Scientific Technical Conference on Low Temperature Plasma During the Deposition of Functional Coatings (LTP Coatings 2017)

Journal of Physics: Conference Series
Volume 1058

Kazan, Russia
5 – 8 November 2017

ISBN: 978-1-5108-6849-6
ISSN: 1742-6588

Printed from e-media with permission by:

Curran Associates, Inc.
57 Morehouse Lane
Red Hook, NY 12571

CURRAN ASSOCIATES INC.
proceedings
.com

Some format issues inherent in the e-media version may also appear in this print version.

Printed by Curran Associates, Inc. (2018)

For permission requests, please contact the Institute of Physics
at the address below.

Institute of Physics
Dirac House, Temple Back
Bristol BS1 6BE UK

Phone: 44 1 17 929 7481
Fax: 44 1 17 920 0979

techtracking@iop.org

Additional copies of this publication are available from:

Curran Associates, Inc.
57 Morehouse Lane
Red Hook, NY 12571 USA
Phone: 845-758-0400
Fax: 845-758-2633
Email: curran@proceedings.com
Web: www.proceedings.com

Table of contents

Volume 1058

Scientific Technical Conference on Low Temperature Plasma during the Deposition of Functional Coatings (LTP Coatings 2017)

5–8 November 2017, Kazan, Russian Federation

Accepted papers received: 5 July 2018
Published online: 31 July 2018

Preface

Papers

IOP Conf. Series: Journal of Physics: Conf. Series **1058** (2018) 011001 doi:10.1088/1742-6596/1058/1/011001

Preface

The IX Republican Scientific Technical Conference "Low-temperature plasma during the deposition of functional coatings" took place from 5 to 8 November 2017 at the Academy of Sciences of the Republic of Tatarstan and the Kazan Federal University. The conference was chaired by a Member of the Academy of Sciences of the Republic of Tatarstan Nail Kashapov – Professor, Doctor of Technical Sciences– a member of the Scientific and Technical Council of the Ministry of Economy of the Republic of Tatarstan.

At the conference, the participants discussed a wide range of issues affecting the theoretical and computational aspects of research problems in the physics and technology of low-temperature plasma. A series of works were devoted to the study of thin films obtained by low-temperature plasma. This year work dedicated to the related field of heat mass transfer in multiphase media and low-temperature plasma was also presented. Of special interest were reports on the exploration of gas discharges with liquid electrolytic electrodes and the study of dusty plasmas.

Kashapov Nail, D.Sc., Professor (Kazan Federal University)

Peer review statement

All papers published in this volume of *Journal of Physics: Conference Series* have been peer reviewed through processes administered by the proceedings Editors. Reviews were conducted by expert referees to the professional and scientific standards expected of a proceedings journal published by IOP Publishing.

Interaction of bubbles in an ultrasonic running wave

I A Aganin and A I Davletshin

Institute of Mechanics and Engineering - Subdivision of the Federal State Budgetary Institution of Science "Kazan Scientific Center of the Russian Academy of Sciences" (IME - Subdivision of FIC KazanSC of RAS) 2/31, Lobachevsky str., Kazan 420111 Russia

E-mail: aganel@gmail.com

Abstract. The influence of the initial sizes of two bubbles on their dynamics is studied under the action of a plane ultrasonic wave travelling along a straight line passing through the centers of the bubbles. The bubbles are filled with air, the liquid is water in room conditions. Initially the total volume of the two bubbles is equal to two volumes of a 5 μm bubble, the distance between the bubble centers is 30 μm. The wavelength is 5 mm, its amplitude is 0.3 bar. A mathematical model used is of fourth order of accuracy in terms of the ratio of the radii of bubbles to the distance between them. It allows for the small non-sphericity of the bubbles, the liquid viscosity and compressibility, the surface tension. It is shown that under the considered conditions three scenarios of the interaction of bubbles are possible: their collision, their divergence, and destruction of one of them due to large deformations of its surface. The ranges of a parameter characterizing the ratio of the initial radii of bubbles in which these scenarios are realized are revealed. The interaction scenarios are shown to depend on the arrangement of the bubbles.

1. Introduction

Dynamics of bubbles in acoustic fields attracts considerable scientific and applied interest. Its features are associated with such phenomena as luminescence [1], neutron emission [2], acoustic streamers [3]. Bubbles in acoustic fields are used for ultrasonic cleaning of surfaces of bodies from various contaminants [4], intensification of physical and chemical processes [5], diagnostics of diseases, drug delivery [6, 7]. In this paper, we study the features of the interaction of two gas bubbles of micron sizes in liquid in a plane ultrasonic wave propagating along a straight line passing through the centers of the bubbles. Similar problems may arise in medical applications. A mathematical model [8] is used, which is a small modification of the models of [9, 10], in which the wave acting on the bubbles is considered to be standing.

2. Problem statement and numerical technique

The dynamics of two gas (air) bubbles in liquid (water) in a plane ultrasonic running wave is considered (figure 1). The wave propagates in liquid in the positive direction of the z axis, on which the bubble centers are located. It is assumed that at the initial time $t = 0$ the bubbles are spherical, and their surfaces are motionless. The initial radii of the bubbles $R_{0,1}$ and $R_{0,2}$ are varied so that the sum of their volumes remains unchanged and equal to the sum of the volumes of two identical bubbles of radius $R_0 = 5$ μm. The initial distance between the centers of the bubbles is $6R_0$. Radii $R_{0,1}$ and $R_{0,2}$ are defined as follows

$$R_{0,1} = (2-\alpha^3)^{1/3}R_0,\ R_{0,2} = \alpha R_0, \quad (1)$$
$$R_{0,1} = \alpha R_0,\ R_{0,2} = (2-\alpha^3)^{1/3}R_0. \quad (2)$$

The parameter α varies in the range $0.2 \le \alpha \le 1$. In the case (1), we have $R_{0,1} > R_{0,2}$, and in the case (2), on the contrary, $R_{0,1} < R_{0,2}$.

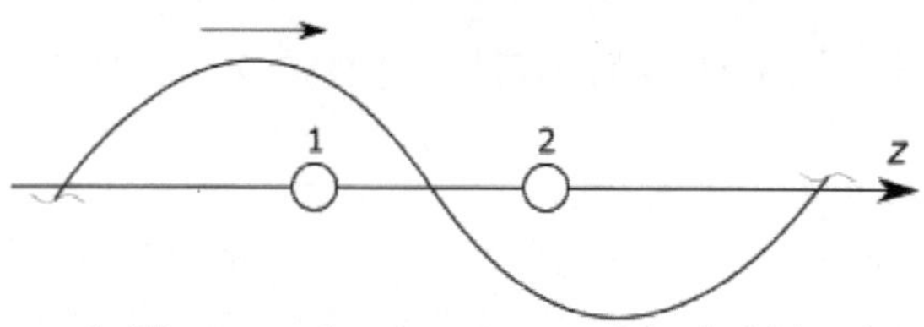

Figure 1. The mutual arrangement of the bubbles 1 and 2 relative to the z axis and the traveling ultrasonic wave at the initial instant of time.

The room conditions are assumed: the static liquid pressure $p_0 = 1$ bar, the liquid density $\rho = 1000$ kg/m^3. The length of the traveling wave is $\lambda = 2\pi\ c_0 / \omega$ where $\omega = 0.6\pi$ MHz, $c_0 = 1500$ m/s is the sound velocity in liquid, the wave amplitude is $p_a = 0.3$ bar.

The surfaces of the bubbles are presented in the form of the series of spherical harmonics

$$r_i = R_i(t)\left[1+\sum_{m=2}^{N}\varepsilon_{m,i}(t)P_m(\cos\theta_i)\right],$$

where $i = 1, 2$ is the bubble number, r_i, θ_i are the radial and angular coordinates of the spherical reference system r_i, φ_i, θ_i with the origin at the center of the i-th bubble, P_m is the Legendre polynomial of degree m, $\varepsilon_{m,i} = a_{m,i}/R_i$ is the dimensionless amplitude of the deviation of the shape of the bubble from the spherical shape in the form of $P_m(\cos\theta_i)$, $a_{m,i}$ is a similar but dimensional amplitude, N is the number of harmonics in the expansion.

The non-sphericity of a bubble is assumed to be small, so that $|\varepsilon_{m,i}| << 1$, it is accepted that $N = 5$. The bubble dynamics model [8] utilized in the present paper is a system of ordinary differential equations of the second order in the radii of the bubbles R_i, the coordinates of their centers z_i on the z axis, and the dimensionless amplitudes of the deviations of their shapes from the spherical one $\varepsilon_{m,i}$. It is of the fourth order of accuracy in terms of the ratio of the radii of the bubbles to the distance between them. The gas in the bubbles is assumed to be homobaric, the effects of the iquid viscosity and compressibility are taken into account approximately.

3. Results

The results of calculations have shown that with the initial data and the conditions of the action of a plane ultrasonic traveling wave on two bubbles, taken in this paper, three scenarios of their interaction are realized: collision of bubbles, destruction of bubbles due to large deformations of the surface of one of them, and unlimited divergence of bubbles. In the literature, one more scenario is known in which the bubbles travel for a fairly long time in the direction of propagation of the wave, being at an equal distance from each other [9], but under the conditions of this work it is not implemented. It should be noted that here, unlike other papers, possible destruction of bubbles due to large deformations of their surfaces is analyzed.

Figure 2 illustrates the first scenario of bubble interaction, when bubbles eventually approach and collide. In the presented case, they behave almost identically, so that the influence of the wave phase on their interaction is unessential. Distortions of the bubble surfaces remain small until collision.

Figure 3 illustrates the second scenario of bubble interaction, when one of the bubbles (the first one) is destroyed. The destruction is due to the increase in the distortion of its sphericity in the form of the second harmonic. An increase in the magnitude of the dimensionless amplitude of the bubble sphericity distortion $\varepsilon_{m,i}$ in the form of any harmonic to the value 1 is accepted as a criterion for destruction.

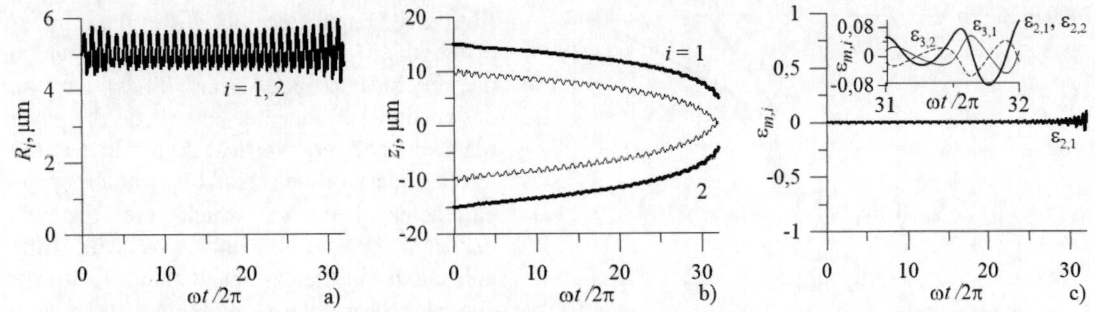

Figure 2. Changes in the radii of the bubbles R_i (the curves for the two bubbles graphically coincide) (a), in the positions of their centers z_i on the z axis (thick solid curves) (b) and in the dimensionless amplitudes of the deviations of their shapes from the spherical one $\varepsilon_{m,i}$ (c) in the case of bubble collision ($R_{0,1} = R_{0,2} = 5$ μm). In figure (b): thin solid curves are the change of the closest-to-each-other points of the surfaces of bubbles located on the z axis.

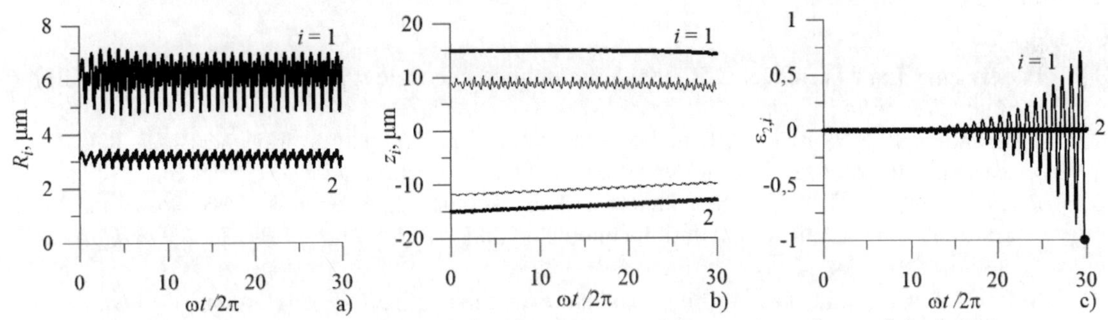

Figure 3. The same as in figure 2, but for the case of destruction of one of the bubbles ($\alpha = 0.625$, $R_{0,1} \approx 6.032$ μm > $R_{0,2} = 3.125$ μm).

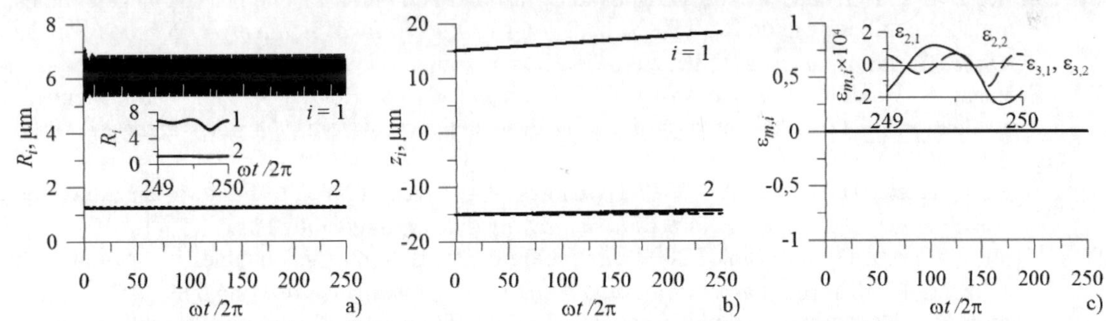

Figure 4. The same as in figure 2, but for the case of bubble divergence ($\alpha = 0.25$, $R_{0,1} \approx 6.283$ μm> $R_{0,2} = 2.25$ μm). In figure (b): the dashed lines (for bubble 1 it graphically coincides with the solid line) are the displacements of similar but single bubbles.

Figure 4 characterizes the third scenario of interaction, when the bubbles diverge (the distance between them increases unlimitedly). It can be seen that the effect of the interaction on the larger bubble is insignificant.

Figure 5 characterizes the ranges of the parameter α in which each of the three identified scenarios of bubble interaction is realized. It is seen that for $0 < 1-\alpha < 0.438$ the "rearrangement" of bubbles at the beginning of their interaction does not affect the scenario of their interaction. At

$0.438 < 1-\alpha < 0.6$, the bubble collision is realized earlier if $R_{0,1} < R_{0,2}$. At $0.6 \leq 1-\alpha < 0.8$, the bubbles collide in the case of $R_{0,1} < R_{0,2}$ and diverge at $R_{0,1} > R_{0,2}$.

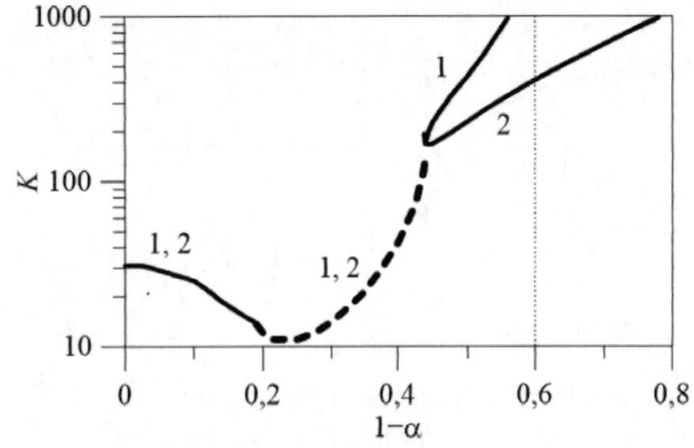

Figure 5. Dependence on $1-\alpha$ of the number of lengths of the traveling wave K, at which the bubbles collide (solid lines) or the destruction of one of them takes place (dashed line). The vertical dotted line is the left boundary of the region of variation of the parameter $1-\alpha$, in which the distance between the bubbles increases with time unlimitedly in the case (1). Curves 1, 2 were obtained for α determined by (1) ($R_{0,1} > R_{0,2}$) and (2) ($R_{0,1} < R_{0,2}$), respectively.

Acknowledgments
The work was supported by the Russian Science Foundation (grant No. 17-11-01135).

References

[1] Putterman S J and Weninger K P 2000 Sonoluminescence: How Bubbles Turn Sound into Light *Annu. Rev. Fluid Mech.* **32** 445

[2] Taleyarkhan R P, West C D, Cho J S, Lahey R T (jr), Nigmatulin R I and Block R C 2002 Evidence for Nuclear Emissions During Acoustic Cavitation *Science* **295** 1868

[3] Parlitz U, Mettin R, Luther S, Akhatov I, Voss M and Lauterborn W 1999 Spatio-temporal dynamics of acoustic cavitation bubble clouds *Phil. Trans. R. Soc. Lond. A* **357** 313–34

[4] Mason T J 2016 Ultrasonic cleaning: An historical perspective *Ultrasonics Sonochemistry* **29** 519–23

[5] Suslick K S 1990 Sonochemistry *Science* **247** 1439–45

[6] Miller D L and Quddus J 2000 Diagnostic ultrasound activation of contrast agent gas bodies induces capillary rupture in mice *Proceedings of the National Academy of Sciences of the United States of America* **97** 10179–84

[7] Seemann S, Hauff P, Schultze-Mosgau M, Lehmann C, and Reszka R 2002 Pharmaceutical evaluation of gas-filled microparticles as gene delivery system *Pharmaceutical Research* **19** 250–57

[8] Aganin I A and Davletshin A I 2017 Dynamics of two gas bubbles in liquid in an ultrasonic traveling wave *Proceedings of the Mavlyutov Institute of Mechanics* **12** no 1 33–9

[9] Aganin A A and Davletshin A I 2009 Simulation of interaction of gas bubbles in a liquid with allowing for their small asphericity *Matematicheskoe modelirovanie* **21** no 6 89–102

[10] Aganin A A, Davletshin A I and Toporkov D Yu 2014 Dynamics of cavitation bubbles arranged in a line in an intense acoustic wave *Computational Technologies* **19** no 1 3–19

IOP Conf. Series: Journal of Physics: Conf. Series **1058** (2018) 012002 doi:10.1088/1742-6596/1058/1/012002

Hydrogen isotope trapping and retention in graphite and boron carbide under consecutive irradiation by deuterium and hydrogen plasma

A.A. Ayrapetov, L.B. Begrambekov, S.S. Dovganyuk, A.V. Grunin, A.S. Kaplevskiy, N.A. Puntakov
National Research Nuclear University MEPhI (Moscow Engineering Physics Institute)
31 Kashirskoe shosse, Moscow, 115409, Russia

E-mail: lexxair@plasma.mephi.ru

Abstract. The results of study on hydrogen and deuterium trapping and retention in MPG-8 grade graphite and boron carbide coating under consecutive irradiation by deuterium and hydrogen plasma in varying hydrogen irradiation conditions are presented in this paper. It is shown that deuterium content decreases both in graphite and boron carbide under irradiation by hydrogen plasma. It is also shown that the main mechanism of deuterium removal is sputtering for graphite, and isotope exchange for boron carbide.

1. Introduction

The problem of hydrogen isotope trapping and retention in graphite materials is important when considering their implementation as plasma-facing materials in thermonuclear facilities [1-4]. Uncontrollable desorption of hydrogen from graphite and graphite materials during plasma irradiation is of huge concern for such devices. Graphite has to be exposed to temperatures of about 800-900 K for prolonged periods of time for its thermal degassing. However, heating for such high temperatures is unacceptable for first walls of thermonuclear facilities. Degassing of graphite elements for devices using tritium poses a significant challenge, as the isotope becomes trapped in the plasma-facing elements, leading to the halt in operating the device and removing tritium from graphite elements.

One of the measures for protecting graphite materials is using renewable protective boron carbide (B_4C) coating [5-7]. Properties of boron carbide as a plasma-facing material are significantly different than those of graphite and other carbon-containing materials used in thermonuclear research. At the same time, accumulation of hydrogen isotopes in protective coating can negatively affect the device's operation. Thus, studies on hydrogen isotope trapping and detention in graphite and boron carbide coating and the possibility of their removal are important.

2. Experimental facilities and methods

Samples made from full-body MPG-8 graphite with dimensions of 7x7x2mm^3 were used in this work. The samples were consecutively polished using sandpaper with surface roughness of up to P2500 (ISO-6344 standard), washed in ultrasonic bath and annealed in the Multifunctional Investigation Complex for Mass Analysis (MICMA) [8] at temperature of 1450 K. A number of the samples were coated with boron carbide coating 4-5 μm thick. Boron carbide coating was conducted on the Coating Deposition and Material Testing (CODMATT) stand [9,10] by depositing atoms sputtered from boron-carbon target by 10-12 keV energy argon ions on a sample.

Experiments on deuterium implantation, sample conditioning and hydrogen isotope trapping analysis were conducted on the "MICMA" stand. Residual vacuum in plasma chamber before discharge initiation was better than 2×10^{-5} Torr. Deuterium implantation was done by irradiating the sample by D_2^+ ions from gas discharge plasma. The discharge was initiated between the heated cathode and the anode. Anode potential was 150 V, discharge current was 1,6 A. Working gas pressure in the chamber during irradiation was 6×10^{-3} Torr. Deuterium ion energy during irradiation was 100 and 650 eV/at., ion flux was $1,2\times10^{16}$ at./cm^{2}s и $1,6\times10^{16}$ at./cm^{2}s, respectively, fluences were ranging from $0,5\times10^{19}$ at./cm^2 to 1×10^{20} at./cm^2, sample's temperature during the experiment was 630 K.

The content of trapped hydrogen isotopes and their thermal desorption spectrum was measured using thermal desorption spectroscopy (TDS) analysis in MICMA's TDS chamber after irradiation in plasma chamber without exposition to atmosphere. The samples were heated during TDS analysis at a constant rate of 5 K/s in the temperature range of 300-1450 K. H_2, D_2 and HD molecules' spectra were analysed (desorption of other hydrogen isotope-containing molecules was negligible).

3. Results and discussion

Preliminary experiments on measurement of trapped hydrogen isotope content under irradiation by deuterium ions at varying fluences were conducted to determine the fluence for the consecutive irradiation experiments. Saturation fluence for graphite was 7×10^{19} at./cm^2, trapped deuterium content was 1×10^{17} at./cm^2. For boron carbide, saturation fluence was equal to that of graphite, with trapped deuterium content being $1,55\times10^{17}$at./cm^2. This implantation fluence was used in all consecutive experiments.

To determine the effect of hydrogen isotope emission on samples irradiated by deuterium under different conditions, a series of experiments was conducted with duration and temperature characteristic to that of hydrogen isotope plasma irradiation: exposition of the samples to vacuum and to hydrogen atmosphere at elevated temperature; irradiation of the samples by hydrogen atoms. These experiments showed absence of deuterium desorption from graphite under these conditions. Analogous experiments for boron carbide lead to insignificant decrease in trapped deuterium content after exposure to vacuum and to hydrogen atmosphere. Approx. 30% of deuterium trapped under ion irradiation is desorbed from boron carbide when the sample was irradiated by hydrogen atoms.

3.1. Hydrogen plasma irradiation of graphite

A series of experiments to determine the dependence of the efficiency of trapped deuterium removal with conditioning hydrogen discharge on irradiation fluence for two implantation energies (100 and 650 eV/at.) were conducted on graphite (Fig.1). For graphite with pre-implanted deuterium irradiated by hydrogen plasma ions with the fluence $0,5\times10^{19}$ at./cm^2, deuterium content in the sample decreases and becomes 40-45% of initial value. The amount of trapped hydrogen, at the same time, increases by ~1.5 times. Deuterium desorption and hydrogen trapping continued, albeit with decreasing rates, until the maximum irradiation fluence was achieved. At the fluence of $3,5\times10^{19}$ at./cm^2, deuterium content decreased to 19% of the initial value for the irradiation energy of 100 eV/at., and to 13% of the initial value for the irradiation energy of 650 eV/at., and the hydrogen content increased. Nevertheless, total amount of hydrogen isotopes trapped in the sample was less than the initial value. It is worth noting that after irradiation by hydrogen ions, deuterium content in the samples implanted at the energy of 650 eV/at. was higher ($1,5\times10^{16}$ at./cm^2) than that for the samples with implantation occurring at 100 eV/at. ($0,7\times10^{16}$ at./cm^2). However, the proportion of deuterium removed relative to the initial values is comparable for both samples. It can be assumed that desorption of deuterium trapped in graphite at other implantation energies would undergo analogously.

Increase of hydrogen irradiation energy to 100 eV/at. did not affect the value of deuterium left in the sample, or the desorption spectra. For altered values of ion flux density (from $1,5\times10^{15}$at./cm^{2}s to $7,5\times10^{15}$at./cm^{2}s), there was no change in the value of desorbed deuterium under the same fluence of hydrogen irradiation. The amount of deuterium desorpbed increases twofold if the temperature is increased from 400 K to 630 K.

Deuterium removal from graphite under hydrogen ion irradiation can be partially due to sputtering of surface layer containing deuterium. Experiments were conducted to determine sputtering coefficient of investigated samples prepared using same techniques and under same conditions as the ones irradiated by hydrogen. Sputtering coefficient for hydrogen ion energy 50 eV/at., ion flux density $7,5\times10^{15}$ at./cm^{2}s and sample temperature 630 K was 0,08±0,02. One can calculate that approx. 280±20 nm is sputtered under these conditions with the fluence of $3,5\times10^{19}$ at./cm^2, which is significantly higher than the characteristic depth of deuterium implantation. This explains the mechanism of deuterium removal from graphite under hydrogen ion irradiation: graphite layer, in which deuterium is contained, is sputtered by hydrogen ions. This assumption is confirmed by the matching of the dependences of deuterium desorption and graphite sputtering coefficient on ion flux density, ion energy and temperature. However, 15% of deuterium is still present in the sample even for irradiation fluence of $3,5\times10^{19}$ at./cm^2, which can indicate the intensification of trapped deuterium transport into the bulk of the sample under hydrogen ion irradiation.

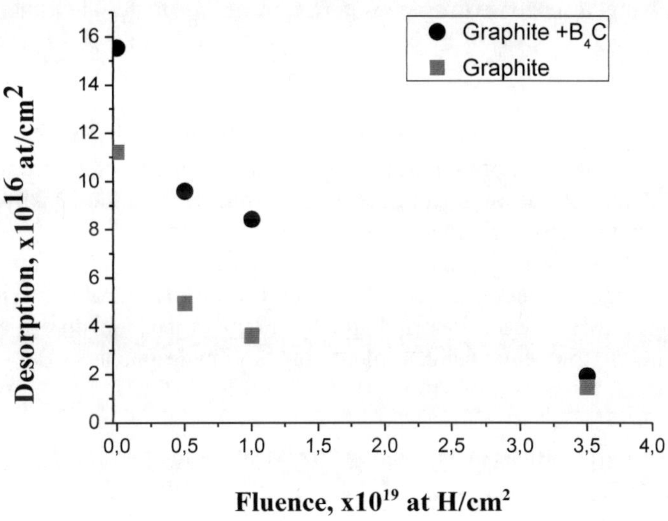

Figure 1 Deuterium desorption from graphite and graphite with boron carbide after hydrogen plasma irradiation (hydrogen ion energy - 50 eV/at., temperature - 630 K, ion flux density - $7,5\times10^{15}$at./cm^{2}s), deuterium implantation ion energy - 650 eV/at.

3.2. Hydrogen plasma irradiation of boron carbide

Irradiation of graphite with boron carbide coating, implanted with deuterium, by hydrogen plasma ions with energy of 50 eV/at., ion flux density of $7,5\times10^{15}$ at./cm^{2}s, and a fluence $0,5\times10^{19}$ at./cm^2, leads to the decrease in the amount of deuterium in the sample to 60% of the initial value (Fig.1). At the same time, the amount of hydrogen trapped is increased by ~2.7 times. With the increase of fluence to $3,5\times10^{19}$ at./cm^2,the amount of deuterium drops to 13% of the initial value, and the amount of hydrogen trapped increases. However, total amount of hydrogen isotopes left inside the sample is 30% lower than the initial value. With further increase of fluence to 7×10^{19} at./cm, the amount of deuterium in the sample becomes 9% of the initial value, with further increase in the amount of trapped hydrogen. At the same time, total amount of hydrogen isotopes is still lower than the initial value. Comparison of these experiments with analogous experiments on uncoated graphite shows that the effectiveness of conditioning of graphite with boron carbide coating is the same as for uncoated graphite.

Increase of hydrogen ion energy to 100 eV/at. does not affect the amount of deuterium left in the sample, trapped hydrogen and desorption spectra. The decrease of ion flux density (from $7{,}5\times10^{15}$ at./cm^{2}s to $1{,}5\times10^{15}$ at./cm^{2}s) lead to the increase of deuterium emission from the sample from 52% to 60% at a fluence of $0{,}5\times10^{19}$ at./cm^2. With the decrease of temperature from 630 K to 400 K the amount of deuterium desorbed from the sample at the same dose decreased almost twofold.

It is known that sputtering coefficient of boron carbide in such conditions is lower than that of graphite for almost an order of magnitude, being 0,002 at./ion [11]. Thickness of surface layer sputtered by hydrogen ions with the energy of 50 eV/at. at a dose of $3{,}5\times10^{19}$ at./cm^2 ($\approx$5 nm) is significantly lower than implantation depth. Therefore, the removal of 87% of deuterium from the sample can't be explained by sputtering. One has to admit that most deuterium atoms trapped in the surface layers of the coating, leave them and desorb due to isotope exchange unrelated to kinetic interaction between hydrogen and deuterium atoms. This is also supported by the fact that all the cases of removal of deuterium from graphite and boron carbide on coated samples was followed by the increase in hydrogen content by the same amount.

As such, the experiments conducted show that the application of protective boron carbide coating can allow removing trapped hydrogen isotopes from the material without its significant sputtering.

4. Conclusion.

It was shown in this work that exposition to vacuum, hydrogen atmosphere or hydrogen atom irradiation does not lead to desorption of deuterium implanted into graphite by plasma ion irradiation. A significant amount of deuterium desorbs from the graphite under irradiation by hydrogen ions. The main mechanism of deuterium removal from graphite is sputtering.

Exposition of graphite with boron carbide coating to vacuum and hydrogen atmosphere leads to insignificant decrease in implanted deuterium content; a significant amount of deuterium desorbs from the sample under irradiation by hydrogen ions and atoms. The main mechanism of deuterium removal from boron carbide coating is isotope exchange of deuterium and hydrogen atoms.

Acknowledgement

This work was supported by the Russian Federation President Grant for the state support of young scientists № 14.Y30.17.2913-MK

References

[1] Skokov G., Sergeev V.Yu., Bykov A.S., et.al. 2014 *Fusion Eng. Des.* **89** Issue 12 P2816
[2] Melnikov A. V., Sushkov A. V., Belov A. M., et al. 2014 *Fusion Eng. Des.* **96–97** P306
[3] Nygren R.E., Youchison D.L., Wirth B.D., et.al. 2016 *Fusion Eng. Des.* **109–111** Part A P192
[4] Mendelevitch B., Boscary J., Peacock A.,et.al. 2015 *Fusion Eng. Des.* **98–99** P1235
[5] Begrambekov L.B., Buzhinskij O.I. 2007 *Plasma Devices and Operations* **15** № 3 P193
[6] Buzhinskij O.I., Semenets Yu.M *Fusion Eng. Des.* 1999 **45** № 4 P343
[7] Buzhinskij O.I., Barsuk V.A., Otroshchenko V.G. 2010 *Fusion Eng. Des.* **85** № 7–9 P1615
[8] Airapetov A., Begrambekov L., Bremond S., et al. 2011 *J. of Nucl. Mater.* **418** P1042
[9] Azizov E., Barsuk V, Buzhinskij O. et al. 2015 *J.of Nuc. Mat.* **463** P792
[10] Sadovskiy Y, Begrambekov L, Ayrapetov A, et al 2016 *J. Phys.: Conf. Ser* **748** 012003
[11] Eckstein W., Garcia-Rosales C., Roth J., Ottenberger W. Sputtering data. IPP report **9/82** February **1993**.

IOP Conf. Series: Journal of Physics: Conf. Series **1058** (2018) 012003 doi:10.1088/1742-6596/1058/1/012003

Modification of basalt fibers by low-temperature plasma

M V Antonova[1], I V Krasina[2], S V Ilyushina[3], R R Mingaliev[4], and A S Parsanov[5]

[1] Associate professor, the Department of Technology of Chemical, Natural Fibers and Products, Kazan National Research Technological University, Kazan, Russia
[2] Head of the Department of Technology of Chemical, Natural Fibers and Products, Kazan National Research Technological University, Kazan, Russia
[3] Associate professor, the Department of Technology of Chemical, Natural Fibers and Products, Kazan National Research Technological University, Kazan, Russia
[4] Senior lecturer, the Department of Technology of Chemical, Natural Fibers and Products, Kazan National Research Technological University, Kazan, Russia
[5] Associate professor, the Department of Technology of Chemical, Natural Fibers and Products, Kazan National Research Technological University, Kazan, Russia

E-mail: marisha.10@list.ru

Abstract. In the article the characteristics of basalt fibers and threads are given, their advantages over other mineral fibers are considered. Materials and methods of research are described. The results of a study of the properties of filaments, such as the breaking load force of a basalt thread before and after its treatment in a low-temperature plasma are presented. The regime of plasma treatment is revealed at which the strength of the fibers of the basalt thread increases, it established that modification of low-temperature plasma allows increases the characteristics strength of basalt filaments and transforming the structure of their surfaces.

1. Introduction

In Russia and abroad, the production of fibrous materials based on mineral, glass, kaolin and other fibers is intensively developing. However, it's also increases technical requirements for these materials, and as well as the shortage of raw materials restrains the growth of their production volumes.

In connection with this, the industrial production of basalt fibers and materials based on them has become widespread, where rock as a raw material - basalts, gabbro-debase porphyries and others - were used as an initial one-component raw material.

Basalt fibers have high heat, sound insulation and structural properties. In terms of temperature resistance, as well as in acid, alkali and vapor resistance, they go beyond mineral and glass ones.

The high physical and mechanical properties of basalt fibers determine the creation and growth of new high-efficiency construction, technical, heat-sound-insulating and as well other composite materials and products for various industries. The use of basalt fibers as raw materials makes it possible to produce materials that replace asbestos, metal, wood and others [1].

The aim of the research is to modify the basalt threads with a low-temperature plasma and to study its effect on fiber strength.

2. Materials and methods

IOP Conf. Series: Journal of Physics: Conf. Series **1058** (2018) 012003 doi:10.1088/1742-6596/1058/1/012003

The objective of the study selected basaltic twisted thread BS 10-68Z40-KV-12. The plasma treatment was carried out on an experimental high-frequency plasma plant of reduced pressure, described in [2], using as a plasma-forming gas, and a mixture of argon / propane-butane gases in a ratio of 70/30. Determination of the breaking load of the basalt thread was carried out on the Shimadzu AGS-X testing machine. The microstructure of the basalt fiber was examined by confocal laser microscopy using the Olympus LEXT 4000 microscope.

3. Results of experiments and discussion

Threads from basalt fibers are used for the production of various textiles. However, in the process of weaving products, the fibers of the yarns break due to various mechanical factors, which in turn leads to a deterioration in the quality of the finished product and difficulties in the production process.

Based on the previous work and their results in the study of the influence of low-temperature plasma on the properties of various types of fibers and filaments [3-6], studies of the effect of plasma basalt thread processing parameters on its strength are presented.

The processing of the basalt thread in a high-frequency plasma installation was carried out in two stages: with the voltage variation and the current intensity at the anode of the generator lamp; with a variation in processing time.

At the first stage of the studies, processing of the samples of the basaltic filament was performed in an industrial plasma installation with a plasma-forming gas argon / propane-butane in the ratio of 70/30. The value of plasma gas flow (G) was 0.04 g/s, a generator frequency (f) was 13.56 MHz, and pressure (P) in the vacuum chamber was 26.6 Pa, a processing time (t) 3 min. The voltage of the anode (Ua) varied from 4 to 7 kV, and the anode current (Ia) was from 0.5 A to 0.85 A. Based on the results of the treatments, the most optimal values of the voltage at the anode and the forces current at the anode, at which the best results for the force, applied load were achieved. The results of the studies presented in Table 1.

Table 1. Values of the maximum breaking load force of the basalt thread at the selected plasma processing parameters

Voltage (Ua), kV	Current strength (Ia), A	Maximum breaking load force, N
Control sample		30.07
6.5	0.8	36.05

The maximum value of the breaking load force for the selected parameters is increased by 20% compared to the control sample.

At the second stage of the studies, the basalt fiber samples were processed in an experimental plasma installation with a plasma-forming gas flow rate G = 0.04 g/s, f = 13.56 MHz, P = 26.6 Pa, Ua = 6.5 kV, Ia = 0.8 A. The processing time (t) then ranged from 1 to 9 minutes. The results of the studies presented in Table 2.

Table 2. The dependence of the maximum breaking load force of the basalt thread on the plasma treatment regimes

Voltage (Ua), kV	Current strength (Ia), A	Processing time (t), min	Maximum breaking load force, N
Control sample			30.07
		1	33.21
6.5	0.8	3	36.05
		5	38.36

	7	38.32
	9	38.82

Based on the data in Table 2, it can be concluded that the modification of basalt fibers with a low-temperature plasma increases the maximum breaking load force, and the maximum value is achieved with a processing time of 9 minutes. The best mode of plasma treatment at which the maximum strength of the basalt thread is maximized, and the mode with parameters: G = 0.04 g/s, f = 13.56 MHz, P = 26.6 Pa, Ua = 6.5 kV, Ia = 0.8 A, t = 9 min, and the value of its maximum breaking load increases to 29%. A study of the dependence of the tensile strength on the duration of plasma treatment in the direction of increasing the latter is being carried out at the present time.

In Figure 1 are microphotographs of samples of fibers of a basaltic filament of a control and processed low-temperature plasma.

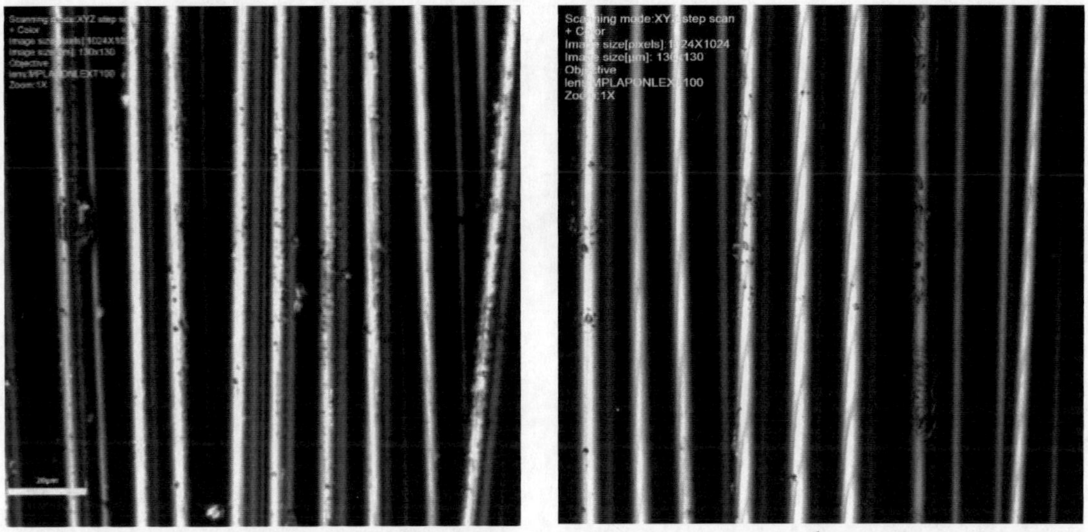

a b

Figure 1. Microphotographs of fibers of a basalt thread (x100): a – control sample; b – sample treated in argon/propane-butane gas (70/30), G = 0.04 g/s, f = 13.56 MHz, P = 26.6 Pa, Ua = 6.5 kV, Ia = 0.8 A, t = 9 min

The above diagram shows microphotographs of a sample that has undergone plasma treatment, more uniform structure of the fibers of the basalt thread and less irregular surface are observed in comparison with the untreated sample. Due to these changes in the surface of the fibers, an increase in the maximum bursting force of the basalt thread treated with low-temperature plasma is likely to occur as compared to the untreated yarn.

4. Conclusion

Thus, based on the results of study, it can be concluded that modification in conditions of low-temperature plasma is upcoming method of increasing the characteristics strength of basalt filaments and transforming the structure of their surfaces.

References

[1] D Jigiris et al 1989 *Basalt fiber materials* (Moscow: VNIIMSM) pp. 1-3
[2] E Sergeeva et al 2010 *Optimization of modes of low-temperature plasma processing of high-modulus polyethylene fibers* (Kazan: KSTU) pp. 94-98

[3] S Ilyushina et al 2015 *Investigation of the influence of the plasma of the high-frequency discharge of a low pressure on the fixation of fire retardant in the surface layer of textile materials* (Kazan. KNITU Publishing House) pp. 78-80
[4] S Ilyushina et al 2017 J. Phys.: Conf. Ser. 927 012035
[5] M Shaehov 2006 *Physics of high-frequency discharge of low pressure in the processes of capillary-porous and fibrous materials* (Kazan) p. 350
[6] E Sergeeva 2010 *Regulation of the properties of synthetic fibers, yarns, fabrics and composite materials on their basis with the help of nonequilibrium low-temperature plasma* (Kazan) p. 363

IOP Conf. Series: Journal of Physics: Conf. Series **1058** (2018) 012004 doi:10.1088/1742-6596/1058/1/012004

Plasma spray coating with iron oxides for manufacturing of the corrosion-resistant electrodes

P A Katasonov, R I Valiev, A A Khafizov, Yu I Shakirov and R A Valiev

Kazan Federal University Naberezhnye Chelny Institute, 68/19 Mira Street, Naberezhnye Chelny, 423810, Russian Federation

E-mail: comrade.hammond@gmail.com

Abstract. Magnetite anodes were manufactured by plasma spray coating of titanium with the iron oxide powder. The powder was obtained by plasma-electrolytic atomization of carbon steel. Porosity, adhesion, specific resistivity of these coatings were measured in relation to size distribution and chemical composition of the powder used for spray coating.

1. Introduction

Magnetite is characterized by relatively high conductivity and corrosion resistance, which makes it suitable for producing anodes for various electrochemical applications.

Bulk magnetite anodes are the most common ones. These anodes suffer from significant drawbacks such as brittleness, difficulty of manufacturing, considerable mass and size as well as increased resistivity. By making an anode with plasma spray coating by magnetite on an inert and conducting substrate such as titanium, niobium, zirconium, tantalum or stainless steel, it is possible to create electrodes that are free of these drawbacks [1].

In order to produce high quality magnetite anodes by plasma spray coating it is important to use magnetite powder of particular properties [2, 3]. Low-temperature plasma is a good tool for surface treatment of materials [4 – 11]. One method of creating magnetite powder suitable for plasma spraying is plasma-electrolytic atomization of carbon steel [12 – 14]. This method is easy to implement and it makes it possible to produce spherical particles of magnetite with diameter of 0.01 – 0.50 mm that has low amount of impurities. This study aims to determine operational characteristics of anodes made by plasma spraying of magnetite powder on a titanium substrate. The powder used was obtained by plasma-electrolytic atomization of carbon steel.

2. Materials and methods

Magnetite powder was produced using the plasma-thermic reactor with electrolytic cathode and metal anode [12]. The anode was made of SAE AISI 1120 steel rod 10 millimeters in diameter. Solution of 0.01% sodium chloride in distilled water was used as the electrolyte.

During the atomization process the distance between electrodes has been maintained at 1 mm and voltage of 600 V was applied. This allowed for the current of 1.9 A to pass through the gap between the electrodes. The anode was atomized and thus obtained product was carefully collected, washed and dried under 90 °C until a constant mass was obtained. A number of fractions by size was isolated by sifting through a set of sieves with a mesh of 40 – 160 μm.

The resulting powder consists of two unbound components: one is ferrimagnetic (Fe304) and the other is paramagnetic (FeO), which are easily separable by a magnet. The ratio of these components allows for Fe(II):Fe(III) relation of 0.43 – 0.52 to be achieved.

Plasma spray coating was performed using an UPU-3D device with argon as a working gas at a flow rate of 20-30 liters/min. The arc electrical power was set at 18-30 kW with spraying distance of 100 – 130 mm. A coated electrode formed on a titanium SAE AISI Grade 2 substrate with homogeneous and tough coating having a thickness of about 300 micron was produced for further study.

The coated electrodes were studied using techniques of microscopy analysis on Micromed MET metallographic microscope, Vickers hardness test on PMT-3M hardness tester and adhesive peel strength test on a Elcometer 108 hydraulic tester. Anodic dissolution rate was studied by electrolysis in 5% solution of sodium chloride in distilled water at a current density of 5 A/dm^2.

3. Results and discussion

Hardness and adhesion strength of the coatings produced are given in Table 1, and microphotograph of a cross section is presented in Figure 1.

Figure 1. Microphotograph of the magnetite coating, magnification 200^x

Table 1. Hardness and adhesive strength of the coatings.

Fraction by size, μm	Hardness, HV 0.05	Adhesive strength, MPa	Porosity, %
40 – 50	537	11.4	5 – 8
50 – 71	491	16.3	6 – 10
71 – 90	471	17.2	7 – 10
90 – 120	468	15.9	8 – 12
120 – 160	453	9.8	12 – 18

Using of coarse particles with the diameter of more than 120 μm resulted in deposition of highly porous coatings with poor adhesion. Only a small part (about 50%) of this size fraction took part in coating formation because coarse particles failed to heat through and gain enough momentum in a plasma jet.

Size fraction of 40 – 50 μm forms non-porous coatings with strong adhesion and high coalescence rate. The only downside is that small particles can partially oxidize during the deposition process, which leads to cracking due to the elevated structural stress.

The highest quality coatings were produced by deposition of the size fraction of 50 – 120 μm. Adhesion strength of these coatings is consistent with the known values for coatings of the same type [15].

Measured parameters for coatings with particle size fraction of 72 – 90 μm of various Fe(II):Fe(III) relations used as a spray coating source are given in Table 2. Porosity increases and adhesion strength decreases as the relation increases.

Table 2. Parameters of coatings obtained from 72-90 μm size fraction with various Fe(II):Fe(III).

Fe(II):Fe(III)	Hardness, HV 0.05	Adhesive strength, MPa	Porosity, %
0.43	485	12.6	9
0.45	471	13.8	8
0.47	463	14.2	6
0.49	450	14.7	6
0.51	446	15.0	5

The most important parameters of magnetite anodes are specific resistance ρ and anodic dissolution rate ΔM_A. Direct current specific resistance of the coatings in logarithmic scale is plotted versus the Fe(II):Fe(III) relation and shown in Figure 2.

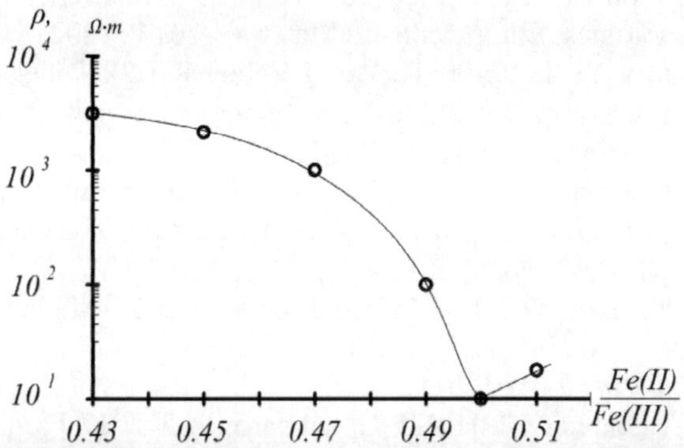

Figure 2. Specific resistance plotted versus Fe(II):Fe(III) relation.

Anodic dissolution rate in linear scale is plotted versus the Fe(II):Fe(III) relation and shown in Figures 3.

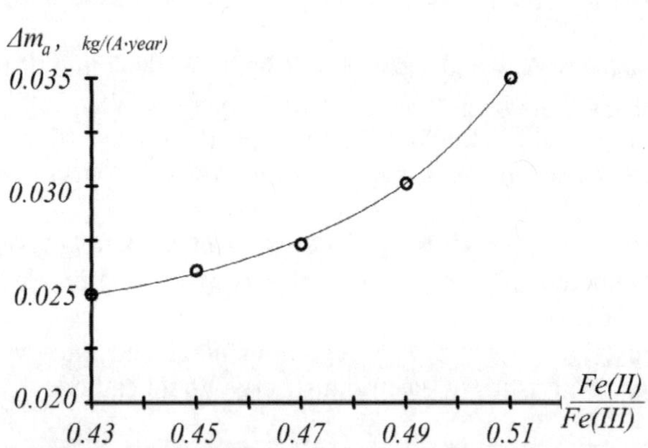

Figure 3. Anodic dissolution rate (b) plotted versus Fe(II):Fe(III) relation.

Anodic dissolution rate of these coatings is consistent with the known values for bulk magnetite anodes [16]. ΔM_A increases with the decreasing amount of Fe(II) ions, but specific resistance increases exponentially below stoichiometric point.

4. Conclusion

The study shows that coated electrode formed on a titanium substrate with homogeneous and tough magnetite coating having a thickness of about 300 μm produced through the above-described spray coating treatment has the same electrochemical durability as a bulk magnetite anode, but at the same time has decreased mass, size and brittleness. The best quality coatings can be produced when magnetite powder of size fraction of 50 – 120 μm with Fe(II):Fe(III) relation of 0.495 – 0.510 is used as a spray coating source.

References

[1] Hayes M and Kuhn A T 1978 *Journal of Applied Electrochemistry* **8**(4) 327

[2] Hideo I, Setsuo Y and Sakae I 1992 *US patent* 5143746

[3] Sigaev N P, Zaharov V F, Travin A L, Horishko B A and Rumjancev E M 2003 *Izvestija VUZ Himija i himicheskaja tehnologija* **46**(7) 27

[4] Kashapov N F and Sharifullin S N 2015 Hardening of the surface plasma jet high-frequency induction discharge of low pressure. *IOP Conference Series: Materials Science and Engineering* "International Scientific and Technical Conference "Innovative Mechanical Engineering Technologies, Equipment and Materials-2014" 12021

[5] Khafizov A A, Shakirov Yu I, Valiev R A and Valiev R I 2017 Determination of regression materials mircohardness, processed by low-temperature plasma dependence on process conditions *J. of Phys.: Conf. Ser.* **789** 012024

[6] Gabdrakhmanov A T, Israphilov I H, Galiakbarov A T, Samigullin A D and Gabdrakhmanov A T 2016 Improving the efficiency of plasma heat treatment of metals *J. of Phys.: Conf. Ser.* **669** 012014.

[7] Bashmakov D A, Israphilov D I and Samigullin A D 2017 Mathematical modeling of the crystallization process to obtain a fine fraction aluminates *IOP Conference Series: Materials Science and Engineering* **240** 012011.

[8] Gabdrakhmanov A T, Israphilov I. H and, Galiakbarov A T 2017 Preparation of metal surfaces for application of functional coatings *J. of Phys.: Conf. Ser.* **789** 012009.

[9] Gabdrakhmanov A T, Shafigullin L N, Galimov E R and Ibragimov A R 2017 Surface thermohardening by the fast-moving electric arch. *J. of Phys.: Conf. Ser.* **789** 012010.

[10] Tazmeev A K, Tazmeeva R N and Sarvarov F S 2016 The features of high-current gas discharge in a narrow gap between the liquid electrolyte and solid electrode *J. of Phys.: Conf. Ser.* **669** 012056

[11] Gibadullina G R, Tazmeev A K and Tazmeeva R N 2017 Recycling of polymer waste by liquid electrolyte plasma generators *Journal of Fundamental and Applied Sciences* **9** pp 1779-1789

[12] Valiev R A, Gaisin F M and Shakirov Yu I 1991 Special traits of powder obtained in discharge between steel electrode and electrolyte *Soviet Powder Metallurgy and Metal Ceramics* **30**(6) 448

[13] Khafizov A A, Shakirov Y I, Valiev R A, Valiev R I and Khafizova G M 2016 Study of thermal and electrical parameters of workpieces during spray coating by electrolytic plasma jet *J. of Phys.: Conf. Ser.* **669** 012030

[14] Valiev R I, Shakirov Yu I, Khafizov A A, Valiev R A and Nuriev I M 2017 Generalized current-voltage characteristics of electric discharge liquid cathode *J. of Phys.: Conf. Ser.* **789** 012067

[15] Borisov Ju S, Harlamov Ju A, Sidorenko S L and Ardatovskaja E N 1987 *Gazotermicheskie pokrytija iz poroshkovyh materialov. Spravochnik* (Kiev: Naukova dumka) p 544

[16] Wakabayashi S and Aoki T 1977 *Journal de physique. Colloque C1, supplement au* **38**(4) 241

IOP Conf. Series: Journal of Physics: Conf. Series **1058** (2018) 012005 doi:10.1088/1742-6596/1058/1/012005

Influence of intermetallide coatings on the high-temperature corrosion resistance of martencite steel in various structural conditions

E L Vardanyan, K N Ramazanov, R D Agzamov, I I Yagafarov, A R Khamzina and A Yu Nazarov

Ufa State Aviation Technical University

E-mail: vardanyaned@gmail.com

Abstract. The effect of coatings deposited from a vacuum-arc plasma on the high-temperature corrosion resistance of martensitic steel in ultrafine-grained and coarse-grained state is studied. It is proved that coatings obtained from a vacuum-arc discharge plasma increase the high-temperature corrosion resistance of martensitic steel. The analysis of high-temperature corrosion resistance was carried out using the method of accelerated cyclic tests developed by the All-Russian Institute of Aviation Materials.

1. Introduction

In gas turbine engines (GTE), the most loaded parts are compressor blades and turbines. They work in conditions of high temperatures, aggressive environment and heavy loads. To improve the mechanical properties of GTE parts, it is perspective to use materials with ultrafine-grained structure (UFG) together with traditional materials with a coarse-grained structure (CG). However, materials with a UFG structure are subject to a higher corrosion rate compared to samples with a CG structure, this is due to a significant number of defects and the presence of a longer grain boundary, which are active centers of corrosion damage [1].

Thus, work is underway to develop new methods and coating processes in the direction of creating multi-layer composite coatings that allow increasing the life and reliability of the GTE parts [2].

Article [3] illustrated the corrosion properties of Ti-Al-N / Ti-Al two-layer coating deposited by magnetron sputtering on magnesium alloy AZ31 are presented. Coatings were produced by spraying composite Ti / Al composite targets. Corrosion tests were carried out in an aqueous solution of 3.5% NaCl. The result showed that the coating effectively works in aggressive environments.

In the work [4], the authors of A.M. Smyslov, A.D. Mingazhev and others consider the possibility of obtaining vacuum-plasma nanolayer coatings for blades of turbomachines made of titanium alloys by combining the processes of electric arc deposition and ion implantation. The vacuum-plasma deposition of a metal sublayer and nanolayers of aluminum nitrides, carbides and / or aluminum carbonitrides or titanium, zirconium and aluminum compounds with nitrogen and / or carbon is produced on the blade surface. Allowed to increase in comparison with the existing coating (ionic nitrogen implantation and stabilizing annealing, ion-plasma coating of titanium nitride is carried out after ion implantation) resistance to salt corrosion (up to 1.5 times), drip-shock erosion (up to 2.1

times), gas-abrasive erosion (by 2.5 times). Thus, research aimed at developing technology to increase the corrosion resistance of aviation materials is relevant.

Thereby. this article is devoted to the generalization and identification of the regularities of the corrosive behavior of various materials with the UFG structure with their CG analogs and the increase in their corrosion resistance at high temperatures by the deposition of coatings from the vacuum-arc discharge plasma.

2. Material and methods

As initial materials, were used samples of EI 961 stainless steel with a coarse and ultrafine-grained structure.

To form a UFG structure, samples with a diameter of 20 mm and a thickness of 1 mm were subjected to intense plastic deformation by torsion (IPDT) at a pressure of 6 GPa, with a speed of 10 turns.

The experiments were carried out on a modernized installation for vacuum ion-plasma deposition of NNV-6,6 I1. In the vacuum chamber, samples were installed from martensitic steel EI 961c CG and UFG structure. When the pressure in the chamber is P = 1.5-2 * 10-3 Pa, the TiN coating was deposited on the first sample. The second sample was deposited with a multi-layer composite coating based on the intermetallics of the TiAl (TiAl N) system developed at the USATU laboratory. The third sample was nitrided at the first stage with a high-current plasma source with a cathode for 60 min, and a multilayer composite coating of the intermetallide system TiAl (TiAl N) was deposited in the second stage. The currents of the arc evaporators are set in the range I = 60-120 A, the current of the cathode is Ical = 90-120 A, the arc discharge current of the plasma source is Iugus = 40-60 A, the temperature of the part is t = 550 ° C. Within 60 minutes the process of deposition of the multilayer composite coating of the TiAl / TiAlN and TiN system occurs.

Corrosion resistance tests were carried out according to RTM 1.2.189 - 2006 (VIAM) using the accelerated cyclic test method. During the course of this procedure, the samples with coatings are held for 1 hour at the test temperature in an atmospheric oven and then air-cold for 1-2 minutes and cooled in a 3% NaCl solution to room temperature, then held for 22 hours in a wet desiccator.The duration of the test is 10 cycles.

3. Results and discussion

On samples of steel EI 961 with a CG structure with coatings, TiAlN and nitriding + TiAlN compositions provided better corrosion resistance. Specific change in the mass of samples coated with TiAlN, nitriding + TiAlN decreases in comparison with uncoated samples by 4 and 3 times. Researches on the corrosion resistance of TiN coated samples revealed that the specific gravity losses are 4,1353 g / m^2, which is 2 times lower than in the initial state. (Fig.1)

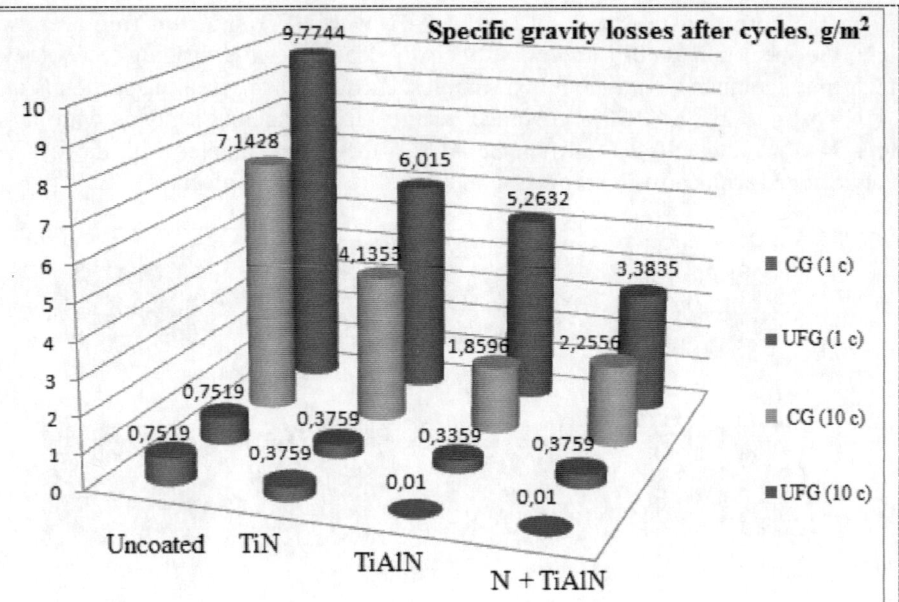

Figure 1. Histogram of the corrosion rate of steel EI 961 in CG and UFG condtions.

Visual inspection (Fig. 2) of the samples after 10 cycles of the accelerated cyclic corrosion tests (ACCT) at a temperature of 500 ° C showed that the surface of the uncoated samples of EI961 steel is powdered with loose corrosion products. On samples with TiAlN coatings, nitriding + TiAlN, a partial coating peeling is seen, no corrosion uclers are detected. It is obvious that cracks arise and develop mainly along the grain boundaries of the matrix under the coating, while the coating deforms plastically, and cracks appear in it after they are formed in the matrix. It has been established that TiN coated samples have a coating peel of about 70%, small pittings with a diameter <0.1 mm.

Figure 2. Appearance of samples from steel EI 961 (CR) before the tests: a) uncoated; b) coated with TiN; c) coated with TiAlN; d) coated with nitriding technology + TiAlN;
After 10 ACCT cycles at a temperature of 500°C: e) uncoated; f) coated with TiN; g) coated with TiAlN; h) coated with nitriding technology + TiAlN

Tests on the corrosion resistance of steel EI961 with UFG structure (Fig. 3) also revealed a decrease in the specific mass of samples with TiAlN coatings, with nitriding technology + TiAlN in 1,8 and 2,8 times compared with the initial samples. As a result of visual inspection (Fig. 2), samples with a TiN coating have a 80-90% coverage peeling, the remaining samples with TiAlN coatings, nitriding + TiAlN - about 10-20%. After the ACCT, the initial samples with the UFG structure are completely covered with corrosion products, and there are corrosive ulcers.

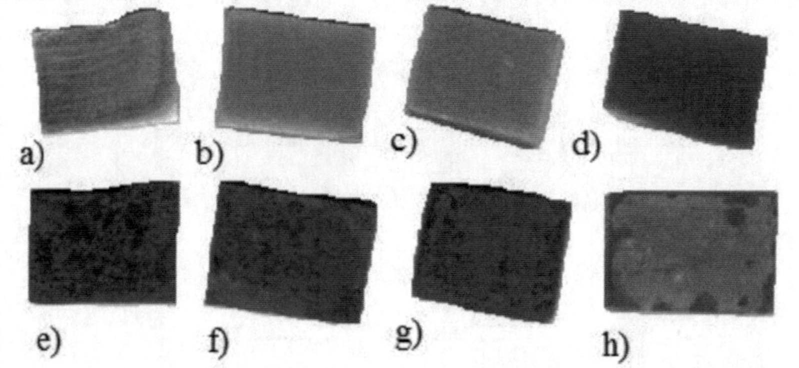

Figure 3. Appearance of samples from steel EI 961 (UFG) before the tests: a) uncoated; b) coated with TiN; c) coated with TiAlN; d) coated with nitriding technology + TiAlN;
After 10 ACCT cycles at a temperature of 500°C: e) uncoated; f) coated with TiN; g) coated with TiAlN; h) coated with nitriding technology + TiAlN

As seen from (Fig. 4a.), after the corrosion test the entire surface of the sample was subjected to continuous corrosion. On the surface of the sample ulcers formed, as well as the places of exfoliation of the material.

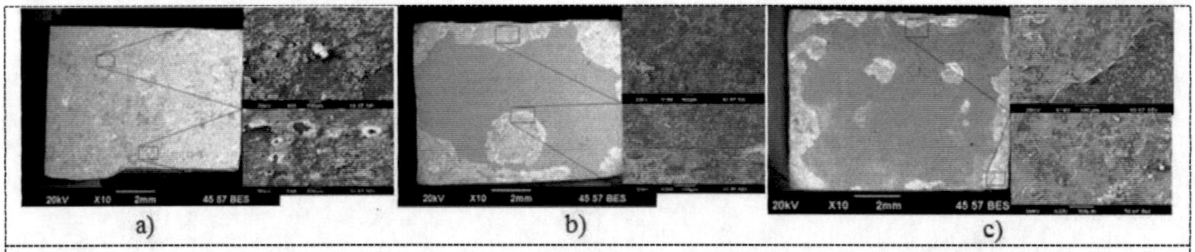

Figure 4. Surface structure of UFG without coating after corrosion resistance test.

(Fig. 4b) shows a sample coated with TiN after the corrosion test. As seen from the photo that the coating peeled off mainly along the edges of the sample, and one large piece has peeled closer to the middle of the sample. The peeled and corroded areas account for 37% of the total sample area.

(Fig 4c.) shows a sample coated with TiAlN after the corrosion test. As seen from the photo that the coating has peeled off mainly along the edges of the sample, and also there are several small areas in the middle of the sample. The peeled and corroded areas account for 22% of the total area of the sample.

Corrosive destruction of coatings having high chemical and thermal stability can be contributed by various disruptions in their structure. These include: pores, depressions, delamination, lamination, separation from the substrate, chips, cracks, submicrochannels between columnar crystals, scaly, droplets and sprays of hardened metal on the surface of coatings, uneven thickness. All these structural defects can cause, to varying degrees, accelerated local corrosion damage to the coatings.

4. Conclusions

The conducted researches of corrosion tests of martensitic steels with coarse-grained and ultrafine-grained structure allowed to establish that:

- The corrosion resistance of martensitic steels with a CG structure is 30% higher than that of the UFG structure, and therefore, when using martensitic steels with a UFG structure, the effect of the structure on high-temperature corrosion resistance should be taken into account.
- Deposition of TiAl3 / TiAlN protective coatings on the surface of martensitic steels makes it possible to increase the high-temperature corrosion resistance for materials with a short-circuit structure by a factor of 4, and for the UFG structure in 3 times;
- The introduction of ionic nitriding prior to the application of protective coatings allowed to improve the result of corrosion resistance only for materials with UFG structure (3,2 times).

References

[1] N A Amirkhanova, R Z Valiev, A F Razyapova, G I Raab. "*Vestnik UGATU*" T. 10, № 1 (26). P. 91-95
[2] V A Belousov. *Kharkov Scientific Assembly*, 2006. P. 60-73.
[3] GuosongWu et al. *Materials Characterization* 60 (2009) 803–807
[4] A M Smyslov, A D Mingazhev, M K. Smyslova, K S Selivanov, A A Mingazhev. *"Vestnik UGATU"* T.15 # 1 (1) 2011 p. 109-112

IOP Conf. Series: Journal of Physics: Conf. Series **1058** (2018) 012006 doi:10.1088/1742-6596/1058/1/012006

Development of anti-adhesive coatings based on nitrides and oxides of zr by deposition from vacuum-arc plasma

E L Vardanyan, K N Ramazanov, A N Abramov, A Yu Nazarov and R Sh Nagimov
Ufa State Aviation Technical University

E-mail: vardanyaned@gmail.com

Abstract. Anti-adhesive coatings based on Zr nitrides and oxides deposited form the vacuum-arc plasma were developed. Laboratory tests that determine tribotechnical properties were conducted. Proved, that coatings based on nitrides and carbides increase the anti-adhesive properties

1. Introduction

During the exploitation of machine parts in the sliding contact zone, a complex of physical, physico-mechanical, electrochemical, acoustical and other appearances, which affect the friction and metal wearings [1]. The most effective method of modification the surface properties of the tool is the deposition of functional coatings. Currently, the processes of physical deposition of coatings (PVD processes) on various products, including die tools, are becoming oftenly used because of high reliability, versatility, the possibility of obtaining coatings of virtually any architecture, composition, structure with ensuring environmental cleanliness of processes in the production of tools.

Article [2] studies of the mechanism of interaction of aluminum alloys and tool material (with and without coatings) are given. According to the results of the research, established that the mechanism of aluminum adhering at hard coatings occurs in two phases: primary adhering due to micro unevenness of the coating; secondary transfer of aluminum on the surface. Coatings of CrN and CrAlN showed similar mechanism of interaction. The resistance to aluminum adherence of CrN coatings is lower than of CrAlN. Aluminum particles adhere to the surface of the coating and thereby are formed large dimensions, which leads to an increasing of the friction coefficient, metal wear and adhering on the surface of the coatings. To increase the resistance to adhering, it is necessary to improve the quality of the contactable surface (roughness).

Article illustrated the wear resistance and frictional characteristics of various hardening coatings at sliding friction without a lubricant. As a samples were used disks with a diameter of 50 mm and height of 35 mm, made of an aluminum alloy D16T (roughness Ra = 0,11 – 0,14 μm) with different hardening coatings: TiN, a carbon diamond-like film (CDLF), an anode and chemical oxidation. As a result of the researches established, that the formation of hardening vacuum-plasma coatings provides a significant, more than 10 times, increase in wear resistance of products in frictionless sliding without lubricant.

In the work dedicated to improve the resistance against the setting of aluminum alloys, as a protective coatings are considered coatings based on nitrides and zirconium oxides (ZrN and ZrO_2), because Zr has the least tendency to interact with an aluminum melt [5].

Article [4] contains results of researches and comparative analysis of various hardening coatings of resistance to adhering and scuffing at interaction with different working materials. Coatings were deposited to high-alloyed tool steel for cold working.

Regardless of the surface hardening technology used (nitriding or coating), the adherence of the material from the aluminum alloy to the surface of tool steel occurred at the beginning of the tests. Coatings of TiB_2 and TiC showed an increase in the friction coefficient of more than 0.5 immediately after the start of the tests. DLC coatings also showed fast increase of friction at a load of about 300 N, while TiN and VN coatings showed a more moderate increase in the friction coefficient. For all the surfaces studied, the increase in friction corresponded to the formation of a thick layer of the transferred aluminum alloy.

The results of the literature review showed that in order to increase the resistance to aluminum adherence, it is necessary that the roughness of the contactable surface is minimal; Solid coatings based on nitrides and oxides of Zr, CrAlN can be used as anti-adhesive coatings (aluminum alloys that are resistant to adherence). Because the friction coefficient of these coatings when interacting with aluminum alloys is minimal.

Thus, studies aimed at developing anti-adhesion coatings by deposition from a vacuum-arc plasma are actual.

2. Material and methods

Deposition of the coatings will be carried out at the upgraded plant NNV-6,6-I1, intended for condensation of the substance from the metallic plasma and surface modification by the non-self-sustaining high-current diffusion, arc and glow discharges, which is equipped with a gas supply system, a plasma generator with a filament cathode – "PINK", evaporators. The technological modes of applying coatings at the installation of NNV-6,6-I1 varied within the following limits: pressure in the chamber $P = 10^{-1} \div 10^{-2}$ Pa; arc current I = 40 – 120 A; processing time 60-75 minutes; The temperature of the processed samples did not exceed 520°C. Scheme of the experiment (Fig. 1).

For the structure analysis of the instrumental material and the transition zone of the coatings-substrate will be used modern metallographic analysis techniques (Zeiss Axiotech 25HD Optical Microscope) and scanning electron microscopy (SEM JEOL JSM-6490LV).

To assess the effectiveness of the new method for modifying the surface of the instrument will be used laboratory stands, in addition to the generally accepted methods for determining the tribotechnical properties of surface modified layers, will be realized typical processes of cold working of metals by pressure.

ZrO and ZrN coatings were applied to the samples in the form of a ring with dimensions 380x280x20 from the tool material (Fig. 2), to study the antiadhesive properties when contacting with aluminum alloys.

The following types of coatings were applied to the samples: Zr-ZrN; Zr-ZrO.

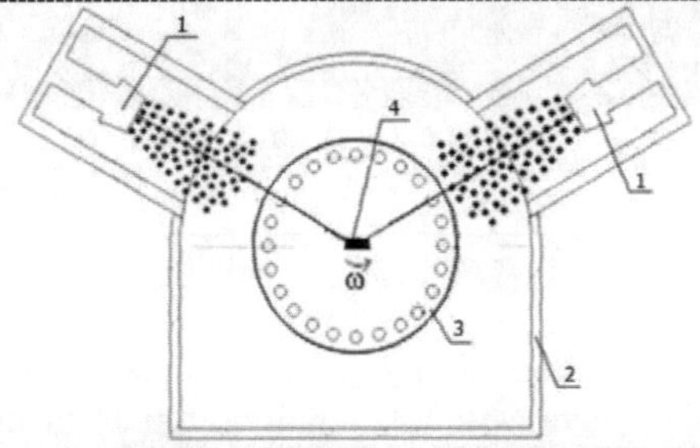

Figure 1. Scheme of the experiment:
1 - cathodes, 2 – vacuum chamber, 3 – worktable, 4 – sample

IOP Conf. Series: Journal of Physics: Conf. Series **1058** (2018) 012006 doi:10.1088/1742-6596/1058/1/012006

Figure 2. A sample coated with ZrN

During the experiments, the samples were tested in the initial state without coating, with ZrN and ZrO coatings. A sample of the tool material is mounted on a laboratory bench, specimens of a cylindrical shape of aluminum alloys are placed in the holder, a load of 10 kg is fed to the holder through a load lever. The ring-shaped sample starts to rotate by contacting at the top with the aluminum alloy. During the rotate on surface there are formed scuffs and aluminum is adhered to the surface of the tool. When the friction force increases and the tool is subjected to a critical load, the motor automatically stops. The laboratory stand is shown in (Fig. 3.).

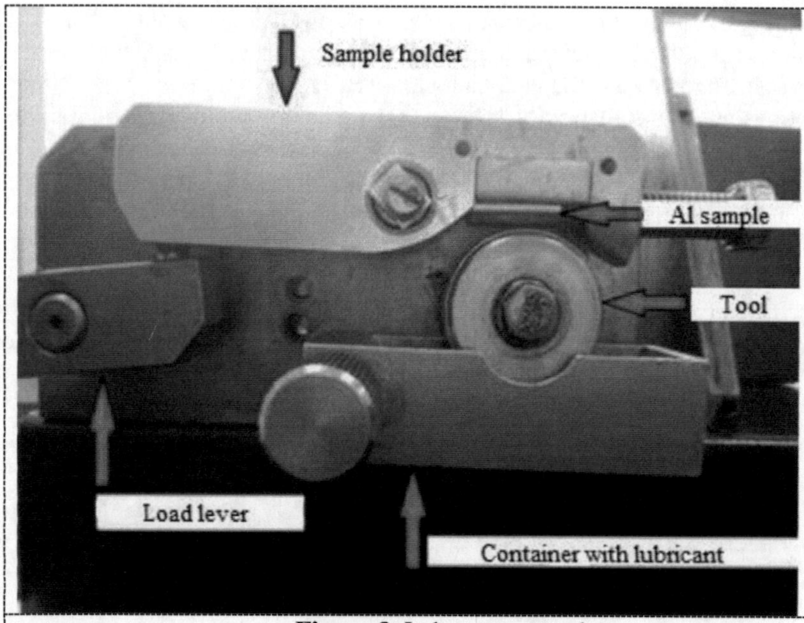

Figure 3. Laboratory stand

3. Results and discussion

Tool samples after the test are presented in (Fig. 4.). As seen, the width of the track on the surface of the rings formed as a result of contact with aluminum samples is different. In the course of the experiments, it was found that the stopping time of the motor for the original uncoated instrument sample was 1.5 seconds, for the sample with ZrO coating the value was slightly more than 2.3 seconds, and for the sample with ZrN coating - 7.2 seconds.

Figure 4. Tool samples after the test

The tracks of the clad aluminum on the surface of the tool samples are shown in (Fig. 5.)

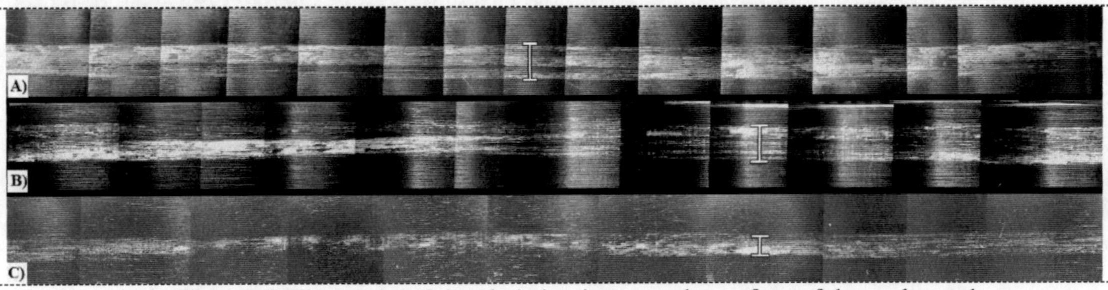

Figure 5. The tracks of the clad aluminum on the surface of the tool samples
A) Uncoated, Б) with ZrO coating, B) without ZrN coating.

The results of measuring the maximum track width showed that the original uncoated sample and the ZrO coating are approximately the same and are 4.98 and 4.90 mm, respectively and 2.5 mm on the ZrN coating.

4. Conclusions

Thus, the results of the literature review showed that to increase the resistance to aluminum adherence, it is necessary that the roughness of the surface to be contacted is minimal; Solid coatings based on nitrides and oxides of Zr, CrAlN can be used as anti-adhesive coatings (resistant to adherence of aluminum alloys). The carried out researches have shown that the deposition of zirconium nitride based coatings improves the antiadhesive properties of the tool by 3 times compared to the uncoated tool material.

Acknowledgments

The work was carried out in the framework of project No. 17-48-020688 with the support of the Russian Foundation for Basic Research and the Academy of Sciences of the Republic of Bashkortostan

References

[1] J A Mrochek [et al.]. *UE "Technoprint"*, 2004. - 369 p.
[2] Ivashchenko S A, Koida S G. "*Vestnik BNTU*", No. 6, 2008 30-34
[3] M. Berger, S. Hogmark / *Wear* 252 (2002) 557–565
[4] B. Podgornik et al. / *Wear* 261 (2006) 15–21
[5] Graschenkov D V: *VIAM* / 2010-205320, viam.ru/public/

Formation of the hydrophobic coating on polymeric textile materials in nonequilibrium low-temperature plasma

O V Vishnevskaya[1], E F Voznesensky[1], V V Vishnevsky[1], Y V Kharapudko[1], M D Gavrilov[1], F S Sharifullin[1], N V Tihonova[2], L Y Mahotkina[2]

[1] Department of Plasma Technology and Nanotechnology of High Molecular Weight Materials, Kazan National Research Technological University, Kazan, 420015, Russia
[2] Department of Designing clothes and shoes, Kazan National Research Technological University, Kazan, 420015, Russia

olesya-zef@yandex.ru

Abstract. The use of low-temperature plasma for drawing a hydrophobic coating among argon-acetylene (70:30) on polyether textile materials is investigated in work. As a result of the modification, a granular coating with a thickness of 5–10 μm is formed. This hydrophobic covering has allowed to increase values of a contact angle of wetting to 138 degrees.

1. The Introduction

Modification by drawing hydrocarbonic coatings 0,01–50 microns thick in the conditions of nonequilibrium low-temperature plasma (NLTP) with the set structure, chemical and physic-mechanical properties which allows to hydrofobisate the surface of textile materials was widely adopted. Such coatings allow to change significantly mechanical, hygienic and operational properties of initial material.

The most widespread methods of receiving volume hydrocarbonic plasma is use smoldering, radio frequency (RF) and microwave of discharges and their modifications. Advantages of RF of the discharge are: a possibility of drawing hydrocarbonic coatings on dielectric substrates, high concentration of plasma (10×10 cm^3) and smaller working pressure (0,1–100 mTorr), in comparison with the glow discharge.[1].

In this regard the possibility of formation of a hydrophobic coating on the surface of polyester textile materials when processing in nonequilibrium low-temperature plasma is considered in this work.

2. Materials, methods and equipment

As objects of a research the textile is used art. 80304 Climate 150 RS, 100% polyether. It has a polyurethane membrane coating Climate 3 and oil–repellent impregnation and textile (art. 87015 Climate 260, 78% polyether, 22% viscose) with a polyurethane membrane coating Climate 3+, a polyurethane lining and water-repellent impregnation of production of LLC «Tchaikovsky Textiles».

Modification was carried out in the experimental radio frequency capacitive (RFC) plasma generator, description explicitly of which is provided in operation [2]. The mode of plasma modification was regulated by change of parameters in the following limits: the expenditure of gas (G) of 0–0,04 g /s with, tension on the anode of the oscillator tube (U_a) of 5–7 kV, working pressure in the working camera (P) of 21–27 Pa; frequency of the generator (f) is 13,56 MHz, a processing time (τ) 1–10 min. As plasma-forming gas the compound of argon and acetylene in the ratio 70:30 was used.

Coating of polyester textile materials was created on the surface of plasma argon–acetylene (70:30) after modification. For visualization of the created coating on the surface of f textiles, the research of surface morphology and cross sections of the studied materials by means of the confocal laser scanning 3D microscope of Olympus LEXT OLS 4100 was conducted [3].

Influence of the hydrocarbonic coating received by means of plasma modification on hydrophilic and hydrophobic properties of the chosen fabrics was investigated with the help of method "lying drop" on the Kruss Easy Drop DSA 20E device [4].

Hygienic properties after RFC-plasma modifications of polyester textile materials were estimated in parameters of vapor permeability (GOST 22900–78), air permeability on the A0002D Digital device of production IDM Instruments according to the ASTM D737 standard and water resistance on the FX 3000 HYDROTESTER III device of production of Textest AG according to the EN 20811-1992 standard [5–10].

3. Results

Results of confocal microscopy (CM) of the created hydrophobic coating on textile art. 80304 Climate 150 RS are presented in Fig. 1.

On the basis of results of microscopy (Fig. 1), it is possible to draw a conclusion that after modification in nonequilibrium plasma on the surface of fibers of textiles art. 80304 Climate 150 RS and art. 87015 Climate 260 the granular coating (Fig. 1 b and d) was formed out of acetylene polymerization products. Thickness of this coating was estimated according to micro photos of cross sections (Fig. 1 g and h) was also 5–10 microns.

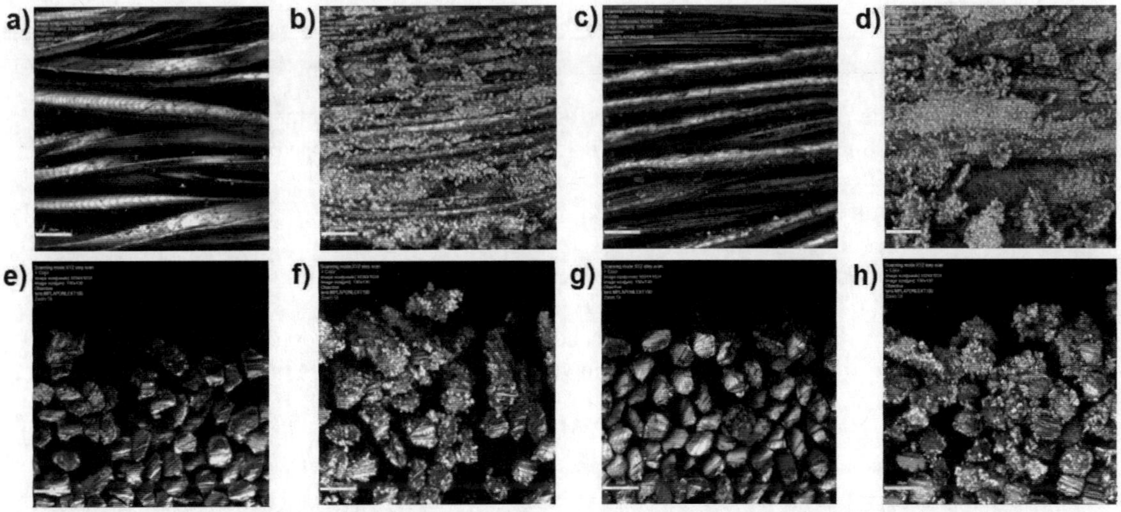

Figure 1. CM images (×100) of surfaces before (a, c) and after (b, d) modification in NLTP and cuts before (e, f) and after (g, h) modification in NLTP of textile fibers art. 80304 Climate 150 RS pictures (a, b, e, f) and art. 87015 Climate 260 in pictures (c, d, g, h)

Contact angle of wetting textile surfaces art. 80304 Climate 150 RS and textiles art. 87015 Climates 260 was measured for change research superficial and hygienic properties of the studied materials after formation of a hydrophobic coating in low-temperature plasma. Also the vapor permeability, water absorption, air permeability and water resistance was measured. Results of a research are presented in table 1.

Apparently effect of a hydrophobization of a surface of the studied materials increase at modification in plasma of argon–acetylene (70:30). It comes from the results given in table 1. The value of contact angle of wetting after plasma modification increases by 37,1% for textile art. 80304 Climate 150 RS and for 41,0% of textile art. 87015 Climate 260.

Results of a research of hygienic properties of textile materials after formation of a hydrophobic coating demonstrate that there is an increase in their water absorption by 17% and more. Also it should be noted that after formation of a coating on the surfaces of textiles, there is an insignificant decrease in their vapor- and air permeability. Fabric water absorption decreases considerably art 80304 Climate 150 RS – for 28,6%, where as textile water absorption art. 87015 Climate 260 doesn't change. Decrease in values of parameter of water absorption can have an adverse effect on the term of operation of products from such textiles.

Table 1. Values of hygienic properties of textile polyester materials before and after plasma hydrophobization

Parameter	Sample	Textile art. 80304 Climate 150 RS	Textile art. 87015 Climate 260
Contact angle of wetting, degree	Control	97	98
	Modification	133	138
Water resistance, mbar	Control	981	853
	Modification	> 1000	> 1000
Vapour permeability, g/m^2 in 24 hours	Control	1740	2500
	Modification	1739	2497
Water absorption, %	Control	1,4	0,6
	Modification	0,4	0,6
Air permeability, mm/s	Control	2,6	5,2
	Modification	2,5	5,2

4. Conclusions

Thus, as a result of the conducted researches on formation of a hydrocarbonic coating on the surface of textiles art. 80304 Climate 150 RS and art. 87015 Climate 260, it is possible to draw the following conclusions:

1. The created hydrocarbonic coating out of products argon–acetylene plasma has uneven granular character at a thickness of 5–10 microns.
2. The value of a contact angle of wetting of textiles after modifications on NLTP increases to the 138 degrees.
3. Influence of the created hydrocarbonic coating on hygienic properties of polyester textile materials has positive effect which was shown in values of water resistance increase.

Acknowledgments
Authors thank professor I. Sh. Abdullin for help in discussing the results of work.

References

[1] Krivobokov V P 2011 *Plazmennye pokrytiya (svojstva i primenenie)*[Plasma coatings (properties and applications)] Tomsk (Izdatel'stvo Tomskogo politekhnicheskogo universiteta) 137 p (In Russian)

[2] Abdullin I Sh, Zheltuhin V S, Kashapov N F 2000 *Vysokochastotnaya plazmenno–strujnaya obrabotka materialov pri ponizhennyh davleniyah. Teoriya i praktika*

primeneniya [Radio-frequency plasma-jet treatment of materials at low pressures. Theory and practice of application] Kazan (Izdatel'stvo Kazanskogo universiteta) 348 p (In Russian)

[3] Ibragimov R G, Voznesensky E F,Vishnevskaya O V, Vishnevskaya E V, Svetikov K A 2017 *Mikroskopicheskoe issledovanie plazmomodificirovannyh tekstil'nyh materialov s membrannym pokrytiem posle mnogokratnyh stirok* [Microscopic examination of plasmo-modified textile materials with membrane coating after repeated washing]. Kazan (Izdatel'stvo KNITU) pp. 67-70 (In Russian).

[4] Abdullin I Sh, Ibragimov R G, Zaitseva O V, Vishnevsky V V, Osipov N V 2014 *Gidrofobizaciya poverkhnosti materialov s membrannym pokrytiem s pomoschyu plazmennoy obrabotki* [Hydrophobization of the surface of materials with membrane coating by plasma treatment]. Kazan (*Vestnik tekhnologicheskogo universiteta* vol 19) pp. 47-49 (In Russian).

[5] Abdullin I Sh, Ibragimov R G, Zaitseva O V, Vishnevsky V V, Osipov N V 2014 *Tkani s membrannym pokrytiem* [Fabrics with membrane coating]. Saint-Petersburg (*Dizayn. Materialy. Tekhnologiya.* vol 5) pp. 25-29 (In Russian).

[6] Ibragimov R G, Voznesensky E F, Grevtsev V A, Vishnevskaya O V, Vishnevsky V V, Osipov N V 2017 *Vliyanie plazmennoy obrabotki na morozostoykost membrannykh materialov* [Effect of plasma treatment on frost resistance of membrane materials]. Kazan (*Vestnik tekhnologicheskogo universiteta* vol 3) pp. 88-90 (In Russian).

[7] Abdullin I Sh, Ibragimov R G, Vishnevskaya O V, Vishnevsky V V, Osipov N V 2016 *Regulirovanie vozdukhopronicz aemosti dvukhsloynykh membrannykh materialov s pomoschyu plazmy* [Regulation of air permeability of two-layer membrane materials by means of plasma]. Kazan (Izdatel'stvo KNITU) pp. 123-126 (In Russian).

[8] Abdullin I Sh, Ibragimov R G, Vishnevskaya O V, Vishnevsky V V, Osipov N V, Gorelisheva V E 2016 *Issledovanie izmeneniya svoystv poverkhnosti dvukhsloynykh membrannykh materialov posle plazmennoy obrabotki* [Investigation of changes in surface properties of two-layer membrane materials after plasma treatment]. Kazan (*Vestnik tekhnologicheskogo universiteta* vol 8) pp. 75-78 (In Russian).

[9] Abdullin I Sh, Ibragimov R G, Vishnevskaya O V, Vishnevsky V V, Osipov N V 2016 *Povyshenie vodoupornosti materialov s membrannym pokrytiem* [Increase in water resistance of materials with membrane coating]. Ekaterinburg (Izdatel'stvo UrFU) pp. 326-327 (In Russian).

[10] Voznesensky E F, Ibragimov R G, Vishnevskaya O V, Sisoev V A, Lutfullina G G, Tihonova N V 2017 *J. Phys.: Conf. Series* **927** 012075

Study of the behavior of fine tungsten crystallic and amorphous particles in varying conditions

L B Begrambekov, A N Voityuk, A M Zakharov

National Research Nuclear University MEPhI (Moscow Engineering Physics Institute), 31 Kashirskoe highway, Moscow 115409, Russia

Corresponding author e-mail: lbb@plasma.mephi.ru

Abstract. Formation of dust particles and their agglomerates with sizes ranging from tenths to hundreds of microns is observed in many modern plasma facilities. Dust accumulation can significantly affect discharge parameters in future thermonuclear facilities and lead to the accumulation of dangerously high amounts of tritium. Experiments on the generation of tungsten and tungsten-aluminium dust made of flakes formed during exfoliation of coating deposited on chamber walls from tungsten and aluminium targets sputtered by ions are described in this paper. A comparative analysis of the behavior of such dust and dust (figure 1), formed by crystal particles with simple geometric shapes ("crystal particles") in electric field, their degassing and change occurring in temperatures of up to 2000 K is conducted.

Figure 1. SEM photograph of the tungsten dust crystal particles with simple geometric shapes.

1. Generation of amorphous tungsten dust during exfoliation of deposited layers.

Amorphous tungsten dust was formed in the "DECOR" plasma coating deposition stand [1]. Gas discharge was induced in the atmosphere of argon with hydrogen impurity. Tungsten and aluminium targets were places in the area of gas discharge, which were sputtered by plasma ions when negatively biased towards plasma potential. Collectors made of 12X18H10T grade stainless steel (closest analogue AISI 321) sheets were placed along the plasma chamber walls, on which atoms sputtered during discharge formed the layers.

A 6A current discharge was initiated in the chamber before the target was biased to clean the surface of oxidized layers. Discharge parameters during the target sputtering were the following: working gas Ar+10% H_2, pressure $3x10^{-1}$ Pa, discharge current (4-5) A, target bias -600V, current on the target (150-250) mA, sputtering duration (4-5) hours. Sputtering of aluminium target was conducted before tungsten target to generate dust particles containing aluminium.

Deposited layers were formed on stainless steel sheets during deposition which partially exfoliated and formed dust. Approx. 0,5 g of dust was collected after each discharge. The particles were in the form of flakes ranging in size from a couple of microns to 500 μm (figure 2a), and 2-6 μm in thickness (figure 2b). As such, their sizes correlated to the sizes of dust particles formed in the chambers of modern plasma facilities [2]. Particle content measured using Energy Dispertion Spectroscopy (EDS) analysis, included W (up to 90% at.), Al (up to 60% at.), and O (up to 50% at.). Inclusion of a small (<10% at.) Fe, Cr and Ni impurity is apparently due to surface oxides on stainless steel exfoliating with flakes of deposited coating. The particles were calibrated by size using a 35x35 μm^2 cell net.

a b

Figure 2. Dust particles: (a) general view; (b) thickness of a dust particle.

2. Investigation of gas desorption during heating of dust particles.

Thermal Desorption Spectroscopy (TDS) analysis of heated crystal and amorphous particles registered desorption of mostly H_2O, H_2 and –OH hydroxyl group molecules (figure 3). Total number of registered particles per unit weight of powders is shown in Table 1.

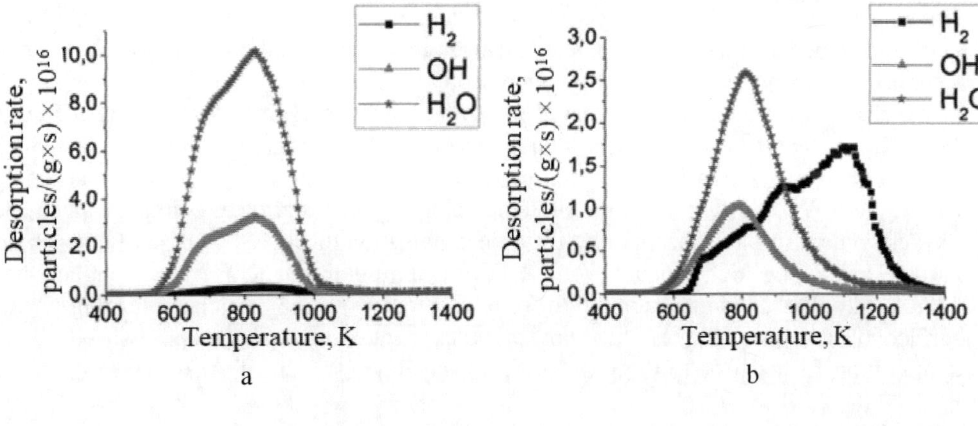

a b

Figure 3. TDS spectra for: (a) crystal particles; (б) amorphous particles.

Large desorption of water molecules from the surface of crystal particles can be explained by their surface area being larger than the surface area of amorphous particles at the same total mass since the average size of amorphous particles is 4-5 times larger. At the same time, hydrogen TDS spectrum form (figure 3b) for amorphous particles, as well as a higher amount of hydrogen desorbed from amorphous particles compared to crystal particles indicates that most of hydrogen came from the bulk of flake particles. The reason behind high amounts of hydrogen in amorphous particles is, supposedly, trapping of hydrogen prom plasma during deposition.

Table 1. Desorption from particles of both types per 1 gram of weight.

	Crystal particles	Amorphous particles
H_2O (particles/g)	5.7×10^{18}	1.3×10^{18}
OH (particles/g)	1.8×10^{18}	5.1×10^{17}
H_2 (particles/g)	2.0×10^{16}	1.2×10^{18}

3. Study on the behavior of dust particles in electric field.

It was found that both types of particles detach from surface under applied electric field. To deduce minimal electric field strength under which the particles detach from the surface, experiments were conducted, in which dust particles were placed on a horizontal metal surface. Above the surface, parallel to it, a biased plate was located. It was found that the emission of amorphous particles with sizes $\leq 35\times35$ μm^2 began when the electric field strength between the plates was $E \approx 230$ V/mm, and particles of larger size were emitted at $E \approx 150$ V/mm. Crystal particles detached from the surface at $E \approx 350$ V/mm. Emission of amorphous particles at lower electric field strength can be explained by lower adhesion strength with the surface of the plate, and a higher electric charge formed on particles due to larger surface area-to-volume ratio.

4. Study on the structural change of dust particles during.

During the heating experiments dust particles were placed on the box-shaped heater element made of thin tungsten sheet, which was mounted between electrodes in the vacuum chamber of a VUP-2K stand. Electric current was fed through the heater, the temperature was controlled using a tungsten-rhenium thermocouple. Dust particles were heated incrementally with a 100 K step, changes in structure and composition of particles were deduced using scanning electron microscopy and EDS.

No change in structure was observed for amorphous particles for temperatures of up to 1073 K. Starting from 1473 K, which corresponds to minimum temperature at which tungsten recrystallization occurs, metal grains and pores along grain boundaries appeared (figure 4a). With further increase of temperature, grains started growing in size, and oxygen content in flakes decreased due to decomposition of oxides. In case of amorphous particle containing high amounts (>55% at.) of aluminium, at temperatures higher than 1473 K, droplets of molten aluminium were secreted from the particle, spreading on the surface of a heater (figure 4c). For particles containing two layers, with different composition for each layer (85% at. of W on layer 1, 60% at. of Al on layer 2), cone-shaped structure growth could be observed along the edges of a dust particle (figure 4b), which can allude to the development of internal stress in the layers. If a particle had a low Al content (<25% at.), no such phenomena occurred for temperatures specified above, and it is presumed that aluminium evaporated without forming droplets.

In case of crystal dust particle heating, larger crystals were baked together along the edges, and smaller crystals baked together into clusters (figures 4d, 4e).

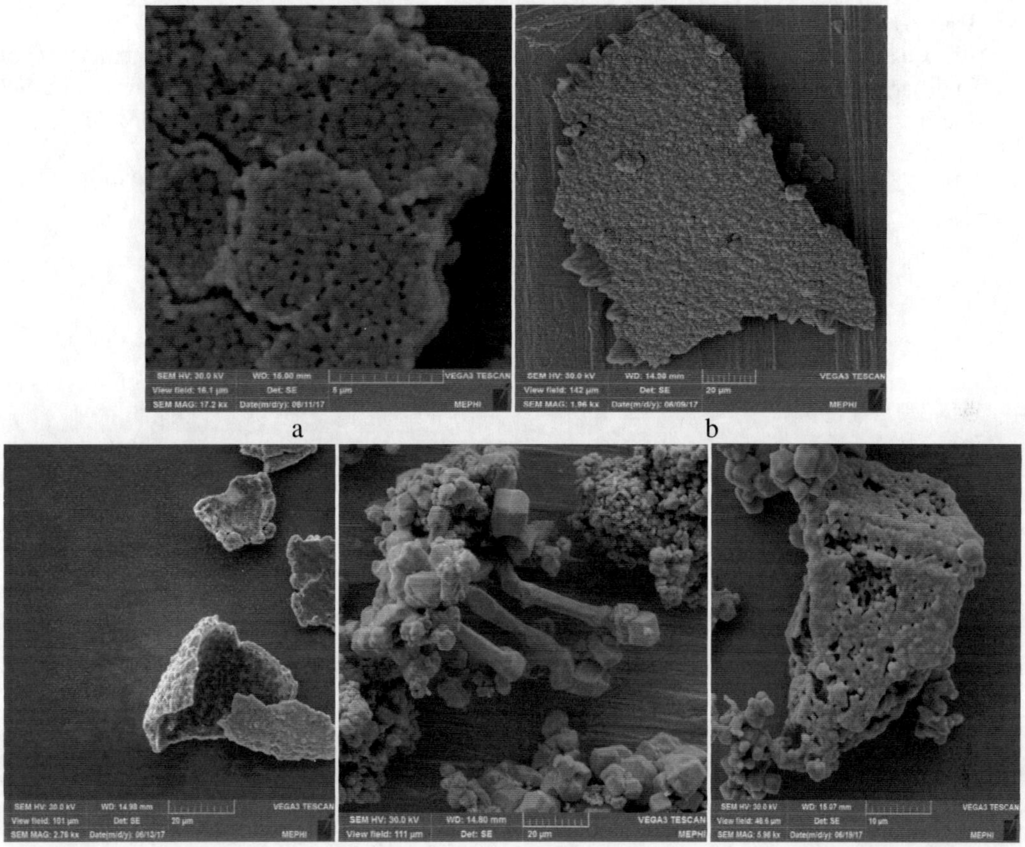

a b c d e

Figure 4. Different structural changes of dust particles during heating. a) recrystallisation, б) cone structures, c) molten Al droplet secretion, d) baking of larger crystal particles, e) baking of smaller crystal particles into clusters.

5. Conclusion.

Experiments on the generation of amorphous dust particles are conducted. A comparative study of the behavior of such dust with crystal particle dust during heating and subjection to electric field is conducted.

It was shown that gas desorption occurs from the surface and the bulk of dust particles during heating, as well as change in their structure: recrystallisation and metal grain growth, formation of cone-like structures due to internal stress, melting and evaporation of low melting point components (Al) in case of amorphous dust particles, baking of larger particles along the edges and baking of smaller particles into clusters in case of crystal particles.

It was shown that both types of particles are emitted from the surface under applied electric field. Minimal values of electric field strength for emission are deduced for each

IOP Conf. Series: Journal of Physics: Conf. Series **1058** (2018) 012008 doi:10.1088/1742-6596/1058/1/012008

type of particle, them being E ≈ 230 V/mm for amorphous particles of size ≤ 35×35 μm^2, E ≈ 150 V/mm for larger amorphous particles and E ≈ 230 V/mm for crystal particles.

Acknowledgments
This work was carried out with the financial support of Ministry of Science and Education of Russian Federation under the agreement No. 14.575.21.0169 (RFMEFI57517X0169).

References

[1] Begrambekov L.B., Gordeev A. A., Grunin A.V., Evsin A.E., et al. The Peculiarities of hydrogen trapping and retention in zirconium with chromium coating // Materials of VII International School-Conference of young researchers and specialists IHISM'11 (24-28 October, 2011, Zvenigorod), p. 323-340

[2] S.I. Krasheninnikov et al., // Plasma Phys. Controlled Fusion 53(2011) 083001.

The metal surface cleaning using a vapor-gas discharge

Az T Gabdrakhmanov, I H Israphilov, L N Shafigullin
Kazan Federal University, Naberezhnye Chelny, Russian Federation

E-mail: ATGabdrahmanov@kpfu.ru

Abstract. In this paper, the study results of the surface of samples after their purification using a vapor-gas discharge are presented.

1. Introduction
The undesirable or harmful contaminants and foreign substance arise in the process of production (manufacture, processing, storage) on the surface of metal parts.
The metal surface cleaning is the procedure for removing contaminants from the metal surface to a certain level of its purity. The cleaning is carried out by various methods - mechanical, physical, chemical, physico-chemical and chemical-thermal.

In papers [1-10], an efficient and environmentally pure method of the metal surfaces cleaning using the vapor-gas discharge is proposed. The discharge region lies is in the air space between the liquid electrolyte and the aluminum anode [1-2]. A plasma stream flows from the discharge region into the surrounding atmosphere. The electrolyte flow (processing water) circulates inside of the cathode unit, which creates favorable conditions for reducing heat losses at the cathode. The cleaning part is used as the cathode.

2. Experimental studies
For study the surface after cleaning using the vapor-gas discharge, two samples from steel 08U without treatment (No. 1) were taken and after purification (No. 2). The studies were carried out at the following environmental parameters: air temperature is 25°C; - relative humidity is 60%. At external examination on the surface of sample No. 2, matte light and dark areas were detected in accordance with Fig. 1.

sample No. 1 sample No. 2

Figure 1. Visual appearance of samples

IOP Conf. Series: Journal of Physics: Conf. Series **1058** (2018) 012009 doi:10.1088/1742-6596/1058/1/012009

The studies were carried out on a scanning electron microscope VEGA 3 LMH from Company TESCAN, Czech Republic, using the X-ray energy dispersive microanalysis system AZtecEnergy Advanced with the X-max 20 Standart detector from Company Ochford Instruments.
The study results are presented in Figures 2-4.

×39 ×1300

Figure 2. Visual appearance of the surface of the sample plate No. 1

Figure 3. The surface of the sample plate No. 2 with an increase in × 42

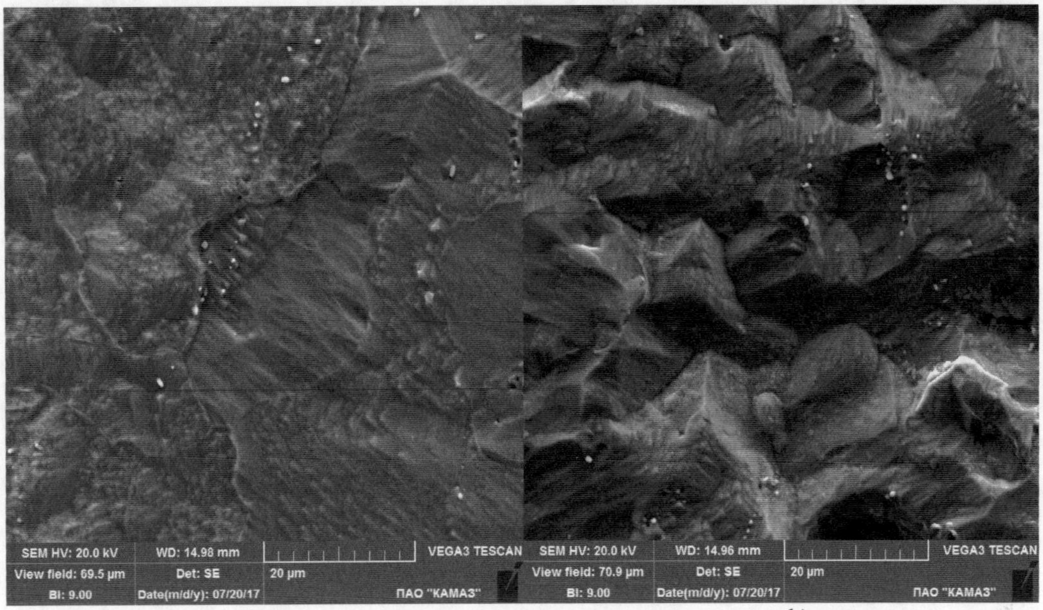

a) b)

Figure 4. Visual appearance of the surface of the sample plate No. 2 with an increase of × 1300; a) section 1, b) section 2

In the electron-microscopic investigation of the samples plates and parts revealed:
- the sample surface No. 1 with factory finish;
- the sample surface No. 2 with increased etching;

3. The conclusion

The study results show the proposed method provides high-performance and high-quality the surface cleaning of parts from virtually any contamination - mineral and organic preservative greases, rust, scale, remnants of old galvanic and paint coatings, enamel insulation from electric wires. When cleaning with the use of steam and gas discharge, the surface roughness rises, which increases the adhesion to the functional coatings applied in the following process operation.

References

[1] Gabdrakhmanov A. T., Israphilov I. H., Galiakbarov A. T. *Preparation of metal surfaces for application of functional coatings* // IOP Conf. Series: Journal of Physics: Conf. Series. Volume 789 (2017). Article number 012009.

[2] Gabdrakhmanov Az.T., Gabdrakhmanov Al.T., Galiakbarov A.T. *Investigation of the characteristics of a gas-vapor discharge with an aluminum anode and a liquid cathode.* Online electronic scientific and technical journal "Socio-economic and technical systems: research, design, optimization". №3 (70) - Nab. Chelny: K (P) FU, 2016.

[3] Shakirov Yu. I. , Valiev R.I., Khafizov A.A., Valiev R.Aю and Khakimov R.G. Erosion of electrode metal in the electric discharge under the exposure of the electrolyte stream J. of Phys.: Conf. Ser. 2016, 669, 012064.

[4] Khafizov A.A., Shakirov Y.I., Valiev R.A., Valiev R.I., Khafizova G. M. *Study of thermal and electrical parameters of workpieces during spray coating by electrolytic plasma jet.* J. of Phys.: Conf. Ser. 2016, 669, 012030.

[5] Khafizov A.A., Valiev R.I., Shakirov Yu.I., Valiev R. A. *Steel surface modification with plasma spraying electrothermal installation using a liquid electrode*. J. of Phys.: Conf. Ser. 2014,

567, 012026.
[6] Denisov D.G., Kashapov N.F., Kashapov R.N. *The appearance of shock waves in the plasma electrolytic processing.* Iop conference series: materials science and engineering, 2015, №012005.
[7] Gabdrakhmanov Az.T. *Features of the current-voltage characteristics of pulsed plasma generator* / Online electronic scientific and technical journal "Socio-economic and technical systems." №2 (59) - Nab. Chelny: INEKA, 2011.
[8] Israfilov I.H., Saubanov R.R., Rakhimov R.R. *Prospective application of highly concentrated energy for the surface heat treatment products* // Socio-economic and technical systems: research, design, optimization. № 1, 2011, p. 25-30.
[9] Zvezdin V.V., Galiakbarov A.T., Nugumanova A.I., Gabdrakhmanov Az.T., Saubanov R.R. *The investigation of the influence the parameters the pulsed plasma generator on quality process* // Bulletin KSTU. AN Tupolev. Number 2, 2010, s.50-52.
[10] Denisov D.G., Kashapov N.F., Kashapov R.N. *The appearance of shock waves in the plasma electrolytic processing* // Iop conference series: materials science and engineering – 2015. – №012005.

Modeling of glow discharge in longitudinal gas flow in a plane channel of variable cross section

I G Galeev and T Ya Asadullin

Kazan National Research Technical University named by A.N.Tupolev
420111 Russia Kazan, K.Marx str., 10

galeev.physics@kstu-kai.ru

Abstract. A two-dimensional model of glow discharge positive column in gas flow has been suggested in this paper. It is considered that the discharge maintained between the permeable mesh electrodes installed in the initial and end sections of the plane channel. In general case the channel width changing along the stream is accounted. The positive column in gas flow is described by integro-differential system of equations with an approximation of constant gas flow density. The border conditions for gas ionization degree and temperature in the initial section of channel are given by any functions. An analytical solution of that system was obtained and special cases of the boundary conditions analyzed.

The glow discharge in gas flow is an important object of study in relation to the creation of the active medium of lasers, obtaining and maintaining plasma of glow discharge plasmatron for various purposes [1,2]. The organization of gas flow through the discharge region allows significantly improve its characteristics such as spatial discharge homogenity, limit energy input etc. Therefore theoretical and experimental study of the glow discharge is an actual task. The authors [3] investigated the glow discharge pressed by magnetic field to surface of liquid hydrocarbons for the production of carbon nanotubes. In a work [4] the discharge stability was increased by organization of a supersonic gas flow, numerical simulation the discharge characteristics was done.

An analytical model of the glow discharge in the gas flow is present in this paper. The accepted model describes the glow discharge maintained between two mesh electrodes spaced at the beginning and end of the plane channel. For instance the mesh electrodes AA* and BB* can consist of a system of rods or water-cooled tubes arranged parallel to each other (Figure 1). As in the first approach the uniformity is assumed in the perpendicular direction to the flow so the two-dimensional model is used. The origin of coordinate system is chosen in the middle of the initial section of camera.

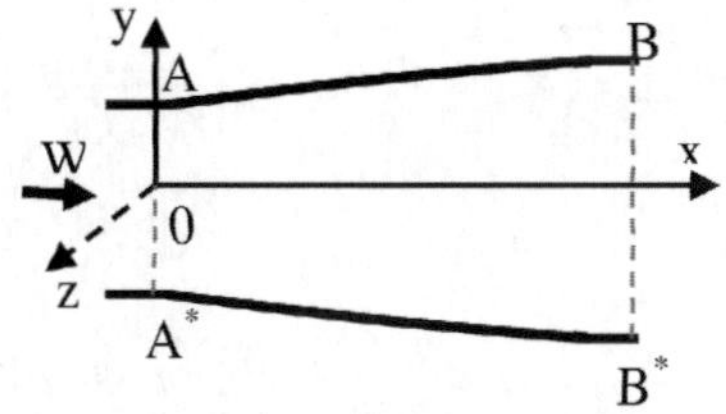

Figure 1. The scheme of the discharge chamber of variable section.

The accepted system of equations has the form

$$c_a \frac{\partial^2}{\partial y^2}\alpha - \frac{g_x}{m}\frac{\partial}{\partial x}\alpha - \frac{g_y}{m}\frac{\partial}{\partial y}\alpha + c_v E^2 \alpha = 0,$$

$$I = 2e\mu_e b_z E \int_0^{b(x)} \frac{n}{N} dy,$$

$$c_p g_x \frac{\partial T}{\partial x} + c_p g_y \frac{\partial T}{\partial y} = \lambda \frac{\partial^2 T}{\partial y^2} + \delta\, e\mu_e n \frac{E^2}{N},$$

$$P = NkT.$$

$$G = 2b_z \int_0^{b(x)} g_x dy.$$

It is written in approximation of the constancy of gas flow density in any cross section perpendicular to the flow - $g_x = \rho W_x$=const.

The solution of this system was obtained under the conditions:

-at the entrance to the discharge zone (at x = 0)) $\tilde{\alpha}_e = \frac{\tilde{n}}{\tilde{N}} = \varphi_\alpha(\tilde{y}),\ \tilde{T} = \tilde{\vartheta},$

- on the discharge chamber walls (at y = ±1/f(x)) $\tilde{\alpha}_e = 0,\ \tilde{T}(\tilde{x}, \pm 1/f(\tilde{x})) = \tilde{\theta}(\tilde{x})_\omega,$

- on the plane of symmetry of the discharge chamber (y = 0) $\tilde{g}_y = 0, \frac{\partial \tilde{\alpha}_e}{\partial \tilde{y}} = 0,\ \frac{\partial T}{\partial \tilde{y}} = 0.$

Here marked with: D_a – the ambipolar diffusion coefficient, $C_a = N D_a$, N - is the concentration of neutral particles, g_x, g_y –the gas flux densities in the respective directions. δ - a share of inserted power going into gas heating, $\alpha = \frac{n}{N}$ - degree of ionization of the gas, the function f(x) defines the shape of the channel, $b = \frac{b_{|x=0}}{f(\tilde{x})} = \frac{b_0}{f(\tilde{x})}$, I –the total electric current, G –the complete mass flow of the gas. μ_e - mobility of electrons, C_P – the heat capacity of gas at constant pressure. Any variable with a sign ~ over symbol corresponds to the dimensionless value.

The system of equations includes the continuity equation for electron gas, the Ohm's law in integral form, the equation of heat balance, state equation of the ideal gas, the equation of full gas flow constancy in any cross section of the channel. The system of equations is written taking into account the plasma quasineutrality condition of glow discharge positive column. The continuity equation for the electron gas accounts for the ambipolar diffusion of electrons to discharge chamber walls, convective transport and birth of particles due to ionization. It is assumed that the discharge is maintained at a sufficiently low gas pressure in the channel so the volume recombination of charged particles is neglected.

The solution can be represented in the form

$$E = \left[2c_E \Phi^2(\tilde{x})\right]\left[\int_0^{\tilde{x}} \frac{c_E}{f(\tilde{x})\Phi^2(\tilde{x})} d\tilde{x} + \frac{\Phi^2(\tilde{x})}{E^2(0)\Phi^2(0)}\right]^{-0,5},$$

$$\tilde{\alpha}_e = \frac{I\sum_{m=1}^{\infty} A_m \exp\left[-\frac{\pi^2(2m-1)^2}{4Pe_d}\int_0^{\tilde{x}} f(\tilde{x})d\tilde{x}\right]\cos\left[\frac{\pi}{2}(2m-1)f(\tilde{x})\tilde{y}\right]}{2e\mu_e b_z b_0 \alpha_e(0,0)E(\tilde{x})\Phi(\tilde{x})},$$

$$\tilde{T} = \tilde{\theta}_\omega(\tilde{x}) + \sum_{m=1}^{\infty} \tilde{\theta}_m(\tilde{x})\cos\left[\frac{\pi}{2}(2m-1)f(\tilde{x})\tilde{y}\right].$$

$$\tilde{\theta}_m(\tilde{x}) = \exp\left[-\frac{\pi^2(2m-1)^2}{4Pe_d}\int_0^{\tilde{x}} f(\tilde{x})d\tilde{x}\right]\left\{\frac{\delta I E(0)}{2g_0 c_p T_0 b_z}\times\right.$$

$$\left.\times\int_0^{\tilde{x}} \frac{E d\tilde{x}}{f(\tilde{x})\Phi(\tilde{x})} A_m \exp\left[-\frac{\pi^2(2m-1)^2}{4Pe_d}\left(\frac{1}{Pe_d} - \frac{1}{Pe}\right)\int_0^{\tilde{x}} f(\tilde{x})d\tilde{x}\right]d\tilde{x} - \right.$$

$$-\int_0^{\tilde{x}} \frac{4(-1)^{m+1}}{\pi(2m-1)} \frac{d\tilde{\theta}_\omega}{d\tilde{x}} \exp\left[\frac{\pi^2(2m-1)^2}{4Pe}\right]\int_0^{\tilde{x}} f(\tilde{x})d\tilde{x} + B_m\Bigg\},$$

here indicated $\Phi(\tilde{x}) = \sum_{m=1}^{\infty} A_m \frac{2(-1)^{m+1}}{\pi(2m-1)f(\tilde{x})} \exp\left[-\frac{\pi^2(2m-1)^2}{4Pe_d}\int_0^{\tilde{x}} f(\tilde{x})d\tilde{x}\right]$, $Pe, Pe_d = \frac{g_0 b_0}{c_a m}$ - the Pecle number and the diffusion Pecle number, $c_E = \frac{c_v m b_0}{g_0}$. The constants of integration B_m are determined from the conditions on the degree of ionization and temperature at the inlet to the discharge chamber

$$A_m = 2\int_0^1 \varphi_{\alpha(\tilde{y})} \cos\left[\frac{\pi}{2}(2m-1)\tilde{y}\right] d\tilde{y}.$$

$$B_m = 2\int_0^1 \tilde{\vartheta}(\tilde{y}) \cos\left[\frac{\pi}{2}(2m-1)\tilde{y}\right] d\tilde{y} - \frac{4(-1)^{m+1}}{\pi(2m-1)} \tilde{\theta}_\omega(0).$$

In the particular case when the shape of the channel have the saturation downstream and of assignment of initial condition in the form $\tilde{\varphi}_\alpha = \cos\left(\frac{\pi\tilde{y}}{2}\right)$ distribution of the electric field and ionization gas, respectively have the form

$$E = \frac{\frac{f(\tilde{x})}{f_*} E_*}{\left\{1 - \left[1 - \left(\frac{1}{f_*}\frac{E_*}{E_0}\right)^2\right] \exp\left[-\frac{\pi^2}{2Pe_d}\int_0^{\tilde{x}} f(\tilde{x})d\tilde{x}\right]\right\}^{0,5}},$$

$$\tilde{\alpha}_e = \frac{E(0)}{E(\tilde{x})} f(\tilde{x}) \cos\left[\frac{\pi}{2} f(\tilde{x})\tilde{y}\right],$$

here f_*, E_* - the corresponding values in the steady-state section of the channel.

The obtained relations allow to calculate the degree of ionization, gas temperature inside the channel under various conditions at the inlet to the discharge area.

The work was done with support of Grant of Ministry of education and science of the Russian Federation 3.6564.2017/8.9.

References

[1] Raizer Yu P 2009 *Physics of gas discharge* (Moscow: Publishing house: Intellect) 736 p
[2] Timerkaev B A, Sofronitskiy A O, Andreeva A A 2016 Carbon nanotubes formation in the decomposition of heavy hydrocarbons creeping along the surface of the glow discharge *J. of Phys. Conf. Ser.* **669(1)** 012062.
[3] Saifutdinov A I, Timerkaev B A, Zalyaliev B R 2016 Control of the glow discharge parameters at low pressures by means of a transverse supersonic gas flow *High Temperature* **54(5)** 632-8
[4] Timerkaev B A, Ahmetov M M, Zalyaliev B R, Petrova O A, Israfilov D I 2014ongitudinal distribution of electrical parameters in normal glow discharge. *J. of Phys. Conf. Ser.* **567(1)** 012036

IOP Conf. Series: Journal of Physics: Conf. Series **1058** (2018) 012011 doi:10.1088/1742-6596/1058/1/012011

Method calculation of a temperatures field of gas flow, in a cylindrical channel in the presence of internal sources of non-uniformly distributed heat in the axial direction

Alexander Gerasimov, Alexander Kirpichnikov, Leonid Rachevsky, Farida Sabirova, Igor Yakimov

Kazan National Research Technological University 420015, Karl Marx str., 68 Kazan, Russia
Kazan National Research Technological University 420015, Karl Marx str., 68 Kazan, Russia
Kazan National Research Technological University 420015, Karl Marx str., 68 Kazan, Russia
Kazan National Research Technological University 420015, Karl Marx str., 68 Kazan, Russia
Kazan National Research Technical University named A. N. Tupolev 420111, Karl Marx str., 10 Kazan, Russia

gerasimov@kstu.ru

Abstract. The calculation of a temperature field of gas flow with an internal heat source is generalized in case of a non-uniform axisymmetric heat distribution. The distribution is approximated by a set of segments of various lengths. The source function in each segment is considered constant, in the axial direction, while its radial profile is specific only to that segment. Thus, gas flow sequentially moves through $n+2$ areas: input z < 0, output z > 1 and the area of the internal heat source $0 \leq z \leq l$, presented by n segments. The obtained analytical solutions provide us with a straightforward way to calculate the heat balance for a wide class of complex problems related to describing air flow in the presence of non-uniform heat source distribution in the axial direction.

1. Introduction

The majority of theoretical papers dealing with convective heat transfer in gas flow with an internal heat source [1-6] considered the energy balance equation only for the area in which the heat source was present. The area was assumed infinite in the direction of the gas flow. An attempt to improve the accuracy in determining gas temperature (especially for short areas) was undertaken in article [7], in which heat source distribution was assumed uniform in the axial direction, even though this is a strong restriction.

The authors of this paper propose to get rid off the aforementioned restriction by approximating the heat source distribution with n segments of various lengths. Solving the energy balance equation for each segment is simple since its power density q_{vk} is given, the heat source function is assumed constant in the axial direction and the radial profile $F_k(r)$ is known as well. The final solution of the

IOP Conf. Series: Journal of Physics: Conf. Series **1058** (2018) 012011 doi:10.1088/1742-6596/1058/1/012011

temperature field is obtained by "stitching" the per-segment solutions at the boundaries of the segments. Obviously, the required accuracy can be reached by increasing the number of segments.

2. Equation

Similar to [7, 8] we assume the physical properties of a gas unchanged for the considered interval of temperatures and equal to some average values. The processes of vibrational relaxation are ignored because gas is treated as mono-atomic. Axisymmetric internal heat source distribution in the channel is presented in fig. 1 and is described by the function $E(r, z)$:

$$E(r,z) = \begin{cases} 0, & z<0, \quad z>\dfrac{L}{R} \\ F_k(r)\cdot q_{v_k}, & z_{k-1}\le z\le z_k \end{cases}$$

where z_k are boundary-determining segments in the axial direction, by which the whole area of an internal heat source is divided.

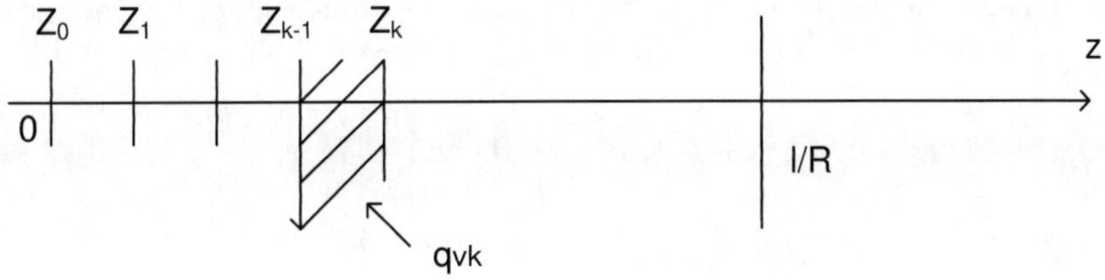

Fig. 1. Area of an internal heat source

(layers with various values of power density)

U designates the averaged (by time and radial direction) axial velocity.

Based on the above assumptions, the process of convective heat transfer can be described by the following differential equations

$$\mathrm{Pe}\frac{\partial\theta}{\partial z} = \frac{1}{r}\frac{\partial}{\partial r}\left(r\frac{\partial\theta}{\partial r}\right) + \frac{\partial^2\theta}{\partial z^2}; \; z<0, \quad z>\frac{l}{R};$$

$$\mathrm{Pe}\frac{\partial\theta}{\partial z} = \frac{1}{r}\frac{\partial}{\partial r}\left(r\frac{\partial\theta}{\partial r}\right) + \frac{\partial^2\theta}{\partial z^2} + \frac{q_{v_k}R^2}{\lambda}F_k(r);\; 0<z\le\frac{l}{R} \tag{1}$$

Here $\mathrm{Pe} = \dfrac{UR}{\mathrm{a}}$ is the Pe number, $\theta = T - T_W$, where T_W is the temperature on the wall, a and λ are temperature conductivity and heat conductivity coefficients respectively; q_{vk} is thermal-power density on the channel axis in a *k-th* segment. Gas temperature on the wall T_W is assumed constant, and its profile in the radial direction symmetric.

It is important to emphasize the continuity conditions for T and $\dfrac{\partial T}{\partial z}$ on the boundaries of the segments.

3. Solution method

The arbitrary radial function of a heat source $F_k(r)$ can be expanded into a Fourier-Bessel series according to functions $J_0(\mu_i r)$, $0 \le r \le 1$, where μ_i is a *i-th* root of the equation $J_0(x) = 0$. Therefore,

$$F_k(r) = \sum_{i=1}^{\infty} a_{ik} J_0(\mu_i r)$$

Where

$$a_{ik} = \frac{1}{J_1^2(\mu_i)} \int_0^1 F_k(r) J_0(\mu_i r) r dr$$

Thus, the distribution function of excess temperature $\theta(r, z)$ can be written as:

$$\theta(r, z) = \sum_{i=1} f_i(z) J_0(\mu_i r) \tag{2}$$

By analogy with [7], let us apply the Kantorovich-Galerkin method [9] to equations of convective heat transfer (1) by computing the inner product of these equations with basic functions $J_0(\mu_i r)$. As a result, we get a system of ordinary differential equations of the second order with constant coefficients:

$$\begin{cases} Pe \cdot f_i'(z) = -\mu_i^2 f_i(z) + f_i''(z), \quad z < 0, \quad z > l/R \\ Pe \cdot f_i'(z) = -\mu_i^2 f_i(z) + f_i''(z) + \dfrac{q_{v_k} R^2}{\lambda} \cdot a_{ik}, \quad z_{k-1} \le z \le z_k \end{cases} \tag{3}$$

where a_{ik} are expansion coefficients of a radial function of the source $F_k(r)$ in each of the segments $k=1,2... n$.

As seen above (3), the first equation can be derived easily from the second one if the power density factors, q_{v0} and q_{vN+1} are zeroed in the input and output areas, respectively.

Let us designate $f_i(z)$ as $f(z)$, μ_i as μ, and the last term (source) in the second equation of the system (3) as S_k, for simplicity, i.e.

$$S_v = \frac{q_{v_k} R^2}{\lambda} \cdot a_{ik}$$

Then the second equation of the system (3) takes the form

$$f'' - \mu f - Pe \cdot f' + S_k = 0, \tag{4}$$

Where $S_0 = S_{n+1} = 0$, $S_k = const$ is in the interval $z \in [z_{k-1}, z_k]$.

Let us look for the solution of equation (4) in the form

$$f(z) = A_k \cdot e^{\alpha z} + B_k \cdot e^{\beta z} + \frac{S_k}{\mu^2},$$

where

$$\alpha = \frac{Pe + \sqrt{Pe^2 + 4\mu^2}}{2} > 0, \tag{5}$$

$$\beta = \frac{Pe - \sqrt{Pe^2 + 4\mu^2}}{2} < 0. \tag{6}$$

A_k and B_k are unknowns.

Let us find them by "stitching" $f(z)$ and $f^{'}(z)$ using their aforementioned continuity requirements on the segment boundaries z_k:

$$\begin{cases} A_k \cdot e^{\alpha z} + B_k \cdot e^{\beta z} + \dfrac{S_k}{\mu^2} = A_{k+1} \cdot e^{\alpha z} + B_{k+1} \cdot e^{\beta z} + \dfrac{S_{k+1}}{\mu^2} \\ \alpha A_k \cdot e^{\alpha z} + \beta B_k \cdot e^{\beta z} = \alpha A_{k+1} \cdot e^{\alpha z} + \beta B_{k+1} \cdot e^{\beta z} \end{cases} . \quad (7)$$

Let us enter the designations: $\Delta A_{k+1} = A_{k+1} - A_k$, $\Delta B_{k+1} = B_{k+1} - B_k$. Then the system of linear algebraic equations takes the form

$$\begin{cases} e^{\alpha z} \Delta A_{k+1} + e^{\beta z} \Delta B_{k+1} = \dfrac{S_k - S_{k+1}}{\mu^2} \\ \alpha e^{\alpha z} \Delta A_{k+1} + \beta e^{\beta z} \Delta B_{k+1} = 0 \end{cases} \quad (8)$$

and can be easily solved for ΔA_{k+1} and ΔB_{k+1}.

As a result

$$\Delta A_{k+1} = \frac{\beta}{\beta - \alpha} \cdot \frac{S_k - S_{k+1}}{\mu^2 \cdot e^{\alpha Z_k}}$$

and

$$\Delta B_{k+1} = \frac{\beta}{\alpha - \beta} \cdot \frac{S_k - S_{k+1}}{\mu^2 \cdot e^{\beta Z_k}} .$$

It is necessary to obtain A_k and A_{k+1} from ΔA_{k+1} and B_k and B_{k+1} from ΔB_{k+1}. It is apparent from formulas (5) for α and (6) for β that they have real and opposite-in-sign values. Let us assume

$$\left[\begin{array}{l} \alpha > 0 \\ \beta < 0 \end{array} \right. .$$

As seen from the above formulas, the temperature field in each segment is computed using the two unknowns (A_k, B_k). The exception takes place in the areas $z < z_0 = 0$ and $z > z_n = \frac{l}{R}$, in which one of these unknowns disappears. Otherwise, the function *f(z)* would approach infinity, which makes no physical sense.

So, from

$$f(z) = A_0 \cdot e^{\alpha z} + B_0 \cdot e^{\beta z} + \frac{S_0}{\mu_2} = 0, \; z < z_0 = 0$$

It follows that $B_0 = 0$.
Similarly, from

$$f(z) = A_{n+1} \cdot e^{\alpha z} + B_{n+1} \cdot e^{\beta z} + \frac{S_{n+1}}{\mu_2} = 0, \; z > z_n = \frac{l}{R}$$

It follows that $A_{n+1} = 0$.

Using the fact that $A_{n+1} = 0$ and ΔA_{n+1} are known, as are B_0 and ΔB_1, we apply the following recurrent relations $A_k = A_{k+1} - \Delta A_{k+1}$, $B_{k+1} = B_k + \Delta B_{k+1}$ with initial values $A_n = -\Delta A_{n+1}$, $\Delta B_1 = B_1$ to obtain a final solution to the problem.

4. Results and conclusion

The method described above can be easily applied to obtain the solution in [7] by letting n=1 i.e. where the area of the internal heat source is uniform. The presented model enables us to calculate the influence of a non-uniform heat source distribution on a temperature profile in both the presence of gas flow (Pe$\neq$0) and its absence (Pe=0).

The considered approach seems promising. First, it allows for the approximation of a wide class of non-uniform heat source distribution functions in order to compute local and integral energy balance for laminar and turbulent modes of gas flow in a cylindrical channel.

Secondly, it enables us to investigate the inverse (synthetic) problem, i.e. what heat source distribution should be in order to synthesize a desired temperature profile with a predefined accuracy. Besides, it gives us an opportunity to solve a related problem pertaining to heat input optimization; specifically, how to arrange power sources to obtain the desired distribution of temperatures in the zone of maximal interest, for example, in an area of chemical reaction.

References

[1] Zhukov M.F., et al. Electric arc gas heaters. Novosibirsk: Science. 1973.
[2] Slattery, J. C., *Momentum, Energy and Mass Transfer in Continua*, McGraw-Hill, New York, USA, 1971
[3] Sergeev V. L.//Low temperature plasma generators.- Minsk, 1986 p. 19
[4] Gerasimov A. V., Kirpichnikov A. P.//Reports of Academy of Sciences. 2003. V. 389. No. 2. p. 177.
[5] Gerasimov A. V., Kirpichnikov A. P., Rachevsky L. A. //Thermal Science. 2005. Vol. 9.No 1. P. 131.
[6] Gerasimov A. V., Kirpichnikov A. P.//Thermophysics and Aeromechanics. 2013. No. 4. p. 439.
[7] Gerasimov A. V., Kirpichnikov A. P., Rachevsky L. A.//Achievements of applied physics. 2013. No. 1 p. 30.
[8] Gerasimov A. V., Kirpichnikov A. P., Rachevsky L. A // Thermal Science. 2015. Vol. No. 2. P. 735
[9] Kantorovich L. V., Krylov V. I. Approximate Methods of Higher Analysis. M.-L.: Gostekhizdat (State Technical Publishing). 1949.

IOP Conf. Series: Journal of Physics: Conf. Series **1058** (2018) 012012 doi:10.1088/1742-6596/1058/1/012012

On the possibility of low-temperature removal of hydrogen isotopes from tungsten in conditions of thermonuclear and plasma facilities

L B Begrambekov, S S Dovganyuk, A E Evsin and A S Kaplevskiy.

National Research Nuclear University MEPhI (Moscow Engineering Physics Institute), 115409 Kashirskoe shosse 31, Moscow, Russia Federation

e-mail: ssdovganyuk@plasma.mephi.ru

Abstract. Acceleration of hydrogen isotope trapping and desorption has been observed in the experiments with stainless steel (Activated Surface Penetration - ASP) under irradiation by hydrogen atoms and hydrogen plasma ions with oxygen impurity. ASP is considered in this paper for a method of low-temperature hydrogen isotope removal from tungsten and aluminum-coated tungsten under irradiation by hydrogen plasma ions with oxygen impurity. It is shown that periodic irradiation by hydrogen plasma ions with oxygen impurity allows preventing accumulation of implanted deuterium ions in tungsten.

1. Introduction

Beryllium and tungsten are chosen as the materials of the first wall and the divertor of the ITER tokamak, respectively. Due to the possible presence of oxygen impurity in ITER plasma, existence of oxides on their surfaces, as well as the surface of beryllium layer deposited on tungsten. The development of the low-temperature hydrogen isotope removal method from tungsten and tungsten with beryllium layers is relevant for safety reasons as tritium trapping in the first wall and the divertor of the tokamak is limited as 700g.

Barrier properties of oxide layers relative to hydrogen diffusion are well known [1-4]. It was shown in work [1] that barrier properties of the oxide layer on stainless steel relative to hydrogen isotope diffusion change under ion and plasma irradiation. It was found that irradiation of 12X18H10T-grade steel (closest analogue is AISI 321) by hydrogen atoms and hydrogen plasma ions with oxygen impurity drastically improve hydrogen isotope diffusion through the surface of stainless steel (Active Surface Penetration – ASP), accelerating hydrogen isotope trapping and desorption at low temperatures ($\leq$ 500 K).

The goal of this work was to study the possibility of using ASP phenomenon for developing the method of low-temperature removal of hydrogen isotopes from tungsten and aluminum-coated tungsten under the irradiation by hydrogen plasma with oxygen impurity. As beryllium is toxic, tungsten samples with aluminum coating were used in the experiments due to the simplicity of aluminum's usage in laboratory conditions, not imposing additional safety measures, and its physical and chemical properties closely resembling those of beryllium.

IOP Conf. Series: Journal of Physics: Conf. Series **1058** (2018) 012012 doi:10.1088/1742-6596/1058/1/012012

2. Sample preparation and experimental setup

Tungsten samples of 7x7x1 mm^3 were mechanically polished, washed in an ultrasonic ethanol bath and then annealed at pressure of $\leq 5\times10^{-6}$ Pa and temperature of 1600 K to remove "technological" hydrogen. For a part of these samples, 200±10 nm thick aluminum coating was deposited on one side in argon plasma with hydrogen impurity. Residual gas pressure did not exceed 5×10^{-4} Pa. Deposition rate was 0.055 nm/s. Temperature of the samples was 450-500 K. EDS analysis showed formation of an oxide layer on aluminum coating exposed to the atmosphere.

Experiments were conducted in the «MIKMA» stand [6]. Deuterium plasma ions were implanted into tungsten samples and tungsten side of aluminum-coated samples. The samples were then irradiated by hydrogen plasma ions with oxygen impurity. The cycles of irradiation repeated and deuterium content was measured using thermal desorption spectrometry (TDS) analysis after each cycle.

Residual gas pressure in the plasma chamber of the stand did not exceed 3×10^{-3} Pa. Deuterium, and then hydrogen plasma was initiated between a glow cathode and an anode. Plasma ion flow irradiating the sample included mainly molecular ions of the respective hydrogen isotope (82-85%), with the rest being mono- and triatomic ions. During the analysis of the results it was assumes that the sample was irradiated exclusively by H_2^+ or D_2^+ ions. The temperature of the sample was regulated by radiation from a flat tungsten spiral behind the back of the sample. The energy of irradiating ions was E=650 eV/at., the ion flow on the sample was $j=3.7\times10^{19}$ at/m^2s, the irradiation dose was $\Phi=1.3\times10^{23}$ at/m^2, the temperature of the samples during deuterium implantation and hydrogen ion irradiation was T=500 K.

Residual gas pressure in the TDS camera was $1.3\text{x}10^{-5}$ Pa. During the TDS measurements, samples were heated at a rate of 5 K/s, the desorption of H_2, HD, D_2, H_2O, HDO and D_2O molecules was registered. Desorption of other hydrogen- and deuterium containing molecules was negligible. The amount of deuterium and hydrogen in the samples after each stage of the experiment was calculated based on these results.

The first sample implanted with deuterium was used for a TDS measurement of hydrogen and deuterium content after the implantation. The second sample was exposed to vacuum for 1 hour at 500 K i.e. for the duration of hydrogen plasma with oxygen impurity irradiation. Hydrogen and deuterium content was then measured in the sample. This allowed deducing whether particles trapped would desorb during exposure to vacuum without plasma irradiation. After deuterium implantation, the third sample was irradiated by hydrogen plasma with 1% oxygen impurity; in case of aluminum-coated sample, the film was irradiated. The parameters of irradiation by H_2+1% O_2 plasma were as follows: E=50 eV/at., $j=3.7\times10^{19}$ at/m^2s, $\Phi=1.3\times10^{23}$ at/m^2, T=500 K. TDS was then used to measure the change of hydrogen and deuterium content after irradiation by H_2+1%O_2 plasma ions. Experiments were conducted with one, two, three and five repetitions of deuterium implantation and H_2+1%O_2 plasma irradiation

3. Results of the experiments for tungsten samples

Deuterium and hydrogen atom content in tungsten samples before and after the experiments are shown in Table 1. Interaction of deuterium ions during implantation (Table 1 row 2) with surface oxide layer initiated hydrogen trapping from residual gas molecules on the surface in the quantities roughly equal to the quantity of trapped deuterium. Exposure of tungsten samples to vacuum (Table 1 row 2, 3) did not lead to desorption of implanted deuterium atoms. Irradiation by H_2+1%O_2 plasma (Table 1 row 2, 4) initiated desorption of ≈40% implanted deuterium atoms, and the hydrogen in the ample increased to about 2.8 of the initial value. The results of the experiment showed that ASP develops on the surface of tungsten and reduces the quantity of previously implanted deuterium. However, to make a conclusion on the possibility of using ASP to remove hydrogen isotopes from tungsten, it was necessary to make sure that the effect is repeated for multiple cycles of deuterium implantation and H_2+1%O_2 plasma irradiation.

Table 1. Hydrogen and deuterium content in tungsten samples after deuterium implantation, exposition to vacuum, exposition to atmosphere and H_2+1%O_2 plasma irradiation of the side deuterium was implanted in.

№	Experiment	Deuterium content, $\times 10^{20}$ at/m^2	Hydrogen content, $\times 10^{20}$ at/m^2
1	Before deuterium implantation	-	-
2	After deuterium implantation	2.2	2.3
3	After implantation and exposition to vacuum	2.2	2.3
4	After irradiation by H_2+1%O_2 plasma	1.3	4.8

4. Results of the experiments for aluminum-coated tungsten samples.
Deuterium and hydrogen content in aluminum-coated tungsten before and after deuterium implantation into tungsten, exposition to vacuum and H_2+1%O_2 plasma irradiation are shown in Table 2.

Table 2. Hydrogen and deuterium content in aluminum-coated tungsten samples after deuterium implantation, exposition to vacuum, exposition to atmosphere and H_2+1%O_2 plasma irradiation of aluminum coating.

№	Experiment	Deuterium content, $\times 10^{20}$ at/m^2	Hydrogen content, $\times 10^{21}$ at/m^2
1	Before deuterium implantation	-	3.6
2	After deuterium implantation into tungsten	2.9	3.1
3	After implantation and exposition to vacuum	3.1	3.3
4	After irradiation of Al by H_2+1%O_2 plasma	1.5	4.0

It can be seen that a significant amount aluminum was absorbed into the sample from the residual gas (Table 2 row 1). Its decrease in content during deuterium implantation (Table 2 row 2) can be explained by sputtering of the surface layer by deuterium ions and isotope exchange. Exposition of the sample implanted from tungsten side to vacuum did not lead to any changes in deuterium content (Table 2 row 2, 3), which correlates with the results obtained for a non-coated tungsten sample (Table 1 row 2, 3). Irradiation by H_2+1%O_2 plasma ions of the aluminum coating initiated the desorption of approx. half (48%) of trapped deuterium (Table 2, row 2, 4), as with tungsten sample (Table 1, row 2, 4); hydrogen content increased by 11%. The conclusion can be made that ASP develops on the surface of the aluminum coating on tungsten.

5. Multiple tungsten irradiations by deuterium and hydrogen ions.
The results of the experiments with multiple iterations of tungsten irradiation by deuterium and hydrogen ions are shown on Fig. 1. It can be seen that after the second cycle deuterium concentration

inside the sample was set to 3.1×10^{20} at/m^2, which remained almost unchanged until the fifth cycle. It can be seen that hydrogen is trapped at a much higher rate than deuterium, however, the increase of total hydrogen isotope content inside the sample slows down after the second iteration of the cycle, which means that periodic irradiation by H_2+1%O_2 plasma ions prevents deuterium accumulation in tungsten.

As the first wall and the divertor of a thermonuclear reactor will be irradiated not only by fast neutrals from the center of plasma, but also slow ions and atoms from the periphery, it can be assumed that oxygen impurity sorbed on the surface of elements will assist ASP. For that reason, it can be assumed that tritium removal from tungsten and beryllium plasma-facing elements will occur at a constant rate during the operation of the reactor.

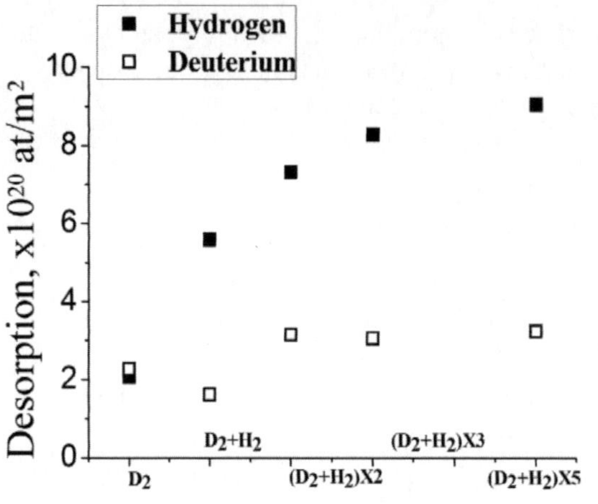

Figure 1. Dependence on the number of deuterium and hydrogen atoms in tungsten under multiple cycles of irradiation by deuterium and hydrogen ions, where D_2+H_2 (H_2+1%O_2); ×2, ×3, ×5 – number of cycles.

6. Conclusion

It was shown in this work that ASP develops on the surfaces of both tungsten and aluminum-coated tungsten. The features of the development of ASP on tungsten and aluminum film on tungsten allow concluding that the growth of tritium trapping in the ITER tungsten divertor tiles during the operation of the reactor can be halted by periodic irradiation by low-ionized deuterium plasma with small oxygen impurity.

Similarity of beryllium and aluminum parameters permits halting tritium trapping in beryllium-coated divertor tiles, as well as beryllium first wall tiles, in a similar fashion.

This work was conducted with financial support from Ministry of Science and Education of Russia under the contract № 14.575.21.0169 (RFMEFI57517X0169).

References

[1] Begrambekov L., Grunin A., Kaplevsky A. et al. / J. of Surf. Inv. X-ray, Synch. and Neut. Tech. (2015). V. 9. P. 190–195.
[2] Kurokawa H, Oyama Y et al. / J. of Electrochem. Soc. (2004). V. 151 (8). P. 1264-1268.
[3] Ishikawa Y., Nemanic V. / Vacuum. (2003). V. 69. P. 501-5123.
[4] Zhang G., Wang X., Xiong Y. et.al. / Int. J. of Hydr. En. (2013). V. 38. P. 1157-1165.
[5] Marot L., Linsmeier Ch., Eren B. et al. / Fus. Eng. And Des. (2013). V. 88. P. 1718-1721.
[6] Airapetov A., Begrambekov L. et al. / J. of Nucl. Mater. (2011). V. 415. P. 1042-1045.

IOP Conf. Series: Journal of Physics: Conf. Series **1058** (2018) 012013 doi:10.1088/1742-6596/1058/1/012013

Formation of Composite Coatings by Sub- and Supersonic Plasma Spraying under and on Special Coating Materials

S A Ilyinikh, V A Krashaninin, B R Gelchinsky M N Zakharov, A V Dolmatov and A A Ponomarenko

Institute of Metallurgy of Ural Branch of Russian Academy of Sciences
620137, 101 Amundsen st., Ekaterinburg, Russia

E-mail: mr.mizani@mail.ru

Abstract. A technology was developed for the formation of a multilayer composite coating obtained by the sub – and supersonic plasma spraying method based on Al, Ti, Ni, and Mo. The results of the investigation of a coating of Al-Ti-Ni-Mo powder are deposited by subsonic plasma deposition on a steel substrate followed by the application of an anticorrosion coating from a paint coating material (PCM) of the CINEP brand, which is protected by the supersonic spraying powder Al-Ti-Ni-Mo. This coating layer is designed to protect the anticorrosive layers of the main multicomponent coating of products, which can be subjected to external mechanical stress during operation. It is shown that the resulting composite coating is sufficiently strong and withstands strains in the zone of elastic deformation of the matrix. Carried out corrosion tests showed a high resistance of these coatings.

Keywords: composite coating, plasma spraying, plasma

1. Introduction

The combination of technologies of supersonic plasma spraying, airborne plasma subsonic spraying and PCM allows to obtain a new generation of composite coatings with improved performance characteristics, maximize the service life of existing equipment and facilities while preserving the reliability of their operation, at minimal cost [1].

Optimization of technological modes of plasma spraying was carried out using modern numerical methods and available software tools. The choice of the parameters of the plasma flow made it possible to determine the velocity and maximum acceleration of the particles, the temperature of their heating, the residence time, the evaporation rate of the particles, and also the particle trajectory in the plasma jet.

Coatings used in environments with high corrosive activity are effective only in the complex of the coating system. It is proposed to apply polyurethane and epoxy varnish materials based on scaled pigments to plasma coatings. In this case, the covering layers enhance the barrier properties of the coating, shield it from the influence of aggressive factors, which slows down the oxidation of zinc present in the plasma coating and prolongs its protective action. In addition, they impart greater hardness, resistance to abrasion.

The use of high-speed plasma (with velocities of approaching the deposited particle to the surface of the substrate v = 600-1500 m/s) opens up new possibilities for realization the multi-purpose protection of metal structures due to the following reason i.e. the possibility of a strong sputtering on products made from almost any material.

The task was to create conditions and obtain a coating that, when deformed, would retain its operational properties for as long as possible. In the literature there is the concept of "multiple cracking" [2, 3], in the manifestation of which the onset of cracking does not immediately lead to the destruction of the layer or its detachment from the substrate. The cover is fragmented, but remains on the basis and continues to perform, though not fully, its task. There are many common methods for testing coatings [4, 5]. However, in most cases researchers and technologists are interested in strength (hardness, adhesion, cohesion) and tribological (resistance to abrasive and erosion wear, sliding friction parameters, etc.) coating properties. Adhesion characterizes the bond strength of the coating with the substrate [6].

2. Methods and materials

Intermetallic alloys based on titanium, aluminum, nickel and molybdenum have increased performance characteristics. The alloy was obtained by aluminothermic reduction of oxides in a Tamman furnace. Subsequently, it was subjected to grinding in a vibratory shredder to a particle size of 40-160 μm.

Using the method of supersonic plasma spraying using air and its mixture with methane as a plasma-forming gas, laboratory samples of Al-Ti-Ni-Mo coatings were obtained on plates of steel St45 with a size of 100x15x1 mm.

Three-point bending tests (Fig. 1) were carried out on a universal Zwick / Roell Z050 test machine (BT1-FR050THW / A1K). The distance between the supports was set at 20 mm. The moving speed of the movable crosshead was set to 4 mm/min and three tests of each coating were performed.

3. Results and discussion

Photographs of coated plates after a test with characteristic fractures for each coating and application conditions are shown in Figure 1. Tests have shown that the samples do not have fracture with a central crack.

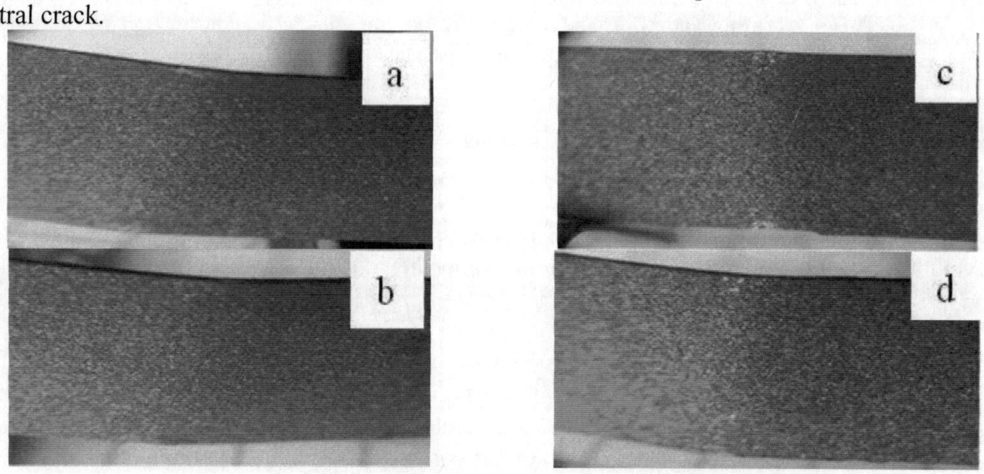

Fig. 1. Characteristic fractures of the Al-Ti-Ni-Mo coating.

Plasma-forming gas: a, b - air, methane; c, d – air.

Rockwell hardness was also measured. The measurement was carried out by ultrasonic-measurer UZIT-3 in Rockwell in HRC units (measurement limits from 20-70 HRC, error of measurement no more than ± 2.0 HRC). The results of the fixed hardness values of HRC coatings on Al-Ti-Ni-Mo samples obtained in air and air + methane are presented in Table 1. The scatter of the values of hardness is related to the features of the structure of the coating, its lamellar structure, and also to the multiphase nature of the applied material containing regions of different hardness. The load increase graph, shown in Figure 2, represents the measurements obtained with a three-point bend [7]. There are different areas of deformation, including the peeling of the coating.

Table 1. Hardness values of HRC coatings on Al-Ti-Ni-Mo samples obtained in air and air + methane.

Measurement №	Methane + air	Air
1	77.5	88
2	81.4	79.5
3	87.2	84.5
4	85.2	78.8
5	84.1	87.7
6	80.1	80.0
7	79.6	87.2
8	84.2	71.9

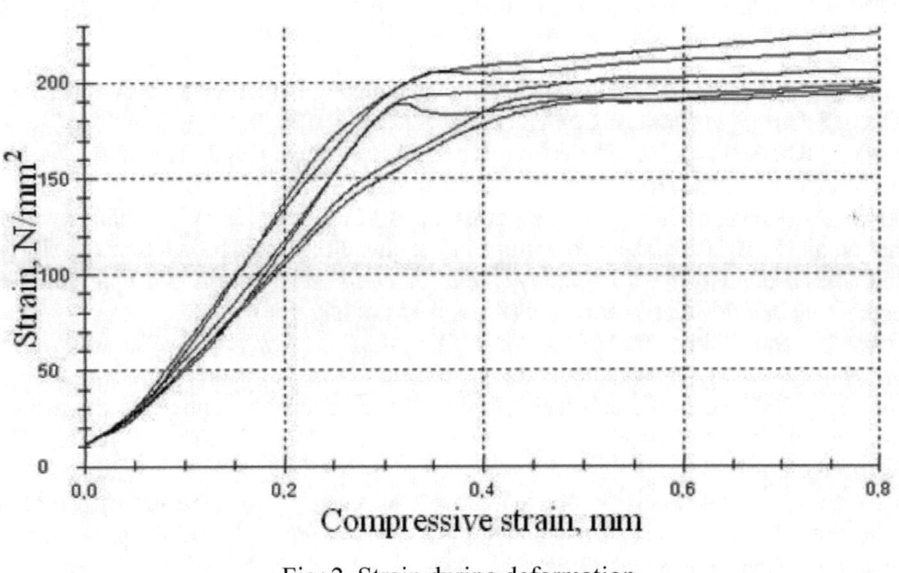

Fig. 2. Strain during deformation.

To investigate the corrosion resistance of combined coatings, tests were carried out in a salt mist chamber Q-FOG / SSP / 600 according to ASTM B117 standard test. This test consists in the continuous spraying of salt mist for a predetermined time. To create salt fog, a 5% aqueous solution of NaCl, pH 6.5-7.2 was used. The test temperature is 35 ° C. The duration of the tests is until the appearance of visible damage to the coatings. For the criterion of the quality of the coatings, the appearance of the coatings, characterizing the protective properties, is adopted. During the tests, the appearance of the coatings was periodically monitored and evaluated according to GOST 9.407. As test specimens, plates of steel grade 08ps (GOST 16523) measuring 100x150x1 mm were used. The total thickness of coatings was up to 170 and more microns. The test duration was 883 hours before the destruction, which is a good result.

4. Conclusions

It was possible to connect the technology of plasma (subsonic and supersonic) evaporation and the technology of applying coatings to create multifunctional composite coatings obtained by the method of plasma spraying of the Al-Ti-Ni-Mo system with the application of special paint and varnish materials to reduce porosity and enhance corrosion protection.

The properties of the Al-Ti-Ni-Mo coating applied to the steel substrate have been studied. It is shown that the coating is sufficiently strong and withstands strains in the zone of elastic deformation. It is shown that in the bending test there is a peeling of the coating without cracking.

Tests have been conducted to assess the protective properties of combined coatings consisting of a plasma-dust coating and a paint coating layer (zinc-primed CINEP primer). The test duration was 883 hours. During the tests, the main types of damage were pitting corrosion. The influence of the total thickness of the coating on the protective properties was found. A coating with a larger thickness showed higher protective properties. The coating based on Al-Ti-Ni-Mo has good strength properties and, taking into account good anticorrosion properties, can be recommended for use in various engineering fields.

Acknowledgements

This work was financially supported by the Ural Branch of the Russian Academy of Sciences (grant No. 15-17-3-41). The research was carried out using the equipment of the Ural-M Center for Collective Use.

References

[1] Zakharov M N, Rybalko O F, Romanova O V, Gelchinskiy B R, Il'Inykh S A and Krashaninin V A 2017 *Journal of Physics: Conference Series* **789 1** 012070

[2] Borisov Y S, Kharlamov Y A, Sidorenko S L and Ardatovskaya E N 1987 *Gas thermal coatings from powder materials* (Kiev: Naukova Dumka) 543

[3] Ilyinikh S A, Krasikov S A, Ponomarenko A A, Sitnikova O A, Gelchinsky B R and Krashaninin V A 2013 Phase formation in the metallothermal production of Al-Ti-Ni-Mo alloys *Proceedings of the 13th International Scientific and Technical Conference "Surface Engineering and Product Renovation* (Kiev: ATM of Ukraine) 116

[4] Tushinsky L I and Plokhov A V 1986 *Investigation of the structure and physico-mechanical properties of coatings* (Novosibirsk: Science) 216

[5] Astakhov E A 2005 Scientific and technological fundamentals of controlling the properties of detonation coatings Disertation of Dr. Tech. Sciences (Kiev: NAS of Ukraine, Institute of Electric Welding) 382

[6] Ulyanitsky V Y, Nenashev M V, Kalashnikov V V, et al. 2010 Experience in research and application of detonation coating technology. *Izvestiya of the Samara Scientific Center of the Russian Academy of Sciences* **12** 569

[7] Rech S, Trentin A and Vezzu S 2014 Different Cold Spray Deposition Strategies: Single- and Multi-layers to Repair Aluminium Alloy Components. *Journal of Thermal Spray Technology* **23 8** 1237

Plasma Recycling of Mechanical Processing Wastes from Titanium Alloys Stock Materials

M N Zakharov, O V Romanova, O F Rybalko, S V Zhidovinova and B R Gelchinsky

Institute of Metallurgy of Ural Branch of Russian Academy of Sciences
620137, 101 Amundsen st., Ekaterinburg, Russia

E-mail: mr.mizani@mail.ru

Abstract. The article shows the possibility of processing and involving in production of a powder, made from wastes of the VT-22 alloy ingots mechanical processing. These powders are applied as protective coatings by plasma spraying method. From VT-22 powder there were obtained protective coatings on a steel substrate by plasma spraying method. Coating samples were tested for adhesion with the substrate by the three-point bending method and on the corrosion resistance in a salt mist chamber (combined coatings: metallic + paintwork material). When tested in a salt mist chamber, pitting corrosion was quickly detected on the combined coatings, which indicates low protective properties. Three-point bending tests showed that coatings have a good adhesion to the substrate. After the work on the application of plasma coatings, a powder passed through the argon plasma was collected to study the effect of the plasma-forming gas on the properties of the VT-22 powder. Research have shown that even after passing through the plasma, the powder practically does not change physical-processing properties, chemical composition, or morphology of the particles. Based on the data obtained, it can be concluded that it is possible to reuse the powder.

1. Introduction

During mechanical treatment of metal stock materials (shaping, slotting, milling, drilling, etc.) and parts manufacturing by traditional methods (casting, forging, stamping, etc.) [1] a large amount of man-made waste is produced, which must be disposed of and used in production [2 -4].
Powder metallurgy methods make it possible to reduce the material consumption and the volume of products mechanical treatment, as well as the amount of wastes in the production of the final product [5-7]. With the development of additive technologies (AT) appeared additional methods of obtaining details from powders [8-11].

Powders of titanium and its alloys can be obtained by various methods from ingots, by HDH method from standard waste and titanium sponge, and others [6,7,9,12]. AT methods based on two main requirements for powder materials: sphericity (for better fluidity) and particle size (20 to 100 μm). To obtain such powders, a number of methods have been developed [8, 9, 12], among them the most frequent are: atomization of liquid metal by an inert gas jet under pressure (gas atomization) and plasma sputtering of a metal electrode. The first method is considered quite costly, and is acceptable for obtaining powders from metals having a not too high melting point. The second one is more technological for a wide range of metals, including refractory metals, and is more effective in industrial applications. This way was obtained the powder material corresponding to GOST 19807-91 from the mechanical treatment wastes of VT-22 alloy ingots at OJSC “Composite” (Korolev, Moscow Region).

The aim of this work is to investigate the possibility of processing and involving in the production of powder made from the mechanical treatment wastes of the alloy VT-22 ingots.

2. Methods and materials

Powder VT-22 was applied to create combined protective coatings (metal, metallic + paintwork material) using the technology developed by authors [13]. The application of powder to steel substrates (Steel. 3) was carried out with the help of the "Corvette-4" plasma spraying installation in a protective atmosphere of argon. Some of the samples with metallic coating were sent to determine the adhesion properties by the three-point bending method on the universal testing machine ZWICK BT1-FR050THW / A1K. Combined coatings (metal + paintwork material) were tested for corrosion resistance in a salt mist chamber Q-FOG / SSP / 600 according to GOST R52763-2007.

After the application of plasma coatings, the powder, passed through the plasma, was collected. Scanning Electron microscope Carl Zeiss EVO 40 was used to study the morphology of the VT-22 powder particles before passing through the plasma and after. Chemical analysis of the powder was carried out using S4 X-ray fluorescence spectrometer (Bruker-AXS, Germany), X-ray diffraction analysis using XRD- 7000 (Shimadzu Corp., Japan).

Bulk density, fluidity and tap density were determined according to the corresponding GOSTs (GOST 19440-94, GOST 20899-98, GOST 25279-93).

The average size, the degree of sphericity (characterizes the degree of proximity of the particle shape to the sphere) and the degree of symmetry (characterizing the degree of deviation of the particle shape from the centrally symmetric) of powder were measured on the CAMSIZER-XT by dynamic image analysis (according to ISO 13322-2: 2006).

3. Results and discussion

The powder material VT-22 was applied to the steel substrate by plasma spraying method. Studies of the obtained samples on a three-point bend showed that plastic deformation of the coating starts at 391-460 N / mm^2. Coatings have good adhesive properties, are strong enough and withstand the bending test. On the other part of the samples, a paintwork material coating of the CINEP brand was applied according to [13,14]. Samples survived the test in a salt mist chamber for 580 hours and showed the presence of pitting corrosion, which began to appear after 197 hours of testing, which indicates their low corrosion resistance.

Studying the powder collected after passing through the plasma, it was established that particles shape does not change (Fig. 1) [15].

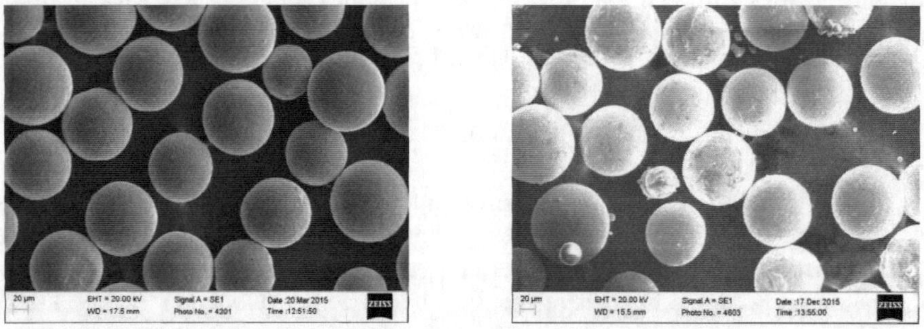

a b

Fig. 1. Morphology of VT-22 powder: before passing through the plasma (a), magnification - 500; after passing through the plasma (b), magnification - 500.

Conducted chemical and X-ray phase analysis of powders before and after plasma treatment did not reveal any significant changes. Data on chemical composition of the VT-22 alloy in accordance with GOST 19807-91, as well as before and after plasma treatment of powder from this material, are given in Table 1.

Table 1. Chemical composition of VT-22 powder before and after plasma spraying

	Ti	Al	V	Mo	Fe	Cr	Zr	Si	C	O	N	H	Other
VT-22 alloy GOST 19807 - 91	Basis	4.4 … 5.7	4.0 … 5.5	4.0 … 5.5	0.5 … 1.5	0.5 … 1.5	Up to 0.3	Up to 0.15	Up to 0.1	Up to 0.18	Up to 0.05	Up to 0.015	0.3
Powder before plasma spraying	82.50	5.75	4.88	4.56	0.85	0.93	0.11	0.09	0.018	0.157	0.018	Not found	0.137
Powder after plasma spraying	82.64	5.54	4.79	4.44	1.04	0.96	0.14	0.1	0.015	0.179	0.024	0.0045	0.129

Physical-processing properties of the collected powder were determined and compared to the similar properties of the original powder (Table 2).

Table 2. Physical-processing properties of powders before and after passing through the plasma

	Bulk density, g/sm^3	Fluidity, g/s	Tap density, g/sm^3	Sphericity degree	Symmetry degree
Before plasma, fraction (-100)	2.72	2.94	3.02	0.722	0.904
After plasma, fraction.(-100)	2.70	2.91	3.07	0.754	0.916

As follows from the table, physical-processing properties remained practically unchanged, which indicates the possibility of powder material further use not only for plasma spraying of coatings, but also for other additive technologies.

4. Conclusion

Tests of plasma sprayed coatings samples on a three-point bend showed that obtained coatings are strong enough and have good adhesion with the substrate.

When testing in a salt mist chamber, pitting corrosion was quickly detected on combined coatings (with paintwork material), which indicates low protective properties.

Investigation of the VT-22 powder before and after passing through the plasma showed that in the protective atmosphere of argon, using it as a plasma-forming gas, physical-processing properties, chemical composition, and morphology of the particles practically do not change. This indicates that this powder can be reused.

Acknowledgements
This work was financially supported by the Ural Branch of the Russian Academy of Sciences (grant No. 15-17-3-41). The research was carried out on the equipment of the Ural-M Center for Collective Use.

References

[1] Kolobov G A, Pruttskov D V, Pozhuev V I and Telin V V 2011 *Titan* **1** 9 - 15.

[2] Kolobov G A 2006 *Secondary titanium. Part 1* (Zaporozhye: Zaporozhye State Engineering Academy) 124.

[3] Zakharov M N, Romanova O V, Rybalko O F, *et al.* 2017 Possibilities of processing and involving in the production of mechanical processing wastes made of titanium and its alloys *Proceedings of the Technogen-2017 Congress*. (Ekaterinburg: UB RAS) 394-396.

[4] Ivashchenko V P, Kuris Yu V and Kolobov G A 2011 *East-European Journal of Advanced Technologies. Applied physics and materials science.* **52** 4/5 23-25.

[5] Smetkin A A 2013 Trends in the development of processes for the production of titanium materials by powder metallurgy *PNIPU. Mechanical engineering, materials science* **15** 3 26-32.

[6] Neikov O. 2014 Non-ferrous powder production: Manufacturing methods and properties of copper, aluminum, titanium and nickel powders *Powder Metallurgy Review* 65-87.

[7] Qian M., Froes F H 2015 *Titanium Powder Metallurgy. 1st Edition Science, Technology and Applications* (Butterworth-Heinemann) 648.

[8] Juan F, Isaza P and Aumund-Kopp C 2014 Additive Manufacturing with metal powders: Design for Manufacture evolves into Design for Function *Powder Metallurgy Review* 41-51.

[9] Zlenko M A, Nagaitsev, Dovbysh V M 2015 *Additive technologies in mechanical engineering. Manual for engineers.* Moscow NAMI. 220

[10] Dzhugan A A, Olshanetsky V E and Ovchinnikov AV 2016 *New materials and technologies in metallurgy of machine-building* **2** 77- 81.

[11] Zakharov M N, Rybalko O F, Romanova O V, Gelchinskiy B R, Il'Inykh S A and Krashaninin V A 2017 *Journal of Physics: Conference Series* **789** **1** 012070

[12]Leyens C, Peters M 2003 Titanium and Titanium Alloys. Fundamentals and Applications (WILEY-VCH Verlag GmbH & Co. KG. KGaA, Weinheim) 513.

[13] Krashaninin V A, Ilyinykh S A, Chusov S A ,*et al.* 2015 *Proceedings of the International Scientific and Technical Conference "Modern Metal Materials and Technologies" (SMMT'15).*1103-14.

[14] Ilyinykh S A, Krashaninin V A, Gelchinsky B R, *et al.* 2016 Investigation of the corrosion resistance of combined coatings of Zn-Me (Al, Mo, Ti, Ni) powder systems *Proceedings of the XIV International Scientific and Technical Conference.* (Moscow, MIR) 91-94.

[15] Zakharov M N, Rybalko O F, Ilyinikh S A, *et al.* 2016 *Reports of the V International Conference-School on Chemical Technology HT'16 Satellite Conference of the XX Mendeleev Congress on General and Applied Chemistry* (VSTU Volgograd) **V 2** 214-215.

IOP Conf. Series: Journal of Physics: Conf. Series **1058** (2018) 012015 doi:10.1088/1742-6596/1058/1/012015

Mössbauer effect studies of thin iron films synthesized by ion beam assisted deposition technique

A L Zinnatullin[1], N M Lyadov[1,2], I A Faizrakhmanov[1,2] and F G Vagizov[1]

[1] Kazan (Volga Region) Federal University, Kremlevskaya 18, 420008 Kazan, Russia
[2] Zavoysky Physical-Technical Institute, Sibirsky Trakt 10/7, 420029 Kazan, Russia

almaz.zinnatullin@gmail.com

Abstract. The paper presents the results of Mössbauer effect studies of thin iron films obtained by ion-assisted deposition technique. It was found that the films obtained consist of several phases and have a perpendicular magnetic anisotropy. Magnetization direction can be changed by external magnetic field in sample plane.

1. Introduction

To increase the storage capacity of magnetic memory, it is necessary to use new magnetic materials that have certain properties. Such materials should have high coercivity at room temperature. The size of the magnetic domain should be as small as possible. Finally, it is important that the materials have perpendicular magnetic anisotropy (PMA). This feature of magnetization gives us possibilities to extremely increase the density of domains and, consequently, to increase the storage capacity of magnetic memory [1].

Thin films which show perpendicular magnetic anisotropy have been studied in the 70s of the last century [2]. The first magnetic memory devices with perpendicular recording appeared in 2005 [3]. Such devices have a relatively big capacity, more performance then hard disks with longitudinal recording. Furthermore, materials with perpendicular magnetic anisotropy can be used in magneto-optical and spintronic devices [4].

The development of a method for the production of such films is a significant practical task. The key factors here are the reproducibility, control of the elemental and phase composition of the films, as well as the low cost of the materials produced. Films with perpendicular magnetic anisotropy can be produced by magnetron sputtering [5], chemical deposition from the gas phase [6], and electrolytic deposition [7]. Ion beam assisted deposition (IBAD) technique is of a great interest. Main feature of IBAD method is that films in unusual, metastable crystalline and amorphous states can be synthesized under the continuous action of low-energy ions (<1 keV) [8].

In this paper, we present the experimental results on investigations of thin iron films synthesized by ion beam assisted deposition. To analyze the phase composition and magnetic properties of the obtained samples, a gamma-resonance spectroscopy (Mössbauer spectroscopy of conversion electrons) was used.

2. Experimental part

Thin iron films on silicon and glass substrates, called as "Sample 1" and "Sample 2", respectively, were synthesized by ion beam assisted deposition technique. Xenon ions with energy of 0.7 keV sputtered the metallic iron target enriched with ^{57}Fe isotope up to 60 %. The xenon ions incident on the surface of the substrate at small angles caused amorphization of the synthesized layer and stimulated the growth of the iron clusters under conditions which are different from the thermodynamic equilibrium. The density of the ion current was about 200 $\mu A/cm^2$. Films were deposited for 20 minutes. The film thickness, estimated from the image of scanning electron microscopy ("Zeiss EVO 50 XVP"), was about 160 nm for the "Sample 1' (Figure 1).

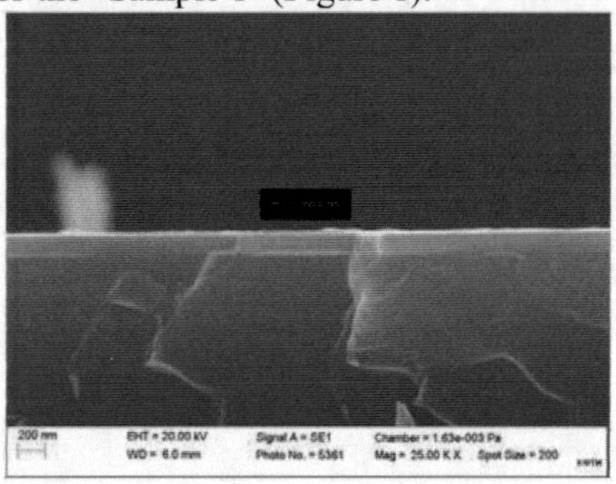

Figure 1. Scanning electron microscopy image of iron thin film deposited on the glass substrate.

Mössbauer spectra of conversion electrons were measured at room temperature on a conventional constant-acceleration spectrometer (WissEl, Germany), equipped with a commercial source of gamma-radiation (Co-57 in the rhodium matrix, Retverc). To detect the conversion electrons, a gas flow detector, operating on a mixture of helium (He) and methane (CH_4, 5% vol.), was used. Spectrometer was calibrated using the metallic iron absorber at room temperature. The isomer shifts of the spectrum components were determined relatively to the center of gravity of the metallic iron spectrum.

3. Results and discussion

Conversion electron Mossbauer spectra of iron thin films at room temperature and the results of fitting are shown in Figure 2. A model of four components was used for the fitting. Mathematical processing revealed the presence of three sextets corresponding to magnetically ordered phases of iron atoms and one doublet that correspond to paramagnetic state of the iron.

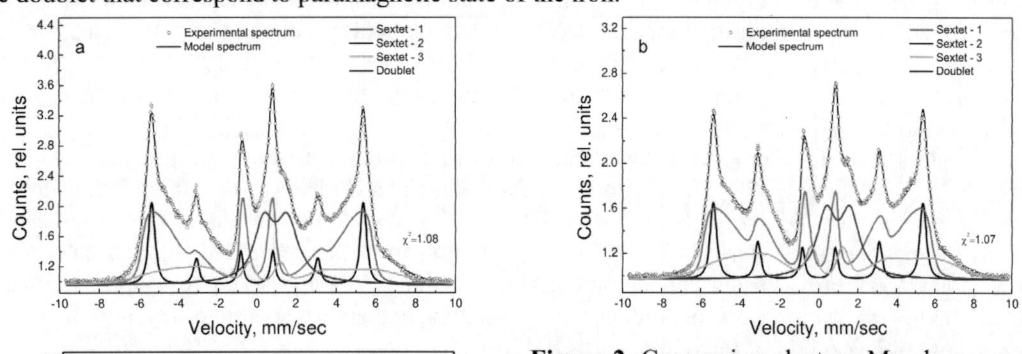

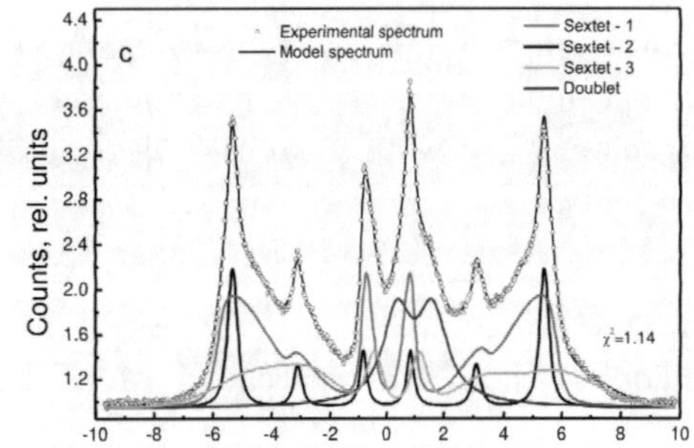

Figure 2. Conversion electron Mossbauer spectra of thin iron films on silicon substrate without external magnetic field (a) and with magnetic field of 700 Oe (b); conversion electron Mossbauer spectra of thin iron film on glass substrate (c) without external magnetic field. Model spectra and components are shown. The solid lines represent the resulting fit curve and the components of the spectrum.

The hyperfine parameters of the spectrum components are given in Table 1. It should be noted that an asymmetric Gaussian distribution of hyperfine magnetic fields on resonant iron nuclei was assumed for more accurately approximation of spectra. Asymmetric distribution of hyperfine magnetic fields is typical for nanostructured materials [9].

Table 1. Hyperfine parameters* and relative content of the components.

	IS, mm/sec	QS, mm/sec	HF, kOe	S, %	A_2
Sample 1					
Sextet-1	0.03	-0.04	348	43.4	0.5
Sextet-2	0.01	0.02	333	13.1	0.9
Sextet-3	0.60	0.36	325	22.2	1.5
Doublet	0.99	1.21	-	20.2	-
Sample 1 in magnetic field of 700 Oe					
Sextet-1	0.03	-0.07	346	47.3	1.15
Sextet-2	0.01	0.02	333	12.4	1.29
Sextet-3	0.56	0.28	315	22.5	1.26
Doublet	0.99	1.21	-	17.7	
Sample 2					
Sextet-1	0.02	-0.05	347	42.4	0.5
Sextet-2	0.02	0.02	332	13.7	0.8
Sextet-3	0.52	0.31	313	25.3	1.0
Doublet	0.99	1.25	-	18.5	-

IS – isomer shift;
QS – quadrupole splitting;
HF – average hyperfine magnetic field on iron nuclei;
S – relative content of component;
A_2 – intensity of second (fifth) lines of sextets relative outer lines.

*Estimate errors for isomer shift and quadrupole splitting are ~0.01 mm/sec, for hyperfine magnetic field - ~1 kOe.

The hyperfine parameters of the sextet, indicated in Figure 2 as "Sextet-2", are close to the well known values of bulk metallic iron, α-Fe. However the parameters for other two sextets are slightly different relate to the parameters for bulk metallic iron and iron oxides. They can be attributed to small particles of metallic iron, as well as to amorphous iron [10]. The values of the doublet parameters are close to the values of the hyperfine parameters of iron (II) oxide.

Values of hyperfine parameters of components, called as Sextet – 2 on model spectra, close enough to the well known parameters for bulk metallic iron. However parameters for other two sextets are slightly different relate to the parameters for metallic iron and iron oxides. We assumed that these components correspond to the small metallic iron particles as well as to particles of amorphous iron [10]. Values of hyperfine parameters for the doublet are close to parameters for iron (II) oxide.

From the analysis of the Mössbauer spectra of iron films deposited on the silicon and glass substrate (Figure 2a, c) and the hyperfine parameters of the spectrum components (Table), it can be assumed that the substrate material has no significant effect on the phase composition of the resulting films under the same deposition conditions. The ratio of the intensity of the second (fifth) sextet lines relative to the intensity of the third (fourth) lines depends on the angle between the direction of the hyperfine magnetic field on the nucleus and the direction of the wave vector of the resonance photon:

$$A_2 = \frac{4 \cdot \sin^2 \theta}{1 + \cos^2 \theta},$$

where θ is the angle between the direction of the magnetic field on the nucleus and the wave vector of the gamma-photon. Thus, this ratio gives us information about primary orientation of the magnetic moments [11]. In the case of random orientation of magnetic moments (isotropic case) this ratio should be equal to 2. As can be seen in Figure 2a, this ratio is different for our samples. Figure 2b shows the Mössbauer spectrum of the conversion electrons of the first sample placed in a longitudinal magnetic field with strength of 700 Oe. It can be seen that the intensities of the 2nd and 5th lines have increased noticeably.

In case of our measurements the wave-vector of gamma photons is perpendicular to the sample plane. So, as follow from above equation, such values of the ratio A_2 correspond to the case when the magnetization of the sample does not lie in the plane of the sample. An increasing of intensities of second and fifth lines might be connected with decrease of angle between the magnetization of sample and the film plane. For well-defined metallic iron sextet (Sextet-2) the angle θ between the primary orientation of the magnetic moments and the wave vector of the gamma-photon changes from 23° to 34° for the sample in the magnetic field.

4. Conclusion

The results of Mössbauer effect studies show that the thin iron films synthesized by ion-assisted deposition technique have a pronounced perpendicular component of the magnetic anisotropy. The resultant magnetization of the sample makes an angle of 23° with respect to the plane normal and can vary with the application of an external magnetic field. It was found that the iron film consists mainly of alpha-iron and disordered iron. The nature of the observed anisotropy requires further study.

Acknowledgments
This work was partially funded by the Russian Foundation for Basic Research (Grant No. 18-02-00845-a), the Program of Competitive Growth of Kazan Federal University funded by the Russian Government.

References

[1] Tudu B. and Tiwari A. 2017 *Vacuum* **146**, 329.
[2] Iwasaki S. and Nakamura Y. 1977 *IEEE Trans. Magn.* **13**, 1272.
[3] Iwasaki S. 2009 *Proc. Jpn. Acad. Ser. B Phys. Biol. Sci.* **85**, 37.
[4] Ikeda S, Miura K, Yamamoto H, Mizunuma K, Gan H D, Endo M, Kanai S, Hayakawa J, Matsukura F and Ohno H 2010 *Nature Materials* **9**, 721.
[5] Wu D, Zhang Z, Li L, Zhang Z, Zhao H B, Wang J, Ma B and Jin Q Y 2015 *Scientific Reports* **5**, 12352.
[6] Ootera Y, Shimada T, Kado M, Quinsat M, Morise H, Nakamura S and Kondo T 2015 *Applied Physics Express* **8**, 113005.
[7] Zana I, Zangari G and Shamsuzzoha M 2005 *Journal of Magnetism and Magnetic Materials* **292**, 266.
[8] Lyadov N M, Bazarov V V, Vagizov F G, Vakhitov I R, Dulov E N, Kashapov R N, Noskov A I, Khaibullin R I, Shustov V A and Faizrakhmanov I A 2016 *Applied Surface Science* **378**, 114.
[9] Eibschütz M, Lines M E and Chen H S 1983 *Physical Review B* **28**, 425.
[10] Long G J, Hautot D, Pankhurst Q A, Vandormael D, Grandjean F, Gaspard J P, Briois V, Hyeon T and Suslick K S 1998 *Physical Review B* **57**, 10716.
[11] Wertheim G K 1964 *Mossbauer Effect: Principles and Applications* (New York: Academic Press).

IOP Conf. Series: Journal of Physics: Conf. Series **1058** (2018) 012016 doi:10.1088/1742-6596/1058/1/012016

Long-lived photoinduced absorption in granular molybdenum disulfide thin films

A L Zinnatullin, I R Vakhitov, A M Rogov, I V Yanilkin, S I Nikitin and R V Yusupov

Kazan (Volga Region) Federal University, Kremlevskaya 18, 420008 Kazan, Russia

almaz.zinnatullin@gmail.com

Abstract. We present results of the differential photoinduced absorption spectra investigation of the molybdenum disulfide nanogranular thin films. The films were produced by means of a sulfurization of pre-deposited by magnetron sputtering thin molybdenum films. It has been shown that photoexcitation with the light quanta with the energy higher than the bandgap leads to a modification of the absorption spectra in the visible range. Possible nature of the photoinduced absorption spectra will be discussed.

1. Introduction

Two-dimensional (2D) materials, especially single-layer ones, attracted a lot of attention in the recent time. It is due to the properties cannot be realized in the bulk. However, an absence of reproducible methods for synthesis of such kind of materials inhibited experimental developments. Successful synthesis of graphene drastically increased the interest to 2D materials. Two-dimensional transition metal dichalcogenides represent a special class among them. The use of these materials in nanoelectronic and optoelectronic devices is expected [1].

2. Experimental section

Molybdenum disulfide thin films have been prepared in two steps. At the first step, thin films of molybdenum were deposited by magnetron sputtering technique onto the fused silica and silicon substrates. Film thickness was measured at the edge between the film surface and undeposited region with Bruker DektakXT profilometer and had a value of ~4 nm. Then the films were annealed at 800°C in a weak flow of sulfur vapor (Figure 1). High purity argon gas (99.999%) has been used for a transport of sulfur.

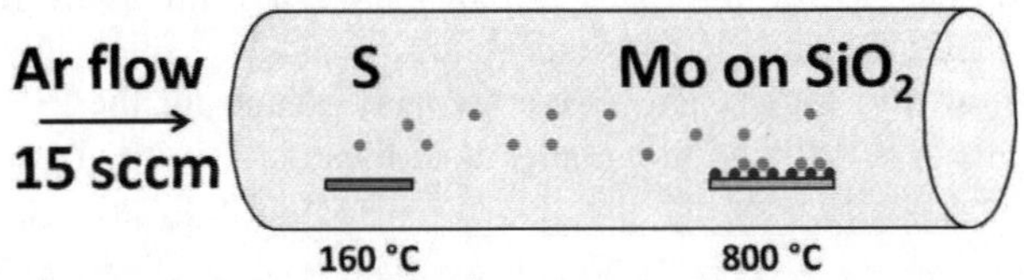

Figure 1. Experimental setup for molybdenum disulfide synthesis by the sulfurization of pre-deposited molybdenum thin films.

Morphology of the produced thin film on silicon substrate was studied by scanning electron microscopy (Merlin, Carl Zeiss). Optical transmission of thin MoS_2 film on the fused silica substrate was investigated with the SPECORD M40 spectrophotometer (Carl Zeiss) at room temperature. Photoinduced absorption measurements were performed with an intent to study long-lived photoexcited states. Excitation (355 nm, 405 nm, 532 nm) was modulated by an optical chopper. Measurements have been carried out with different power and at different frequencies of the modulation.

3. Results and discussion

Scanning electron microscopy image of synthesized thin film of molybdenum disulfide on silicon substrate is shown in Figure 2a. Clearly, the sample is a nanogranular film composed of triangular, truncated-triangular, hexagonal and round-shaped flakes. Thickness of these flakes is about several nanometers, while longitudinal dimension is about 100 nm.

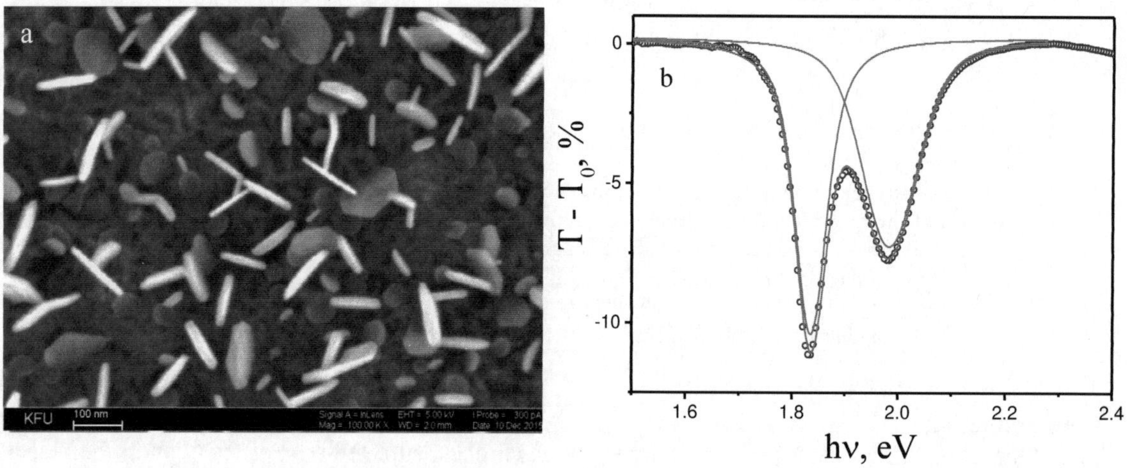

Figure 2. SEM image of the MoS_2 thin film on the silicon substrate (a); transmission spectrum of MoS_2 thin film on the fused silica substrate (b). The term corresponding to scattering processes is subtracted from the raw data in (b). Solid lines indicate the results of the spectrum fit with the sum of the two Voight components.

Optical transmission spectrum is presented in Figure 2b. It is clear that there are two absorption bands at 1.83 eV and 1.98 eV in the spectrum. These bands correspond to the well-known A and B exciton absorption lines, respectively [1]. Excitons are forming at the K and K′ points of the Brillouin zone split by the spin-orbit interaction. The absorption spectrum was fit with the sum of two Voight functions. It should be noted that positions of these bands depends on the thickness of the film [2].

In the simplest approximation, an exciton can be represented as a dipole consisting of an electron and a hole, which are held together by the Coulomb interaction. The exciton is an excited state, therefore relaxation to the ground state is inevitable for it. Thus, excitons have a finite lifetime. However, it is possible to relax to the ground state via some intermediate states [3].

The photoinduced absorption spectra are shown in Fig. 3. They have the same shape for the three excitation wavelengths used. The reason for this is that the photon energies of these lasers are greater than the bandgap, so the formation of excitons is preceded by the relaxation to the bottom of the conduction band.

All spectra consist of five peaks: two positive and three negative. Negative peaks correspond to an increase of the absorption, i.e. the *photoinduced absorption*, and the positive peaks correspond to the decrease of absorption, i.e. *photobleaching* of the sample. The observed spectra qualitatively correspond to the shape of the second derivative of the transmission spectrum without any excitiation

IOP Conf. Series: Journal of Physics: Conf. Series **1058** (2018) 012016 doi:10.1088/1742-6596/1058/1/012016

(Figure 4). This shape of the photoinduced absorption would take place in the case of a broadening of the absorption lines under photoexcitation [4].

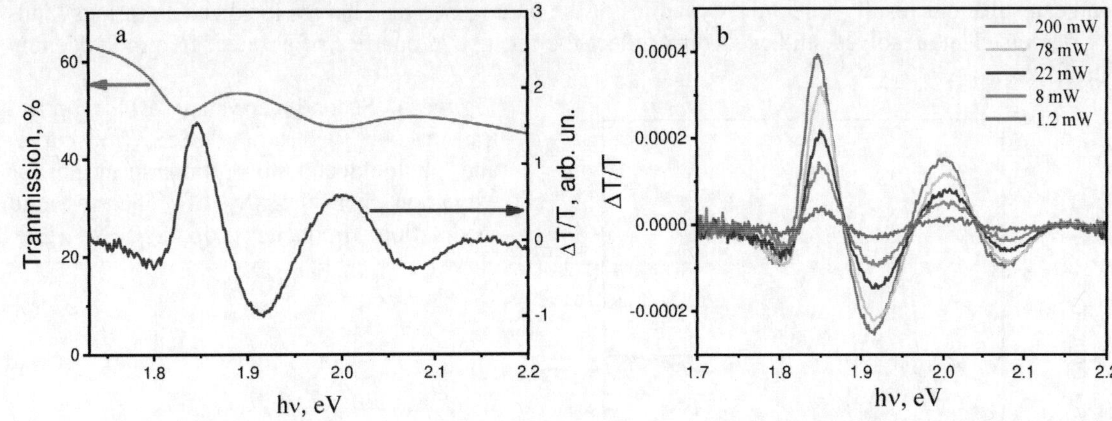

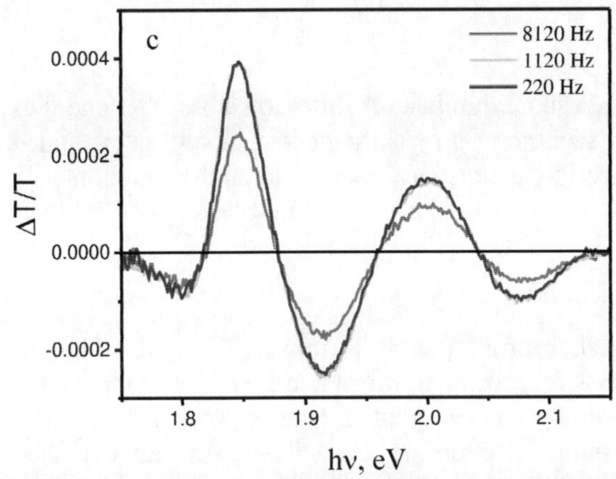

Figure 3. Photoinduced absorption spectra: a - for excitation wavelength of 532 nm and modulation frequency of 220 Hz; optical transmission spectrum is shown with the red line; b – for various intensities of the photoexcitation; c – for various frequencies of the photoexcitation modulation.

In our opinion, a simple explanation for such line broadening is associated with an inevitable presence of a nonzero electric dipole moment of excitons and the appearance of local electric fields upon photoexcitation in MoS_2 particles from photogenerated charge carriers – electrons and holes, as well as excitons. The position of the energy levels of excitons varies under the influence of an external electric field due to the Stark effect. For systems with a nonzero electric dipole moment, a linear Stark effect is observed. In this case, the shift of the line maximum depends linearly on $\vec{E}$, i.e. depends not only on the magnitude, but also on the direction of this field relative to the axis of the dipole. As a consequence, the shift in the energy levels occurs both towards the high and the low energies. Thus, different fractions of the exciton ensemble can experience shifts of different values, and this will be manifested through the broadening of the spectral components. When photoexcitation fluence increases the concentration of the local electric field sources also increases.

An open question is the long lifetime of the photogenerated charges. According to Figure 3c, the amplitude, and hence the change in the concentration of local sources of the electric field, are unchanged at laser modulation frequencies of 220 Hz and 1120 Hz, but fall by approximately 30% with an increase in the frequency to 8120 Hz. In this case, obviously, the lifetime of the charge carriers becomes comparable with the period of laser modulation. Thus, it is of the order of hundreds of microseconds. At the same time, according to the results of laser femtosecond spectroscopy, the relaxation time of the induced change in the transmission coefficient lies within the limits of units -

hundreds of picoseconds [3, 5, 6]. The magnitude of the effect in time-resolved studies reaches tens of percent, that is, orders of magnitude greater than in our observations. In our opinion, such a difference indicates that our results reflect the dynamics of the photogenerated charges localized at various kinds of defects. Time-resolved studies in turn reflect the intrinsic properties of a defect-free molybdenum disulfide.

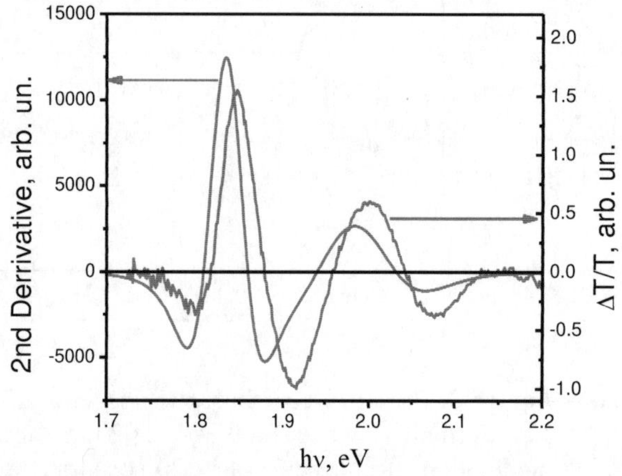

Figure 4. Second derivative of the optical transmission spectrum of MoS_2 (red curve) and photoinduced absorption spectrum for excitation wavelength of 532 nm and modulation frequency of 220 Hz (grey curve).

Also, we note that the photoinduced absorption spectra exhibit small shifts toward higher energies relative to the second derivative of the transmission spectrum. In recent papers such shift is explained by the interactions between A and B excitons. These interactions are referred to as the "exciton self-energy renormalization". They have a repulsive character, thus the absorption peaks shift to higher energies [5].

4. Conclusion

In this paper we present the long-lived photoinduced absorption study of thin nanogranular films of molybdenum disulfide. The observed spectra consist of several positive and negative components, corresponding to an increase and decrease in absorption under photoexcitation with a laser. The qualitative agreement of the photoinduced absorption spectrum to the second derivative of the transmission spectrum in our opinion is an indication that the Stark effect is at the origin of the observed phenomenon due to interaction with the photoexcited charge carriers localized at the defects.

Acknowledgments

The reported study was funded by the subsidy allocated to Kazan Federal University for the state assignment in the sphere of scientific activities, Nos. 3.8138.2017/8.9 (A.L.Z.) and 3.7704.2017/4.6 (R.V.Y.). Film synthesis and characterization were performed with the equipment of the Federal Center for Shared Facilities of Kazan Federal University. Scanning electron microscopy studies were done at the Interdisciplinary Center for Analytical Microscopy of Kazan Federal University.

References

[1] Ganatra R and Zhang Q 2014 *ACS Nano* **8**, 4074.
[2] Frindt R F 1965 *Phys. Rev.* **140**, A536.
[3] Borzda T, Gadermaier C, Vujicic N, Topolovsek P, Borovsak M, Mertelj T, Viola D, Manzoni, Pogna E, Brida D, Antognazza M R, Scotognella F, Lanzani G, Cerullo G and Mihailovic D 2015 *Adv. Funct. Mater.* **25**, 3351.
[4] Aleithan S H, Livshits M Y, Khadka S, Rack J J, Kordesch M E and Stinaff E 2016 *Phys. Rev. B.* **94**, 035445.
[5] Sim S, Park J, Song J, In C, Lee Y, Kim H and Choi H 2013 *Phys. Rev. B.* **88**, 075434.
[6] Vella D, Vega-Mayoral V, Gadermaier C, Vujicic N, Borzda T, Topolovsek P, Prijatelj M, Tempra I, Pogna E A A and Cerullo G 2016 *Journal of Nanophotonics* **10**, 012508.

Disperse admixtures in flame plasma influence on the heterogeneous combustion rate in the electrostatic field

I A Zyryanov, A G Budin, A P Pozolotin

[1] Vyatka State University, Department of Engineering Physics , 610000, Kirov, Moskowskaya Street, 36, Russia

b185@mail.ru

Abstract. The paper presents experimental study results of the disperse admixtures in a flame plasma influence on the polymethylmethacrylate combustion rate change under the action of an electric field in a model hybrid rocket engine. It is shown that increase of disperse admixtures concentration in flame leads to the influence of the field on combustion intensification. An explanation of the results obtained on the basis of the ion-wind mechanism is proposed.

1. Introduction

Hydrocarbon fuels combustion is the most common method of energy obtaining in thermal devices. of Investigation of the electric field action on heterogeneous combustion mechanism makes it possible to develop power plants parameters control methods. Studies conducted on model systems, as well as on a test hybrid rocket engine, show the possibility of the combustion rate controlling by imposing an electric field [1-3].

The dispersed combustion products (soot) presence in the flame allows to consider flame as a dust plasma. The dusty plasma with external fields interactions differ from the classical approaches describing combustion processes. The disperse admixtures introduction into the combustion zone makes it possible to control the dust plasma electrical properties in the flame [4], which affects the resulting combustion rate change under the influence of an electric field.

The aim of this work is the hybrid rocket engine combustion rate change in the presence of disperse admixtures in the gas phase and under the action of an electrostatic field study.

2. Experiment Procedure

Investigation of the disperse additives influence on the net field effect during combustion was performed on an engine with a coaxial arrangement of fuel blocks [5]. The combustion chamber of the engine is formed by two coaxially located fuel blocks: the outer block is a thick-walled pipe with an internal diameter of 32 mm and a wall thickness of 5 mm, inner block - cylinder with a diameter of 12 mm. The length of the blocks is 50 mm. The central block was made by extrusion or carved on the machine. In the gap between the blocks an oxidizer -gaseous oxygen is blown. Electrodes for creating an electric field are located: the first - outside the outer fuel block, the second - a metal rod, placed along the axis of the inner block. With this configuration, the inner fuel block acts as an insulator for the electrode. The outer fuel block combustion rate was studied, the combustion rate determination method is described in [1].

IOP Conf. Series: Journal of Physics: Conf. Series **1058** (2018) 012017 doi:10.1088/1742-6596/1058/1/012017

Disperse admixtures adding into the combustion zone was accomplished by filling the inner fuel block. To create blocks with a given mass concentration of admixtures, a mixture of polyethylene and admixtures was prepared. The mixture was then extruded into the primary preform. The primary preform was extruded three times to ensure uniform mixing of the components, after which the final block was formed. Thus, good mixing was achieved, as evidenced by the lack of own conductivity of the block. In experiments the following fuel block compositions was used: the outer fuel block is made of polymethylmethacrylate (PMMA); inner block - polyethylene (PE) with a disperse admixture (iron oxide). The compositions of the inner blocks are: PE, PE + 20% of the disperse admixture, PE + 40% of the disperse admixture, PE + 60% disperse admixture.

3. Experimental Results

Preliminary experiments conducted with inner blocks of different compositions showed that outer block combustion rate is determined only by the oxidant flow and external electric field intensity. The potential difference between the electrodes in experiments with the field was 5 kV.

The influence of the disperse admixtures concentration in the inner block on the outer block combustion rate change is shown in Fig. 1. The oxidizer flow rate is 5 kg/m2s. The legend shows the central electrode polarity. The disperse admixtures concentration in the inner fuel block is plotted along the horizontal axis.

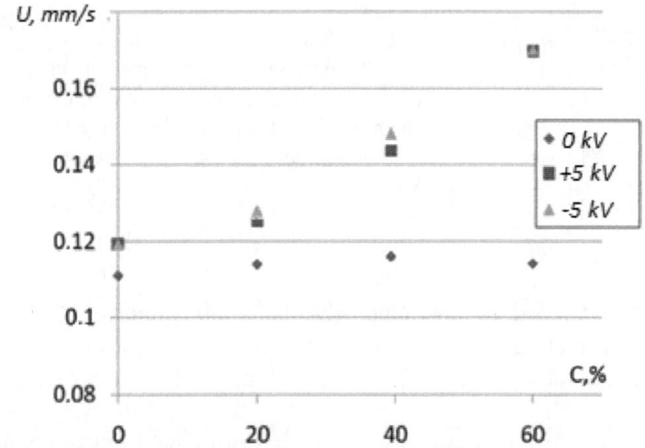

Figure 1. The outer block combustion vs various disperse admixtures concentrations.

As can be seen from the presented results, with an increase in the disperse admixtures concentrations the field effect increases. The direction of the field is not important.

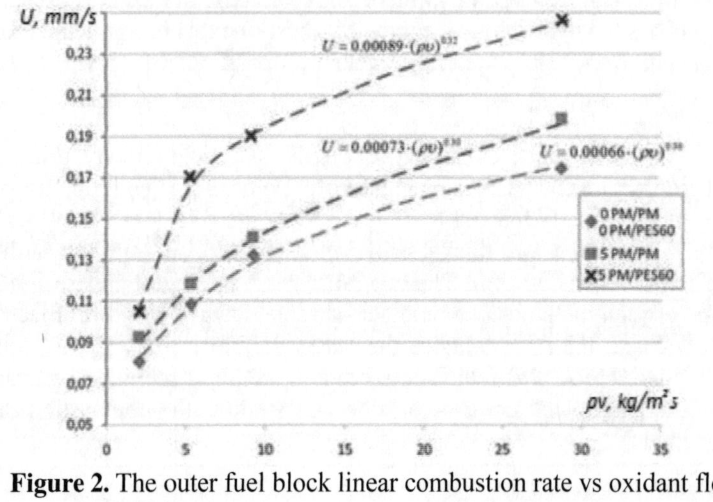

Figure 2. The outer fuel block linear combustion rate vs oxidant flow rate.

Investigation of the combustion law, i.e. the outer fuel block linear combustion rate dependence on the oxidant flow rate, was made with a central fuel block of polyethylene + 60% of the disperse admixture and with a central block of PMMA. The results are shown in Fig. 2. In the legend, the compositions of the blocks are designated: outer and inner blocks of PMMA - PM/PM; outer of PMMA, inner of polyethylene with admixtures - PM/PES60.

The combustion laws obtained from the experimental results:

$U = 0.00066 \cdot (\rho \upsilon)^{0.30}$ $U = 0.00089 \cdot (\rho \upsilon)^{0.32}$ (without field)

$U = 0.00073 \cdot (\rho \upsilon)^{0.30}$ (in the field without admixtures)

$U = 0.00089 \cdot (\rho \upsilon)^{0.32}$ (in the field with admixtures)

On average, the combustion rate in a field without admixtures increases by 10%, and with admixtures by 30%.

4. Discussion of results

The experimental results show that the disperse admixtures introduction into the gas phase of the combustion zone substantially increases the effect of the field influence on the condensed component combustion rate. To explain the combustion rate change in the field, the "ion wind" mechanism is most often used [6]. Under the "ion wind" we mean the charged particles of a flame motion under the action of an electric field. As shown in [7], disperse particles in a flame accumulate an essential electric charge and, therefore, begin to interact with an electric field. The particles, being in the flame, acquire a high temperature, and shifting under the influence of the field, cause the heat flux to the condensed phase change. This leads to a change in the gassing rate. Consequently, the admixtures concentration increase will lead to field effect intensification (Fig. 1) and the pre-exponential factor in the combustion law increase (Fig. 2). In the electric field, with the introduction of dispersed impurities, an insignificant change in the combustion regime is observed: the degree in the combustion law varies by 7%, which also confirms the proposed explanation. The absence of field direction influence on the resulting combustion rate change is due to the fact that the dispersed particle charge depends on the time of its staying in the combustion zone [7,8]. Consequently, an additional heat flux can be created both negatively and positively charged particles of the disperse phase.

5. Conclusion

The paper presents study results of the disperse admixtures influence on the combustion rate change under the action of an electric field in a hybrid rocket engine. It is shown that the polymethylmethacrylate mass combustion rate increases regardless of the electric field direction. The field effect on combustion increases with the disperse admixtures introduction. The combustion laws of the investigated substance are obtained. An explanation for the observed effect is proposed on the ion-wind mechanism basis.

References

[1] Reshetnikov S M, Zyryanov I A, Pozolotin A P, Budin A G, Electrostatic field influence on the combustion rate in hybrid rocket. *Kazan, Vestnik of the Kazan State Technical University,* 2015, V.71, P. 52-57.

[2] Reshetnikov S M, Zyryanov I A, The electrostatic field influence on the combustion macrokinetic of alkanes and kerosene. *Kazan, Vestnik of the Kazan State Technical University,* 2011, V.1, P. 120-128.

[3] Reshetnikov S M, Zyryanov I A, Pozolotin A P, Features of polymers combustion in the electrostatic field. *Taganrog, Southern Federal University Messenger. Technical Science,* 2013, V.8, P. 30-36.

[4] Fairushin I I, Dautov I G, Kashapov N F, Distribution of the potential and concentration of electrons in low-temperature plasma with hollow microparticles, *International Journal of Environmental Science and Technology,* 2016, P. 1-6.

[5] Reshetnikov S M, Zyryanov I A, Budin A G, Pozolotin A P, Heterogeneous mass transfer in HRE in the presence of electrostatic field research, *Journal of Physics: Conference Series. – IOP Publishing,* 2017, V.789, P. 012042.

[6] Park D G, Chung S H, Cha M S, Bidirectional ionic wind in nonpremixed counter flow flames with DC electric fields, *Combustion and Flame,* 2016, V.168, P. 138-146.

[7] Savelev A M, Starik A M, Features of ions and electrons interaction with plasma nanoparticles formed during the hydrocarbon fuel combustion, *St. Petersburg, Journal of Technical Physics,* 2006, V.76, №.4, P. 53-60.

[8] Reshetnikov S M, Zyryanov I A, Pozolotin A P, Budin A G, The distribution of excess charges in the diffusion flame of hydrocarbons, *Journal of Physics: Conference Series,* 2016. V. 669. P. 4.

Low-pressure glow discharge modeling in transverse supersonic gas flow

D I Israphilov

Kazan Federal University

diisraphilov@yandex.ru

Abstract. Article presents a model for the realization of a glow discharge at low pressures due to the organization of supersonic gas flow in a limited region of discharge chamber. A model of flow regime of a supersonic flow in a vacuum chamber is described. Results of experiments on realization of a glow discharge at low pressures due to the organization of a transverse supersonic gas flow are showen.

1. Relevance

Field of glow discharge application for plasma processing of details surfaces continuously expanding. Ion-plasma sputtering of material or modification of surfaces is increasingly used in optic industry, electronics and other fields. One of new area is additive technologies.

In ion-plasma technological process is necessary to organize smooth achievement of the particles to processing surface. To achieve this condition free path length λ of particles should be several times larger than the distance L from the area of movement beginning to processing surface.

One of conditions for glow discharge existence is presence of all near-electrode zones. Number of electron collisions with neutral particles determines length of each zone. With pressure, decreasing length of cathode zones is increasing and certainly, there comes a time that the electrode gap is not enough space for all near-electrode zones. Therefore, at very low pressures, discharge cannot exist. To find the critical pressure, at which it is still possible glow discharge, we assume condition of cold discharge. We assume that inter-electrode distance of 10 cm and number of collisions 10, so mean of free path λ is 1 cm. Critical pressure for glow discharge existing [1]:

(1)

Thus, existence of glow discharge at pressures below 10-2 Torr hampered. Therefore, there is an insoluble problem: on one side discharge existence pressure above 10-2 Torr is necessary and from another glow discharge application in technological processes of coating deposition required pressure below 10-2 Torr.

One solution to this problem is magnetron sputtering application [1, 2, 3]. Magnetic field may increase the probability of collision of electrons with neutral atoms to order more times at low pressures by the joint action of electric and magnetic fields on moving charges. All this gave opportunity to magnetron discharge application at a pressure of about 10-3 Torr. That led to a widespread of magnetron systems in installations for ion-plasma treatment.

In works [4-7] is proposed a fundamentally new solution to problem. Namely, they proposed the creation of zones of different concentrations of neutral particles in electrode gap. For example, in cathode area to create a high vacuum with a minimum concentration of neutral particles, in anode area to provide a relatively high concentration of neutral particles. Electrons in region with high concentration of neutral particles can experience dozens of collisions with neutral particles, while the processing of the material will take place at the maximum possible low pressures. For this condition, it is necessary to create a stream with particles of the working gas in the interelectrode space and at the same time to maintain the vacuum in the discharge chamber below 10-2 Torr. To achieve such a separation of zones in the electrode gap may allow the supersonic gas flow. The density of pumped gas through the working area should provide more than a dozen collisions of electrons with neutral particles, which will allow maintaining the glow discharge, keeping chamber pressure below 10-2 Torr.

To maintain different particle concentrations in chamber device shown (Fig.1), providing the organization with a continuous supersonic flow and its pumping from the vacuum chamber using a supersonic nozzle and mixing chamber. In mixing chamber, flow restore and removed after by vacuum pump. At certain conditions device can operate as a jet pump that will not only keep the initial chamber pressure, but also lowering it.

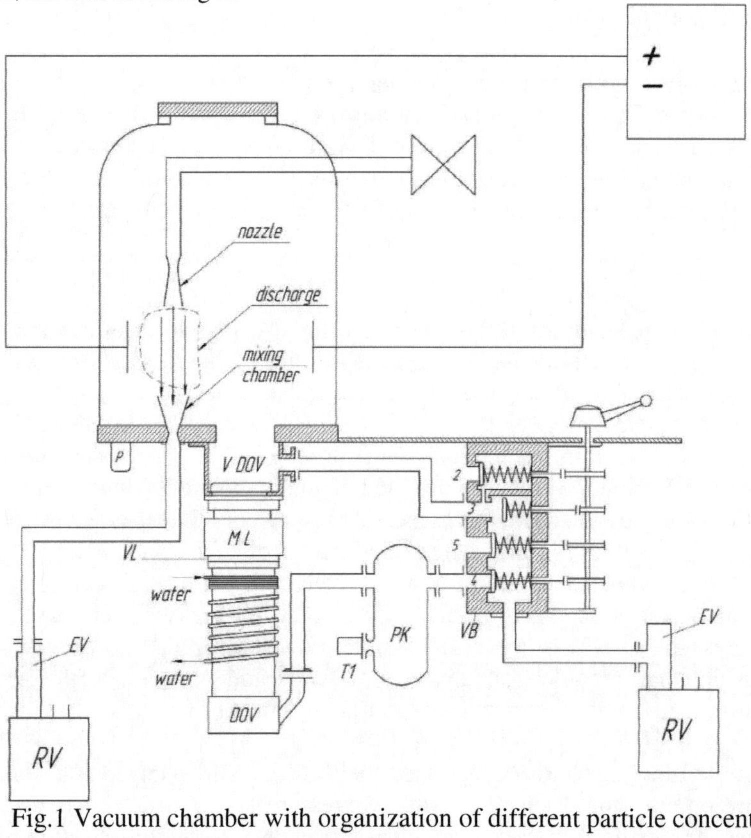

Fig.1 Vacuum chamber with organization of different particle concentration.

RV- rotary vacuum pump, EV-valve, VB- valves box, PK- prev vacuum chamber, DOV - diffusion vacuum pump, ML - oil trap, VL - nitrogen trap, R – vacuum chamber, T1, T2 - thermocouple vacuum gauges, P - high vacuum gauge.

The aim of the work is to create a model of supersonic flow and carry out experimental studies on the implementation of a glow discharge at pressures of 0.03 Torr.

Nozzle and mixing chamber consists of a diffuser and confusor parts and passage of critical cross-section flow velocity reaches critical speed, with the further expansion of flow occurs the growth rate. In calculation of nozzle, main dependencies deduced from analysis of equations of flow continuity and the first two laws of thermodynamics [7].

IOP Conf. Series: Journal of Physics: Conf. Series **1058** (2018) 012018 doi:10.1088/1742-6596/1058/1/012018

Calculating mixing chamber geometric parameters and operating gas flow regimes in supply system, field of continuous flow and pumping [8] allowed carrying out modeling and pilot studies of implementing low-pressure glow discharge in transverse supersonic flow.

2. Simulation

Based on the obtained model [8], geometric dimensions of confuser-mixing chamber elements were calculated. Pressure parameters were as follows: before nozzle P_1 = *1 atm*, in vacuum chamber P_2 = *0.03 Torr*, after mixing chamber in vacuum pumping sustem P_3 = *0.2 Torr*, gas air. Simulation results (Fig. 2) indicate the possibility of creating a continuous flow in a vacuum to produce zones with different particle concentrations.

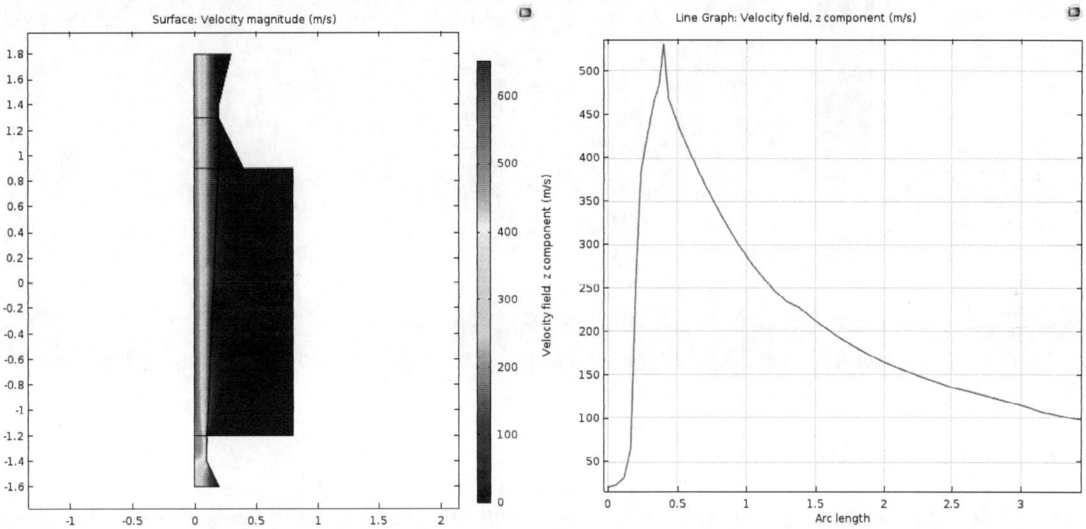

a) b)

Fig. 2. Distribution of gas particles velocity in a vacuum chamber with supersonic pumping of the working gas: a) axisymmetric model; b) velocity distribution along vacuum chamber axis of symmetry.

When supersonic flow is organized in a vacuum chamber, pressure remains unchanged (Fig 3), flow is continuous [9-10]. Gas flow is restored after passing mixing chamber

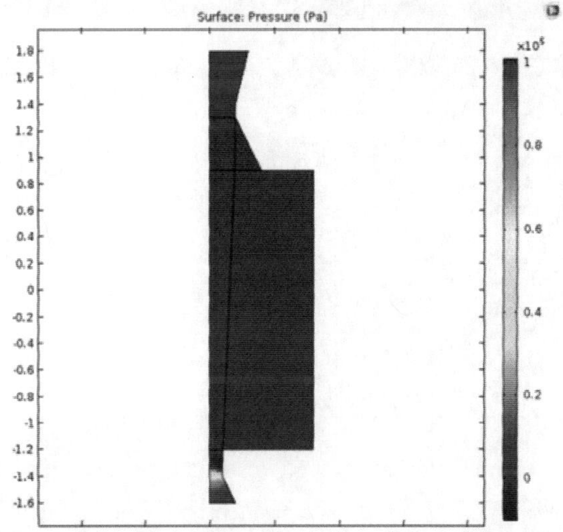

Fig. 3 Pressure distribution in a vacuum chamber with supersonic pumping of working gas

Simulation of glow discharge were carried out using following form (Fig. 4a). Between cathode and anode potential difference is *1250V*. In interelectrode gap an area of high particle concentration is organized through the organization of continuous flow described above.

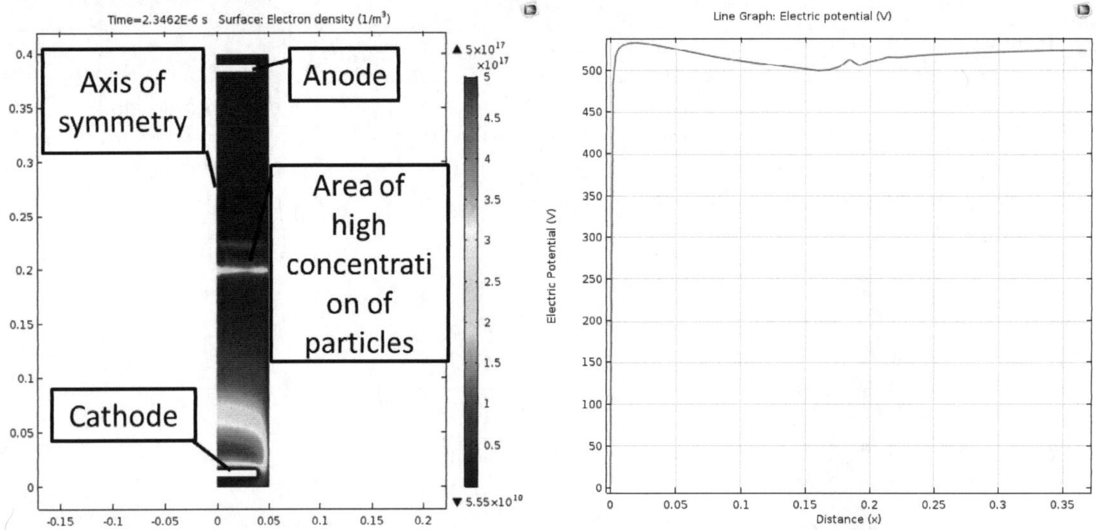

a) b)

Fig.4 Result of simulation: a) surface electron density pattern; b) electric potential in interelectrode gap.

3. Experimental studies

Experimental studies were conducted to determine accuracy of modeling

Gas consumption defined through characteristic of the used vacuum pump 2NVR-5DM with ultimate total pressure $Pn=5*10^{-3}$ *Torr*, pumping speed $G = 5.5$ *l/s*. By use coefficient *0.9* at *1 Torr* pressure get $G = 4.1$ *l/s*.

As a result of experimental studies (Fig. 5a), a volt-ampere characteristic of discharge was obtained (Fig. 5b)

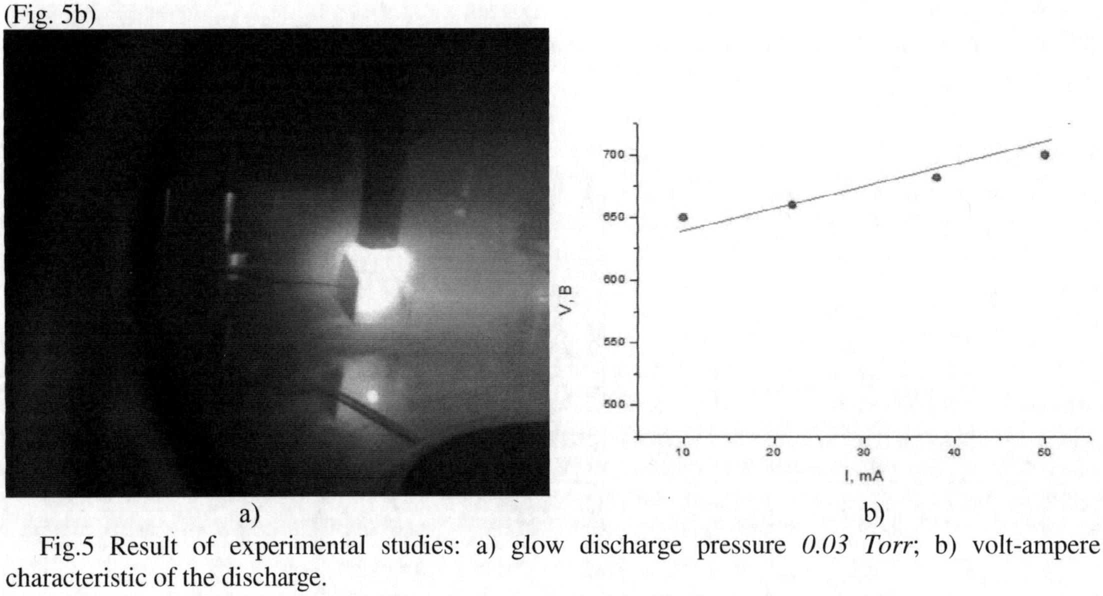

a) b)

Fig.5 Result of experimental studies: a) glow discharge pressure *0.03 Torr*; b) volt-ampere characteristic of the discharge.

4. Conclusion

As a result of simulation and experimental studies, parameters of supersonic flow in vacuum chamber and characteristics of the glow discharge with transverse supersonic pumping of working gas flow were determined. The steady-state pressures in the gas flow, behind the confuser and in the cathode region are determined. The specific values of the parameters for this case were as follows:

Supersonic flow velocity was from 250 to 550 m / s in the gap between the nozzle and the mixing chamber. Pressure in vacuum chamber remained uniform and amounted to 0.03 Torr.

Distribution of the potential of glow discharge indicates a high voltage along the discharge of 550-650 V, a high concentration of electrons is observed in the near-cathode zone and region of high particle concentration. Results of the experiments confirm implementation of discharge at a pressure of 0.03 Torr and model accuracy in the region of 8%.

References

[1] Rayzer J.P. Physics of gas discharge. 2009. — 736 p.

[2] S.B. Nesterov et.all. Methods of vacuum systems calculation. — M.:,2004. — 220 p.

[3] E.V. Berlin, L.A. Seydman. Ion-plasma processes in thin-film technology. Moscow: 2010.- 528p. ISBN: 978-5-94836-222-9

[4] B.A.Timerkaev, B.R. Zalyaliev. Glow Discharge in a Transverse Supersonic Gas Flow at Low Pressures. High Temperature, 2014, Vol. 52, No. 4, pp. 471–474.

[5] B.A.Timerkaev, B.R. Zalyaliev, B.R.Karimov, D I Israfilov. Behavior of a glow discharge in plasma deposition installations in distributed supersonic gas flow / Vestnik KGTU. 2013. №4. P. 198-291.

[6] B A Timerkaev, B R Zalyaliev, A I Saifutdinov. Glow Discharge Characteristics in Transverse Supersonic Air Flow. Journal of Physics: Conference Series 567 (2014) 012032.

[7] A.I.Saifutdinov, B.A.Timerkaev, and B R Zalyaliev. Control of glow discharge parameters using transverse supersonic gas flow - numerical experiment Journal of Physics: Conference Series 567 (2014) 0120313.

[8] G.J. Dautov, B A Timerkaev. Nonequilibrium gas-discharge plasma generators. 1996, 200p.

[9] Saifutdinov A.I. Drift model of glow discharge with nonlocal value of the electric field in the ionization source / A.I. Saifutdinov, B.A.Timerkaev - Journal of Engineering Physics, T. 85, №5, 2012 S.1104-1109. (A.I. Saifutdinov, B.A. Timerkaev - Journal of Engineering Physics and Thermophysics, Vol. 85, No. 5, 2012, P. 1202-1207).

[10] B A Timerkaev, M M Ahmetov, B R Zalyaliev, O A Petrova and D I Israfilov. Longitudinal distribution of electrical parameters in normal glow discharge. Journal of Physics: Conference Series 567 (2014) 012036.

IOP Conf. Series: Journal of Physics: Conf. Series **1058** (2018) 012019 doi:10.1088/1742-6596/1058/1/012019

Application of six- jet plasmatron in science and technology

F G Karikh, I M Arslanov

Kazan Federal University

Arslanov_IM@mail.ru

Abstract. The design and operation of the six-jet plasmatron are considered. Experiments using a six-jet electric arc plasmatron are described and their results are shown.

1. Introduction.

To get a low-temperature plasma, an electric-arc two-jet plasmatron is widely used at present as a source of excitation of the spectrum of solutions and powder materials, as well as for the thermal treatment of metals and alloys.

The disadvantage of using an electric arc two-jet plasmatron in spectroanalytics is unsuitable for monitoring the composition of gas streams and powders with a complex fractional composition.

2. The contruction of a six-jet electric arc torch plasmatron.

This disadvantage is eliminated by using a six-jet plasmatron, assembled using three two-jet plasmatrons, powered from a three-phase current network [1]. It's device shown in the plan in Fig. 1, where the plasma-forming copper heads 1 mounted on the dielectric plateaus 2 are rigidly attached to the arms 3 with the ability to move along the axes of the heads 1 in a direction perpendicular to the tubular posts 4. Above them, a tubular chamber 5 feeding into the argon heads protecting the electrodes from oxidation and the working gas (air) distribution chamber 6 by means of flexible hoses 7. The supply of protective gas to the copper anodes with contacts A1, A2, A3 and a tungsten cathode m with contacts K1, K2 and K3 is carried out by means of flexible hoses 8. Above the posts 4, the cooling water inlet chamber 9 and the water discharge chamber 10 are inserted into channel 19 in axial chambers 5 and 6. The cooling water inlet in the head section is made from the vertical channel 17 (Fig.2, which shows a vertical section of the plasmatron along a plane passing through the axes of a pair of heads, powered from one of the phases of the three-phase rectifier used, in particular connected to an anode head with an electrode A2 with an auxiliary cathode VK2 and a cathode head with a cathode m K2 with auxiliary anode A2). Next to the annular chamber 9 is a water discharge chamber 10 in the duct 19. The connection of the chambers 9 and 10 to the head sections is carried out by means of flexible hoses 11 and 12. The water stream 19 heated by the operation of the plasmatron is directed to cool into the radiator (in Fig. 2, it is not shown) from which the cooled water returns to the channel 16. The posts 4 are located on the mounting table 20, between which is rigidly mounted a branch pipe 14 forming the analyzed gas flow or the processed powder material 15. A cylinder 13 is mounted axially to the nozzle 14, synchronous change in the angle of convergence of the six heads by means of a hinge system 21 providing a synchronous change in the angle of inclination of the heads with respect to the plane of the table 20. The system 21 comprises a plateau 22 with movable arms 3 mounted on them, providing a change in the magnitude of the interelectrode gap between the plasmatron plasma-forming heads that are mounted on dielectric plateaus rigidly connected to brackets with the possibility of moving heads along their axes in a direction perpendicular ohm with respect to the tubular posts, above which an annular chamber for supplying argon to the plasma-forming heads is annularly disposed, a working gas distribution chamber and a water-cooling chamber connected to the heads by means of flexible hoses. Between the posts located on the mounting table, a stiff pipe is assembled to form the analyzed gas flow or the processed powder material and a cylinder providing the synchronism of the angle of convergence of the axes of the six heads by means of the system shown in Fig. 2, which comprises a plateau with movable brackets , providing a synchronous change in the magnitude of the interelectrode gap of each pair of plasma-forming heads.

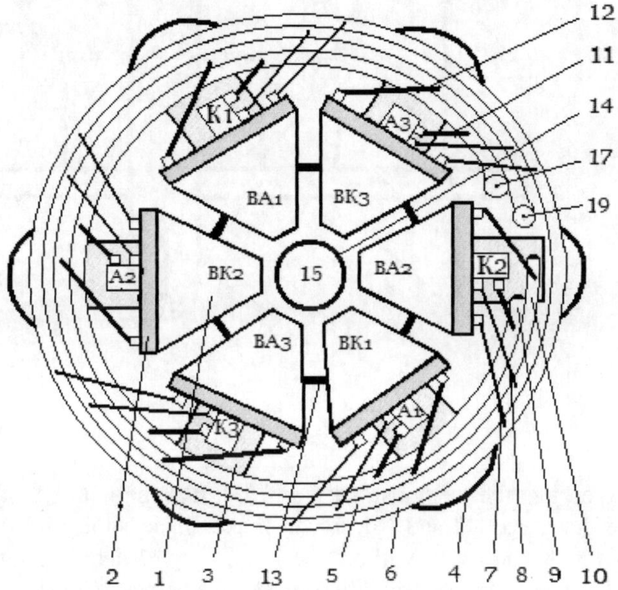

Fig.1

3. Application of a six-jet electric arc torch plasmatron in engineering and results

Production tests of the ability to control the smelting conditions in real time by the composition of waste gases carried out at the foundry of JSC "KamAZ" established that the application of the method [2] carried out in 28 batches on an electric arc furnace of the DSP-50 grade allowed:

- adjust the composition of the alloy in the process of it's preparation by pre-washing the required components;
- Reduce electricity consumption by 10% with a reduction in the consumption of electrodes by 3%;
- control the temperature of the liquid metal, both during the melting period and during the period of the furnace standstill;
- reduce metal losses associated with the fumes of remelting disperse waste by more than two times;
- reduce the time constant of the process of measuring the chemical composition of the alloy being prepared by an order of magnitude.

In the tests carried out using the method [3], the example of obtaining castings of aluminum alloys during the operation of the Dozamat-400 dispenser by monitoring the content of the modifier in the melt, reduced the marriage from 22% to 15%.

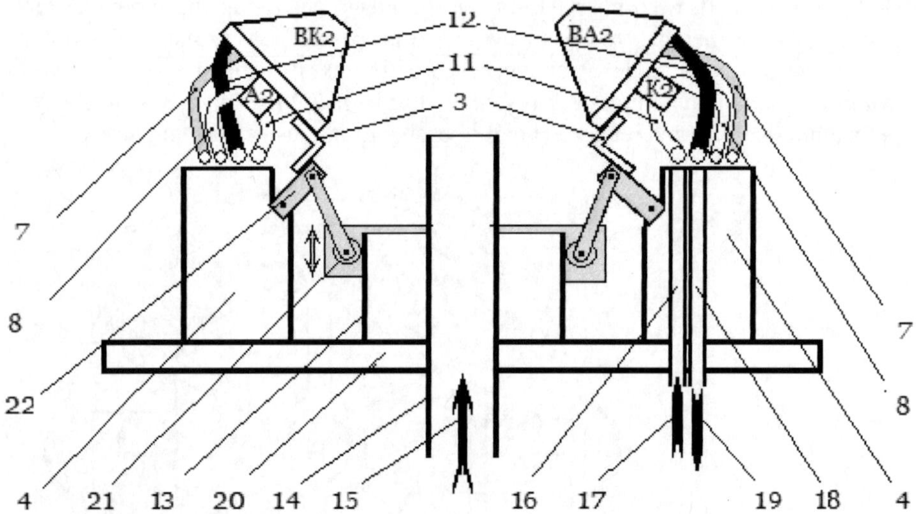

Fig.2.

It should be noted that the methods of [2] and [3], like the whole work [4], were realized using a high-voltage flare discharge [5] and [6], which makes it possible to obtain spectral lines only with excitation energies less than 5 eV, without the possibility of determining the total This meltdown is eliminated only by using an electric arc six-jet plasmatron. Considering the undesirability of a significant flow of argon in the operation of a six-jet plasmatron, a switch has been developed for the ratio of the flow rates of two gas streams, which makes it possible to reduce argon consumption by two orders of magnitude, allowing the possibility of its considerable consumption only during the launch phase.

4. Application of a six-jet electric arc torch plasmatron in science

It is interest to use a six-jet plasmatron in combination with other sources of excitation of the spectrum, for example, in the invention of patent RF No. 2303255 [7], the source of excitation of the spectrum in which is laser radiation. The joint application of a plasma torch and a laser is a scientific task. Now in science the question is raised on the possibility of diagnosing the state of human health. This task was carried out in the invention patent No. 2119768 [8]. However, its using associated with using of labor-intensive and expensive chemical methods for analyzing hair composition. Elimination of these drawbacks was achieved using the method of laser atomic emission spectral analysis of hair [9] in combination with the use of a six-jet plasmatron, which allows reducing the cost of hair composition analysis by more than an order of magnitude compared with the method [8].

References

[1] Karikh F.G. and others. Electric arc six-jet plasmatron. Patent RF, No 2529740, BI, 2014. Byul. No. 27.

[2] Karikh FG, Mukhametzyanova GF, Petrov DM Method of spectroanalytical control of the determination of the content of trace elements in gas flows with a hydrocarbon matrix. Patent

RF, No. 2229114, BI, 2004. Bul. No. 14.
[3] Karikh FG, Karikh AF Method for determining the composition of gas streams. Patent RF, No.2231776, BI, 2004. Bul. No. 18.
[4] Karih F.G. Methodology for determining the physical and chemical parameters of melting metals on the basis of spectroanalytical data. / / Thesis for a degree of Doctor of Technical Sciences. Chelyabinsk-2004, 310 with.
[5] Karikh FG Device for excitation of the gas flow spectrum. Patent RF, No. 2085871, BI, 1997. Bul. No. 21.
[6] Karikh F.G. Method of spectroanalytical determination of smoke composition. Patent RF, No. 2090857, BI, 1997. Bul. No. 26.
[7] Savilov S.V. and others. The laser atomic emission spectrometer "ЛАЭС". Patent RF, No.2303255, BI, 2007. Bul. No. 20.
[8] Chernozubov E.I. The method of diagnosing the state of human health. Patent RF, No. 2119768, BI, 1998, Bul. No. 14.
[9] Karikh F.G., Mukhametzyanova G.F. Method of laser atomic-emission spectral analysis of hair. Patent RF, No.2589960, BI. 2016, Bul. No. 19.

Investigation of amplifying and electrical properties of plasma carbon dioxide lasers

A S Kiselev and E A Smirnov
Department of electronic instruments and devices, Saint Petersburg Electrotechnical University "LETI", 197376, Saint Petersburg, Russia

E-mail: alex.kiselev.epu@gmail.com

Abstract. In this work devices to study the electrical and amplifying properties of the plasma of gas mixtures corresponding to carbon dioxide lasers were designs. The devices to a vacuum system that allows them to be pumped out and filled with a working gas mixture are connected. Experiments to study the current-voltage characteristics of the discharge at various pressures of the gas mixture and the diameters of the discharge gaps were carried. The discharge voltage decreases with increasing discharge current, which is characteristic of a discharge with a longitudinal positive column. Large number of experiments to study the enhancement of the emission of carbon dioxide laser in the discharge of various gas mixture were carried: CO_2, $CO_2 : N_2$, $CO_2 : N_2 : He$. The obtained dependences have a maximum at a certain pressure. The three-component mixture has the greatest amplification. The results of the studies were used when replacing the active gas mixture of a carbon dioxide laser, as well as in the educational process when studying the amplifying properties of plasma.

Gas-discharge lasers form the most extensive class of lasers. Gas-discharge lasers currently occupy a leading position in the fields of application. Gas-discharge lasers are beyond competition in terms of the degree of coherence of radiation, the developed ranges of generation wavelengths and output power levels. The unique properties of radiation from gas-discharge lasers, and the prospects for their development, allow maintaining relatively high volumes of production, development and application in such areas as metrology, recording, processing and transmission of information, technology, medicine and other [1, 2]. Gas-discharge lasers, including glow discharge lasers (GDL), are not only widespread but also quite complex devices in structure and manufacturing technology. From this point of view and taking into account the economic reality, it is quite logical to find ways to increase the service life of the gas-discharge lasers used, and to restore active elements in the event of complete loss of generation. The relatively short life of lasers makes this problem even more urgent. Possible reasons for the complete or partial failure of gas-discharge lasers are: a disturbance in the operation of the power source; increase in the level of losses in the optical resonator; reduction of amplifying in the active element of the laser. The service life of power supplies is much higher than that of active elements. In optical resonators, the main causes of a noticeable reduction in radiation power are the degradation and contamination of mirrors and Brewster's windows. The main reason for the failure of the active element is the degradation of the active medium due to the leakage of atmospheric gases, gas desorption by elements of construction, the absorption of the working gas by the shell, by the electrodes and their spray products, and the diffusion of the working gas through the envelope.

The effect of the leakage process on the electrical parameters of the laser was studied experimentally and theoretically by comparing the potential gradient E_z in the positive column for the pure active media of a CO_2–laser and media containing additives of atmospheric gases [3]. An automated calculation of the electron temperature Te required to find E_z was carried out for recombination conditions on the walls of a longitudinal cylindrical tube with an internal radius R. In this work was assumed Maxwell distribution electron velocity, and approximation of the ionization cross section was a linear [4].

For multicomponent gas mixtures [5], on the left side of the ionization balance equation, the sum of the frequencies $\nu_i(j)$ of ionization of all sorts of gases j entering into the mixture was recorded:

$$\nu_\mathrm{i} = \frac{b_\mathrm{i} k (2.405)^2 T_\mathrm{e}}{R^2 e}, \tag{1}$$

The average mobility of gas ions was calculated using the Blank law [6]:

$$\frac{1}{b_\mathrm{i}} = \sum_j \frac{p_j}{b_{\mathrm{i}0j}}, \tag{2}$$

where p_j is the partial pressure of the gas of type j; b_{i0j} is the unit (for p = 1 Pa) mobility of ions of a gas of type j in its own gas.

To calculate the electron temperature T_e, expression (1) was transformed to the form:

$$\sum_j \nu_\mathrm{i}(j) - \frac{23.1 b_\mathrm{i} k T_\mathrm{e}}{e d^2} = 0, \tag{3}$$

The longitudinal gradient of the potential in the positive column was found to be [7]:

$$E_z = \frac{3kT_\mathrm{e}\sqrt{\chi}}{2e\lambda_\mathrm{e}}, \tag{4}$$

where χ - coefficient of accommodation of electrons; λ_e is the mean free path of electrons in a mixture of gases.

In the case of an GDL, the quantity χ can be determined from the experimental data. The average mean free path of electrons in the gas mixture with sufficient accuracy for practice can be determined from the expression:

$$\frac{1}{\lambda_\mathrm{e}} = \sum_j \frac{p_j}{\lambda_{\mathrm{e}0j}}, \tag{5}$$

where λ_{e0j} is the unit (for p = 1Pa) the mean free path of electrons in a gas of type j.

In the case of strong leakage of atmospheric gases into the active medium of the laser, there remains a cardinal solution-replacement of the gaseous active medium. Of all the ways to increase the life of active elements of gas-discharge lasers, the replacement of the gas active medium is the most laborious solution, since it assumes the depressurization of the active element. This raises a number of problems: the need for a vacuum pumping system and a set of working gases, the selection of optimal gas ratios in the working mixture for each model of the laser, the search for an optimal operating mode for the new mixture and others [8].

The regeneration of the generation ability of active elements of CO_2-lasers was carried out by the method of replace the gaseous medium. Preliminary, with the help of a specially designed layout (figure 1), were investigated the dependences of the amplifying k_{amp} on the discharge current and pressure for different gas mixtures [9]. The voltage on the discharge decreases with increasing discharge current. The layout had length of active medium 0.33 m with a discharge channel diameter of 10 mm [10]. The ends of the layout were covered with Brewster windows of gallium arsenide. Below are the results obtained for pure CO_2 (figure 2), double (figure 3) and triple (figure 4) gas

IOP Conf. Series: Journal of Physics: Conf. Series **1058** (2018) 012020 doi:10.1088/1742-6596/1058/1/012020

mixtures. All dependencies were obtained at a constant cooling water flow rate (0.53 gallons in minute).

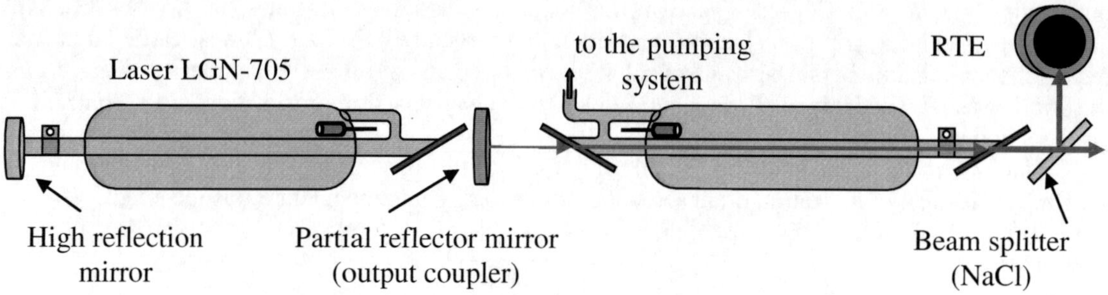

Figure 1. Scheme of investigation of the amplifying properties of the active medium of a carbon dioxide laser

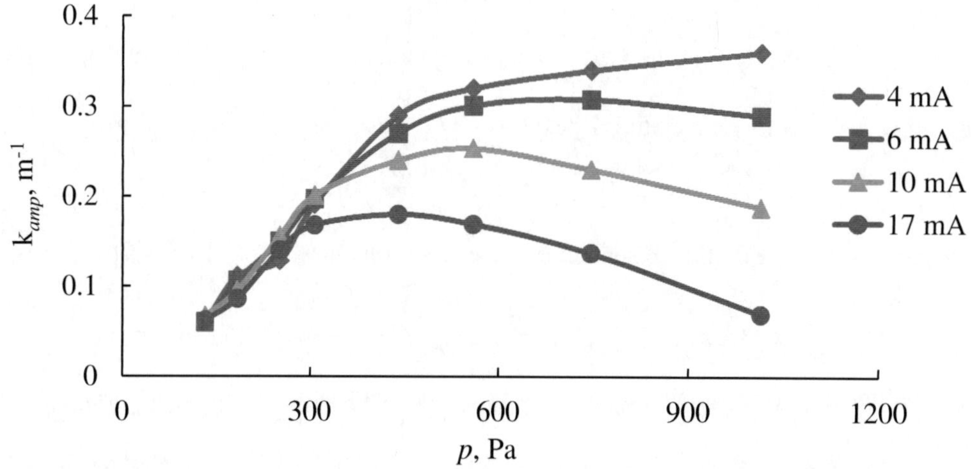

Figure 2. Dependence of the amplifying on the pressure of carbon dioxide at different discharge currents.

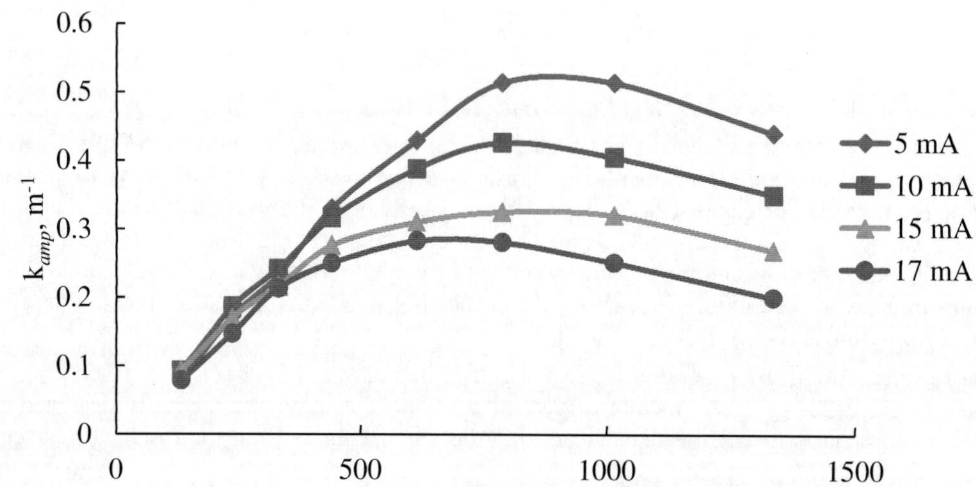

Figure 3. The dependence of the amplifying on the total pressure of the double gas mixture $CO_2 : N_2 = 1 : 1$ at different discharge currents

IOP Conf. Series: Journal of Physics: Conf. Series **1058** (2018) 012020 doi:10.1088/1742-6596/1058/1/012020

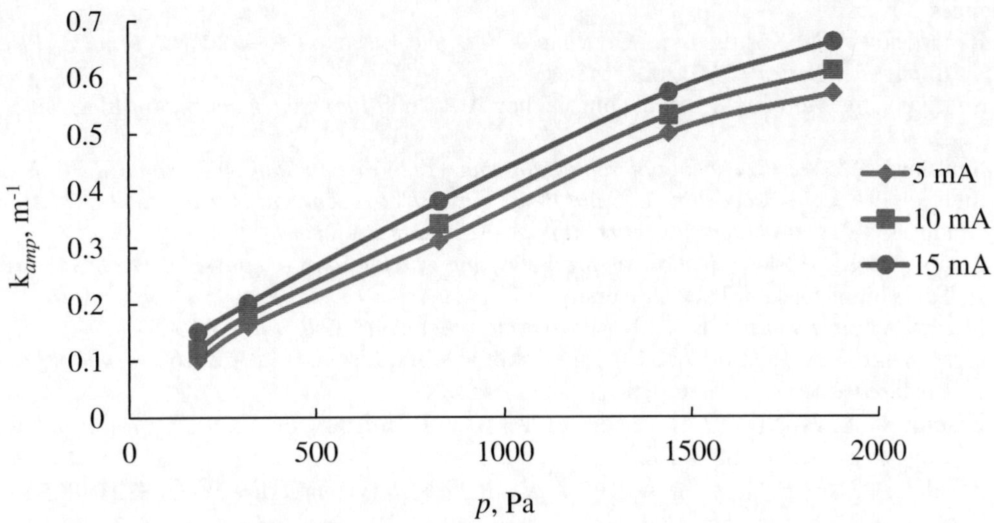

Figure 4. The dependence of the amplifying on the total pressure of the triple gas mixture $CO_2 : N_2 : He = 1 : 1 : 6$ at different discharge currents

The study of the current-voltage characteristics was carried out on a layout having 4 discharge tubes of different diameters. Current-voltage characteristics have a falling character, which corresponds to lasers of a glow discharge. Figure 5 shows the CVC of the discharge for different diameters of the discharge gap and pressures of gas mixture.

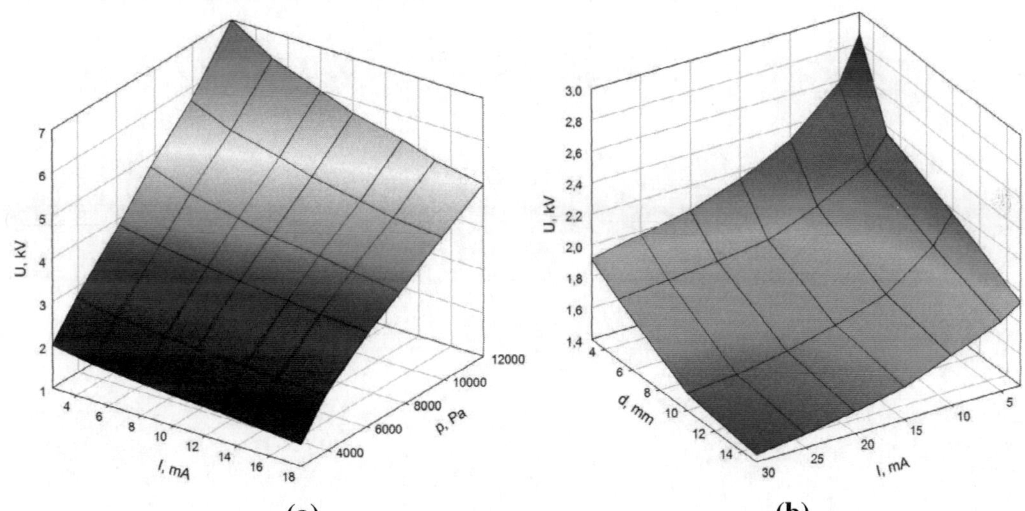

(a) **(b)**

Figure 5. 3D-plots of current-voltage characteristics of the discharge: **(a)** in a carbon dioxide; **(b)** in a gas mixture $CO_2:N_2:He=1:1:6$

The dependences of k_{amp} on the pressure have a maximum. The maximum value of (0.2 ... 0.3 m^{-1}) increases with the transition from pure CO_2 to a mixture of CO_2 : N2 : He (0.6 ... 0.7 m^{-1}). Similar results are useful both in the selection of the gas mixture of the CO_2-laser, and in the study of the CO_2-amplifier itself, for example, in the educational process. Connecting the amplifier to the vacuum system facilitates periodic over-filling with the active mixture in the event of its leakage. In order to ensure the maximum time between overflows in the laboratory version of the layout, it is possible to avoid using expensive helium and use either pure CO_2 or CO_2 with a N_2 [11].

References

[1] Ramazanov A N, Kostrin D K, Goncharov V D and Lisenkov A A 2016 *Journal of Physics: Conference Series* **729** 012004
[2] Smirnov E A, Morozov S S and Timoshichev A A 2008 *Petersburg Journal of Electronics* **2–3** 39–48
[3] Smirnov E A 2005 *Proceeding of Saint Petersburg Electrotechnical University "LETI"* **2** 43–8
[4] Smirnov E A 2013 *Proceeding of Saint Petersburg Electrotechnical University "LETI"* **10** 6–10
[5] Granovsky V L 1971 *Electrical current in gas* (Moscow: Science)
[6] Levitsky S M 1964 *Collection of problems and calculations on physical electronics* (Kiev: Publishing House of Kiev University)
[7] Kiselev A S and Smirnov E A 2014 *Vacuum technique and technology* **1** 56–9
[8] Martsinukov S A, Kostrin D K, Chernigovskiy V V and Lisenkov A A 2016 *Journal of Physics: Conference Series* **729** 012023
[9] Kiselev A S, Kostrin D K, Lisenkov A A and Smirnov E A 2017 *Journal of Physics: Conference Series* **789** 012027
[10] Kiselev A S and Smirnov E A 2017 *Journal of Physics: Conference Series* **872** 012050
[11] Kiselev A S and Smirnov E A 2017 *Journal of Physics: Conference Series* **872** 012053

IOP Conf. Series: Journal of Physics: Conf. Series **1058** (2018) 012021 doi:10.1088/1742-6596/1058/1/012021

Methodic of the computer calculation of the plasma impedance of the glow discharge lasers

A S Kiselev, E A Smirnov and A Yu Chizh
Department of electronic instruments and devices, Saint Petersburg Electrotechnical University "LETI", 197376, Saint Petersburg, Russia

E-mail: alex.kiselev.epu@gmail.com

Abstract. In this work an expression for the dynamic resistance of a glow discharge in long tubes was obtained and analyzed. The expression describes the physical processes taking place in the positive column of a glow discharge. The analysis of the frequency dependences of the active and reactive constituents, as well as the dynamic resistance module for discharge conditions corresponding to glow discharge lasers, was performed. On the basis of the expressions obtained a computer program for modeling the dynamic resistance of discharge was developed. The program allows you to calculate and display the plots of the frequency dependences of the active and reactive constituents, as well as the module of the dynamic resistance of the discharge gaps of lasers. Also, the program provides a function for constructing hodographs of dynamic resistance.

Among the widest range of devices for plasma electronics, devices with a longitudinal positive column (PC) of discharge occupy a special place, in particular, glow discharge lasers (GDL). GDL is actively used for technological and metrological purposes. However, due to physical and structural features, the laser radiation has a temporary instability caused by the influence of various factors. This implies the use of GDL in conjunction with systems of stabilization power of radiation [1].

One of the options for stabilizing the power of laser radiation is to adjust the pump level of active medium. In glow discharge lasers, this process reduces to a forced antiphase change in the discharge current. An important factor is the knowledge of the discharge behavior in dynamics. The discharge gap of lasers as an element of an electrical circuit has an AC resistance, which is determined by the capacitance and inductance of the plasma.

The discharge conditions realized in LTE correspond to the region of sharp drop in the current-voltage characteristic (CVC) of the discharge gap, when there is a transition stage from free electron diffusion to ambipolar one. Since GDL operate on direct current, one can use the method of small perturbations to obtain expressions for the total dynamic resistance. In accordance with it, the current and voltage of the discharge, accordingly, have the form: $U = U_0 + u$, $I = I_0 + i$. In these expressions I_0, U_0 are stationary values of the current and voltage on the discharge, corresponding to the decreasing current-voltage characteristic; i, u are the variable components of the discharge current and voltage, with $u << U_0$, $i << I_0$. Then the dynamic resistance will be determined by solving the equation:

$$\frac{1}{I_0}\frac{\partial^2 i}{\partial t^2}+\frac{1}{I_1}\frac{di}{dt}+\frac{i}{I_2}=\frac{1}{U_0}\frac{\partial^2 u}{\partial t^2}+\frac{1}{U_1}\frac{du}{dt}+\frac{u}{U_2}. \quad (1)$$

Solution of equation (1) was sought in the form

$$Z_d = \frac{u_0}{i_0} e^{j\varphi} = \frac{I_0^{-1}(j\omega)^2 + I_1^{-1} j\omega + I_2^{-1}}{U_0^{-1}(j\omega)^2 + U_1^{-1} j\omega + U_2^{-1}}, \tag{2}$$

After the transformation of equation (2), an expression is obtained for the dynamic resistance of the GDL discharge, as the total resistance to AC

$$Z_d = R + jX = \frac{(I_2U_2)^{-1} + \omega^2\left[(I_1U_1)^{-1} - (I_2U_0)^{-1} - (I_0U_2)^{-1} + \omega^2(I_0U_0)^{-1}\right]}{U_2^{-2} - 2\omega^2(U_0U_2)^{-1} + \omega^2U_1^{-2} + \omega^4U_0^{-2}} +$$

$$+ j\frac{\omega\left[(I_1U_2)^{-1} + \omega^2(I_0U_1)^{-1} - (I_2U_1)^{-1} + \omega^2(I_1U_0)^{-1}\right]}{U_2^{-2} - 2\omega^2(U_0U_2)^{-1} + \omega^2U_1^{-2} + \omega^4U_0^{-2}}, \tag{3}$$

where R and X are the active and reactive components of the PS impedance; ω is the circular frequency. The remaining parameters in (3) depend on the discharge conditions [2].

An important role in the analysis of the dynamic discharge resistance is played by the static CVC. As was said above, for the GDL conditions the CVC has a decreasing character [3]. In this case, the CVC can be approximated by an expression that includes the length of the discharge gap L and the longitudinal gradient of the potential E_z:

$$U_0 = U_k + aE_zLI^b,$$

where U_k is the cathode voltage drop, which is determined by the pair of gas-cathode material; a, b - constants, depending on the kind of gas and discharge conditions. In this case, it is possible to introduce a static slope of the current-voltage characteristic [4]:

$$\rho = dU / dI = abLE_zI^{b-1}.$$

In this case, the anode voltage drop was assumed to be zero. The longitudinal gradient E_z of the potential in the positive column depends on the nature of the gas, the working pressure, and also the diameter of the discharge gap, and is related to the electron temperature Te through the electron accommodation coefficient χ [5]. The electron temperature is a complicated function of the discharge conditions and can be solved by solving the ionization balance equation [6] - [8].

To automate the process of analyzing the discharge impedance behavior, a program was developed in the MS Visual C # programming environment.

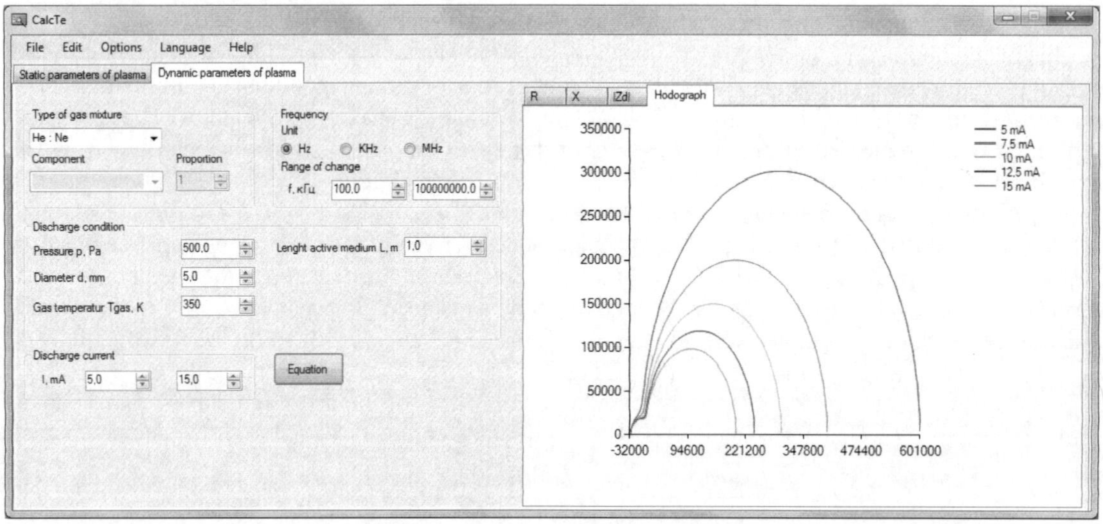

Figure 1. Program for modeling the dynamic resistance of glow discharge lasers

Program makes it possible to calculate the active, reactive constituent, the impedance module, and also construct a hodograph of impedance. Before calculating the user, it is possible to select the type of gas mixture, specify the discharge conditions, the range of the discharge current and the frequency of the signal. The result of the calculation is the family of dependences of the impedance components on the frequency for different discharge currents. Each dependency family is located on a separate tab. The technique based on the use of expression (3) is the basis of the technique for modeling the impedance of plasma of glow discharge lasers.

The results of calculating the dynamic resistance of the plasma obtained by using the developed program was analyze. At low modulation frequencies, the discharge plasma exhibits weakly inductive properties, and the impedance has a purely active value. Since under these conditions the VAC is of a falling nature, the active component of the discharge resistance $R(f, I)$ will have a negative value, and be determined by the operating point on the characteristic (figure 1). With increasing frequency, the active resistance of the discharge reaches a constant value equal to the static discharge resistance. The dependence of the reactive component of the resistance $X(f, I)$ has a resonance character (figure 2). At high modulation frequencies, the plasma does not have time to respond to external influences, so the reactive component of the impedance tends to zero. Figures 3 and 4 show respectively the families of dependencies of the impedance module of the discharge $Z_d(f, I)$, and its travel time curves. They reflect the general trends in the dependencies $R(f, I)$ and $X(f, I)$. The advantage of the proposed method for calculating the impedance of the discharge of laser tubes is that it allows one to correctly assess the nature of the reactivity of the discharge, the presence of active resistance of both signs [9].

The results of the simulation were compared with the experimental data of the dynamic resistance of the most widely distributed types of glow discharge lasers: a He-Ne-laser and a carbon dioxide laser (figure 2). Good agreement between theory and experiment is obtained.

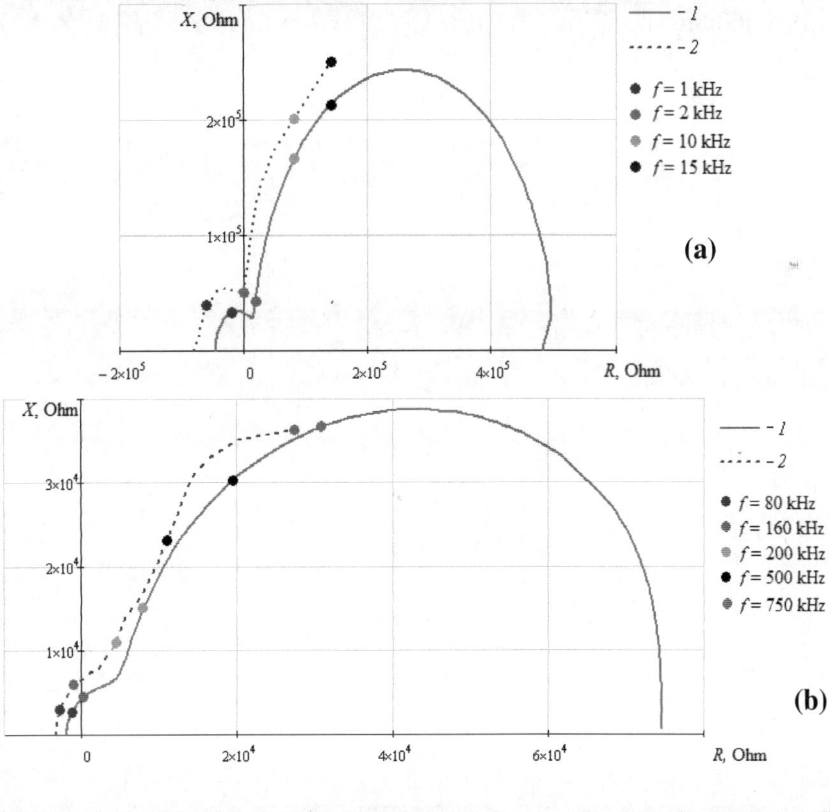

Figure 2. Comparison of experimental data and simulation results: **(a)** – He-Ne-laser; **(b)** – carbon dioxide laser.

The dynamic properties of the discharge in the range of measurement parameters corresponding to the working conditions of glow discharge lasers [10] are described quite adequately by a model that takes into account the inertia of the direct ionization process, the finite time of space charge establishment, and the falling character of the current-voltage characteristics as the main factors. The resulting expressions for impedance can be used in analyzing the stability of power supplies of gas-discharge lasers based on the principles of current, passive or active power stabilization, when knowledge of discharge behavior in non-stationary mode is required.

References

[1] Smirnov E A, Morozov S S and Timoshichev A A 2008 *Petersburg Journal of Electronics* **2–3** 39–48
[2] Smirnov E A 2013 *Proceeding of Saint Petersburg Electrotechnical University "LETI"* **10** 6–10
[3] Kiselev A S and Smirnov E A 2015 *Proceeding of Saint Petersburg Electrotechnical University "LETI"* **6** 3–6
[4] Granovsky V L 1971 *Electrical current in gas* (Moscow: Science)
[5] Kiselev A S and Smirnov E A 2015 *Proceeding of Saint Petersburg Electrotechnical University "LETI"* **8** 13–5
[6] Kiselev A S and Smirnov E A 2014 *Vacuum technique and technology* **1** 56–9
[7] Levitsky S M 1964 *Collection of problems and calculations on physical electronics* (Kiev: Publishing House of Kiev University)
[8] Kiselev A S, Kostrin D K, Lisenkov A A and Smirnov E A 2017 *Journal of Physics: Conference Series* **789** 012027
[9] Kiselev A S and Smirnov E A 2017 *Journal of Physics: Conference Series* **872** 012050
[10] Kiselev A S and Smirnov E A 2017 *Journal of Physics: Conference Series* **872** 012053

Electrical characteristics of pulsed atmospheric generator of plasma flows based on a localized exploding wire

L Y Volodin[1], E Y Kotovrasov[1] and V E Fomin[1]

[1]Bauman Moscow State Technical University

E-mail: e.kotovrasov@yandex.ru

Abstract. Description of the designed impulse plasma torch with localized exploding wire generating high-energy plasma flows are considered. Electrical characteristics of capillary discharge for the two modes of energy input were experimentally studied. In capacitive storage was stored ~ 4 kJ, the charging voltage was ~ 3 kV. The reached peak power exceeded 165 MW at a maximum current of 65 kA and the efficiency of energy transfer to the load is about 80%.

1. Introduction

Based on exploding wire pulsed high-current discharges are an effective sources of plasma flows with controlled gas dynamic parameters and chemical composition [1]. The usage of exploding wire makes it possible to reach about unlimited energy of the discharge and to obtain high instantaneous discharge characteristics. The current passing through the conductor blasts it into vapor at high values of the power, generating discharge plasma injected into the atmosphere.

The generator construction is based on a capillary discharge. Metal electrodes separated by a dielectric insert with a channel (10 mm inner diameter, 35 mm length). The plasma jet flows through one of the electrodes with a nozzle. Unlike the traditional capillary discharge with evaporating walls [2], the plasma-forming substance is not the ablating material of the channel wall, but a cylindrically folded aluminium foil with a diameter and length corresponding to the channel geometry. After installation in the capillary, the foil galvanically connects the electrodes. This method of organizing the discharges assumes the presence of two stages: metallic conductivity and a pulsed arc in vapor, localized by the walls of the channel. This is in different from the most common modes of organizing an electric explosion - the use of flat foils [3] or cylindrical wires [4].

In that work, we measured the time dependences of the discharge circuit current and the generator voltage to determine the discharge electrical specification in two energy input modes: subcritical and near-critical (damping ratio $\gamma = 0.39$ and $\gamma = 0.73$, respectively) in order to select the nominal mode of the plasma torch operation with increased energy content in the generated flows.

2. Experimental Setup

The scheme of the experimental setup is shown at Figure 1. The generator (1) was powered by a capacitor bank (2) of a common capacity $C_0 = 962$ μF, which was charged from a high-voltage source (3). To initiate the discharge, a special ignition system (4) was used, controlling the trigatron air gap (5). To measure pulsed discharge currents, the Pearson 4418 current probe (6) with a weakening attenuator (7) was used. Voltage control was carried out by a high voltage probe Pintek HVP-39 Pro

IOP Conf. Series: Journal of Physics: Conf. Series **1058** (2018) 012022 doi:10.1088/1742-6596/1058/1/012022

(8) was used. The B-dot probe (9) made it possible to register the derivative of the current and determine the moment of the explosion. All signals were recorded with a digital oscilloscope Tektronix TDS 2014C (10).

Battery (2) was charged to a voltage of 3 kV, which corresponded to the stored energy of ~ 4 kJ. A foil weighing 150 mg was rolled up in the form of a cylinder and placed in the channel of the generator (1), shorting the discharge gap. The main discharge was initiated by a spark discharge with a voltage of 12-14 kV in the gap of the trigatron (5).

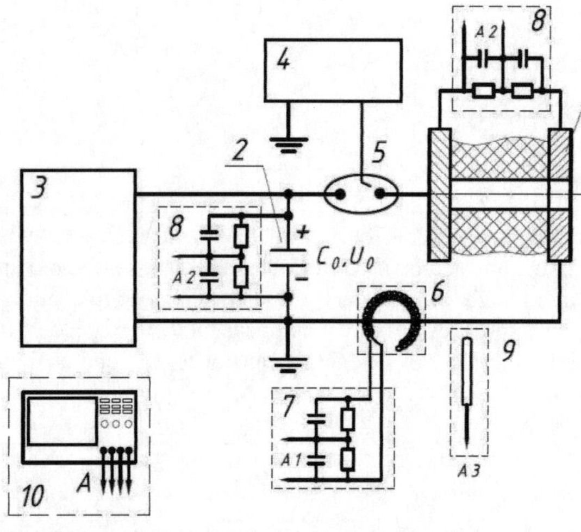

Figure 1. Experimental setup scheme.
1 – pulse plasma torch; 2 – capacitor bank (C_0 = 962 μF); 3 – high-voltage source; 4 – ignition system; 5 – trigatron; 6 – current probe; 7 – attenuator 1:20; 8 – frequency-compensated voltage divider (К = 1000); 9 – B-dot probe; 10 – digital oscilloscope.

3. Results

The circuit parameters (inductance and efficient resistance) for each mode were determined from the oscillograms of the current without inclusion the plasma load in the circuit upon short-circuiting. L_{cir} ~ 3.2 μH and R_{cir} ~ 16 mΩ corresponded for underdamped regime; for subcritically damped regime the values were somewhat lower: L_{cir} ~ 1.32 μH and R_{cir} ~ 4.5 mΩ.

Figure 2 shows the time dependences of the discharge current and voltage on the generator for the described regimes, and in Figure 3 - the time dependences of the input power and energy.

The kink of the current derivative (signal from the B-dot probe) determined the moment of the explosion.

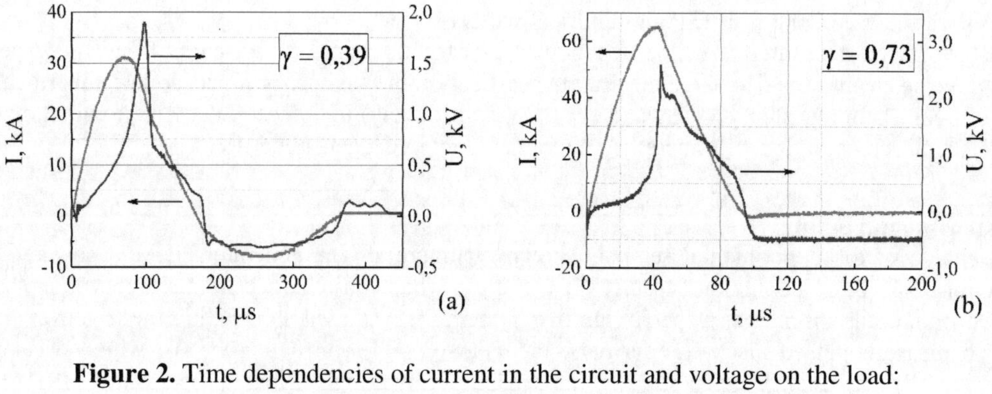

Figure 2. Time dependencies of current in the circuit and voltage on the load:
(a) underdamped regime; (b) subcritically damped regime.

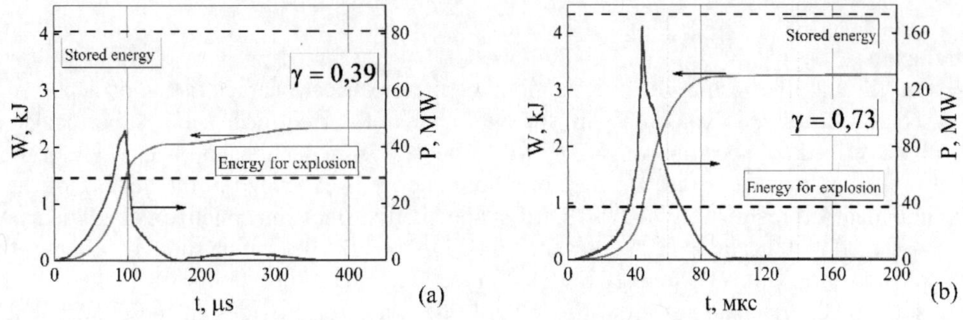

Figure 3. Time dependencies of input power and energy:
(a) underdamped regime; (b) subcritically damped regime.

For underdamped regime, the characteristic amplitude value of the discharge currents was 30 kA, the duration of the current first half-cycle was ~ 170 μs, the time of reaching a maximum is ~75 μs. The peak voltage on the load exceeded 1.5 kV. The maximum electric power reached 45 MW. The explosion occurred and about 1.5 kJ was transferred to the load after ~ 100 μs. In total, ~ 2.3 kJ was invested in the load, which corresponds to the efficiency of energy transfer to the discharge of ~ 60% and to the plasma ~ 20% of the stored energy.
In subcritically damped regime, the maximum current is reached by ~ 40 μs and is equal to 65 kA. The duration of the quasi-half-period is ~ 100 μs. The total energy release time was ~ 100 μs. The peak voltage value on the load exceeded 2.5 kV.
The peak electric power was ~ 165 MW. The load was transferred to ~ 3.3 kJ, which corresponds to an efficiency of energy input of ~ 80%. The explosion costs amounted to ~ 1 kJ, which corresponds to the efficiency of the energy input to the plasma ~ 60% of the stored energy.
It was found that during the discharge with a damping ratio $\gamma = 0.39$, the B-dot probe together with the current derivative records the HF oscillations, whose oscillogram is shown in Figure 4. In the other regime, no oscillations were observed.

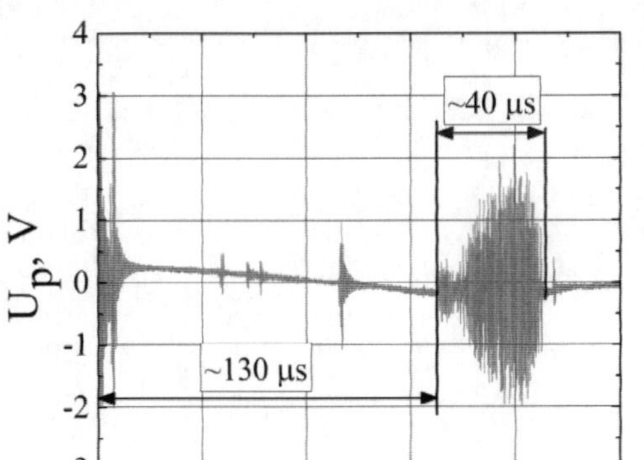

Figure 4. The signal from B-dot probe.

The occurrence of high-frequency oscillations is a common effect in discharges with an unstable discharge gap [5]. It can be assumed that in this case, HF oscillations occur at the moment of the conductor dispersion, as a result of which a heterogeneous system with charged "large" drops of foil material is formed in the channel, which cause local fluctuations of the current in the circuit. It characterizes underdamped regime as inefficient. Issue of the plasma flow in this regime is accompanied by a strong sparking.

4. Conclusion

This report presents the results of investigations of the electrical characteristics of a pulsed plasma torch based on a localized exploding wire for two modes of current decay. For subcritically damped regime characterized by a peak power of 165 MW, the explosion occurs at a lower value of the energy invested in the conductor (~ 1000 J), therefore more energy is expended in the heating of the plasma than in underdamped regime - 2300 J vs. 800 J, respectively. Since the conductors had the same mass, it can be assumed that the value of the power supplied to the load determines both the time of the onset of the explosion and the level of energy invested in the load at that time.
Thus, in subcritically damped regime with $\gamma = 0.73$, the generator provides an energy efficiency in the plasma of ~ 60% of the stored energy.

Refences

[1] Zharnikov M H, Kamrukov A S, Kozhevnikov I V, Kozlov N P and Roslyakov I A 2008 *Appl. Phys. Letters* **34**(1) 74
[2] Bobrova N A, Bulanov S V, Farina D, Pozzoli R, Razinkova T L, Sakai J I, Sasorov P V and Sokolov I V 2000 *Laser and Particles Beams* **18** 623
[3] Anderzhanov E K, Divnov I I, Zotov N I and Hristoforov B D 1989 *J. Appl. Phis.* **59**(8) 17
[4] Lodes A, Curry R D, Rudroff M E J, Schmidt M, Bauer A J R and Brown W 2011 *IEEE Pulsed Power Conf.* (Chicago: IEEE) pp 1077-80
[5] Tazmeev G H, Timerkaev B A, Tazmeev H K, Sarvarov F S and Arslanov I M 2015 *Appl. Phis.* **2** 58

IOP Conf. Series: Journal of Physics: Conf. Series **1058** (2018) 012023 doi:10.1088/1742-6596/1058/1/012023

Effect of RF plasma modification on electrical characteristics of collagen containing polymer materials surface

L Yu Makhotkina, V I Khristoliubova
Kazan National Research Technological University, 420015, Russian Federation, Kazan, 68 Karl Marks st.

E-mail: valllerrriya@mail.ru

Abstract. Artificial materials that used in the manufacture of textile industry products have increased static characteristic and electrical resistivity. The large electrical charge accumulated by them negatively affects the human body. The use of natural collagen containing polymer materials is the safest way to avoid unacceptable risk to human health. With the help of nonequilibrium low temperature plasma (NLTP) treatment of collagen materials under reduced pressure it is possible to achieve an additional reduction of the electric potential by scattering charges on the surface of genuine leather. Under certain conditions the main reason for the generation and accumulation of static electricity is the high hygroscopicity of the material. Hydrophobization of the surface of collagen containing materials is achieved through the use of a hydrophobizator and volumetric NLTP treatment. Plasma treatment of collagen containing polymer materials also increases their biological compatibility and antimicrobial properties, improving mechanical characteristics by changing the internal structure of the material.

1. Introduction

Capillary-porous collagen containing polymer materials which include genuine high molecular tanning materials are in a stable high demand for the production of textile, medicine, petrochemical and other industries products [1,2]. One of the main consumer properties of collagen containing polymer products which determine their value and competitiveness in the international market are reliability and safety in consumption. Reliability and safety of collagen-containing polymer materials are complex indicators of the first level. Reliability in turn is determined by such properties as reliability, durability, maintainability and storageability. Second level complex indicators of the afety of consumption are mechanical, electrical, chemical and biological safeties.

Electrical safety of textile industry products is associated with the use of artificial materials for its production. The large electric charge accumulated by such materials negatively affects the human body. Genuine collagen containing high molecular materials, unlike artificial materials, have significant advantages, because electrification of products made of genuine materials is much lower than that of products of other materials. With the help of nonequilibrium low temperature plasma (NLTP) processing of collagen materials under reduced pressure it is also possible to reduce the electrical potential by scattering charges on the surface of the material which will allow the use of products as special-purpose products in industries with increased danger.

2. Materials and Methods

The studies were carried out on experimental samples of collagen containing polymeric materials made using two technologies: nanostructured samples (genuine materials, which at the final stage of finishing production were treated with low pressure NLTP in a mode of W_p = 1.5 kW, τ = 7 minutes), nanomodified samples (genuine materials , where a biocide agent, a solution of silver nanoparticles of 0.2%, was injected twice in the fattening stage with coating dyeing and at the stage of application of the coating film fixer; NLTP loe pressure treatment in a mode of W_p mode = 1.5 kW, τ = 7 minutes was included within the finishing process).

Plasma torch with flat electrodes for obtaining a plasma flow of a capacitive type is represented by two water cooled copper plates with the following dimensions: 20x30 cm or 40 cm in diameter. The electrodes are placed in a vacuum block. A device for fixing the samples with the device for their rotation during processing is placed between the electrodes. The base of the vacuum unit is mounted in the form of a welded frame. There are two rotor pumps and a water cooling system of the units of the installation on the frame of the block [3-5].

Measurements of the zeta potential control samples of collagen containing polymeric materials, processed in NLTP and modified in NLTP with silver addition are performed with the help of the analyzer of electrophoretic mobility to evaluate the electrical properties of the material [6].

In dispersed systems a double electric layer appears on the surface of solids. The double electrical layer is a layer of ions formed on the surface of the particle as a result of the adsorption of ions from the solution or the dissociation of surface compounds. The surface of the particle acquires a layer of ions of a certain sign uniformly distributed over the surface and creating a surface charge on it. These ions are called potential determining ions. Ions of the opposite sign are attracted to the surface of the body from the liquid medium, they are called counter ions. Thus, the double electric layer consists of potential-determining ions and a layer of counter ions is located in the dispersion medium. The layer of counter ions consists of two layers:

• Adsorption layer (dense layer), adjacent directly to the interfacial surface. This layer is formed as a result of electrostatic interaction with potential determining ions and specific adsorption.

• Diffuse layer where counter ions are located. These counter ions are attracted to the particle due to electrostatic forces. The thickness of the diffuse layer depends on the properties of the system and can reach large values.

When the particle moves, the double electric layer breaks. The place of discontinuity when moving the solid and liquid phases with respect to each other is called the slip plane. The slip plane lies on the boundary between the diffuse and adsorption layers or in the diffuse layer near this boundary. The potential on the slip plane is called the electrokinetic or zeta potential (the ζ potential). In other words, the zeta potential is the difference between the potentials of the dispersion field and the fixed layer of fluid surrounding the body. The zeta potential is not equal to the adsorption potential or the surface potential in the double electric layer. However, the zeta potential is often the only available way to evaluate the properties of the double electrical layer.

3. Results and Discussion

The results of the measurements using the analyzer of the electrophoretic mobility of the zeta potential of control collagen containing polymer materials samples, treated in NLTP and modified in NLTP with silver addition are shown in Fig. 1.

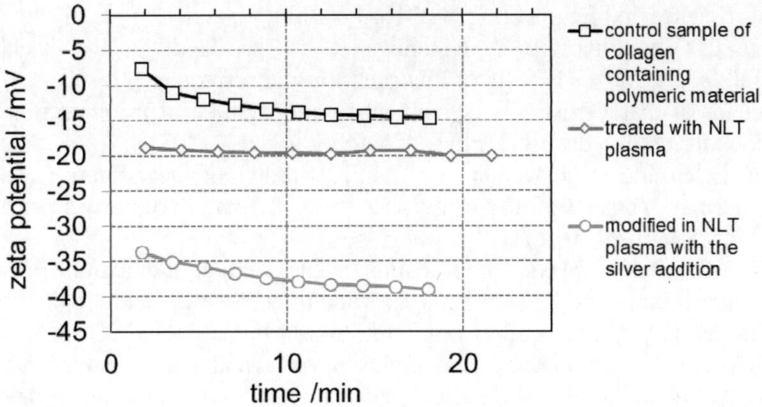

Fig. 1. Zeta potential of control samples of collagen containing polymeric materials, processed in NLTP and modified in NLTP with silver addition.

Under certain conditions the main reason for the generation and accumulation of static electricity is the high hygroscopicity of the material. Hydrophobization of the leather surface, i.e. giving water-repellent properties, is achieved through the use of a hydrophobizator and volumetric NLTP treatment [7].

In turn, plasma processing of collagen containing materials also increases their biological compatibility and antimicrobial properties [8], mechanical properties [9,10] are improved without changing the chemical composition of the material. Processing of collagen-containing materials in NLTP allows to increase the operational indices of light industry products and prolong their service life.

4. Conclusions

On the basis of the conducted experiments it can be concluded that the NLTP treatment of collagen containing materials under reduced pressure leads to a decrease in the electric potential by scattering charges on the surface of the material. The magnitude of the zeta potential determines the degree and nature of the interaction between the particles of the system. The value of the zeta potential equal to 30 mV (positive or negative) can be considered as a characteristic value for the conditional separation of low-charged surfaces and high-charged surfaces. The more electrokinetic potential, the more stable the system. Thus, the potential of a sample subjected to a plasma modification by silver is much lower than a sample from a batch lot or processed in a NLTP, which determines the safety and reliability of the final products during operation.

References

[1] Makhotkina , L.Yu. Design of special purpose products made of nanomodified leather / L.Yu. Makhotkina, V.I. Khristoliubova, L.R. Khannanova-Fakhrutdinova // IEJME – Mathematics education. – 2016. – V. 11, № 6. – P. 1495-1503.

[2] Osin Yu.N. SEM and X-ray analysis of surface microstructure of a natural leather processed in a low tempraeture plasma / Yu.N. Osin, L.Yu Makhotkina, L.N. Abutalipova, I.Sh. Abdullin // Vacuum: volume 51, №2. – P.221-225.

[3] Makhotkina, L.Yu. Design of special purpose products made of nanomodified collagen-containing materials with radio-frequency discharge / L.Yu. Makhotkina, S.N. Sharifullin // IOP Conference Series: Materials Science and Engineering. – 2016. – V. 134. – 012020

[4] Khristoliubova, V.I. Gas and plasma dynamics of RF discharge jet of low pressure in a vacuum chamber with flat electrodes and inside tube, influence of RF discharge on the steel surface

parameters / V.I. Khristoliubova, N.F. Kashapov, M.F. Shaekhov // IOP Conference Series: Materials Science and Engineering. – 2016. – V. 134. – 012017.

[5] Ladianov V I, Gilmutdinov F Z, Nikonova R M, Kashapov N F, Shaekhov M F and Khristoliubova V I 2017 Effects of low pressure radio frequency discharge on the physical and mechanical characteristics and chemical composition of diffusion coating on a surface of complex configuration details J. Phys.: Conf. Ser v 789, 012031.

[6] Makhotkina, L Yu and Khristoliubova V I 2017 Studies of the influence of nonequilibrium plasma thermal exposure on the characteristics of the capillary-porous polymer material J. Phys.: Conf. Ser v 789, 012031.

[7] Kalimullina G.R. Effect of plasma modification on the creation of a hydrophobic leather surface / G.R. Kalimullina, G.N. Kulevtsov, R.R. Mingaliev // New Technologies and Materials of Textile Industry, VIII Intl. sc.-pr. Conf. - 18, 26-30 May 2012. - P. 266-269.

[8] Grebenshchikova M.M. Parameters of complex plasma treatment of natural leather increasing its biological compatibility / M.M. Grebenshchikova, I.Sh. Abdullin // New Technologies and Materials of Textile Industry, VIII Intl. sc.-pr. Conf. - 18, 26-30 May 2012. - P. 263-266.

[9] Makhotkina L Y and Khristoliubova V I 2015 Variation of footwear wear resistance depending on methods of treatment. International Journal Of Applied And Fundamental Research № 1, URL: www.science-sd.com/460-24749 (11.03.2015).

[10] Makhotkina, L Yu and Khristoliubova V I 2017 Improvement of technical purpose materials performance characteristics with the radio frequency low pressure plasma J. Phys.: Conf. Ser. v. 927, 012032

Development of the arc plasma torch operation mathematical model for spheroidization of fine-dispersed powders

Murashov I., Frolov V., Kadyrov A.

Peter the Great St. Petersburg Polytechnic University, Saint Petersburg, Russia

iuriimurashov@gmail.com

Abstract. Brief review of the fine-dispersed powders production technologies for additive laser technologies is presented in the paper. Non-stationary mathematical model of DC plasma torch operation and stationary model of the reactor operation for powder spheroidization have been developed. Air is used as the plasma-forming gas. The influence of the transporting flow and compressing gases on the characteristics of the plasma flow is analyzed. Verification of the non-stationary mathematical model of DC plasma torch operation is based on experimental studies. Radial temperature distributions obtained from the results of numerical simulation and spectral diagnostics of the plasma torch PN-V1 are presented. The results of the experimental study describing plasma flow behavior for given parameters of the fine-dispersed powders spheroidization process are presented in the paper.

1. Introduction

There is a rapid development of additive laser technologies at the present time. The use of additive technologies has received wide application in the aerospace industry, engineering, medicine and other industries. One of the main problems in the development of this technology is the lack of quality powder materials. The methods of spherical shape powder production can be conditionally divided into two main groups:

- mechanical method;
- chemical and electrochemical method.

A block diagram of the methods classification for the powder production for additive laser technologies is shown in Fig. 1

The most common methods of atomization are [1]:

- gas atomization;
- vacuum atomization;
- centrifugal atomization.

The resulted methods of atomization have a number of essential lacks [1]. An alternative way of obtaining powder is spheroidization in a plasma flow. Two main methods of generating a plasma flow are used in industry [2]:

- application of DC plasma torches;
- application of induction plasma torches.

The most adapted method for the powder spheroidization is the application of inductively coupled plasma, described in more detail in [3]. Despite the advantages of this spheroidization technology, there are significant drawbacks associated with the cost of manufacturing process equipment (power source and induction plasma torch). An alternative method of powder spheroidization for additive laser technologies is the application of DC plasma torches (see Fig. 2). The main technological task in developing the design of the "plasma torch-reactor" system and identifying the parameters of operating conditions is obtaining optimal parameters of the plasma flow. The creation of mathematical models makes it possible to significantly accelerate the process of developing structural elements of equipment and determine the parameters of the technological process, ensuring a reduction in the costs of conducting additional experimental studies.

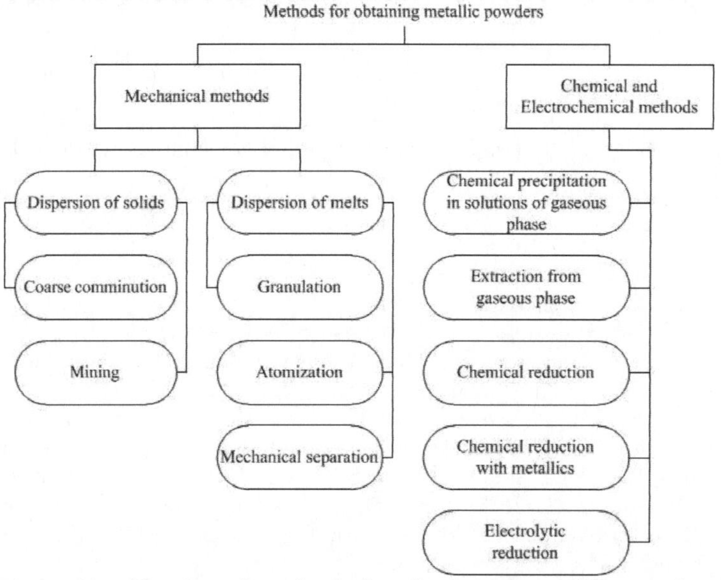

Fig. 1. Classification of methods for obtaining metallic powders

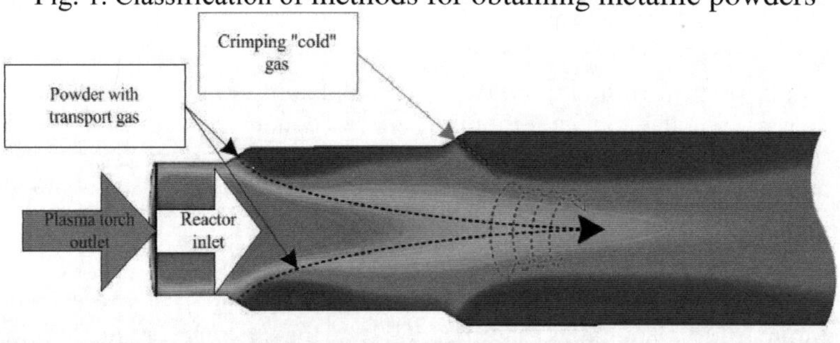

Fig. 2. Scheme of powder spheroidization technological process in the plasma torch jet

2. Mathematical model

Non-stationary mathematical model of DC plasma torch operation is multiphysical task [4], including the equations of electromagnetism, gas dynamics and heat transfer. These equations are interrelated in explicit (the source function of Joule heating) and implicit (thermodynamic and transport properties) form [5].

The mathematical model of DC plasma torch operation contains the following main assumptions:

- plasma is in a local thermodynamic equilibrium (LTE);
- plasma is optically thin;
- two-dimensional axisymmetric;
- thermodynamic and transport properties depend on temperature;
- near-electrode processes are not taken into account.

Basic equations of mathematical model describing nonstationary processes in DC plasma torch:

- Maxwell's system of equations::

$$\begin{cases} \vec{J} = \left(\sigma + \varepsilon_0 \varepsilon \dfrac{\partial}{\partial t} \right) \vec{E} \\ \vec{E} = -\nabla V \\ \sigma \dfrac{\partial \vec{A}}{\partial t} + \nabla \times \vec{H} = \vec{J} \\ \vec{B} = \nabla \times \vec{A} \end{cases} \tag{1}$$

- momentum equation (Navier-Stokes equation):

$$\rho \frac{\partial \vec{\upsilon}}{\partial t} + \rho (\vec{\upsilon} \cdot \nabla) \vec{\upsilon} = \vec{F} + \nabla \cdot \left[-pI + \mu \left(\nabla \vec{\upsilon} + (\nabla \vec{\upsilon})^T \right) \right] \tag{2}$$

- continuity equation:

$$\frac{\partial \rho}{\partial t} + \nabla \cdot (\rho \vec{\upsilon}) = 0 \tag{3}$$

- heat balance equation:

$$\rho C_p \frac{\partial T}{\partial t} + \rho C_p \vec{\upsilon} \cdot \nabla T - \lambda \nabla T = Q_s - Q_{rad} \tag{4}$$

Detailed description of the mathematical model and the calculation technique is presented in [5]. The existing design of the plasma torch PN-V1 was used as the geometry of the computational domain. The time dependences of the radial temperature distributions (see Fig. 3), the axial and radial velocities at the plasma torch outlet (used as boundary conditions for simulation of the plasma reactor) at plasma gas consumption of 0.5 g/s were obtained from the calculation results, which corresponds to the laminar plasma flow (according to the results of the experimental study) (see Fig. 4a).

Verification of the mathematical model is based on experimental studies (see Fig. 4). Techniques for conducting experimental studies on the plasma flow behavior and calculation of the radial temperature distribution on the basis of the results of spectral diagnostics are presented in [5]. Stationary mathematical model describing the reactor operation is developed on the basis of equations (2), (3) and (4) with allowance for $\dfrac{\partial}{\partial t} = 0$.

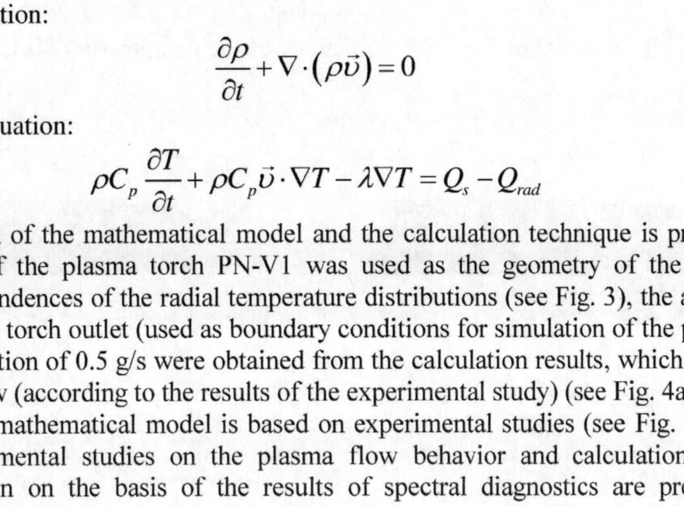

Fig. 3. Dependences of the radial temperature distribution on time and time-averaged radial temperature distribution

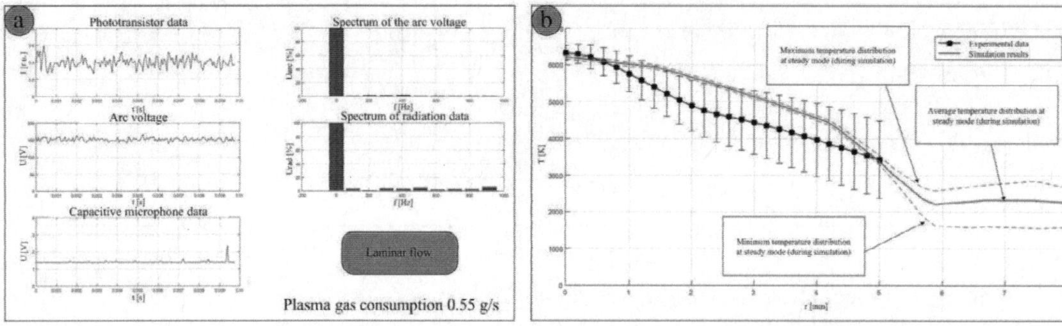

Fig. 4. The results of experimental studies: a - experimental determination of the plasma flow behavior; b - radial temperature distribution based on the results of spectral analysis and numerical simulation

The results of the calculation for different values of the compressing "cold" gas consumption are shown in Fig. 5.

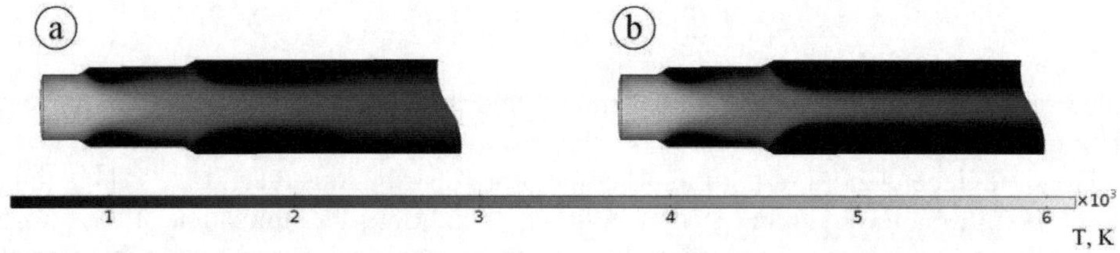

Fig. 5. The temperature distribution in the reactor for the transport gas consumption G = 0.1 g/s and "cold" gas consumption: a - G=0.2 g/s; b - G=0.4 g/s

Conclusions

Non-stationary mathematical model of DC plasma torch operation and stationary reactor operation model are developed. The results of the simulation are consistent with the results of experimental studies. The parameters (I = 190 A, G_{pl} = 0.5 g/s, G_{tr} = 0.1 g/s, G_{com} = 0.2 - 0.3 g/s) of powder spheroidization technological process are obtained based on the results of the simulation, which allows to obtain plasma flow (DC plasma torch) like plasma flow of induction plasma torch. Design elements of technological equipment are developed taking into account the results of simulation.

References

[1] Saprykin A.A. Application of plasma treatment for the metallic powder spheroidization. A.A. Saprykin, V.I. Kuzmin, D.V. Sergachev, D.V. Dudihin. Actual problems in mechanical engineering. Novosibirsk, 2017, vol. 4, no. 2, pp.16-20. (in russian)

[2] Frolov V.Ya. Simulation of the plasma process for obtaining nanopowder materials using a combined plasma torch / V.Ya. Frolov, D.V. Ivanov, M.A. Shibaev. Journal of Applied Physics, 2014, vol. 40, no. 16, pp. 1-7. (in russian)

[3] Dresvin S. Correction for calculation of particle heat transfer in thermal plasma. S. Dresvin, S. Zverev, D. Ivanov. Heat Transfer Research, 2008, vol. 39, no. 3, pp. 183-195.

[4] Murashov I. Numerical simulation of dc air plasma torch modes and plasma jet instability for spraying technology. I. Murashov, V. Frolov, D. Ivanov. ELCONRUSNW 2016, 2016, pp. 625-628.

[5] Murashov I. The development of dc plasma torch for the spraying technology taking into account the phenomena of plasma flow instability. PhD thesis. Saint-Petersburg, 2016.

Characteristics of argon barrier micro-discharge in various operation modes excited by a saw-tooth voltage at atmospheric pressure

A A Saifutdinova[1], **A I Saifutdinov**[2,3]
[1] A.N.Tupolev Kazan National Investigation Technical University
[2] Kazan Federal University, Kazan, Russia
[3] Saint Petersburg State University, Saint-Petersburg, Russia

E-mail:aliya_2007@list.ru, as.uav@bk.ru

Abstract. In this work modeling of atmospheric pressure barrier discharge in argon for saw-tooth voltage signal applied to the electrodes was conducted. The model included balance equations for the densities of charged (electrons, ions) and the excited particles, the electron energy density, and the Poisson equation for the electric potential. The fluxes of charged and excited particles (electrons, ions) and electron energy were given in the drift-diffusion form. Distributions of spatiotemporal parameters of the discharge for different forms of saw-tooth voltage signal were obtained. Numerical experiments have shown that, depending on the growth rate of the external voltage, the one-dimensional model gives three forms of discharge, two of which are slightly different types of Townsend discharge, and one form is a glow discharge.

1. Introduction

In recent years, atmospheric pressure low temperature plasmas have attracted considerable attention due to their extensive applications in various fields of science and techniques. Dielectric barrier discharge (DBD) is one of the most important ways of generating low temperature plasmas with high uniformity under atmospheric pressure [1]. Atmospheric pressure uniform DBD can be divided into two categories by their breakdown mechanisms. One is the atmospheric pressure glow discharge (APGD), and the other is the atmospheric pressure Townsend discharge (APTD) [2]. APGD is characterized by the formation of a cathode fall region which is absent in APTD. The density maxima of electron and ion appear near the anode in APTD. Both APGD and APTD differ themselves greatly in properties from those of the low pressure glow or Townsend discharges. For instance, Townsend discharge is dark under low pressure, while APTD is not dark at all. Another difference is that the electron energy in APTD is higher than that in the low pressure discharge [2,3]. Since atmospheric pressure uniform discharge has proper electron energy and is very uniform, it has extensive application potentials, such as surface modification [4], ozone generation, biological sterilization [5], pollution control [6], etc.

In some works were investigated experimentally and numerically the DBD and found an evolution process from APTD to APGD within a discharge pulse [3, 7-12]. In many works devoted to the study of discharges, a sinusoidal voltage was used. However, recently there have appeared works in which other forms of voltage were used to initiate discharges [12-16].

Therefore, it is important that investigate the discharge characteristics under various form of voltage to excitation. In this paper, the discharge properties excited by a saw-tooth for atmospheric argon DBD with a one-dimensional extended fluid model.

2. Description of the Mathematical Model

The discharge occurs between two parallel-plate electrodes covered by dielectric layers of 1mm thickness ($d_1 = d_2 = 1$ mm). A saw-tooth voltage with amplitude 1.3kV and a repetitive frequency of 100 kHz is applied between the two electrodes. The numerical simulation is based on a one-dimensional extended fluid model [17].

The nonlinear system of partial differential equations comprises the continuity equations for particle densities, the electron energy balance equation determining the spatiotemporal evolution of the electron energy density and Poisson's equation providing selfconsistently the electric potential and the electric field The particle fluxes and the electron energy flux in the *x*-direction are given in drift-diffusion approximation [17].

To achieve an adequate description of atmospheric pressure argon plasmas, atomic and molecular argon ions Ar^+, Ar_2^+, and Ar_3^+ as well as three effective excited atomic: metastable level Ar^m, resonant level Ar^r, effective excited level Ar^* and two effective excimer levels Ar_2^* and Ar_2^{**} were taken into account. The reaction kinetic scheme takes into account 18 electron collision processes. Besides elastic collisions contributing directly to the electron energy balance equation, it includes excitation, deexcitation, ionization and recombination processes with rate coefficients depending on the mean electron energy. In addition, 4 radiation and 16 heavy particle collision processes are considered [17].

3. Numerical simulation and results

Figure 1 shows the calculated waveforms of the electrode voltage V_0, the potential drop V_g across the discharge gap, and the total DBD current density J_0 for different forms of saw-tooth voltage signal (different size of transition zone between rising fronts of the applied voltage): *a)* for 0.1 μs ; *b)* 0.2 μs ; *c)* 0.4 μs .

As can be seen from Fig. 1, APTD mode is observed for all three cases during the first period. On the rising front of the voltage, the conduction current gradually increases from zero to a certain value, depending on the rate of voltage growth, and then remains constant. The discharge current shows a broad plateau per part cycle corresponding on increase front of applied voltage. On a declining front, a situation is almost the same. The difference is that the discharge current consists of some pulses (damped oscillations) superimposed on the plateau.

In the second period, both on the rising and falling fronts of the supply voltage, current pulses occur corresponding to the barrier discharge of the glow type. At the same time, the rate of increase in the magnitude of the voltage on the falling edges of the voltage affects the amplitude of the current pulse. The amplitude of the current pulse is higher in the case when the value of rate of voltage growth *dU/dt* is greater. The periodicity of the formation of current pulses, and hence the initiation of discharges, is established in the third and fourth periods.

Fig. 2 shows the density distributions of charged (electron and sum of all ion types) and excited (sum of all types) particles, as well as the electric field strength at different times 12 mks, 26.7mks and 31.3 mks, respectively for case on fig.1 c. As can be seen at a time point of 12 mks, when current fluctuations are observed in the APTD mode, quasineutral plasma is not observed. Only the charge layers is observed. In this case, the voltage drops across the whole discharge gap. At time points of 26.7 mks and 31.3 mks, classical distributions of the discharge parameters corresponding to the APGD mode are observed.

In the case of APTD mode, a sufficiently high density of excited particles is observed in the discharge gap. The metastable argon atom Ar^m is the predominant excited particle. The predominant ions are the molecular ions $Ar_2{}^+$ and $Ar_3{}^+$. In the case of APGD mode, the predominant ions are also argon molecular ions $Ar_2{}^+$ and $Ar_3{}^+$, and the dominant excited particles are $Ar_2{}^*$ and Ar^m.

IOP Conf. Series: Journal of Physics: Conf. Series **1058** (2018) 012025 doi:10.1088/1742-6596/1058/1/012025

The cathode sheath is followed by the plasma region with the negative glow and Faraday dark space, where the charged particle density first increases to its maximum value near the cathode sheath and then smoothly decreases toward the anode – grounded electrode. This region is characterized by a low (inverse) field with two field reversal points.

After the second reversal point, the plasma field is gradually restored; however, as is seen from the figures, no positive column with a typical homogeneous distribution of the plasma parameters is observed. The Faraday dark space is followed by the anode sheath. The transition to the anode sheath is accompanied by the third field reversal point and the negative anode fall.

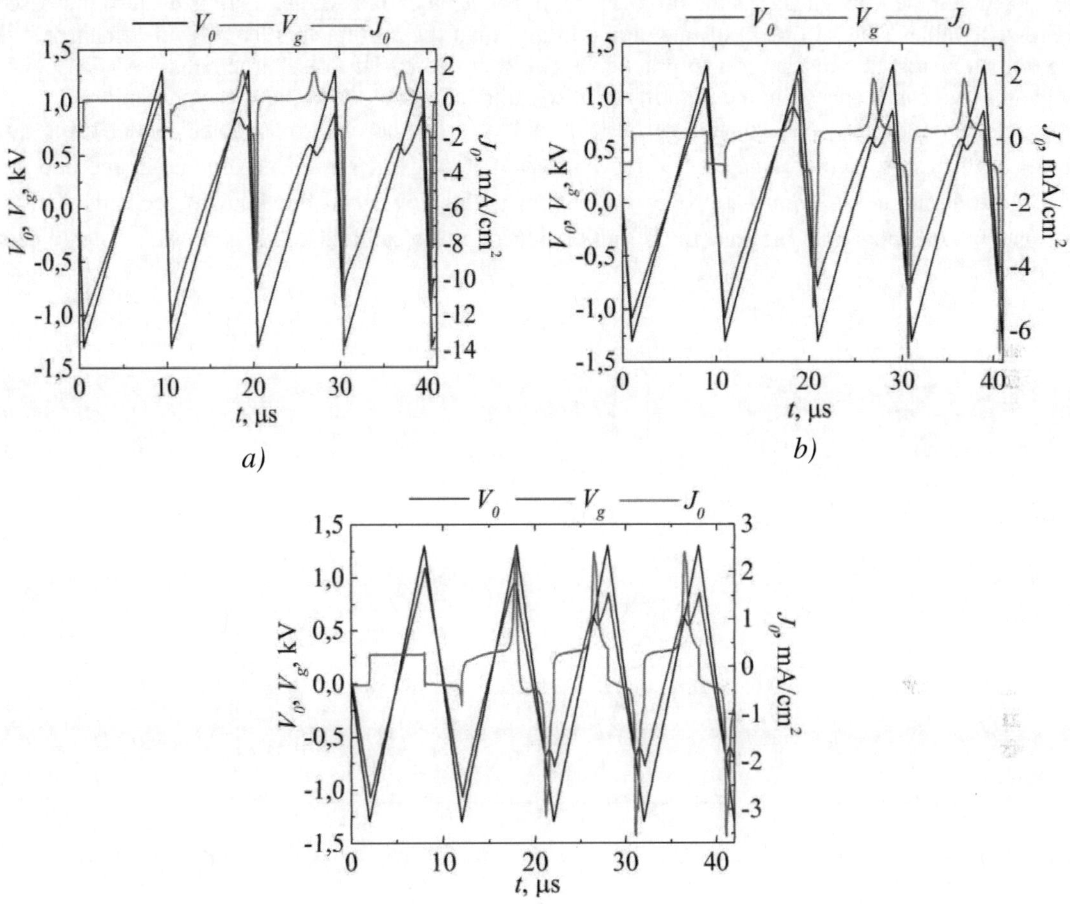

Figure 1. Saw-tooth waveforms of the voltage V0 applied to the electrodes, voltage drop Vg across the discharge gap, and total current density J0 in a dielectric barrier microdischarge for different size of transition zone between rising fronts of the applied voltage): *a)* for 0.1 μs ; *b)* 0.2 μs ; *c)* 0.4 μs .

In addition, a special study was conducted. When using a linearly increasing voltage, one of the parameters regulating the discharge mode is the rate of voltage growth *dU/dt.* As a result of numerical experiments, various discharge modes were detected depending on the *dU/dt.*

At low voltage growth rates (*dU/dt* $\approx 2\cdot10^8 - 3\cdot10^8$ V/s), when the voltage on a discharge gap reaches a certain value, the conduction current increases smoothly from zero to a certain value, depending on the rate of voltage growth. At the same time, the voltage across the discharge gap ceases to increase, since its increase due to the growth of the external voltage is compensated by the charge transfer between the barriers. The discharge in this case is Townsend.

An increase in the value of the rate of voltage growth leads to the appearance of damped oscillations on the conductivity current profile ($dU/dt \approx 4\cdot10^8 - 5\cdot10^8$ V/s). One of the reasons for this is ion-electron emission. In addition, the current oscillations in the Townsend discharge can be due to the distortion of the electric field due to a positive space charge. In this case, they arise only in a narrow transition region between the Townsend and glow discharges and, due to this, are difficult to observe. A discharge of this type is also Townsend, since it does not observe the appearance of a quasineutral plasma and near-electrode regions.

At large values of dU/dt ($dU/dt \approx 1\cdot10^9$ V/s), a qualitative change in the discharge behavior occurs. When the voltage across the discharge gap reaches a certain value, there is a sharp increase in current to values several orders of magnitude higher than the current in a Townsend discharge, with the voltage across the discharge gap practically falling to zero. This discharge behavior is associated with an exponential change in the field in the near-cathode region due to the space charge.

Thus, depending on the growth rate of the external voltage, the one-dimensional model gives three forms of discharge, two of which are somewhat different kinds of Townsend discharge, and one form is a glow discharge. A common property for them is the Townsend breakdown mechanism, which consists in an exponential increase in the initial current at the cathode due to ionization in an electric field.

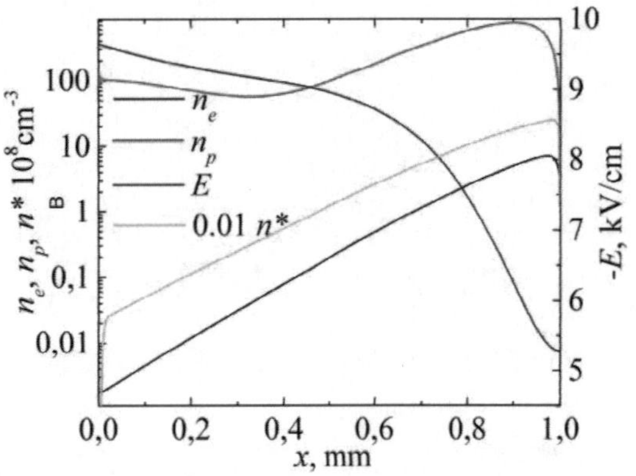

a)

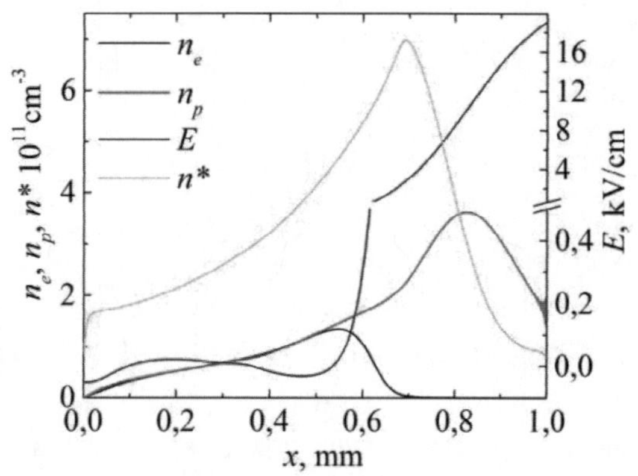

b)

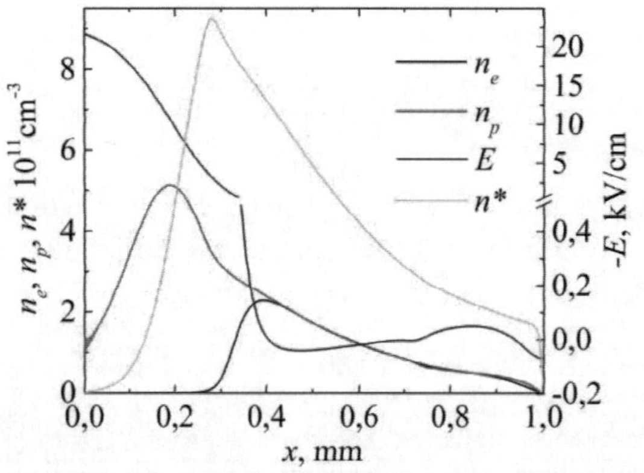

c)

Figure 2. Distributions of the charged particle densities, excited particle densities, and the electric field at the different moments *a)* for 0.1 μs ; *b)* 0.2 μs ; *c)* 0.4 μs for case *c* on fig. 1.

4. Conclusions

Based on the one-dimensional extended fluid model, the characteristics of dielectric barrier discharge excited by saw-tooth voltage are numerically investigated in atmospheric pressure argon. Numerical experiments have shown that, depending on the rate of voltage growth, the one-dimensional model gives three forms of discharge, two of which are slightly different types of Townsend discharge, and one form is a glow discharge. Results show that the discharge mode transits from APTD to APGD when the rate of voltage growth is increase. Distributions of spatiotemporal parameters of the discharge for different forms of saw-tooth voltage signal were obtained.

Acknowledgments

This work was supported in part by the Russian Federation Presidential Grant no. MK-539.2017.1. The development of the numerical model and description of elementary processes in a discharge in argon was supported by the RFBR, project no. 18-31-00098 mol_a.

References

[1] U. Kogelschatz, B. Eliasson, and W. Egli, 1999 *Pure Appl. Chem.* **71** 1819.
[2] F. Massines, N. Gherardi, N. Naude, and P. Segur, 2009 *Eur. Phys. J.: Appl. Phys.* **47** 22805.
[3] F. Massines, N. Gherardi, N. Naude, and P. Segur, 2005*Plasma Phys. Controlled Fusion* **47** B577.
[4] Z. Zhang, Y. Qiu, and Y. Lou, 2003 *J. Phys. D: Appl. Phys.* **36** 2980.
[5] X. T. Deng and J. J. Shi, 2005 *Appl. Phys. Lett.* **87** 153901.
[6] V. Y. Plaksin, O. V. Penkov, M. K. Ko, and H. J. Lee, 2010 *Plasma Sci. Technol.* **12** 688.
[7] S. Kanazawa, M. Kogoma, and T. Moriwaki, 1988 J. Phys. D: Appl. Phys.**21** 838.
[8] A. Sublet, C. Ding, J. L. Dorier, C. Hollenstein, P. Fayet, and F. Coursimault, 2006 *Plasma Sources Sci. Technol.* **15** 627.
[9] H. Y. Luo, Z. Liang, B. Lv, X. X. Wang, Z. C. Guan, and L. M. Wang, 2007 *Appl. Phys. Lett.* **91** 221504.
[10] D. Lee, J. M. Park, S. H. Hong, and Y. Kim, 2005 *IEEE Trans. Plasma Sci.* **33** 949.
[11] Q. Wang, J. Z. Sun, and D. Z. Wang, 2009 *Phys. Plasmas* **16** 043503.
[12] A I Saifutdinov, A A Saifutdinova and N F Kashapov 2016 *Journal of Physics: Conference Series* **669** 012044.
[13] N. Benard and E. Moreau, 2012 *Appl. Phys. Lett.* **100**, 193503.
[14] A. A. Abdelaziz, T. Seto, M. Abdel-Salam, and Y. Otani, 2012 *J. Phys. D* **45**, 115201.
[15] M. Kotsonis and S. Ghaemi, 2012 *J. Phys. D* **45**, 045204.
[16] M. Bogaczyk, G. B. Sretenovic, and H. E. Wagner, 2013 *Eur. Phys. J. D* **67**, 212.
[17] A. I. Saifutdinov, A. A. Saifutdinova, and B. A. Timerkaev 2018 *Plasma Physics Repotrs* **44**, 3.

IOP Conf. Series: Journal of Physics: Conf. Series **1058** (2018) 012026 doi:10.1088/1742-6596/1058/1/012026

Influence of the auxiliary plasma on the deuterium optical spectrum emitted from the dielectric target being irradiated by the e-beam with energy up to 25 keV

Yu S Akishev[1,2], A A Balakirev[1], A V Petryakov[1] and N I Trushkin[1]

[1]SRC RF TRINITI, 108840, Russia, Moscow, Troitsk, Pushkovykh Str., 12
[2]National Research Nuclear University MEPhI, 115409, Russia, Moscow, Kashirskoe sh. 32

Abstract. The dielectric or metal targets isolated from the ground and being irradiated by an electron beam accumulate the electric charge on their surfaces. The amount of a charge is determined not only by the parameters of an e-beam and material of the target and sort and pressure of the gas in which the target is placed but also by the existence or absence of plasma around the target. The paper presents the experimental results about influence of the barrier discharge (BD) plasma on a spectrum of the radiation arising at a surface of a quartz plate irradiated by an electron beam with energy up to 25 keV. The high-current e-beam was generated by the pulsed open discharge in strongly overvoltage regime. The BD-plasma did not influence a shape of the Dα line radiated from the near-surface area of a target that proves a weak impact of the BD plasma on the charge of a target and electric field around the target.

1. Introduction

High-current electron beams of average energy (up to 25 keV) are used in various applications. One of the recent and promising application of such e-beams is the ultrahigh charging of dust particles [1, 2]. High electric charge can be accumulated on dielectric or metal targets isolated from the earth and placed in the path of an electron beam. The amount of charge is defined not only by parameters of a beam, material of a target and sort and pressure of gas surrounding the target but also by the existence or lack of plasma around the target. In case of the particles placed in a vacuum, an e-beam doesn't create plasma around them, and the amount of the charge deposited by a beam on a target will be determined only by parameters of a beam and material of a target. In the gaseous ambiance around a target, plasma is being created by the slow secondary electrons which are beaten out from a target and such electrons have the large cross-section for ionization. On the one hand, the arising positive ions will discharge a negatively charged target. However, on the other hand, this process can play also a positive role.

In papers [1, 2] it is offered to use the deuterium positive ions which can be accelerated in the strong E-field of the negatively charged target up to high energy sufficient for generation of neutrons at their striking the target containing deuterium or tritium. In these works, it was shown by numerical calculations that the preset concentration of ions in the surrounding plasma will quickly diminish due to their leaving onto the negatively charged particles if there is no any source generating the plasma. The rate of depletion of ions around a target depends on the initial density of the plasma. This process reduces the efficiency of generation of neutrons even in the case of the high amount of the charge on the particles. It should be noted that in these calculations the ionization of gas around particles by slow secondary electrons which are beaten out by beam electrons was not considered. The efficiency of formation of slow secondary electrons increases in process of the particle charging close to the potential energy of a beam. In this case, electrons of a beam are being slowed down in the E-field of the charged particle to energy at which the efficiency of knocking-out of secondary electrons approaches unit. In this work the question is raised - whether the plasma generation by a beam in the vicinity of a target is effective in comparison with the barrier discharge creating plasma around the target? As a tester, the radiation of the Dα line of the Balmer series emitted from the near-surface area of a target is used. The shape and intensity of this line react to the electric field and the density of plasma around the target charged by a beam, therefore, comparison of a contour of this line in the presence of the BD plasma and without this plasma will allow us to receive the answer to the question posed.

2. Experimental setup

The scheme of the experimental setup is shown in Fig. 1. The electron beam with the energy of electrons up to 25 keV was generated by the overvoltage open discharge in a deuterium at pressure P= 0.5 - 2 Torr with use of the three-electrode system placed in a quartz tube with an internal diameter of 110 mm and 300 mm in length. The movable electrode system for creation of the auxiliary discharge was also placed in a quartz tube. The dielectric barrier discharge excited by sinusoidal voltage with the frequency of 100 kHz has been used as the auxiliary discharge to generate additional plasma around the target. Two 60x60 mm in size and 1.5 mm thick ceramic (Al_2O_3) plates served as dielectric barriers. The copper foil of 50 microns thick and the size of 40x40 mm has been pasted on the back side of each plate. Each foil was covered by a thick layer (15 mm) of the dielectric material. The distance between ceramic plates is equal to 36 mm. The transparent quartz plate 36x60 mm in size and 1.5 mm thick has been placed in the gap between two electrodes of BD. The plate has been situated in the middle of an inter-electrode gap and oriented perpendicular to the ceramic plates and to an electron beam as well. The electric power consumed by the BD does not exceed of 0.5 W.

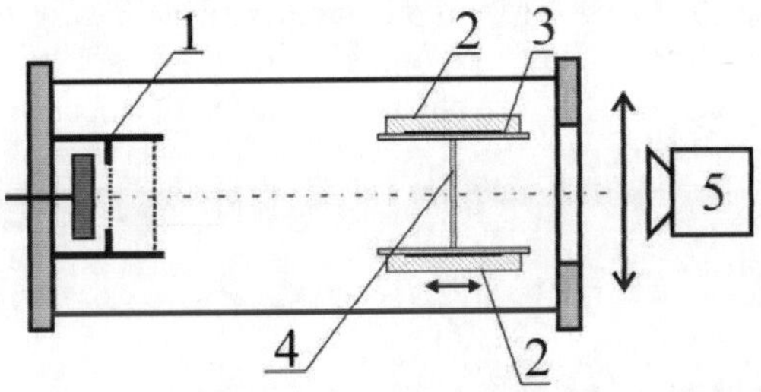

Figure 1. The scheme of experimental setup with the auxiliary dielectric barrier discharge and a quartz target. 1 is the electron gun; 2 are the dielectric plates; 3 are the ceramic plates with metallization (BD electrodes); 4 is the target (thin quartz plate), 5 is the spectrometer.

Dynamic spectral measurements were performed with use of the double monochromator MDR-6 (spectral resolution 0.7 Å), at the exit of which there was a photomultiplier tube. The radiation emitted from the area where the quartz plate interacts with an electron beam was registered by fiber-optical spectrometer AvaSpec-2048L (spectral range λ = 200-1100 nm, spectral resolution Δλ =1.2 nm). The recorded spectrum was integrated over the time of the recording. The currents and voltages of the open discharge and BD were registered by low-inductive current shunts R = 0.024 Ohm and R = 50 Ohm and by the compensated voltage dividers PINTEK HVP-39 (1000:1, 40 kV, 200 MHz), respectively. The electrical signals from the current shunts, voltage dividers and PMT were transferred to digital oscilloscope Tektronix DPO 2024. The images of luminescence formed by the BD, e-beam and plasma in the area of interaction of a quartz plate with an e-beam were taken by Canon EOS 550 and CASIO-EX-F1 digital cameras with a frequency of 1200 shot/s). To maintain high purity of the working gas (99.99% of D_2), all experiments were performed at weak pumping of the gas through the gas-discharge camera.

3. Experimental results and discussion.

The overall view of the e-beam propagating from the right to the left and the target irradiated by e-beam is shown in Fig.2. The distance between the e-gun exit and the target is 18 cm. One may see that the visible cross-section of the gas luminescence produced by the e-beam shrinks with a distance from the e-gun exit. However, the visible diameter of light spot on the target exceeds the visible diameter of the gas luminescence at the target.

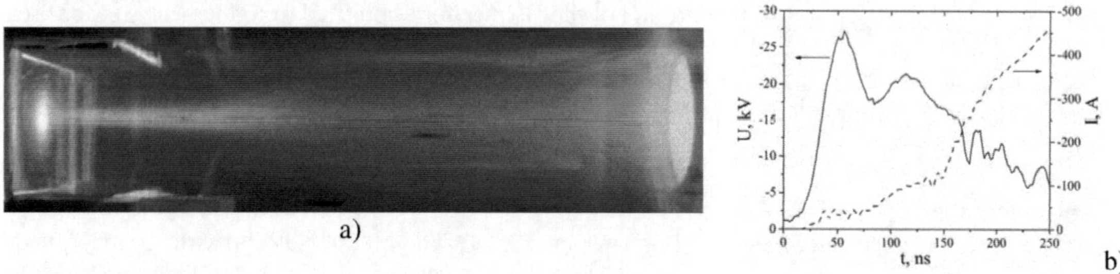

Figure 2. a) The image of a luminescence emitted by both the deuterium along the e-beam trajectory and the target (on the left) area irradiated by e-beam generated by open discharge in the pulsed-periodical mode with a frequency of 150 Hz. The barrier discharge is switched off. The exposure time is 0.6 s. b) The voltage and current waveforms of the overvoltage open discharge generating e-beam. The discharge voltage is 25 kV, gas pressure P = 2 Torr.

The deuterium review spectrum of the light emitted from the target area irradiated by the pulsed-periodical e-beam in the course of 40 pulses is presented in Fig.3a. The spectrum in Fig.3b represents the joint action of the pulsed-periodical e-beam and the BD continuously operating during 40 short pulses of the open discharge pulsing with a frequency of 1 Hz. It means that real exposure time for the BD plasma exceeds the exposure time for the e-beam plasma by factor of 10^6. This is the main reason why the intensity of spectrum in Fig.3b exceeds the intensity of spectrum in Fig.3a. Follows from the presented results that spectrum of the radiation collected from the area of interaction of an e-beam with a target includes both molecular bands and atomic lines. Close examination of these spectra reveals that the BD influences mainly the intensity of molecular bands but the intensity of spectral lines of Balmer series is determined predominantly by the e-beam.

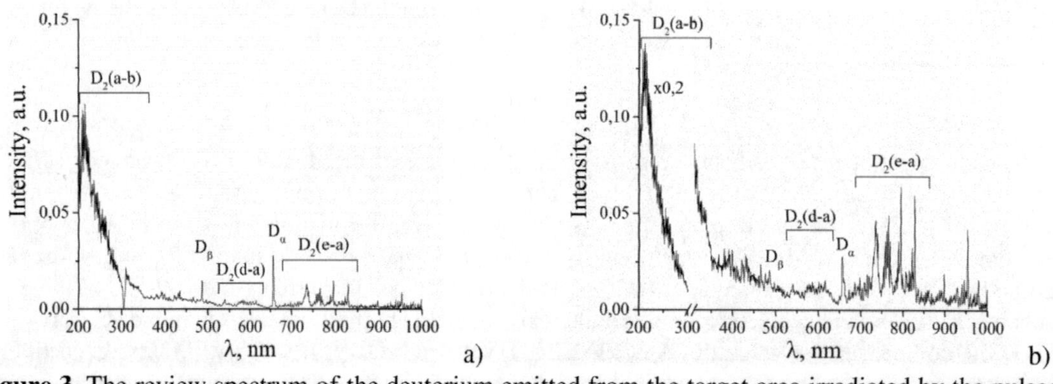

Figure 3. The review spectrum of the deuterium emitted from the target area irradiated by the pulsed-periodical e-beam in the course of 40 pulses (a); the spectrum of the joint action of the e-beam and the BD continuously operating during 40 pulses of the open discharge pulsing with a frequency of 1 Hz. The discharge voltage is 25 kV, gas pressure P = 2 Torr.

It has turned out that in the absence of the BD plasma in the vicinity of the target the radiation of the Dα line arises later about 100 ns after the beginning of charging of a target by an e-beam (see Fig. 4a). The creation of the BD plasma around the target does not change the magnitude of this delay and does not provide the appearance of the Dα line emission before these 100 ns. It means that the plasma quickly created by the e-beam exceeds the influence of the plasma supported stationary by the BD. We suppose that the revealed delay can be attributed to the time that is necessary to charge the target to the potential which is equal practically to kinetic energy of the e-beam. After that, the charged target will strongly decelerate the e-beam. Due to that, the great number of the retarded electrons of the e-beam and the low-energy secondary electrons taken out of the target appear around the target.In contrast to high-energy electrons of the e-beam, the low-energy electrons have ability effectively dissociate a deuterium, excite his atomic and molecular states and also provide intensive ionization. After the termination of the beam, the luminescence of the Dα line continues the increasing that is succeeded by a long decay which is attributed to the discharging of the target by positive ions.

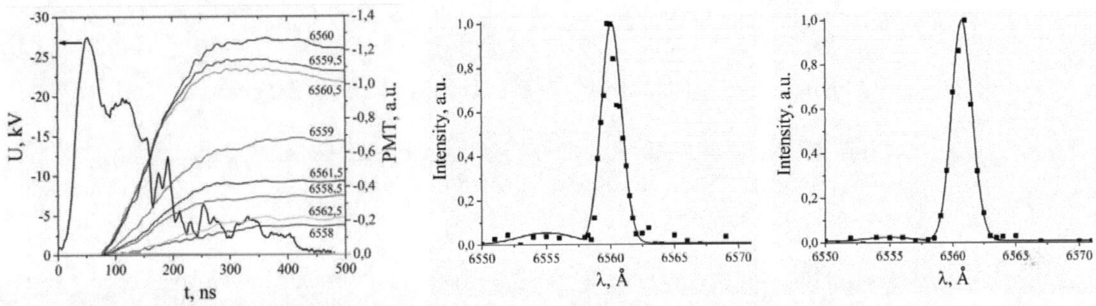

a) b) c)

Figure 4. a) Temporal behavior of the emission intensity of the Dα line for different parts of its contour; the figures at the curves show the wavelength on the contour. The emission happens due to irradiation of the target by the e-beam. This picture shows also the voltage waveform of the discharge generating the e-beam. b) and c) the shape of the Dα line recorded at the moment t=136 ns after applying the voltage; the central part of this line is described by Gaussian approximation with half-width about 1.9 Å (b) and 1.97 Å (c); b) the BD is switched off, c) the BD is switched on.

4. Conclusion.

Approximation of the line form by a Gaussian contour has shown that half-width of lines without and with the BD plasma are almost identical and equal to 1.9 Å and 1.97 Å. It proves high intensity and efficiency of plasma production around the highly-charged target by primary and secondary electrons which are formed due to deceleration of an e-beam by the target. Thus, the formation of the barrier plasma in the area of interaction of an e-beam with a charged target practically changes neither dynamics of Dα line radiation from this area or shape of this line (see Fig.4b, c). The obtained results allow us to hope for the realization of neutron generation due to the ultrahigh charging of targets by an e-beam without the use of any auxiliary discharges.

5. References

[1] Akishev Yu, Karal'nik V, Petryakov A, Starostin A, Trushkin N and Filippov A 2017 *Plasma Physics Reports* **43** 472–479

[2] Akishev Yu, Karal'nik V, Petryakov A, Starostin A, Trushkin N and Filippov A 2017 *Journal of Experimental and Theoretical Physics* **124** 231–243

Acknowledgments

This work has been carried out thanks to full financial support by the Russian Science Foundation (Grant № 16-12-10458).

Features of transition of the open pulse–periodical high-voltage discharge in D2 at low pressure into the low-voltage mode

Yu S Akishev[1,2], A A Balakirev[1], A V Petryakov[1] and N I Trushkin[1]

[1]SRC RF TRINITI, 108840, Russia, Moscow, Troitsk, Pushkovykh Str., 12
[2]National Research Nuclear University MEPhI, 115409, Russia, Moscow, Kashirskoe sh. 32

Abstract. The experimental results on properties of a strongly overvoltage open pulsing discharge being formed in a short gap (3 mm) filled with deuterium at low pressure (P=0.5-2 Torr) are presented. This discharge generates a high-current pulsed-periodical e-beam of 35 mm in diameter with a frequency up to 1.5 kHz and energy up to 25 keV. The repetition frequency influence on the reproducibility of e-beam generation was studied. It was found out that the high-current cathode spots being formed after the transition of the high-voltage regime into low-voltage regime restrict the maximum repetition frequency of a strongly overvoltage open pulsing discharge.

1. Introduction

Generation of the pulsed high-current electron beams with energy up to 25keV by the overvoltage discharges in short gaps (d <1 cm) at a low pressure of the plasma-forming gases (P <10 Torr) is of great interest to many scientific and practical applications. The soft x-ray radiation that can accompany the work with e-beams of average energy doesn't constitute a danger to service personnel as such radiation is easily shielded by the walls of a discharge device. One of the important applications of such e-beams is related to receiving ultrahigh charging of the dust particles containing deuterium or tritium. These particles are surrounded by the plasma created in the deuterium at low pressure. Creation of ultrahigh charged particles by electron beams gives the chance to realize a neutron generator [1] without the accompanying hazardous hard x-ray radiation which necessarily arises when e-beams with energy about 100 keV are used. This is a reason why generators of neutrons without the accompanying hard x-ray radiation are of great interest to biomedicine.

The properties of the overvoltage open discharge in the pulse-periodical mode formed with use of the three-electrode system developed by authors are investigated in this work. The three-electrode system provides a stable ignition of the overvoltage discharge in the pulse-periodical mode. The high-current cathode spots which are being formed after the transition of the discharge from the high-voltage mode into the low-voltage mode can influence the limit opportunities of the open discharge in pulse-periodical regime. The limit opportunities are found out.

2. Experiment

The electric and optical scheme of the experimental setup is presented in Fig. 1. The excitation of the diffusive overvoltage open discharge in a deuterium at a pressure of P=0.5 - 2 Torr was done in a quartz tube in which the gas-discharge system consisting of three electrodes was placed: a solid cathode and two mesh anodes separated from each other. The detailed description of the gas-discharge setup and electrode system is given in [2]. This electrode system allows us to generate two discharges at the same time: the main discharge, i.e. the high-current overvoltage open discharge and the auxiliary discharge, i.e. the low-current discharge which creates the preionization of a gas inside the gap of the main discharge and provides its stable ignition. The pulse-periodical mode of the overvoltage open discharge was organized by using the thyratron operated in the pulse-periodical regime. The auxiliary discharge was working constantly. The existence of the auxiliary discharge increases the frequency and stability of the operation of the main discharge in the pulse-periodical regime. The invented three-electrode system allows us to form the diffusive discharge in a deuterium at the pressure P=0.5 - 2 Torr in the pulse-periodical mode with the voltage amplitude up to 25 kV and the repetition frequency up to 1.5 kHz.

The discharge current and voltage were registered with the low-inductive current shunt R = 0.024 Ohms and the compensated voltage divider PINTEK HVP-39 (1000:1, 40 kV, 200 MHz), respectively, signals from which were recorded by a digital oscilloscope Tektronix DPO 2024.

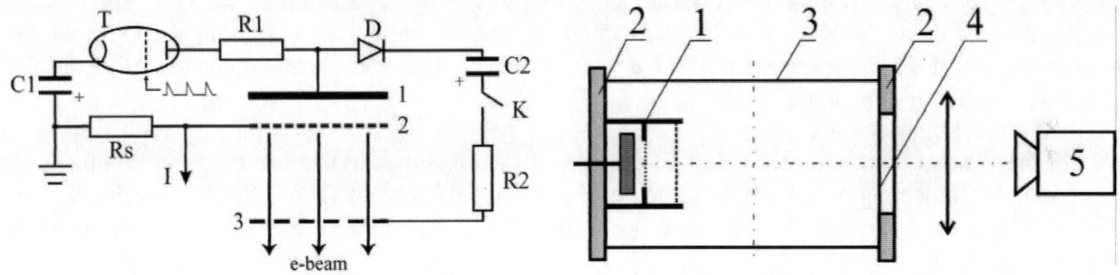

a) Scheme of electrical measurements b) Scheme of optical measurements

Figure 1. The scheme of experimental setup for generation of the overvoltage open discharge in the pulse-periodical mode. a) 1 is the cathode; 2, 3 are the mesh anodes with geometrical transparency about 75%; R1 = 44 Ohm, R2 = 1 MOhm are the ballast resistors, Rs = 0.024 Ohm is the current shunt; D is the diode; C1 =12.5 nF, C2 = 100 μF are the capacitors; K is the key for connection to the auxiliary discharge; T is the thyratron 1000A/25kV which is being switched on by the generator of impulses of the preset frequency. b) 1 is the block in which the three-electrode system of the overvoltage open discharge is mounted; 2 are the flanges which are densely pressed to a quartz tube 3; 4 is a window for observation of the discharge and recording his spectral emission; 5 is a camera or spectrograph.

Spatial-temporal dynamics of both the discharge emission and luminescence of the gas excited by the electron beam in a pipe was registered by Canon EOS 550 and CASIO-EX-F1 digital cameras (the maximum frequency of frames is 1200 a shot/s). The emissive spectrum (λ = 200-1100 nm) of gas excited by an electron beam was registered by a fiber-optical spectrometer of AvaSpec-2048L. To maintain high purity of the working gas (99.99% of D2), all experiments were performed at weak pumping the gas through the gas-discharge camera.

3. Experimental results and their discussion.

Images of the stationary low-current auxiliary discharge and high-current open discharge in the pulse-periodical regime (a look in the direction of a normal to the cathode at a small angle) are given in Fig. 2a and 2b, respectively. It is well visible that the auxiliary discharge is rather uniform. In the case of the periodical open discharge, one can see on the cathode the high-current spots which arise after the transition of the discharge from the high-voltage (HV) mode into low-voltage (LV) one. As a rule, in each impulse, there is no more than one current spot on the cathode (sometimes there is nothing). The existence of a large number of the cathode spots in the photo is explained by a long exposure time in

IOP Conf. Series: Journal of Physics: Conf. Series **1058** (2018) 012027 doi:10.1088/1742-6596/1058/1/012027

comparison with the period of the pulsing discharge, therefore, the image in Fig. 2b corresponds to the superposition of images of many individual discharge impulses. The blurred large spot in the left bottom corner of Fig. 2b corresponds to a luminescence of the gas excited by an e-beam.

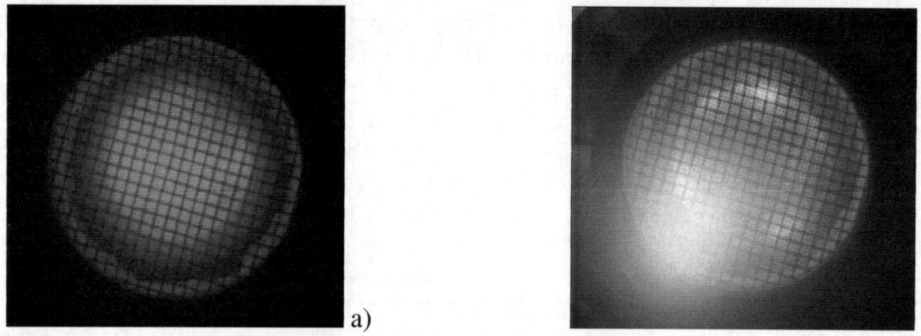

a) b)

Figure 2. Images of the stationary low-current auxiliary category (a) and the pulse- periodical high-current open discharge with a repetition frequency of 150 Hz (b). Deuterium, P = 2 Torr. The pictures are taken in the direction of a normal to the cathode at a small angle The blurred large spot in the left bottom corner of Fig. 2b corresponds to a luminescence of the gas excited by the e-beam. The exposure time is 0.1 s (a) and 0.5 s (b).

The current and voltage oscillograms of the open discharge in a single pulse mode and in the mode of periodically repeating pulses are presented in Fig. 3a and 3b, respectively. After the transition of the HV mode into the LV mode, the voltage drop on the open discharge decreases almost by one order of the magnitude.

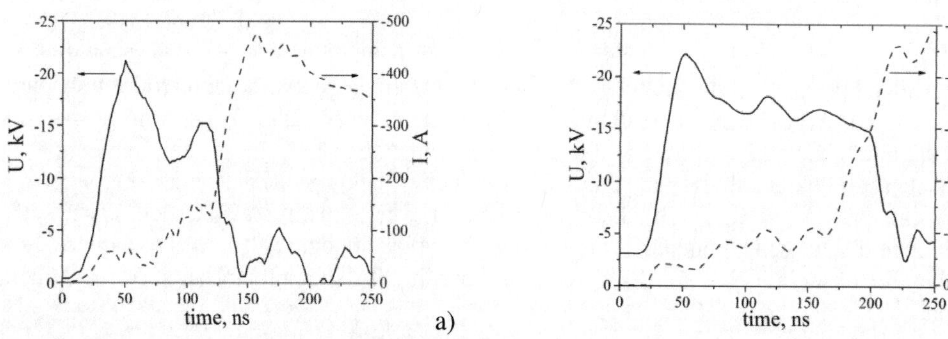

Figure 3. The current and voltage oscillograms of the open discharge in a single pulse mode (a) and in the mode of periodically repeating pulses with a frequency of 150 Hz (b). Deuterium, P = 2 Torr.

A single pulse mode. Exposure time is 0.6 s.

Periodically repeating pulses with a frequency of 10 Hz. Exposure time is 1 s.

Periodically repeating pulses with a frequency of 150 Hz. Exposure time is 0.5 s.

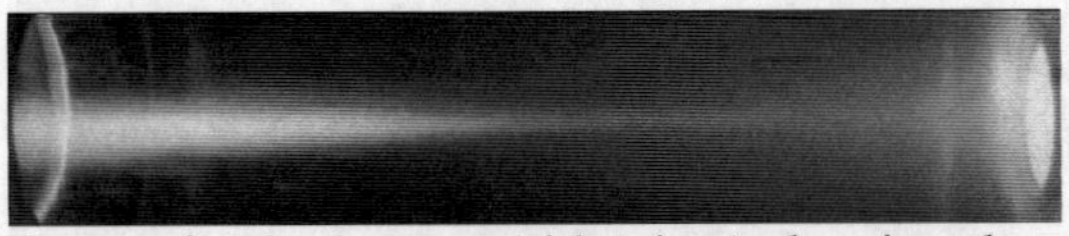

Periodically repeating pulses with a frequency of 1000 Hz Exposure time is 0.5 s.

Figure 4. The images (side view) showing the spatial structure of an e-beam in the direction of his propagation (from the right to the left, length of 23 cm) under different frequencies of the discharge repetition. Deuterium, P =2 Torr. U = 20 kV, I = 100 A.

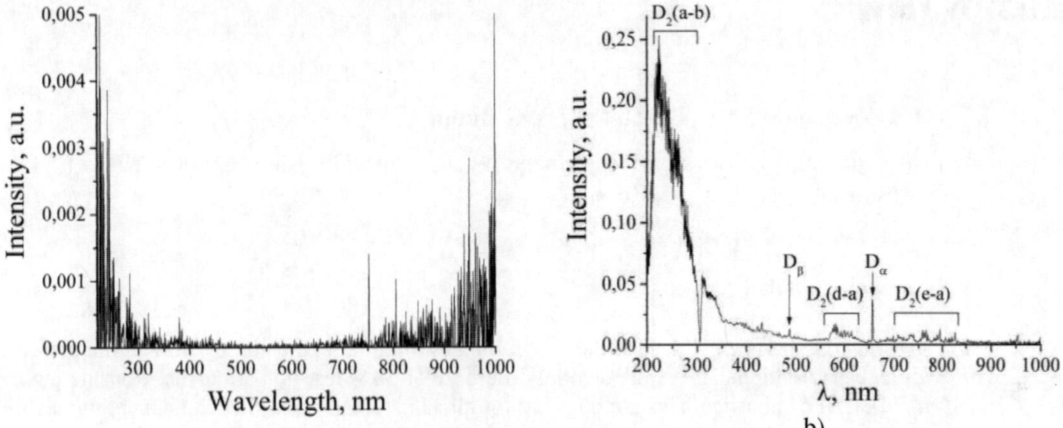

a) b)

Figure 5. The emissive spectra of D2 excited by an e-beam at distance of 20 cm from the HV discharge. a) The spectrum collected from 47 single separated pulses; b) the spectrum collected from a pulse-periodical mode with a repetition frequency of 150 Hz, the time for accumulation of an optical signal is 30 s. One may see that the intensity of the spectral lines significantly increases in the case of discharge in the pulse-periodical regime.

The experiment has shown that in the pulse-periodical mode, the duration of the HV stage of the discharge fluctuates from one impulse to another much stronger compared to that in a single pulse mode. However, the average duration of HV stage increases with the growth of frequency but the average current of the e-beam changes insignificantly. It is possible that increase in duration is caused by the reduction of the gas density in a discharge zone due to his heating. Indeed, at the gas pressure reducing, the duration of the HV stage increases, but the average current of the e-beam decreases. With the growth of frequency, the scattering in the duration of the HV stage is expressed stronger that, perhaps, is connected with the influence of residual plasma from the cathode spots of the previous discharge pulse on development of the discharge in the subsequent pulse. It is obvious that this influence becomes more notable at the reduction of the time between two neighbor discharge pulses. This effect, apparently, limits the maximum frequency about of 1-1.5 kHz for the repetition of pulses of the overvoltage open discharge.

4. References

[1] Akishev Yu, Karal'nik V, Petryakov A, Starostin A, Trushkin N and Filippov A 2017 *Plasma Physics Reports* **43** 472–479

[2] Akishev Yu, Karal'nik V, Petryakov A, Starostin A, Trushkin N and Filippov A 2017 *Journal of Experimental and Theoretical Physics* **124** 231–243

Acknowledgments

This work has been carried out thanks to full financial support by the Russian Science Foundation (Grant № 16-12-10458).

Electrostatic field influence on wood combustion in high-enthalpy flow

I A Zyryanov[1], A P Pozolotin[1], A G Budin[1]

[1] Vyatka State University, Department of Engineering Physics , 610000, Kirov, Moskowskaya Street, 36, Russia

firewcross@mail.ru

Abstract. Experimental data of the wood combustion in a high-enthalpy flow in the presence of an electrostatic field with a potential difference of 5 kV research are given. Considering the field effect combustion laws are obtained for birch, used as a solid fuel component in a hybrid rocket engine.

1. Introduction

The study of the electric fields influence on the hydrocarbon fuels combustion is an important scientific and technical task. The result of the task is creating a contactless management tool for combustion control process. The literature contains a great number of the kinetic and diffusion combustion in electric fields studies. The electrostatic field can deform the combustion front, effect on the phase transformations mechanism, and change the flame temperature [1-5]. The electric field method of combustion control can be used in real power installations, including installations using high-enthalpy flows [6-7].

Wood is a renewable hydrocarbon fuel, and the prospects for its use have not been fully explored. One of the research areas is the high-enthalpy combustion study [8]. This kind of combustion realized in the combustion chamber of hybrid rocket engines [9], in conditions of fires spreading [10].

Therefore, the task of the work is studying of the wood combustion process in a high-enthalpy flow in the presence of an electrostatic field.

2. Experiment Procedure

The experimental setup is an installation with a hybrid rocket engine, described in detail in [7]. As wood, birch with a moisture content of 6% is used. A sample of wood is a cylinder of mass m_0, length = 0.2 m and a channel with a radius of 0.01 m. The combustion is carried out in gaseous oxygen with a flow density [10-40] kg/m^{2}s. The linear combustion rate is calculated by the method given in [11]. Thrust measurement produced by the load cell [7]. The constant electrostatic field in the combustion zone creating system is organized according to [12], the potential difference between electrodes is 5 kV. The heat flux from the flame to the surface of the wood calculation is carried out according to the procedure described in [13].

3. Experimental Results

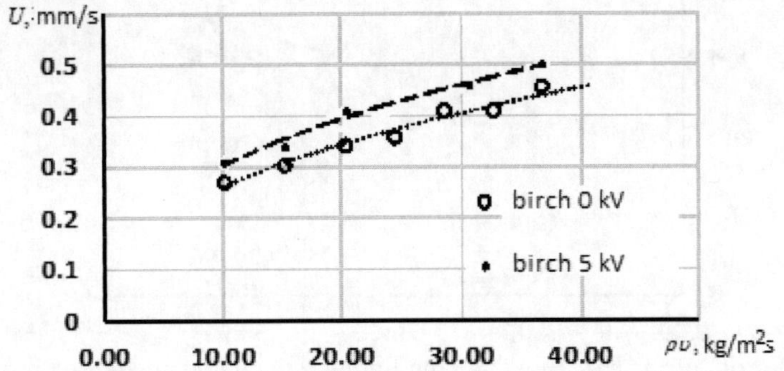

Figure 1. Wood linear combustion rate vs oxygen flow density in the presence of an electrostatic field.

The dependence of the birch linear combustion rate vs oxygen flow density in the presence of an electrostatic field is shown in Fig. 1. The combustion laws have the following forms:

$U = 0.1(\rho\upsilon)^{0.39}$ (Birch 0 kV)

$U = 0.12(\rho\upsilon)^{0.38}$ (Birch 5 kV)

The exponent in the combustion law indicates a combustion mode: in our case for wood it is mixed (diffusion-kinetic). According to the equations obtained, the electrostatic field does not change the combustion mode. The linear combustion rate is increased by 20% with a potential difference of 5 kV. Graph of the time-averaged thrust vs the oxygen flow density in the presence of an electrostatic field is presented in Fig. 2.

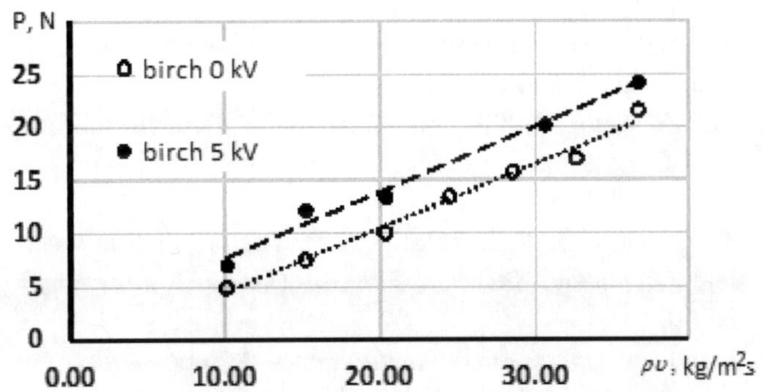

Figure 2. Thrust vs the oxygen flow density.

According to Fig. 2, the thrust in the electrostatic field with a potential difference of 5 kV increases by 40%.

The electrostatic field influence mechanism on the fuels with a condensed phase combustion depends on many factors [2, 3, 7]. It was found that during polymers combustion in the presence of an electrostatic field, the flame temperature increases by an average of 100 K [3], combustion can be accompanied by condensed phase dispersing [3], in the high-enthalpy combustion conditions the polymer gasification centers number in the reaction layer increases [7]. In addition, the flame can change its shape under the action of field mass forces, which leads to a change in the heat flux to the condensed phase [2, 3]. All these factors contribute to the changes in the combustion rate.

According to experimental studies [14], wood combustion rate depends on the density of the heat flux from the flame to the condensed phase.

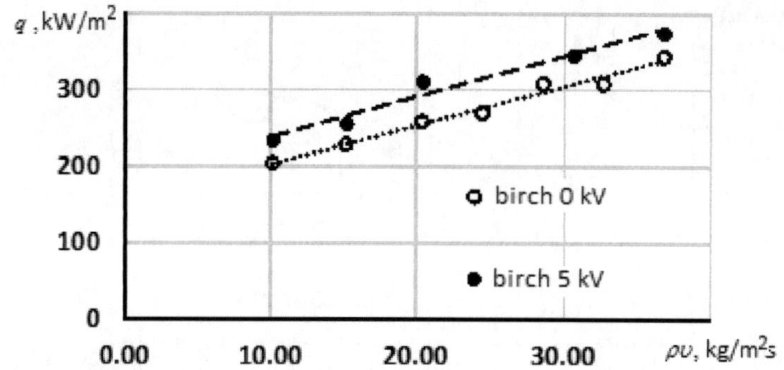

Figure 3. Density of the heat flux from the flame to the wood condensed phase vs the oxygen flow density.

The plot of the heat flux density versus the oxidant flow density is shown in Fig. 3. As can be seen from the graph, the electrostatic field leads to the heat flux increase by 20% according to the calculation technique given in [13]. The reasons for this change in case of wood require further study.

4. Conclusion

Thus, experimental data of the wood combustion in a high-enthalpy flow in the presence of an electrostatic field research are given. It is shown that the electrostatic field increases the wood combustion rate by 20%, which leads to a change in the combustion law and in the hybrid rocket engine thrust by 40%.

References

[1] Kidin N I, External electromagnetic fields influence on combustion processes, http://www.ism.ac.ni/sgv/rtf/l21.rt.

[2] Reshetnikov S M, Zyryanov I A, The electrostatic field influence on the combustion macrokinetic of alkanes and kerosene. *Kazan, Vestnik of the Kazan State Technical University,* 2011, V.1, P. 120-128.

[3] Reshetnikov S M, Zyryanov I A, Pozolotin A P, Features of polymers combustion in the electrostatic field. *Taganrog, Southern Federal University Messenger. Technical Science,* 2013, V.8, P. 30-36.

[4] Lauton D, 1976, *Electrical aspects of combustion*, ed V A Popov (Moscow: Energy) p 296.

[5] Tretyakov P K, Tupikin A V, Zudov V N, Effect of laser radiation and electric field on combustion of hydrocarbon-air mixtures. *Combustion, Explosion, and Shock Waves*, 2009, V.45, №4, P. 413-420.

[6] Afanasev V V, Kidin N I, 2008, *Diagnosis and management of combustion stability in power plants combustion chambers*, (Moscow: Physmathlit) p 176.

[7] Reshetnikov S M, Zyryanov I A, Pozolotin A P, Budin A G, Electrostatic field influence on the combustion rate in hybrid rocket. *Kazan, Vestnik of the Kazan State Technical University,* 2015, V.71, P. 52-57.

[8] Reshetnikov S M, Zyryanov I A, Budin A G, Pozolotin A P, Investigation of wood combustion in the high-enthalpy oxidizer flow, *Journal of Physics: Conference Series. – IOP Publishing,* 2017, V.789, №1.

[9] Gubertov A M, Mironov V V, Gollender R G, 2008, *The processes in hybrid rocket engines*, ed. A S Koroteeva, (Moscow: Nauka), p 405.

[10] Valendik E N, Kosov I V, Thermal radiation of forest fires and its possible effects on the forest stand, *Coniferous boreal zone*, 2008, V.25, № 1-2, P. 88-92.

[11] Hudson M K et al. Guanidinium azo-tetrazolate (GAT) as a high performance hybrid rocket fuel additive, *J. Pyrotechnics*, 2004. V.19, P. 37-42.

[12] Reshetnikov S M, Zyryanov I A, Budin A G, Pozolotin A P, Patent for invention RU2598984, Hybrid rocket engine thrust increase method, applicant and patent holder Vyatka State University - No. 2015100924, publ. 10.10.2016, Bul. № 28, p 6.

[13] Pozolotin A P, Reshetnikov S M, Zyryanov I A, Budin A G, Shaurko P N, Methods of experimental study of wood combustion in a high-enthalpy flow, *Kirov, Society, science, innovation (SPC-2016): Coll. materials*, 2016. P. 1746-1751.

[14] Abduragimov I M, Androsov A S, Batrak M, Ignition and combustion of wood under the influence of heat fluxes, *FGV*, 1986, V.22, №1, P. 10-15.

IOP Conf. Series: Journal of Physics: Conf. Series **1058** (2018) 012029 doi:10.1088/1742-6596/1058/1/012029

Phase transitions and transformation of dust structures in neon dc discharge at cryogenic temperature

D N Polyakov, V V Shumova and L M Vasilyak

Joint Institute for High Temperatures of Russian Academy of Sciences

cryolab@ihed.ras.ru

Abstract. The transformation of dust structures under changes of discharge current in neon dc discharge at a pressure of 20 Pa and the discharge tube temperature of 77 K has been studied. A nonmonotonic dependence of radial and axial sizes of the dust formations on the discharge current have been detected and analyzed. It has been found that dust clouds consisted of a mixture of dust particles and clusters formed from dust particles. The first and second-order phase transitions in dust structures and the mesomorphic state of dusty plasma have been found. The transformation of the dust structure to spherical form has been detected.

1. Introduction

Dusty plasma is an open nonequilibrium dissipative nonideal system with strong Coulomb interaction. Therefore, dust structures can be used to simulate nonideal systems in microworld, including those forming at low and cryogenic gas temperatures. The dust clouds in plasma can exist in forms analogous to the thermodynamic states of matter [1], and phase transitions between these states can be observed. The structure of dust particles can 'melt' and 'evaporate' like a matter. In this case, the long-range order between the dust particles is violated, while the short-range order is maintained. This allows one to describe the dusty plasma in the approximation of the thermodynamic system through the parameter of non-ideality (the kinetic temperature of particles) and to represent its properties in the hydrodynamic approximation through the transport characteristics of a continuous medium [2].

The interest in the studies of characteristics of gas discharge plasma with dust particles cooled to cryogenic temperature is stimulated by the possibility of using such plasmas in novel plasma-chemical technologies. These studies are relevant for the development of the theory of nucleation, growth and agglomeration of nanostructures and coagulation of micro- and nanoparticles in plasma [3-5], and may promote the related sciences such as low-temperature and quantum chemistry [6, 7].

At low and cryogenic gas temperatures in some gases, the structures consisting of dust clusters can be formed [5, 8-11], and the Bose condensation can occur [12]. The process of cluster formation at cryogenic cooling was accompanied by a decrease in the distances between the dust particles [8-11]. As well as individual dust particles, these clusters can form the lattice sites similar to atoms in a solid, and also exist in a 'liquid' and 'gaseous' states. The discharge current is known to exert the greatest influence on the topological parameters of the dust structure in dc discharge. For instance, the influence of the current on the radial size of dust structure was observed in the discharge plasma of a number of gases with various dust particles [13-15]. In this work the influence of the discharge current on the shape, size, composition of complex dust structures consisting of dust clusters, and their phase state has been studied at cryogenic gas temperature.

2. Experimental

The parameters of dust structures formed by spherical plastic particles with a diameter of 4.14 µm, which were injected into dc discharge in neon at a pressure of 20 Pa, have been experimentally studied in a glass discharge tube (1) with an inner diameter of 16.5 mm and a length of 20 cm (figure 1). The discharge current I was varied from 0.6 to 3.2 mA. To measure the voltage drop ΔU_c in the positive column of the discharge and to provide confinement of dust particles (2) at the measurement region, two narrow ring electrodes (3) spaced apart, were located in the discharge tube near the cathode (4). The discharge tube was placed in an optical cryostat (5), where it was cooled to the boiling point of liquid nitrogen T=77 K. Also at different values of the discharge current, the total voltage of the discharge U, and the sizes of the dust structure - the radial D (diameter) and the axial L (length) were measured. Using the electrical measurement data, the values of the linear power input $Q=EI$ and the longitudinal electric field $E=\Delta U_c/l$ in the discharge were calculated. Length l is the distance between ring electrodes. The gas temperature in the discharge was assumed to be equal to the temperature of the discharge tube near the investigated sector containing dust particles. Images of dust structures (6) were recorded with a microscope (7) in the axial section in the reflected light of a plane laser beam (8). According to the obtained images, the composition, shape, degree of order and dynamic stability of the components of dust structures were analyzed. The experimental setup is described in detail elsewhere [9, 10, 16].

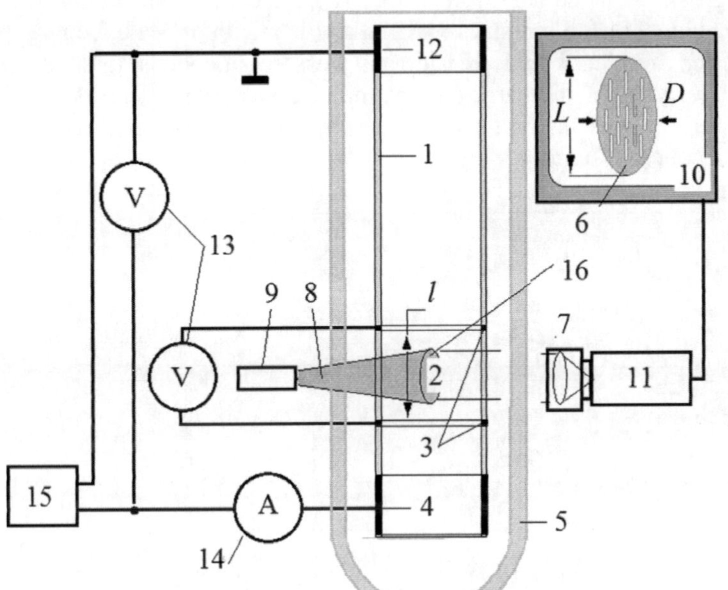

Figure 1. Experimental scheme:
1- glass discharge tube,
2-confinement of dust particles,
3- measuring ring electrodes,
4- cathode, 5- optical cryostat,
6- dust structure image,
7- microscope, 8- laser beam,
9- laser, 10- monitor and video-recorder, 11- video-camera,
12- anode, 13- voltmeter,
14- current meter, 15- high voltage power supply.

3. Results and discussion

At room temperature of the discharge (about 295 K), an increase in the discharge current leads, as a rule, to an increase in the radial D and a decrease in the axial L size of the dust structures [11, 14]. In neon dc discharge at room temperature, the dust structures were formed by individual dust particles (figure 2 (a)). With decreasing gas temperature, the individual dust particles formed clusters (figure 2 (b, c)). At a lower gas temperature, the clusters become ordered in axial direction (figure 2 (c)). When the clusters were formed, typical features of a first-order phase transition, at which the particle density increased sharply and continued to increase with decreasing gas temperature were observed. The process of formation of clusters can be considered as ‘condensation’ and ‘deposition’ of dust particles [11].

IOP Conf. Series: Journal of Physics: Conf. Series **1058** (2018) 012029 doi:10.1088/1742-6596/1058/1/012029

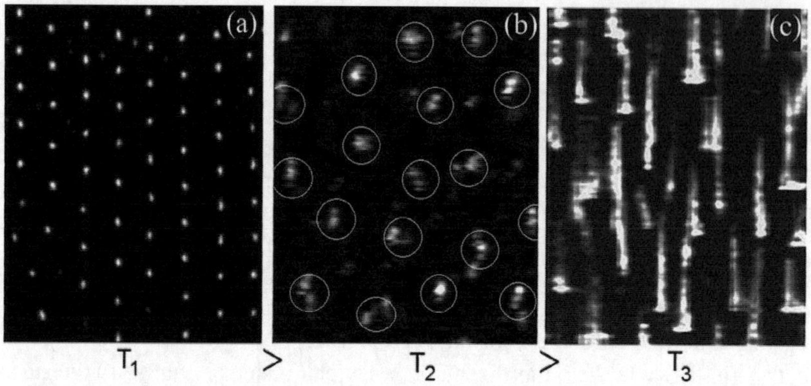

Figure 2. (a-c) Fragments of dust structure at decreasing gas temperature.

At T=77 K the dust structures can be multicomponent [10, 11]. Multicomponent system is represented by a mixture of dust particles and clusters formed by dust particles. The composition of the dust system varies with gas pressure and discharge current. One-component dust systems consist of simple clusters (figure 2 (c)). Simple clusters are threadlike clusters consisting of several dust particles located along the discharge axis. Threadlike clusters form homogeneous structures in the 'liquid' state, which were observed at higher neon pressures $P \geq 120$ Pa [10, 11]. In our case, complex clusters consisting of threadlike clusters, formed with decreasing gas pressure. In contrast to the dust structures at room temperature, upon varying the discharge current, the nonmonotonous dependencies of the radial D and axial L sizes of the dust structure on the discharge current I (figure 3, circles) have been observed in cryogenic multicomponent systems.

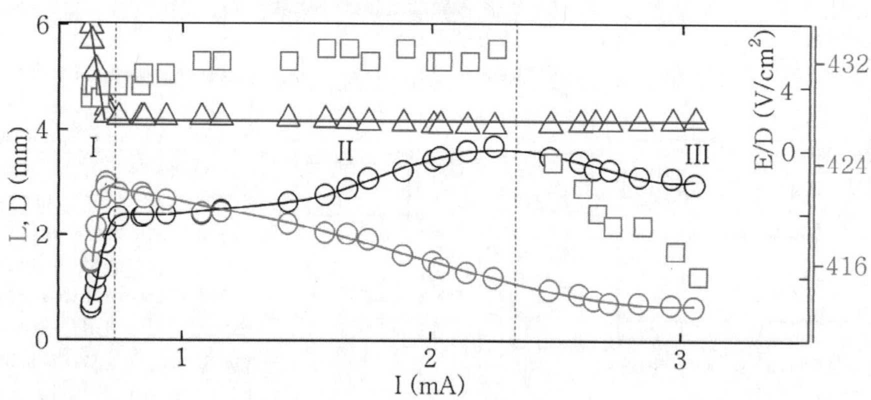

Figure 3. Dependencies of the radial D (black circles) and axial L (red circles) sizes of dust cloud, the longitudinal electric field over the dust cloud diameter E/D (blue triangles), and the total discharge voltage U (green squares) on the discharge current I.

In figure 3, one can distinguish three characteristic regions (I - III), divided with dashed lines. Regions II and III are analogous in the dependences on the current of the shape of the dust structures formed by individual dust particles observed at room gas temperature. The change in the slope of the $D(I)$ curve in region III is caused by the change in the discharge mode. Here, a decrease in the discharge voltage U (squares, figure 3) has been observed; it led to a decrease in the electric field E. Regions II and III are characterized by constant ratio of the longitudinal electric field to the radial size of the dust structure, E/D=const (triangles, figure 3). This result indicates the applicability of one-dimensional model of dusty plasma [17-19] at cryogenic temperatures.

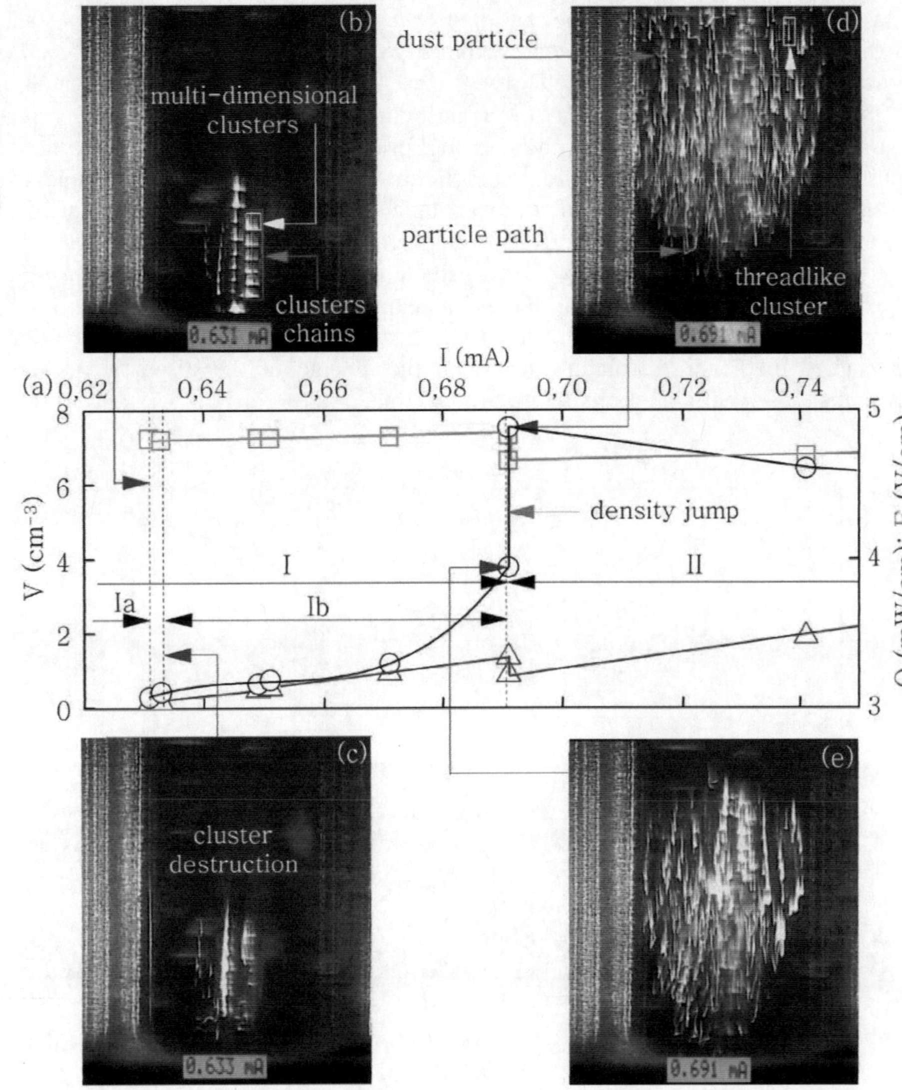

Figure 4. (a) Dependencies of the dust structure volume V (black circles), the longitudinal electric field E (pink squares) and the linear power input Q (blue triangles) on the discharge current I. (b-e) Images of dust structures at phase inversion points.

Figure 4 represents the dependencies of the dust structure volume V, linear power input Q and longitudinal electric field E on the discharge current I. The images of dust structures (figure 4 (b-e)) correspond to the regions I and II in figure 3. These images correspond to characteristic points of phase inversion. From the analysis of the dust structure images and from $V(I)$ dependence, one can conclude that there are the evidences of the first- and second-order phase transitions in dust structures, which follow the changes of the discharge current. At $I = 0.631$ mA (figure 4 (b) (region Ia)), the dust structure is in the 'crystalline' state. The dust 'crystal' is formed by cluster chains, which consist of multi-dimensional clusters. An increase of the current to 0.633 mA (figure 4 (c) (region Ib)) is accompanied by a decrease in the symmetry (up to its disappearance) of the dust structure. Thus, the evidence of a second-order phase transition has been observed. At a higher current (figure 4 (d) (region Ib)), a multicomponent mixture consisting of cluster chains and simple clusters was observed inside the dust structure; this indicated the 'mesomorphic' state of dust system. This is a state where reconstructive transition of allotropic structures from a crystal to an alloy of a crystal in an isotropic liquid and vice versa can occur. At I = 0.691 mA (figure 4 (e) (region II)), there was an abrupt jump in the dust structure volume, i.e. the density of the dust cloud decreased sharply, which indicated that it was a first-order phase transition. Here, the 'melting' of complex dust clusters and partial 'melting' of simple dust clusters were observed. In region II (figure 3 and figure 4), the dust system exists as a mixture consisting of individual dust particles and simple clusters. The unstable dynamic state of the dust particles resembled a 'boiling liquid'. The development of oscillations and rotations of individual dust particles was observed. The intensity of dust particle motion increased with decreasing pressure. As the discharge current increased (region II), the concentration of individual dust particles and the intensity of their motion decreased. The dust system in region II exists in the 'liquid' state. In this region, the change in the dust structure size with the change of the discharge current occurs analogously to that observed in dust structures at room temperature.

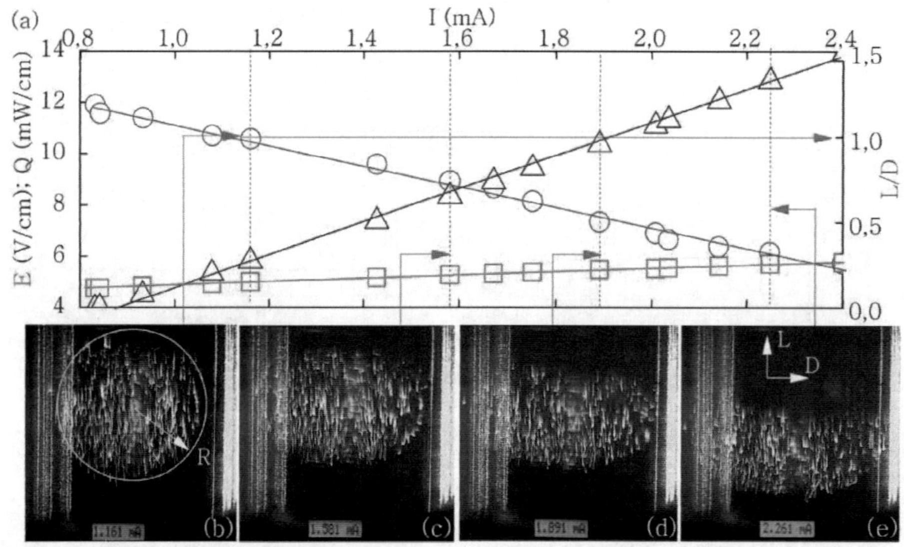

Figure 5. (a) Dependencies of the relative size of the dust structure L/D (green circles), the longitudinal electric field E (pink squares) and the linear power input Q (blue triangles) on the discharge current I. (b-e) Images of dust structures at different values of discharge current.

This behavior is illustrated by the dependence of L/D on I (green circles) in figure 5 (a). At a current of about 1.15 mA L/D=1; in this case, the dust structure takes the spherical shape (figure 5 (b)). The spherical dust clouds were observed in discharges of various types and their formation was associated with various conditions and gas discharge parameters (not only gas temperature). The spherical dust clouds were first observed in neon dc discharge at room temperature of the gas [20]. In dc discharge in air, the spherical dust structures formed under the influence of an external magnetic field [21] (figure 6 (a)). The spherical dust structures also formed in a radio-frequency discharge, when an active and grounded electrodes were shifted in the vertical discharge tube (figure 6 (b)). When a spherical cloud formed, the distances between dust particles decreased from 170 μm to 80-90 μm. At a decrease of gas temperature, an increase in number density of charged dust particles occurs as a result of a decrease in the Debye radius [22].

At cryogenic temperatures, the spherical dust structures were observed for the first time in a glow discharge of direct current in helium at T = 4.2 K [4]. In [23], the transition to the spherical shape of a dust cloud was obtained as a result of a decrease in the temperature of the helium plasma from 175 to 135 K. A sphere-like structure at T=77 K was obtained as a result of decrease of gas pressure and discharge current in helium [24]. The spherical dust structures can simulate the state of a liquid from the point of view of compliance with the Eötvös rule for the magnitude of surface tension force, which increases with increasing the number density and decreasing the temperature of the liquid. The presence of 'surface tension' in the dust structure is indicated by the dynamics of the drop-like shape of the dust structure [25] and the dynamics of its separation into two spherical parts under the influence of an external temperature field [13, 26]. For the dust structure in 'liquid' state, the surface tension can be represented through the binding energy of dust particles [13], which is comparable to the value of the binding energy of a dust molecule [27].

The spatially ordered planar structures of spherical shape formed from charged water droplets above its surface [28]. The round form of such structures, in the opinion of the authors of [29], indicated the presence of a surface tension. Similar structures can exist in atmospheric clouds and fogs [29].

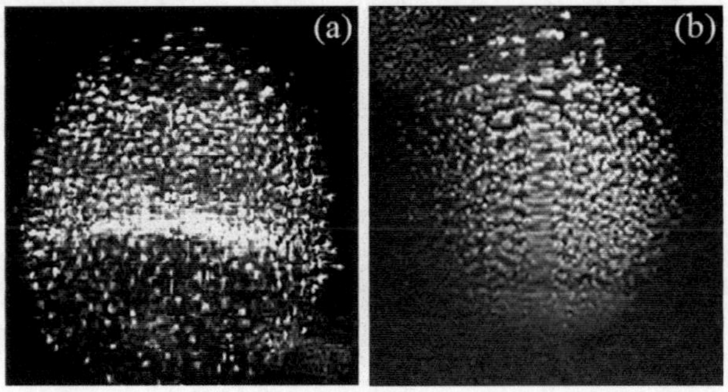

Figure 6. Images of spherical dust structures (axial section) at room temperature: (a) in a dc discharge with a radial magnetic field [20] and (b) in a radio-frequency discharge.

4. Conclusions

The impact of the discharge current on the size, shape and composition of complex Coulomb dust structures consisting of dust clusters at cryogenic temperature has been studied. The first and second-order phase transitions have been found. It has been found that with increasing current, the complex clusters melted, giving a mixture of components consisting of simple clusters, complex clusters, and individual dust particles. In the 'liquid' state of dusty plasma, the formation of a spherical dust structure has been observed.

Acknowledgments

The authors are grateful to the Russian Foundation for Basic Research for the financial support of this study (Grant No. 16-02-00991).

References

[1] Morfill G E and Ivlev A V 2009 Complex plasmas: An interdisciplinary research field *Rev. Mod. Phys.* **81** 1353–403

[2] Polyakov D N, Vasilyak L M and Shumova V V 2013 Positive Column of Glow Discharge with Dust Particles *Surf. Eng. Appl. Electrochem.* **49** 114–24

[3] Huttel Y 2017 Gas-Phase Synthesis of Nano-Particles (Weinheim: Wiley VCH) p 416

[4] Asinovskii E I, Kirillin A V and Markovets V V 2006 Plasma coagulation of microparticles on cooling of glow discharge by liquid helium *Phys. Let. A.* **350** 126–8

[5] Polyakov D N, Vasilyak L M and Shumova V V 2015 Synergetics of Dusty Plasma and Technological Aspects of the Application of Cryogenic Dusty Plasma *Surf. Eng. Appl. Electrochem.* **51** 143–51

[6] Goldanskii V I 1976 Chemical reactions at very low temperatures *Ann. Rev. Phys. Chern.* **27** 85–126

[7] Krems R V 2008 Cold controlled chemistry *Phys. Chem. Chem. Phys.* **10** 4079–92

[8] Polyakov D N , Shumova V V and Vasilyak L M 2015 Electric field on the diagram of phase transitions in cryogenic dusty plasmas *Rom. Rep. Phys.* **67** 1040–8

[9] Polyakov D N, Shumova V V and Vasilyak L M 2014 Formation of structures from dusty clusters in cryogenic plasma of glow discharge *IEEE ICOPS/BEAMS Proc.* 670–4

[10] Polyakov D N, Shumova V V and Vasilyak L M 2016 Formation of structures from dusty clusters in dc discharge at plasma cooling *J. Phys.: Conf. Ser.* **774** 012181

[11] Polyakov D N, Shumova V V and Vasilyak L M 2017 Transformations of dust structures in glow DC discharge in neon: effect of gas temperature and discharge current *Plasma Sources Sci. Technol.* **26** 08LT01

[12] Barmashova T V, Mart'yanov K A, Makhalov V B and Turlapov A V 2016 Fermi liquid-to-Bose condensate crossover in a two-dimensional ultracold gas experiment *Physics-Uspekhi.* **59** 174-83

[13] Vasilyak L M, Vetchinin S P, Polyakov D N and Fortov V E 2005 Formation of Complex Structures in Dusty Plasmas under Temperature Gradients *J. Exp. Theor. Phys.* **100** 1029–34

[14] Shumova V V, Polyakov D N and Vasilyak L M 2015 Transformation of dust structures in a dc discharge in neon *Prikl. Fiz. (Appl. Phys.)* **4** 27–32

[15] Fedoseev A V, Sukhinin G I, Abdirakhmanov A R, Dosbolayev M K and Ramazanov T S 2016 Voids in Dusty Plasma of a Stratified DC Glow Discharge in Noble Gases *Contrib. Plasm. Phys.* **56** 234–9

[16] Shumova V V, Polyakov D N and Vasilyak L M 2017 Neon dc glow discharge at cryogenic cooling: experiment and simulation *J. Phys. D: Appl. Phys.* **50** 405202

[17] Shumova V V, Polyakov D N and Vasilyak L M 2014 Effect of metastable neon atoms in positive column of glow discharge with dust particles *Plasma Sources Sci. Technol.* **23** 065008

[18] Shumova V V, Polyakov D N and Vasilyak L M 2015 Influence of dust particles on ionization and excitation in neon dc discharge *J. Phys.: Conf. Ser.* **653** 012132

[19] Shumova V V, Polyakov D N and Vasilyak L M 2017 Influence of dust void on neon DC discharge *Plasma Sources Sci. Technol.* **26** 035011

[20] Fortov V E, Molotkov V I, Nefedov A P, and Petrov O F 1999 Liquid- and crystallike structures in strongly coupled dusty plasmas *Phys. Plasmas* **6** 1759–1768

[21] Vasilyak L M, Vetchinin S P, Polyakov D N and Fortov V E 2007 Control on particles density in aerosol Coulomb structures in plasma *Herald of Daghestan State University* **4** 23–6 (http://vestnik.dgu.ru/Stat/2007.4.4.pdf)

[22] Fortov V E, Vasilyak L M, Vetchinin S P, Zimnukhov V S, Nefedov A P and Polyakov D N 2002 Plasma-Dust Structures at Cryogenic Temperatures *Dokl. Phys.* **47** 21–4

[23] Petrov O F and Fortov V E 2013 Collective Phenomena in Strongly Coupled Dissipative Systems of Charged Dust: from Ground to Microgravity Experiments *Contrib. Plasma Phys.* **53** 767–77

[24] Antipov S N, Vasiliev M M, Alyapyshev M M, Petrov O F and Fortov V E 2014 Dense dust structures in cryogenic complex plasma *J. Phys.: Conf. Ser.* **511** 012008

[25] Balabanov V V, Vasilyak L M, Vetchinin S P, Nefedov A P and Fortov V E 2001 The Effect of the Gas Temperature Gradient on Dust Structures in a Glow-Discharge Plasma *J. Exp. Theor. Phys.* **92** 86–92

[26] Vasilyak L M, Vetchinin S P, Zimnukhov V S, Polyakov D N and Fortov V E 2003 Dust Particles in a Thermophoretic Trap in a Plasma *J. Exp. Theor. Phys.* **96** 436–9

[27] Tsytovich V N 1997 Dust plasma crystals, drops, and clouds *Phys. Usp.* **40** 53–94

[28] Shavlov A V, Dzhumandzhi V A and Romanyuk S N 2011 Electrical properties of water drops inside the dropwise cluster *Phys. Let. A.* **376** 39–45

[29] Shavlov A V and Dzhumandzhi V A 2010 Spatially ordered structures in storm clouds and fogs *Phys. Let. A.* **374** 2561–5

IOP Conf. Series: Journal of Physics: Conf. Series **1058** (2018) 012030 doi:10.1088/1742-6596/1058/1/012030

Experimental research of the cathode region of the glow discharge in nitrogen

E I Prokhorova, A A Platonov, A G Slyshov and A I Nazarov

Petrozavodsk State University, 33 Lenina Avenue, Petrozavodsk, Russia, 185910

prokhorova@petrsu.ru

Abstract. The main falling of the discharge tension accounts for the cathode region where the main streams of charged particles are concentrated. The research of this region is important for receiving various coverings and processing of surfaces with gas discharge. Authors have presented results of the experimental electric and optics research of the cathode regions of the glow discharge in nitrogen. The research of the potential distribution in the cathode region is conducted; distribution functions of the various electron energy groups are measured; the research of the ions energy distribution near the cathode is executed. The model of formation of the ions energy distribution near the cathode is offered. Criteria for formation of the potential hole for electrons and its influence on a power range of electrons are experimentally defined. The theoretical analysis of the received results is carried out. Nonlocal theoretical model of the cathode region of the glow discharge is offered. Comparison of the received results to an overall picture of a luminescence of the discharge is executed. The agreement of experimental data with theoretical is shown.

1. Introduction

In spite of a large number of publications, so far physical processes in separate parts of the discharge have been insufficiently studied. In particular this holds true for the cathode regions. Two approaches to formation of the cathode region of the glow discharge are still most widespread [1-5]. There is not enough experimental data about distribution of parameters of the plasma in the cathode region and almost no analysis of these data from the point of view of nonlocal kinetics. Meanwhile it is important for the development of the theory of discharge and its practical use. Therefore complex study of physical properties of the near-cathode regions of the glow discharge and creation of the corresponding theoretical model are relevant directions of development of modern physics of the gas discharge.

2. Electrons energy distribution functions

Traditionally the description of the gas-discharge plasma builds on the basis of the hydrodynamic approach operating with average characteristics of particles. However such approach cannot fully describe many phenomena in gas discharges especially when inhomogeneity of plasma is essential [3, 6]. More exact kinetic analysis has been limited to generally homogeneous plasma; recording of its heterogeneity was generally carried out through numerical calculations [7, 8]. Significant progress in the theory of heterogeneous regions of the gas discharge is made thanks to the use of principles of nonlocal kinetics of electrons [2, 3, 5, 6].

IOP Conf. Series: Journal of Physics: Conf. Series **1058** (2018) 012030 doi:10.1088/1742-6596/1058/1/012030

For the purpose of determination of the near-cathode regions structure of short (without positive column) glow discharge in molecular gas (nitrogen) and check of its nonlocal model the discharge tube has been made and measurement of the near-cathode regions of the discharge parameters is executed.

The measurements were performed in a glass tube. The tube (radius 37 mm) contained a mobile flat cold cathode of diameter 56 mm, a flat anode of diameter 60 mm, and a cylindrical 3-mm-long probe with a diameter of 0.2 mm capable of moving along the tube. The anode – cathode distance was 165 mm. The research included measurement of the potential of the space, electron distribution function, concentration of the charged particles and radiation of the discharge as well as calculation of these values, their comparison with experimental data and development of nonlocal theoretical model.

The electron distribution function was determined by using the second derivative of the probe current with respect to the probe potential. The glow discharge conditions were selected so as to provide the absence of ionization waves. The plasma potential was determined at the second derivative zero.

Existence of two types of the structure of the cathode regions is revealed. For small values of the pL parameter (p – the pressure, L – the distance between electrodes) a pit is observed on the profile of the potential. At bigger value of the pL parameter the potential pit apparent experimentally isn't registered. The site of constancy of the potential of the plasma is observed.

The reason of the existence of a potential pit for electrons at low pressure is that the density of the diffusion current exceeds the density of the discharge current. In this connection the fulfillment of the continuity equation requires an existence of a reversed field. At these pressures collisions of electrons with atoms and gas molecules are negligible. Therefore, the diffusion stream is significant. A big reversed field is needed. That is why the potential pit should be rather deep (about the temperature of electrons).

An increase of gas pressure increases the number of collisions. As a result the resistance to the diffusion stream increases. Therefore, the stream decreases. Thus, the reversed field will decrease too and the area of the reversed field be much shorter. In this case $j_{diffusion} \geq j_{discharge}$ and the pit cannot be registered. On the condition that $j_{diffusion} < j_{discharge}$ an existence of a pit is not necessary.

It is established that the measured electron distribution functions in the slow part are Maxwell ones with a temperature less than one eV. This fact is explained by the kinetic model of the glow discharge of a direct current based on nonlocal ionization of gas by the electrons which have gained the energy in a cathode layer. This model predicts presence of three groups of electrons: fast, intermediate and slow (locked in a potential pit). The group of slow electrons consists of the locked electrons. Their energy is lower than the potential of second point of the field address (or anode potential in the short discharge). These electrons do not participate in transfer of current and cool down up to the room temperature. The maximum concentration of the electrons corresponds to a bottom of the potential pit. Electronic current in the Faraday's dark space is transferred by intermediate electrons with energies less than energy of excitement of atoms.

Measurement results of the intermediate electron distribution function along the discharge axis agree in nonlocal rate with the calculations based on the solution of the kinetic equation of Boltzmann with a zero boundary condition on the anode.

Fast electrons are few, therefore for their research the probe technique is ineffective. Information on this part of electron distribution function is obtained by optical methods of a research, namely discharge radiation. It is revealed that the discharge luminescence is as much as possible on border of the layer — plasma. This results from the fact that excitement and ionization in the negative glow are caused by the fast electrons accelerated by the strong field in the cathode layer. Their energy much more surpasses energy of excitement and ionization of atoms. Ionization of atoms exponential grows in process of removal from the cathode, reaching a maximum near border the layer — plasma.

The results of optical and electrical measurements agree well. There is a good recurrence of the results.

Thus, the received experimental results are good in agreement with the model of the near-cathode regions of the glow discharge based on the nonlocal ionization of the gas by the electrons which have gained the energy in a cathode layer.

3. Ions energy distribution functions

Most of the main characteristics of glow discharges, and even their existence, are determined by processes in the near-electrode regions, primarily in the cathode region. It usually accounts for the main part of the discharge voltage. There are significant flows of charged and neutral active particles from the plasma to the cathode surface here. These streams can be used for surface treatment [9]. Therefore the research of the near-cathode region of the glow discharge is important not only for theoretical creation of model of the discharge, but has also important practical application. In this regard of particular interest are the ion fluxes to the cathode, as they mostly initiate different processes on its surface and also distribution of potential in near-cathode region.

For determine of the ion energy distribution function (IEDF) in the near-cathode region of the glow discharge the second discharge tube has been made and has been performed experimental research. The physical processes in the cathode region of a glow discharge were experimental investigated in a glass tube. The layout consists of two round planar electrodes: cathode and anode. Their diameters were 42 mm. The distance between the electrodes was 150 mm. The cathode at its center had a hole (diameter of 4 mm), which was covered by a grid. A flat analyzer-collector of diameter 4 mm was mounted against the hole. The anode had an opening through which a cylindrical probe of diameter 0.2 mm and length 3 mm was inserted into the discharge gap. The probe could be moved along the tube axis. The ion fluxes to the cathode were determined by measuring the collector characteristics (CVC) – dependence of the collector current on the potential difference between the collector and the cathode.

In the same way as during EEDF research, before carrying out measurements careful cleaning of a discharge tube was carried out by its warming up by current with repeated change of working gas and polarity of electrodes.

In all cases, the blocking voltage (the collector current is equal to zero) is much less than the total voltage across the tube, which in the first approximation can be taken as the value of the cathode fall. This means that the ion on the way from the plasma of the discharge to the cathode experiences multiple collisions. This is confirmed by the estimation of the mean free path of ions, which is smaller than the cathode fall. Therefore, the increase in pressure affects the maximum energy of ions reaching the cathode. This corresponds to the received collector characteristics. The change of the discharge current at a constant gas pressure does not affect the blocking voltage.

To calculate the distribution functions of the ions the voltage drop on the cathode layer was taken equal to the voltage across the discharge tube. The molecular nitrogen ions N_2^+ were considerate as the dominant ions.

$$f(\varepsilon)=\frac{I_{c(\varepsilon=0)}}{S_c\cdot e\cdot(1+\gamma)}\cdot\sqrt{\frac{M}{2\varepsilon}}\exp\left(-\frac{L_{sh}}{\lambda}\left(1-\sqrt{1-\frac{\varepsilon}{U\cdot e}}\right)\right)\cdot\frac{L_{sh}}{2\lambda\cdot U\cdot e\cdot\sqrt{1-\frac{\varepsilon}{U\cdot e}}} \quad (1)$$

where $I_{c(\varepsilon=0)}$ is the collector current at $U = 0$, S_c is the effective area of the hole in the cathode, γ – the coefficient of the ion-electron emission [1], M – mass of an ion, ε – energy, e – the ion charge, L_{sh} – the length of the cathode layer, λ – the mean free path determined by charge-exchange, U – the voltage drop across the cathode layer.

Calculations were performed for different ratios of the thickness of the cathode layer to the free path length. The comparison of the calculated and experimental functions allows to determine the optimal value of this ratio.

Figure 1 shows a comparison of an IEDF calculated by formula 1 with an experimental IEDF. The best agreement is observed with $L_{sh}/\lambda=4$. This value is in good agreement with the estimation made on the basis of the charge-exchange process [1]. The value L_{sh} was determined from the axial distribution of the probe floating potential in the tube.

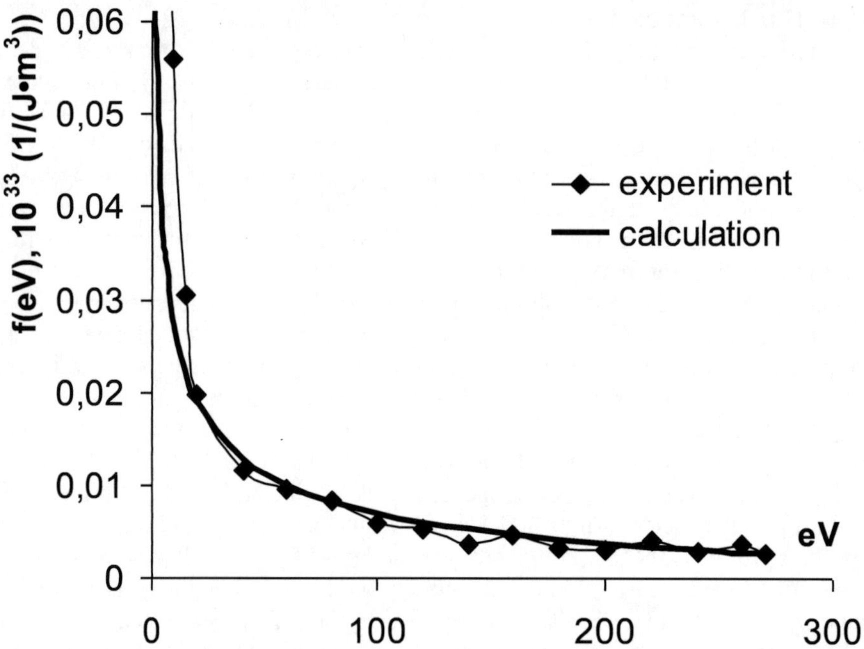

Figure 1. Comparison of the experimental IEDF (nitrogen, p = 0,3 Torr, I = 5 mA, U = 740 V) with that calculated by the formula (1).

References

[1] Raizer Yu P 2009 *Physics of gas discharge* (Moscow: Intellect)

[2] Golubovsky Yu B, Kudryavtsev A A, Nekuchaev V O, Porohova I A and Tsendin L D 2004 *The kinetics of electrons in nonequilibrium gas-discharge plasma* (St. Petersburg: St. Petersburg's University Press)

[3] Kudryavtsev A A, Smirnov A S and Tsendin L D 2010 *Physics of the glow discharge* (St. Petersburg: Publishers Lan)

[4] Granovsky V L 1971 *Electric current in gas. The established current* (Moscow: Nauka)

[5] Kolobov V I and Tsendin L D 1992 Analytic model of the cathode region of a short glow discharge in light gases *Phys. Rev.* A **46** 7837.

[6] Tsendin L D 2010 Nonlocal kinetics of electrons in gas-discharge plasma *Uspekhi Fizicheskikh Nauk* V 180 N 2 pp 139–164

[7] Shveygert V A and Shveygert I V 1989 To the theory of near-cathode region of the glow discharge *Thermophysics of high temperatures* V 27 N 1 pp 23–29

[8] Blinderman M S 1983 *Cathode parts of the glow discharge (the current state of a question)* (Moscow: Institut problem mechaniki Akademii nauk USSR, Preprint N 216)

[9] Michael A Lieberman and Allan J Lichtenberg 2005 Principles of plasma discharge and materials processing (A John Wiley & sons, INC Publication)

IOP Conf. Series: Journal of Physics: Conf. Series **1058** (2018) 012031 doi:10.1088/1742-6596/1058/1/012031

Extraction of neutral particles from a gas discharge

A A Rykov, M I Yurchenkov, A A Lisenkov

Department of electronic instruments and devices, Saint Petersburg Electrotechnical University "LETI", 197376, Saint Petersburg, Russia

E-mail: rikov247@gmail.com

Abstract. In this paper a version of the extraction of neutral particles from a gas discharge using a gas-distributing disk is proposed, which allows creating a pressure drop between the source volume and the working zone of the vacuum chamber.
The gas distributor ensures a uniform supply of gas to the ionization zone and resonant charge exchange, which will make it possible to achieve uniform combustion of the discharge in the entire volume of the source. Also the gas distributor allows to increase the speed of gas flow, due to pressure drop.
An increase in the rate of gas flow in the volume of the source makes it possible to increase the number of ionization events per unit of time. Consequently, it makes it possible to increase the number of ions generated in the gas discharge, that is, the ionization coefficient of the working gas increases. In turn, an increase in the number of ions generated in the discharge leads to an increase in the number of neutral atoms in the output beam.

1. The first section in your paper

The main task in the application of coatings on dielectric surfaces with the use of ion-plasma technological systems is the preliminary preparation of this surface [1-3]. For this purpose, to optimize the process of removing residual gases and activating the surface, a preheating technological operation is used before the coating is deposited. Then ion-plasma cleaning of dielectric substrates on the basis of a controlled gas-discharge source of fast neutral particles is used for final cleaning [4].

If there is a working gas in the source volume and a negative voltage on the main cathode (Udiss), a gas discharge is ignited, occupying the internal volume of the source. In the gas discharge, ions are generated, accelerated by the cathode voltage drop. Neutralization of charged particles occurs mainly in the reflection of ions from the main cathode. As a result of neutralization, an accelerated directed flux of neutral particles is formed.

To maintain the pressure difference between the input volume (1) and the gas source of the neutral particles (2), a gas distribution disk (4) was used (1 mm thick with different, radially located openings with a diameter of 1-2 mm). The total gas conductivity of the holes is an order of magnitude less than the conductivity of the inlet channel. The gas distributor allows to increase the gas flow rate by providing a pressure difference between the volume of the source (2) and the working volume (3) of the vacuum post (neutralizer 5 - dielectric 6). The distribution of the pressure of the working gas in the transportation area (3) decreases (figure 1).

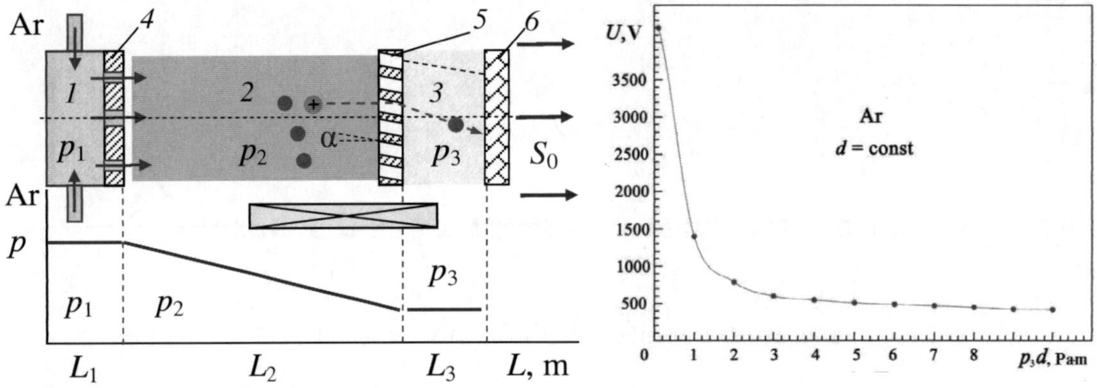

Figure 1. Diagram of variation of working gas pressure (p): 1 - area of argon supply to the volume of gas discharge source, limited by anode and distribution disk 4; 2 - gas discharge region; 5 - cathode with neutralization channels oriented at angle α; 3 - flow transport area neutral particles outside the source volume from the neutralizer 5 to the workpiece 6; S_0 - vacuum pumping out

To obtain the maximum ionization coefficient, the maximum pressure in the volume of the gas-discharge source was provided (pd = 10). On the other hand, for transporting neutral particle beams, the pressure in the working chamber must be minimal in order to minimize the probability of a collision between the flow of fast neutral particles and the atoms of the residual gas.

If the vacuum chamber is continuously evacuated and the gas is allowed to flow into the source, it is possible to create a pressure difference (p_1 - p_3) between the source volume and the vacuum chamber.

In the course of the source study, the dependences of the pressure difference between the vacuum chamber and the source volume were obtained using two different gas-distribution disk samples. The obtained dependences are shown in Figure 2.

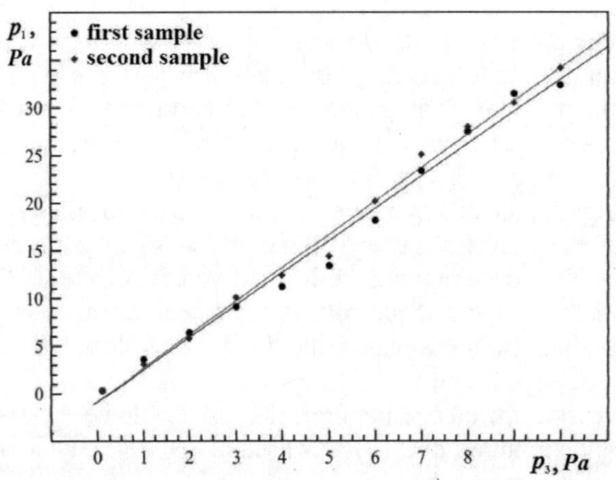

Figure 2. Pressure in the volume of the source from the pressure in the vacuum chamber

Thus, the use of the gas distribution disk allows creating a pressure difference between the source volume and the working area of the vacuum chamber: the pressure in the source is 3.5 times higher than the pressure in the vacuum chamber.

The difference in pressures in the gas supply zone p1 and the transport zone for the neutral particle fluxes p_3 can differ very significantly, with $p_1 > p_3$. The pressure in the ionization and resonance charge-exchange region p_2 changes linearly in the first approximation, and the average pressure value is calculated by the formula $\overline{p_2} = 0.5 * (p_3 + p_1)$. As already mentioned, the working gas supply zone is limited by a gas distribution element, which is a plate with a plurality of radial holes. The total gas conductivity of the holes is an order of magnitude smaller than the conductivity of the inlet openings. The gas distributor ensures a uniform supply of gas to the ionization zone and resonant charge exchange, which will make it possible to achieve uniform combustion of the discharge in the entire volume of the source. Also the gas distributor allows to increase the speed of gas flow, due to pressure drop p_1 and p_3.

An increase in the rate of gas flow in the volume of the source makes it possible to increase the number of ionization events per unit time, and consequently, it makes it possible to increase the number of ions generated in the gas discharge, that is, the ionization coefficient of the working gas increases. In turn, an increase in the number of ions generated in the discharge ultimately leads to an increase in the number of neutral atoms in the output beam.

References

[1] A. I. Kuzmichev, N. A. Babinov, A. A. Lisenkov, Plasma Emitters of Charged and Neutral Particles Sources," Kiev: Avers, 2016. p 181.
[2] Barchenko V.T., Babinov N.A., Lisenkov A.A, et al. Physics and Technology of Plasma Emission Systems. SPb: SPbETU «LETI», 2014. p 262 .
[3] Barchenko V.T., Veresov O.L., Lisenkov A.A., et al. Plasma emission systems with cold cathodes for ion-plasma technologies. SPb: SPbETU «LETI», 2012. p 208 .
[4] Lisenkov A. A., Fadeev A. S., Talanov A. S. et all. Cleaning of the dielectric surfaces us-ing a controlled gas-discharge source of fast neutral particles // IOP Conf. Series: Journal of Physics: Conf. Series 872 (2017) 012021.

Low-temperature plasma as a cleaning tool for pumping and compression pipes from asphalt, resin and paraffin deposits

R.T. Galiakbarov, A. D. Samigullin, A.T. Galiakbarov.

Kazan Federal University, 423800, 68/19 Mira avenue, Naberezhnye Chelny, Russia

samigullin86@mail.ru

Abstract: this article deals with results of experimental research such as volt-ampere characteristics, electric arc traverse speed and end plasmotron operation performance index for cleaning the oil-pipes and melting of side-wall layer of asphalt, resin and paraffin deposits.

1. Introduction.

Pumping and compression pipes are very important components in oil and gas extraction from underground layers. Their strength and reliability ensure the uninterrupted process of oil and gas extraction. The repeated multiple use of pipes after maintenance defines the oil and gas cost price since they are a part of its cost.

There are a lot of means of asphalt, resin and paraffin deposits removal in the world. First, chemical (inhibition, dilution) as least expensive. Besides the chemical method of pipes cleaning the mechanical method is used (scrapers, lowered on the wire and bars). Other methods such as dewaxing by wave action (acoustical, ultrasound, explosive), electromagnetic and magnetic (impact to fluid by magnetic fields), thermal (heating of pumping and compression pipes with fluid or steam, current, thermochemical dewaxing) and hydraulic (line pipes sections choke restriction to initiate gaseous phase discharge) are used more rarely due to their relative expense.

2. Method.

Removal of asphalt, resin and paraffin deposits with low-temperature plasma is a relatively new technology since plasma is quite an energy-intensive tool.

Plasmotron suggested in this research is an end plasmotron i.e. as opposed to linear plasmotrons. The low-temperature plasma will completely envelop the pipe surface without consuming its energy for environmental air warmup. Thus this type of plasmotron gives an opportunity to effectively remove the asphalt, resin and paraffin deposits from pumping and compression pipes without damage to the environment [1].

3. Tasks.

Main task at plasmotron development is getting optimal parameters of plasma flow and plasmotron operation modes which ensure efficiency of technological process and high life cycle of plasmotron structural units.

4. Research part.

To solve the task the test installation was developed. The main components of test installation are power supply, plasma gas delivery system, measurements system (digital oscilloscope, rotameter, thermocouples etc.) and end plasmotron directly.

End plasmotron on Figure 1 presents two parallel circular electrodes, 1 as a pipe fixed in body 2. Plasma gas is delivered to chamber 3 with following enveloping of electric arc 4 and exiting from nose as plasma. Electric arc 4 moves along the ring electrodes by force of its own magnetic field. Since the electric arc rotation speed is high, visually plasma exits the nose as a ring.

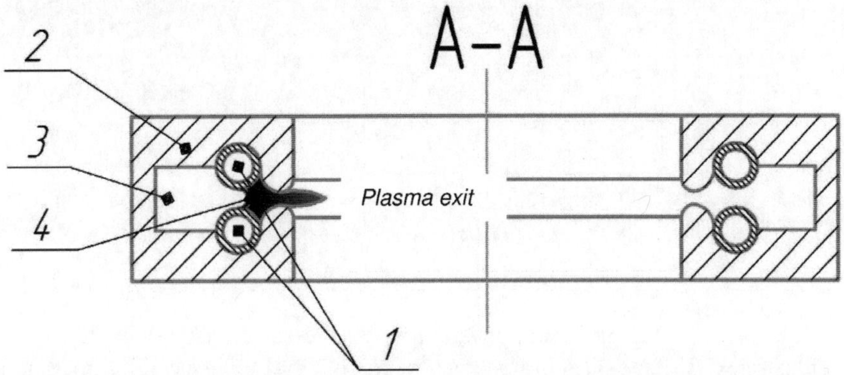

Figure 1. End plasmotron cross section. 1 – Ring electrodes, 2 – plasmotron body, 3 – plasma gas delivery chamber, 4 – electric arc.

Operational characteristics of end plasmotron in test installation for cleaning pipes from asphalt resin and paraffin deposits: diameter of ring electrodes is chosen depending on diameter of pipes to be cleaned from asphalt, resin and paraffin deposits. In this research the diameter of ring electrode was 90 mm and electrode pipes cross-section diameter 6 mm [2]. Interelectrode spacing varied from 3 to 6 mm. Plasma gas consumption in it is regulated from 1 g/sec to 2 g/sec. Cooling liquid consumption on cathode 0,0088-0,0185 kg/sec, at anode 0,0068 - 0,0054 kg/sec [3].

Main energetic characteristics of plasmotron are also the thermal efficiency of device and dependency of arc discharge voltage from current by variation of plasma gas consumption, electrodes diameter, interelectrode gap - voltage-current characteristic (VCC) of arcs.

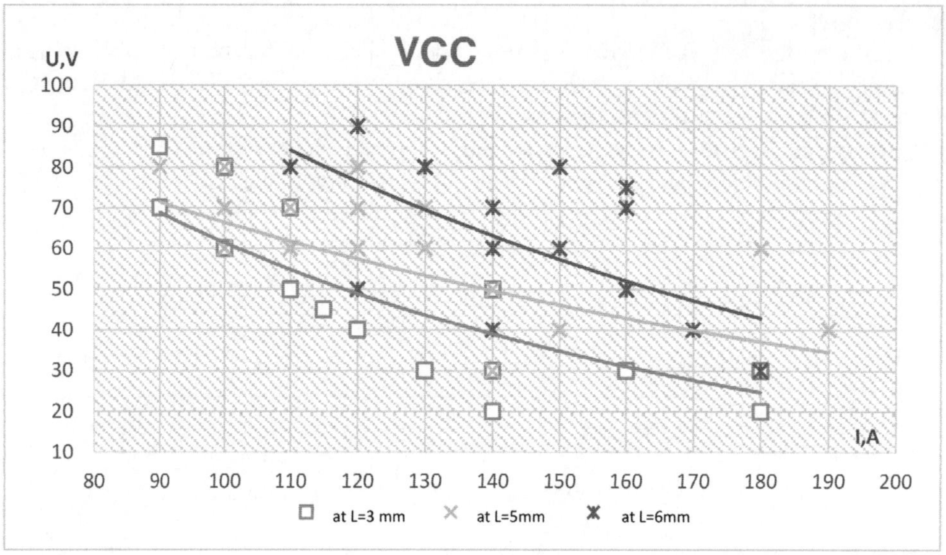

Figure 2. End plasmotron voltage-current characteristic.

Figure 2 illustrates voltage-current characteristic of end plasmotron operation at different interelectrode gaps. In plasmotron with self-centered discharge length of arc VCC has falling view. From analysis of arc VCC the choice of interelectrode gap is defined by many factors (required power, continuous operation resource etc.) for specific technological process [4].

Figure 3 illustrates arc speed graph of behavior depending on arc current at different interelectrode gaps.

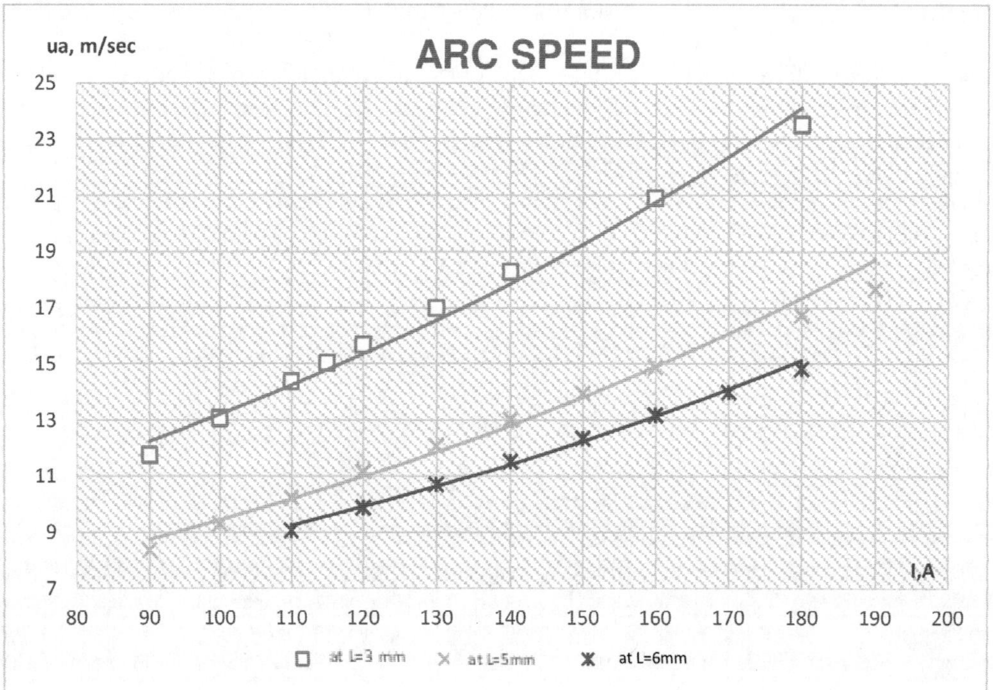

Figure 3. Arc speed graph of behavior depending on arc current at different interelectrode gaps

The graph shows that the highest arc speed happens at interelectrode gap 3 mm. Pic. 2 illustrates that arc voltage at 180 A and at interelectrode gap 3 mm is 40V [5].
Plasmotron efficiency was defined by calorimetric method. Geometry of chamber where plasma is located is designed in a way to decrease losses from emission to environment. Efficiency graph depending on the plasmotron power is illustrated on Figure 4, it is defined that its maximum value is achieved at interelectrode gap 3 mm, at U=70 V and I=110 A, and arc speed at given parameters is 14,5 m/sec. Consumption of cooling liquid for cathode was 18,5 g/sec and for anode 5,4 g/sec [6].

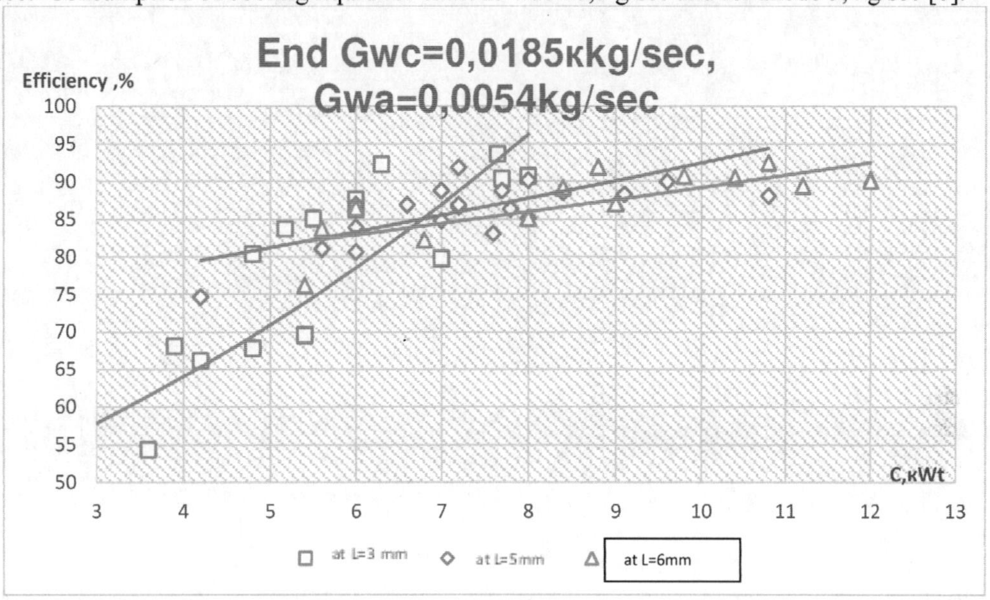

Figure 4. Efficiency definition graph depending on plasmotron power at different interelectrode gaps.

5. Conclusion.
Dependencies of output parameters of plasmotron from its geometrical and energetic parameters at input have been defined. Further research stage will consist of definition by experiment and calculations of effective end plasmotron movement speed along the cleaned pipe.

References
[1] A.D. Samigullin, A. T. Galiakbarov, R. T. Galiakbarov. Study of the petroleum schedules thermal cleaning process from asphalt, resin and paraffin deposits using low-temperature plasma//VII Conference on low temperature plasma in the processes of functional coating preparation. - 2016. - Vol.669, Is. - Art. №12017.
[2] A.D. Samigullin, A. R. Samigullina, A. D. Samigullin, A. T. Gabdrakhmanov. Localized thermal cleaning method for pumping and compression pipes from asphalt, resin and paraffin deposits using plasma. Journal of Physics: Conference Series, 789(1) doi:10.1088/1742-6596/789/1/012047.
[3] A.T. Gabdrakhmanov, I. H. Israphilov, A. T. Galiakbarov, A. D. Samigullin, A. T. Gabdrakhmanov, (2016). Improving the efficiency of plasma heat treatment of metals. Journal of Physics: Conference Series, 669(1) doi:10.1088/1742-6596/669/1/012014.
[4] A.T. Gabdrakhmanov, A. T. Galiakbarov, A. D. Samigullin, A. T. Gabdrakhmanov. Investigation of ring plasmotron for thermal purification of the dismantled pipes of an oil assortment from asphalt-resin-paraffin sediments. Journal of Physics: Conference Series, 789(1) doi:10.1088/1742-6596/789/1/012008.

[5] Khafizov A A, Shakirov Y I, Valiev R A, Valiev R I and Khafizova G M 2016 Study of thermal and electrical parameters of workpieces during spray coating by electrolytic plasma jet J. of Phys.: Conf. Ser. 669 012030.

[6] Khafizov A A, Valiev R I, Shakirov Yu I and Valiev R A 2014 Steel surface modification with plasma spraying electrothermal installation using a liquid electrode J. of Phys.: Conf. Ser. 567 012026.

Asphalt, resin and paraffin deposits from pumping and compression oil pipes removal process modeling

R.T. Galiakbarov, A. D. Samigullin.

Kazan Federal University, 423800, 68/19 Mira avenue, Naberezhnye Chelny, Russia

samigullin86@mail.ru

Abstract: This article deals with temperature calculation results on inner surface of oil pipes and melting of asphalt, resin and paraffin deposits side-wall layer.

1. Introduction.

Asphalt, resin and paraffin deposits, deposited on metal surfaces of field equipment, obstruct the oil extraction and complicate oil-field equipment operation. The main components of asphalt, resin and paraffin deposits are paraffin-naphthenic - and rarely paraffin – naphthenic - aromatic hydrocarbons, condensed in asphaltenic clusters, forming resin-asphaltenic colloids [1]. Asphaltenic associates significantly impact the wells waxing, since on one hand they do not allow paraffin – naphtenes to crystallize and fall out the flow, on the other hand – they initiate the waxing, forming large associates which then coagulate and deposit on pipes surfaces [2].

Research purpose. Due to different component portions in asphalt, resin and paraffin deposits composition the melting temperature and chemical solubility vary in large ranges. This problem requires research of oil pipes cleaning methods from asphalt, resin and paraffin deposits. The purpose of this research is a computer modelling of oil pipes cleaning process from asphalt, resin and paraffin deposits.

Basic data. Computer modelling was performed with CFD-resolvent. For modelling the oil pipe DU 70 was selected with wall thickness 5 mm and asphalt, resin and paraffin deposits thickness 20 mm, the cleaning process of which is described in research works [3] and [4].

Thermophysical properties of asphalt, resin and paraffin deposits which were defined during modelling [5]:

Dynamic viscosity	0.02 Pa·sec
Density	940.0 kg/m3
Latent heat of fusion	150000.0 J/kg
Hardening point	343.0 K
Melting temperature	343.0 K
Thermal conductivity	1.5 W/m·K
Specific thermal capacity	2900.0 J/kg·K
Pipe materials	steel NKT 20

2. Results.

One of the main factors for asphalt, resin and paraffin deposits removal processes is the melting temperature, which allows to evaluate the asphalt, resin and paraffin deposits mobility. For asphalt, resin and paraffin deposits melting the temperature 70 °C - 90°C shall be achieved. Continuous heating of cleaned piped starting from bottom part results in consecutive melting of side-wall layer of asphalt, resin and paraffin deposits along the cleaned pipe which ensures the free exit of generated gases and asphalt, resin and paraffin deposits melting products without causing explosions at gasification and temperature and thermal expansion of asphalt, resin and paraffin deposits side-wall layer in closed space. Melted side-wall layer of asphalt, resin and paraffin deposits can fulfil the role of lubricant at exit of not melted asphalt, resin and paraffin deposits hard mass from cleaned pipe under gravity.

To provide the free exit of generated gases and asphalt, resin and paraffin deposits melting products melting of at least 1 mm of asphalt, resin and paraffin deposits side-wall layer is required. The depth of melted layer of asphalt, resin and paraffin deposits is provided by coaxial movement speed change of circular plasmotron. After melting of side-wall layer with 1 m length the asphalt, resin and paraffin deposits hard mass is separated from main mass under gravity and slides from pipe to the receiving container. I.e. temperature not less than 70° shall be provided at depth 1 mm during passing the 1 m distance by circular plasmotron in 2 minutes [6].

Figure 1 illustrates the temperature distribution scalar scene at the beginning of pumping and compression pipe with asphalt, resin and paraffin deposits at the moment of removal of circular plasmotron 1 m from the beginning of pipe.

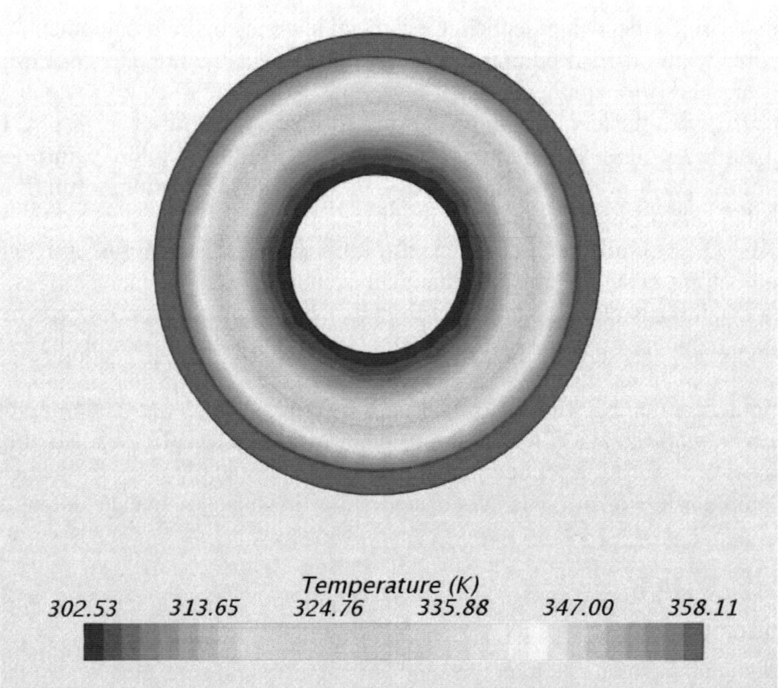

Figure 1. The temperature distribution scalar scene at the beginning of pumping and compression pipe with asphalt, resin and paraffin deposits at the moment of removal of circular plasmotron 1 m from the beginning of pipe.

Figure. 2 illustrates the asphalt, resin and paraffin deposits phase state volume concentration at longitudinal section of the pipe where we can see the asphalt, resin and paraffin deposits melted layer presented in red color.

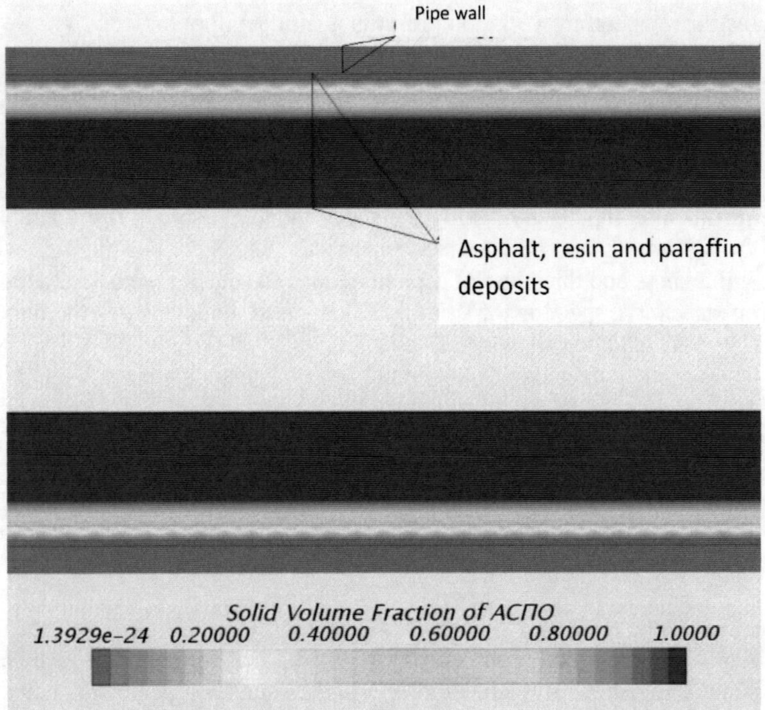

Figure.2. Scalar scene of the asphalt, resin and paraffin deposits phase state volume concentration at longitudinal section of the pipe (red – liquid phase, blue – hard phase).

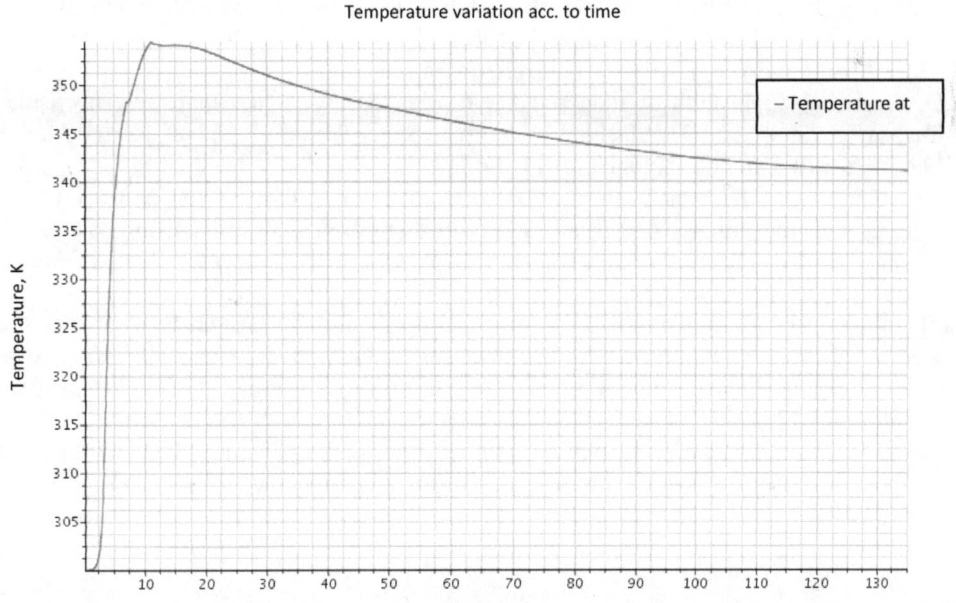

Figure. 3. The temperature variation graph at 1 mm depth point from inner wall of pumping and compression pipe in the asphalt, resin and paraffin deposits.

Figure. 3 illustrates the temperature variation graph in 1 mm depth point from inner wall of pumping and compression pipe in the asphalt, resin and paraffin deposits phase state volume concentration at longitudinal section of the pipe at the beginning of the pipe [7]. Guide line is set to 343 K (i.e. asphalt, resin and paraffin deposits melting temperature 70°C) [8].

3. Conclusion

According to the results of computer modelling the speed mode of asphalt, resin and paraffin deposits cleaning process from pumping and compression pipes was suggested. For the pipe with diameter 70 mm the 5 mm wall thickness and thickness of asphalt, resin and paraffin deposits 20 mm thickness, the circular plasmotron movement speed is 0.51 m/min. This speed mode ensures the removal of asphalt, resin and paraffin deposits main mass in one pass.

References

[1] Oliver C. Mullins, Andrew E. Pomerantz, A. Ballard Andrews, Julian Y. Zuo Asphaltenes Explained for the Nonchemist // Petrophysics: - 2015. - T. 56, № 3. — P. 266-275.

[2] Mansoori Ali. G. Remediation of asphaltene and other heavy organic deposits in oil wells and pipelines // Socar proceedings. — NIPI "Neftegas", 2010. — № 4. — P. 12-23. — ISSN 2218-68.

[3] A.D. Samigullin. Improvement of thermal cleaning process of oil pipes from asphalt, resin and paraffin deposits. A.D. Samigullin, A.T. Galiakbarov. In collected articles: Information technologies. Automation. Actualization and highly-qualified personnel preparation problem solution (ITAP-2015). International research/practice conference digest (distant form). Institute of Naberezhnye Chelny (branch) FGAOU VPO K(P)FU Informational-educational training center for engineering specialists. 2015. P. 54-59.

[4] Pat. 161034 Russian Federation, MPK B08B 9/02. Device for thermal cleaning of dismounted oil pipes from asphalt, resin and paraffin deposits/ A.D. Samigullin, A.T. Galiakbarov, , I. Kh. Israphilov, D.I. Israphilov, A.T. Gabdrakhmanov, applicant and patent holder – Federal FGAOU VO "Kazan (Privolzskiy) Federal University". - 2015134577/05; application 17.08.2015; published on 10.04.2016 Bulletin № 10.

[5] N.D. Shishkin. Application of inductive oil heating during its transportation from the North Caspian oilfields. N. D. Shishkin Astrakhan State Technical University reporter. Series: Marine facilities and technologies. 2011. № 3. P. 52-56.

[6] Samigullin A. D., Galiakbarov A. T., Galiakbarov R. T., Study of the petroleum schedules thermal cleaning process from asphalt, ressin and paraffin deposits using low-temperature plasma//VII Conference on low temperature plasma in the processes of functional coating preparation. - 2016. - Vol.669, Is. - Art. №12017.

[7] Samigullin, A. D., Samigullina, A. R., Samigullin, A. D., Gabdrakhmanov, A. T. Localized thermal cleaning method for pumping and compression pipes from asphalt, resin and paraffin deposits using plasma. Journal of Physics: Conference Series, 789(1) doi:10.1088/1742-6596/789/1/012047.

[8] Shakirova G Yu, Shakirov Yu I, Ilyin V I, Valiev R A, Drogaylova L N 2017 Determination of steel bar dispersed mass in electric discharge with alternative electrode. *J. of Phys.: Conf. Ser.* **789** 012050

Numerical modeling of the plasma processes taking into account the products of transfer processes from aqueous solution of sodium chloride being used as a cathode

N A Sirotkin[1], V A Titov[1] and S A Smirnov[2]

[1]G.A. Krestov Institute of Solution Chemistry RAS, Ivanovo, 153045, Russia
[2]Ivanovo State University of Chemistry and Technology, Ivanovo, 15300, Russia

E-mail: alexsad8@yandex.ru

Abstract. The modeling of the formation processes and further transformations of active particles in atmospheric pressure discharge plasma with a sodium chloride solution used as a cathode is performed. Experimental data on the composition of the gas phase and the physical parameters of plasma were used in the numerical simulation. It is determined that electron collisions with water molecules leads to a change in the form of the electron energy distribution function. As a result, the rates of the processes occurring under the action of the electron impact are altering. In addition, the transfer of water molecules to the gas phase leads to the appearance of new active particles, such as $OH^{\cdot}$, $HO_2^{\cdot}$, H_2O_2. It has been established that the products of the transfer of solutes influence on the balance of charged particles thus the electric field strength are reduced. This results in changing of the rate constants of the processes occurring in the plasma.

1. Introduction

Investigation of gas discharge plasma in contact or in liquids is one of the priority areas in plasma physics [1]. Atmospheric pressure discharges with liquid electrolyte electrodes are of interest as sources of active species for a variety of plasma-chemical applications one of which is the purification and disinfection of water and aqueous solutions [2, 3]. Plasma is the source of various active particles; some of them (for example, atoms of O, ozone, OH radicals) are formed in the gas phase followed by penetration into the solution and participation in the destruction of organic and other water pollutants. For estimation of processes mechanism in the liquid phase, it is necessary to know the active species concentrations in a gas phase and their change under alterations of discharge parameters. Such data are very limited with the exception of ozone concentration. Data on the containing of other active particles in the plasma zone of discharges with an electrolyte cathode are practically absent, which is due to the experimental difficulties in measuring the concentrations of active components. The concentration of hydroxyl radicals ($1.3 - 1.8\times10^{16}$ cm^{-3}) by the method of laser-induced fluorescence spectrometry at currents of discharge from 10 to 28 mA was determined in [4]. The concentration of oxygen atoms by the method of two-photon laser-induced fluorescence spectrometry using a nanosecond laser was measured in [5].

The large number of works devoted to plasma chemistry modeling was published in recent years [6–8]. In particular, the modeling of processes in the discharge of atmospheric pressure in air using the distilled water as a cathode was carry out in [6]. The concentrations of a range of active particles were obtained using the experimental values of the electric field strength in the plasma and the gas temperature. The authors used the value of the mole fraction of water equaled of 0.2%. Taking into account the transfer products in a parametric form, without relying on experimental measurements is a disadvantage of these theoretical calculations.

The aim of the investigation is to determine the composition of the gas phase above the electrolyte surface in discharge - electrolyte cathode system by means of the numerical simulation method using experimental data of physical parameters and transfer processes. The influence of transfer processes on the chemical processes in the plasma and the physical properties of the discharge will be established.

2. Experimental part

The numerical model included 233 reactions which describe the concentrations of the following species: O_2, $O_2(a^1\Delta_g)$, $O_2(b^1\Sigma_g^+)$, N_2, $N_2(A^3\Sigma_g^+)$, $N_2(a^1\Pi_g)$, $N_2(B^3\Sigma_u^-)$, $N_2(C^3\Pi_u)$, $O(^3P)$, $O(^1D)$, $O(^1S)$, O_2^+, O^-, O^+, $N(^4S)$, $N(^2P)$, $N(^2D)$, H_2O, $H_2O(^1A_1)$, $H_2O(^1B_1)$, H_2, NO, $NO(A^2\Sigma)$, NO^+, NO^-, H_2O^+, H^-, O_2^-, $Na(^3P)$, $Na(^4S)$, Na^+, OH, $OH(A^2\Sigma)$, OH^-, H, HO_2, O_3, HNO, H_2O_2, NO_3, N_2^+, H_3O^+, N_2O, Cl, Cl_2, HCl, NaO, NaO_2, NaNO, NaCl, Cl^-, Cl^+ and electrons. The system of equations of chemical kinetics was solved in the stationary approximation using the zero–dimensional model. In the calculation, the experimental values of the mol fractions of the electrolyte cathode components were used. The mol fraction of water molecules in the plasma is 1% and mol fractions of atoms of Na and Cl are 0.003 – 0.006% [9]. The temperature dependences for the rate constants of reactions of heavy particles were taken into account. The calculations were carried out for a gas temperature of 1600 K. The rate constants of processes interaction electrons with heavy particles were obtained using the electron energy distribution function founded by a numerical solution of the Boltzmann kinetic equation. The experimental values of the reduced electric field strength (*E/N*) were used in the calculations: $E/N = (2.2 – 3.0)\times10^{-16}$ V·cm^2 (water used as a electrolyte cathode), $E/N = (1.5 – 2.3)\times10^{-16}$ V·cm^2 (solution of NaCl used as an electrolyte cathode).

3. Results and discussions

The transfer processes alter the composition of the gas phase and it results in the changing cathode voltage drop and electric field strength in plasma [9]. The changes of the reduced electric field strength and the set of electron collisions in the plasma should affect the electron energy distribution function (EEDF). The calculations of EEDF with variation of mole fraction of H2O in the plasma showed that collisions with water molecules lead to a decrease in the average electron energy and, as a consequence, the rate constants of the threshold processes with electrons participation. The electron concentration is $(3 – 6)\times10^{12}$ cm^{-3} at a discharge current of 10 – 70 mA and it depends on the mol fraction of water. The change of the rate constants of processes and the concentration of electrons results in a decrease of the rates of processes occurring under the action of an electron impact. Thus, the rates of low-threshold processes reduce at several times, and the rate constants with high thresholds (ionization, dissociation) decrease in order of magnitude.

The numerical calculation of the plasma composition was carried out using the zero–dimensional model in the stationary approximation in order to estimate the concentrations of the main active particles. At the first stage, the concentrations of the main active components: $O(^3P)$, metastable electronically excited $O_2(a^1\Delta_g)$ and $O_2(b^1\Sigma_g^+)$, ozone were obtained by modeling processes in a system containing initially only N_2 and O_2 molecules. The calculations showed that the concentration of $O(^3P)$ increases with the discharge current and reaches ~ 3×10^{16} cm^{-3}, the concentration $O_2(a^1\Delta_g)$ does not exceed 2.6×10^{16} cm^{-3}, the ozone concentration is ~ $(3.2 – 37.5)\times10^{13}$ cm^{-3}. The nitrogen oxides are formed in the plasma: NO (~ (0.34 – 1.42) × 10^{16} cm^{-3}), NO_2 (~ $5\cdot10^{14}$ cm^{-3}), NO_3 (~ 10^{10} cm^{-3}) and N_2O (~ 10^{14} cm^{-3}).

Radicals OH, HO_2 and hydrogen peroxide are generated in the plasma when water molecules in the gas phase are taken into account. In this case, an additional process for the formation of $O(^3P)$ atom appears: OH + OH → O (^3P) + H_2O. It should be noted that the channel formation rate is commensurate with the rates of formation of atoms upon dissociation of O_2 molecules by electron impact and in collisions with electronically excited nitrogen molecules. The main channels for the formation of hydroxyl radicals in the plasma are the processes: $O(^3P) + H_2O \rightarrow OH + OH$ and the dissociative attachment of electrons to H_2O molecules ($H_2O + e \rightarrow H^- + OH$). Subsequent reactions with OH radicals lead to the formation of hydrogen peroxide ($OH + OH \rightarrow H_2O_2$), HO_2 radicals ($NO_2 + OH \rightarrow NO + HO_2$) and hydrogen atoms. In addition, the reactions affect the balance of nitrogen oxides in the plasma.

The discharge with the solution of sodium chloride used as cathode has low values of the reduced electric field strength in the plasma that is caused by the change in the balance of charged particles. The calculations showed that the ionization frequency of Na atoms exceeds the ionization frequency of the initial plasma components at a mol fraction of Na atoms in the gas phase $\chi > 2\times10^{-6}$. Assuming that chlorine is in the form of Cl_2 molecules in the plasma, the frequency of dissociative attachment of electrons to chlorine molecules ($Cl_2 + e = Cl^- + Cl$) is too small to affect significantly the electron balance (fig. 1).

In the discharge with a solution of sodium chloride, used as a cathode, the electric field strength is reduced. As a result, the form of the energy distribution function of electrons changes and, as a consequence, the rate constants of the threshold processes that occur under the action of electron impacts decrease. In the end, the concentrations of active particles decrease in the discharge. The concentrations of active plasma particles for different initial gas compositions are given in the table 1. The Na and Cl atoms as participants in chemical reactions practically do not affect the balance of the main active particles, because the small molar proportions of sodium and chlorine atoms in the plasma are too small.

Thus, three channels of influence of the transfer processes on the properties of the discharge can be defined. First, the balance of charged particles in the plasma is varied which leads to reduces the electric field strength. Second, the set of electronic collisions is altered, which results in a change in the form of the EEDF. Eventually, this leads to a change of the rate constants of processes involving electrons. Thirdly, there are new reactions of formation and loss of active particles. As a result, the concentrations of the all active components of the plasma are changed.

Table 1. The concentrations (cm^{-3}) of active plasma particles with different initial gas compositions

The initial components	$O(^3P)$, 10^{15}	$O_2(a^1\Delta_g)$, 10^{15}	O_3, 10^{13}	NO, 10^{15}	OH, 10^{14}	HO_2, 10^{11}	H_2O_2, 10^{12}
[a] $[N_2]:[O_2] = 0.79:0.21$	1.4÷30.7	25.9÷2.4	37.5÷3.2	14.2÷3.4			
[a] $[N_2]:[O_2]:[H_2O] = 0.785:0.205:0.01$	4.5÷8.7	9.4÷0.9	20.2÷3.4	10.3÷1.7	0.5÷5.5	5.0÷0.5	1.5÷5.0
[b] $[N_2]:[O_2]:[H_2O]:[Na]:[Cl] = 0.785:0.205:0.01:(3\text{-}6)10^{-5}$	3.4÷5.5	6.9÷0.2	2.4÷0.3	2.7÷0.4	0.7÷1.9	4.2÷0.5	0.8÷1.8

[a] $E/N = (2.2 - 3.0)\cdot10^{-16}$. [b] $E/N = (1.5 - 2.3)\cdot10^{-16}$ $V\cdot cm^2$

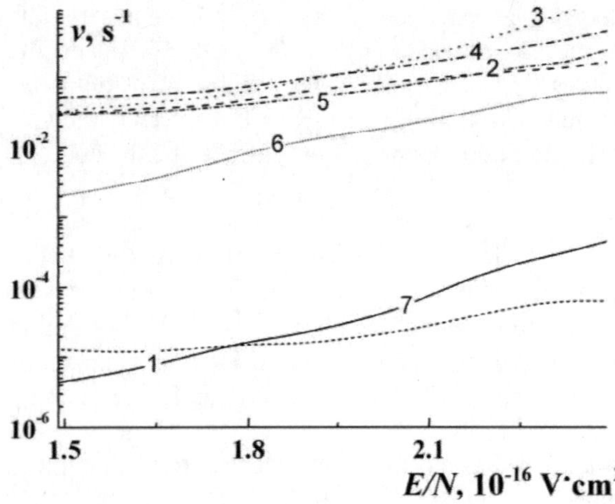

Figure 1. The frequencies of ionization and electron attachment of particles in the plasma:
1 – $N_2 + e = N_2^+ + 2e$ (ε=15.51 eV)
2 – $O_2 + e = O_2^+ + 2e$ (ε=12.08 eV)
3 – $Na + e = Na^+ + 2e$ (ε=5.14 eV)
4 – $H_2O + e = O^- + H_2$ (ε=4.95 eV)
5 – $H_2O + e = H^- + OH$ (ε=5.39 eV)
6 – $Cl^- + e = Cl + 2e$ (ε=3.42 eV)
7 – $Cl_2 + e = Cl^- + Cl$ (ε=2.51 eV)

Acknowledgments
The study was supported by Russian Foundation for Basic Research according to the research project 16–33–60061 mol_a_dk

References

[1] Bruggeman P J, Kushner M J, Locke B R, et.al. 2016 *Plasma Sources Sci. Technol.* **25** 053002
[2] Jiang B, Zheng J, Qiu S, Wu M, et.al. 2014 *Chem. Eng. J.* **236** 348
[3] Du C M, Wang J, Zhang L, et. al. 2012 *New J. Phys.* **14** 013010
[4] Nikiforov A, Xiong Q, Britun N, Snyders R, Lu X P, Leys C. 2011 *Appl. Phys. Express*. **4** 21
[5] Zhang S, Van Gessel A F H, Van Grootel S C, Bruggeman P J 2014 *Plasma Sources Sci. Technol.* **23** 025012
[6] Bobkova E, Smirnov S, Zalipaeva Y, Rybkin V. 2014 *Plasma Chem. Plasma Process*. **34** 721
[7] Matsui Y, Takeushi N, Sasaki K, Hayashi R, Yasuoka K 2011 *Plasma Sources Sci. Technol.* **20** 034015
[8] Mirakami T, Niemi K, Gans T, O'Connel D, Graham W G 2013 *Plasma Sources Sci. Technol.* **22** 01503
[9] Sirotkin N A, Titov V A. 2017 *Plasma Chem. Plasma Process*. **37** 1475

IOP Conf. Series: Journal of Physics: Conf. Series **1058** (2018) 012035 doi:10.1088/1742-6596/1058/1/012035

Research of the influence of the geometry of the discharge chamber on the characteristics of the arc plasmatron.

G Yu Dautov[1], N F Kashapov[2], I G Dautov[1,2] and A O Sofronitskiy[1]

[1] Kazan National Research Technical University named by A.N.Tupolev, 420111, K.Marx str., 10, Kazan, Russian Federation

[2] Kazan Federal University, Institute of Engineering, 420008, Kremlevskaya St. 18, Kazan, Russian Federation

E-mail: artempic8@mail.ru

Abstract. This work shows the possibility of a significant effect on the characteristics of the plasma torch by changing the diameter of the arc chamber along its length.

1. Introduction

Despite the external simplicity of the design of the plasma torch, complex physical processes take place in it: ionization of atoms and molecules, recombination of charged particles, emission and absorption of photons, interaction of charged particles with electric and magnetic fields, transport phenomena (thermal conductivity, viscosity, diffusion), electron emission by cathode and erosion of electrodes.

2. Plasmatron circuit

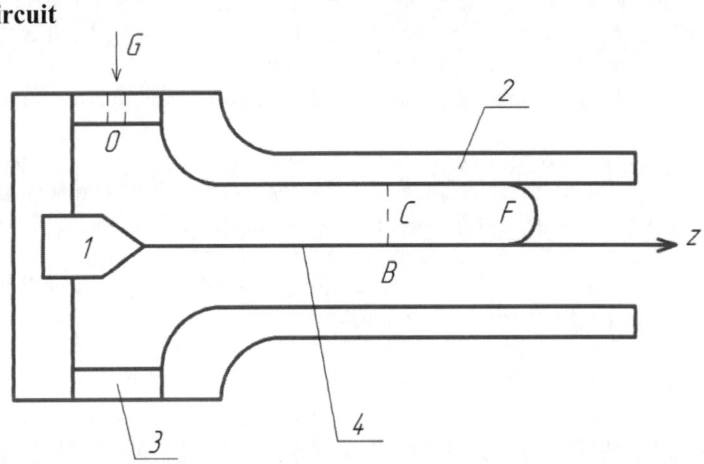

Figure 1. Scheme of single-chamber plasmatron with vortex stabilization of arc.

Figure 1 shows a diagram of a single-chamber plasmatron with vortex stabilization of the arc. The arc chamber is formed by the cathode 1, the anode 2, the insulator 3, which is simultaneously a device for introducing the plasma-forming gas G. The gas inlet G through the tangential hole creates a vortex motion in the arc chamber, which stabilizes the position of the arc 4 on the axis of the plasma torch and rotates the anode spot along the surface of the electrode 2 Under the influence of the axial component of the gas flow velocity, the radial portion of the arc moves along the inner surface of the arc chamber. This increases the length of the arc and the voltage on it. Such an increase in the length of the arc and voltage is limited by the shunting process - an electric breakdown between the positive column of the arc and the anode. along the BC line, the portion of the arc BF dies, the portion BC moves in the z-axis direction to the BF position and again a breakdown occurs. As a result, pulsations of current and arc voltage arise [1-4]. As the current increases, the breakdown voltage decreases and therefore the arc length and voltage also decrease. Therefore, as the current rises, the plasma temperature rises very little, and the efficiency of the plasma torch decreases substantially. These disadvantages can be cleared by installing between the cathode 1 and the anode 2 a sectioned interelectrode insert (SII) 3, as shown in Figure. 2. The SII sections are isolated from each other and the plasma-forming gas G_i is fed through the gaps between them. The rest of the plasma-forming gas G_o enters the arc chamber through a tangential opening in the $z=0$. The characteristics of such plasmatrons at $R=const$ are investigated in [1-5].

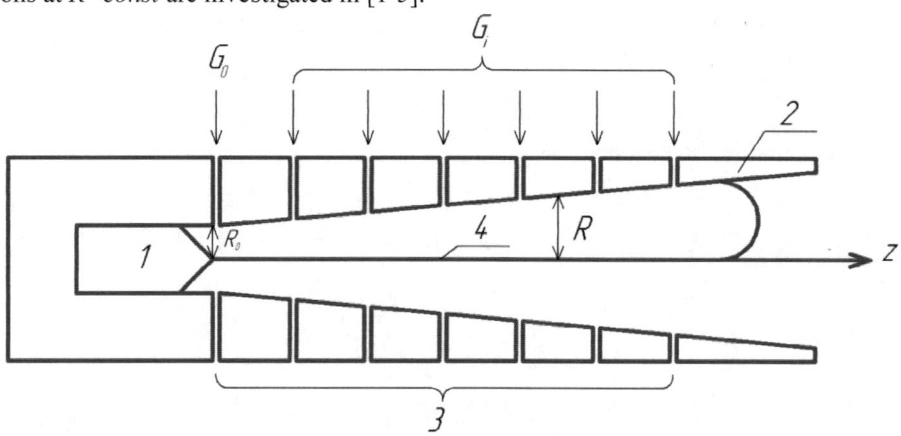

Figure 2. Plasmatron with variable radius of an arc chamber.

In this paper, the characteristics of the plasma torch are investigated, the radius of the arc chamber of which along the z axis varies linearly according to the linear law $R=R_0\ (1 + k_l lx)$, where $lx = z$, l is the SII length (Figure. 2). In the case $k_l>0$, the arc chamber expands along the z axis, and in the case of $k_l<0$ it narrows along the z axis. The experiments were carried out in the range of air flow rate $G=(1\text{-}4)\ 10\text{-}3\ kg\ /\ s$, current $I=(100\text{-}400)\ A$ and at atmospheric pressure at the exit from the arc chamber.

When the radius of the arc chamber does not change along its length, the heat flux through the wall of the arc chamber q in the direction of the z axis only increases [1-4]. Experiments have shown that in the expanding arc chamber the value of q in the direction of motion of the plasma first increases. In the cross section, x_m reaches a maximum and at large x decreases. The numerical value of x_m depends on the gas flow rate and current intensity. The decrease of q for large x plays a very large role, since this leads to an increase in the efficiency of the plasma torch.

3. Distributions of mass average enthalpy of plasma

Figure 3 shows the distributions of the average mass enthalpy of the plasma h_c along the axis of the arc chamber at $G_0 = 3.2\ 10^{-4}\ kg\ s^{-1}$, $G_i = 1.7\ 10^{-4}\ kg^{-1}$. $k_l\ l = 5.25$. In the initial section, h_c increases sharply, reaches a maximum, and then decreases very slowly. This affects the formation of the current-voltage characteristic of the arc, the distribution of the potential and the electric field strength in the arc chamber.

The distribution of the potential φ along the length of the arc was determined by measuring the potentials of the sections. When the plasmatron is directly polarized, electrode 1 in figure 2 is a cathode, and the electrode is an 2-anode. In this case, the section potential is equal to the arc potential in the middle section of the arc [6]. Knowing the potential distribution, the electric field strength can be determined from equation $\mathbf{E} = -grad\ \varphi$.

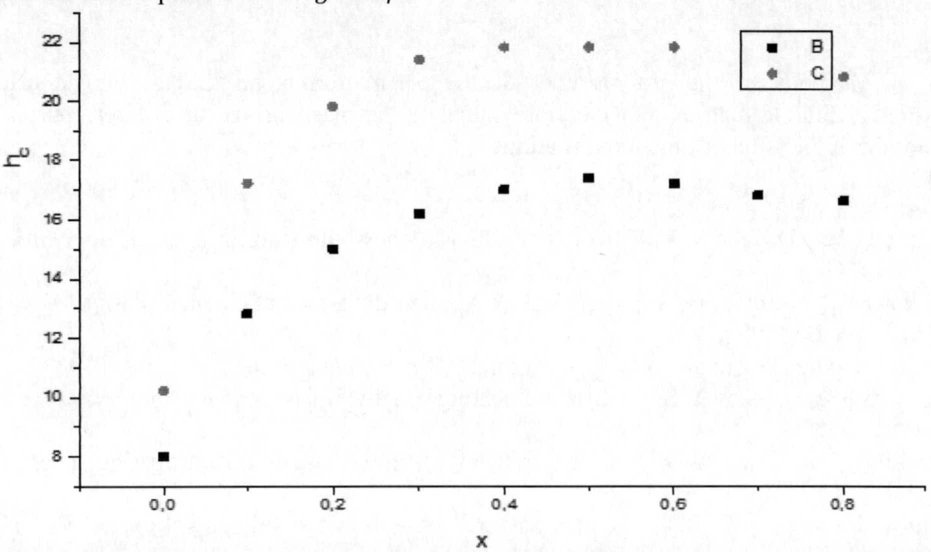

Figure 3. The distributions of the average mass enthalpy of plasma at currents of 150 A (curve B) and 200 A (graph C).

4. Distribution of electric field strength

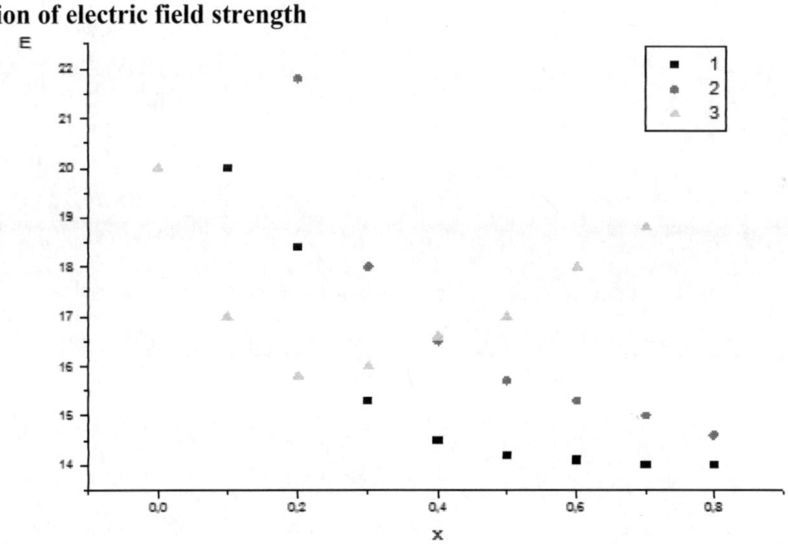

Figure 4. The distribution of the electric field strength along the axis of the arc chamber. *1. k_1=0, G=1.9 10^{-3} kg/s, G_0=0.2 10^{-3} kg/s, l=0.105 m, R=0.005 m. 2. k_1=30 m^{-1}, G=1.9 10^{-3} kg/s, G_0=0.27 10^{-3} kg/with 1_1=0.18 m, R_0 = 0.3 cm, R_1 = 0.75 cm 3. k_1=-6 m^{-1} G=0.8 10^{-3} kg/s, G_0 = 0.2 10^{-3} kg/s, l=0.105 m, R_0=0.0075 m.*

Figure 4 shows the distribution of the electric field strength in arc chambers of various geometries. The point 1 corresponds to a cylindrical arc chamber whose inner radius is 0.005 m nominal. In this case, the temperature and electrical conductivity of the plasma in the direction of its motion increase and therefore the electric field strength. Points 2 correspond to the strength of the electric field in the expanding arc chamber. As can be seen, at the same expense in the case of an expanding arc voltage.

The points 3 correspond to the converging arc chamber. In this case, the electric field strength in the initial section is less than in the first cases. However, at the end of the arc chamber, it is greater than in the two previous cases.

5. Results

Changing the diameter of the arc chamber of the plasmatron along the length can affect the characteristics of the plasmatron, and therefore calculate the plasmatrons for different purposes, for example, for the application of functional coatings [7].

References

[1] Dautov G Yu, Dzyuba V L, Karp I N. Plasma torches with stabilized arc. *Kiev: Nauk.dumka*, 1984.-168 p.

[2] Zhukov M F, Koroteev A S, Uryukov B A Applied dynamics of thermal plasma. *Novosibirsk, Nauka,* 1975, 298 p

[3] Yasko O I. The electric arc in the plasma torch. *Minsk: Science and Technology*, 1977, 151 p.

[4] Yurevich F B, Kulikov V S. Electric arc heating of gas. *Minsk: Science and Technology*, 1973, 189 p.

[5] Mustafin G M. Characteristics of a stabilized arc in a channel with a distributed gas supply. *PMTF*, 1968, no. 4, p. 124-129

[6] Dautov G Yu, Sazonov M I. The electric field strength in a stabilized vortex arc. *PMTF*, 1967, no. 4, p. 127-131

[7] Sadikov K G, Sofronitskiy A O and Dautov I G. Functional plasma sprayed coatings on magnesium ceramic substrates. 2017 *Journal of Physics: Conference Series,* Volume **789**, conference 1, 012043. doi:10.1088/1742-6596/789/1/012043

Thermodynamic analysis of the plasma-chemical conversion of polymers into synthesis gas

A Kh Tazmeev and R N Tazmeeva

Kazan Federal University, Naberezhnye Chelny Institute, 68/19 Mira Street, Naberezhnye Chelny, 423810, Russian Federation

E-mail: tazmeev@gmail.com

Abstract. The composition of the gas formed during the decomposition of polyethylene and polyethylene terephthalate in steam-water plasma was calculated. The database includes thermodynamic information of 32 individual substances that can be formed from carbon, hydrogen and oxygen atoms. The calculated temperature range was chosen from 1000 to 2250 K. The conditions for achieving optimal regimes for the conversion of polymers into synthesis gas was analyzed.

1. Introduction

The world production of polymer materials is increasing every year. In the end, all polymer products are converted into consumer waste and pollute the environment. One of the ways of processing waste polymeric materials is plasma-chemical conversion to synthesis gas. The most effective version of the energy carrier for plasma technologies, including plasma-chemical conversion of polymer waste into synthesis gas, is steam-water plasma [1-11]. It is possible to note other positive effects accompanying the practical application of steam-water plasma [12-15].

Determination of the composition of gases formed during the plasma-chemical decomposition of polymers is extremely difficult. In the first approximation, neglecting of the kinetics of plasma-chemical reactions, one can perform a calculation for the case of thermodynamic equilibrium. Despite the conventionality, such a calculation is widely used in the analysis of high-temperature processes of conversion of hydrocarbon raw materials [16]. In this work, a technique developed for inorganic raw materials was used [17]. The validity of this approach is based on the fact that the current structure firstly, is high-temperature; secondly, homogeneous, consisting only of gas components. Condensed phases, i.e. solid or liquid substances are not considered. At the same time, in the composition of which initial substances (organic or inorganic) there were chemical elements from which individual components are formed in the gas phase, which has no fundamental importance.

2. Mathematical model and numerical calculation

When phase and chemical equilibria are established in any closed system, entropy increases. If internal heat sources operate in the system, the entropy increases until the sources are fully exhaustion. Thus, in the equilibrium state, the entropy of a closed system is always maximal. This is the principle of maximum entropy, on which the technique is based [17].

It should be noted that the system becomes equilibrium only if mechanical and energy equilibria with the environment are achieved, i.e. only if the heat is not supplied from the outside and is not diverted, and the system does not perform mechanical work. Under these conditions, the internal energy is kept constant, and it can not be changed by any phase or chemical transformations within the system itself. Thus, in the absence of internal heat sources for the equilibrium state

$$dS = 0. \tag{1}$$

This condition (1) is the main one in compiling a mathematical model for the maximum entropy principle.

$$S = \sum_{i=1}^{k}\left[S_i^{\circ}(T) - R\ln\frac{RT}{V}n_i\right]n_i. \tag{2}$$

The extremum of this function (2) is found at observance:
- the law of conservation of mass of chemical elements

$$\sum_{i=1}^{k}\nu_{ji}n_i - b_j = 0, \tag{3}$$

- electroneutrality of the mixture

$$\sum_{i=1}^{k}\nu_{ei}n_i = 0, \tag{4}$$

- constancy of volume

$$V = const, \tag{5}$$

- the invariance of internal energy

$$\sum_{i=1}^{k}U_i(T)n_i = const \tag{6}$$

and also under the condition that the steam-gas mixture is described by the equation of state of an ideal gas

$$pV - RT\sum_{i=1}^{k}n_i = 0. \tag{7}$$

Here k – is the number of components of the mixture; $S_i^0(T)$ – the entropy of the i component at the standard pressure p_0 = 0.1013 MPa; R – universal gas constant; n_i – specific molar content of i component (mol/kg); b_j – molar content of the j of the chemical element in 1 kg of working fluid (mol/rg); ν_{ji} – number of atoms j of chemical element in I component (stoichiometric coefficient); ν_{ei} – the multiplicity of ionization of i component; $U_i(T)$ – molar internal energy of i -components.

For numerical calculations the program given in the appendix to [17] was used with small changes. The thermodynamic properties of individual substances are taken from fundamental reference books [18-19]. The database includes thermodynamic information for the following substances: atomic oxygen O, positive ion of atomic oxygen O^+, negative ion of atomic oxygen O^-, molecular oxygen O_2, positive ion of molecular oxygen O_2^+, negative ion of molecular oxygen O_2^-, ozone O_3, atomic hydrogen H, positive ion of atomic hydrogen H^+, negative ion of atomic hydrogen H^-, molecular hydrogen H_2, positive ion of molecular hydrogen H_2^+, positive hydrogen ion H_3^+, hydroxyl OH, positive hydroxyl ion OH^+, negative hydroxyl ion OH^-, hydrogen dioxide HO_2, negative ion of hydrogen dioxide HO_2^-, steam-water H_2O, positive ion of steam-water H_2O^+, hydrogen peroxide H_2O_2, positive ion of trihydrogen oxide H_3O^+, condensed carbon C^*, carbon in gas phase C, positive carbon ion C^+, negative carbon ion C^-, carbon monoxide CO, positive carbon monoxide CO^+, carbon dioxide CO_2, positive carbon dioxide CO_2^+, dicarbone oxide C_2O, methine CH, methylene CH_2, methyl CH_3, methane CH_4, acetylene C_2H_2, ethylene C_2H_4, ethyl C_2H_5, ethane C_2H_6, formyl HCO, formaldehyde H_2CO and methanol CH_3OH.

3. Results of calculations and their analysis

In the actual process, the products of the plasma-chemical reactions pass through the hardening chamber, where the steam-water condenses. In the hardening chamber there is also the formation of molecular gases from atomic gases and the combination of hydroxyl groups into molecules:

$$O + O = O_2; \quad (8)$$

$$H + H = H_2; \quad (9)$$

$$2OH + 2OH = 2H_2O + O_2. \quad (10)$$

Due to these processes (8) - (10), the molar composition of the steam-gas mixture significantly changes. At the exit, the "dry" gas consists only of non-condensed phases of individual substances.

The "dry" gas contains saturated steam-water and electrolysis gases: hydrogen and oxygen. As shown estimated calculations by Faraday's law , at a discharge current of I = 10 A the maximum yield of electrolysis gases does not exceed 0,3 l/min. According to the data presented in [20], the volumetric content of saturated steam at T = 300 K is ~ 3,5 %. Thus, for most regimes of the plasmochemical process in the volume of the "dry" gas, steam-water and electrolysis gases are relatively small, and in the design model they can be neglected. In this paper, it is so done.

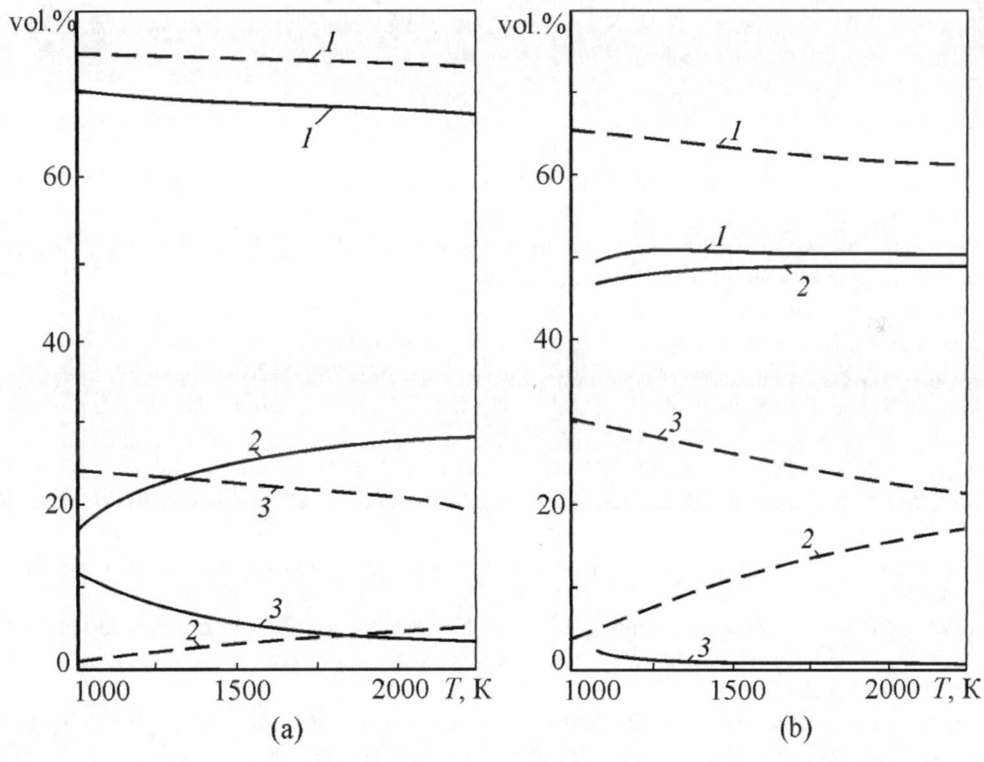

Figure 1. The volumetric content of the main components with the composition of synthesis gas. Raw material: (a) - polyethylene (PE); (b) - PET. The ratio of "raw materials: steam-water": solid lines - 1: 5; the dotted lines are– 1:100. *1* – H_2; *2* – CO; *3* – CO_2.

In the case of polyethylene terephthalate (PET) raw materials, the mass content of the chemical element C in the original composition is reduced by 10%. This accounts for the formation of a solid residue, which mainly consists of carbon [5, 6].

In the figure 1 shows the graphs based on the results of calculations. As can be seen, the composition of gas synthesis varies considerably with increasing temperature. In the case of raw materials from PE, a more significant change in the composition of the synthesis gas takes place. Apparently, such a regularity is due to the fact that PE in steam-water plasma is almost completely converted to gas [5, 6].

4. Conclusions

1. With a decrease in the molar content of carbon in the original composition of the working fluid (PE + steam or PET + steam), the volumetric yield of the "dry" gas is reduced.

2. To obtain final products of plasma-chemical decomposition only in the gas phase, the amount of steam-water should be in excess relative to the amount of raw materials from the polymers, and in the case of PET this excess should be higher.

3. With increasing excess steam, yield of CO_2 and H_2 increases, and the volumetric content of CO is reduced.

4. To reduce the ballast gas of CO_2 and increase the useful component of CO it is necessary to raise the temperature in the reaction zone. The optimum range is 1500 ÷ 2000 K.

References

1. Altovsky G S, Bernadiner M N and Ivanov V V 2011 *Ecology ahd Industry of Russia* No. 2 8
2. Gudim Y A and Golubev A A 2009 *Ecology ahd Industry of Russia* No. 2 4
3. Artemov A V, Pereslavsev A V, Krutyakov Y A, Voschinin C A, Kudrinsky A A, Bulba V A and Ostryi I I 2011 *Ecology ahd Industry of Russia* No. 10 18
4. Gaisin A F and Tazmeev K K 2005 *High Temperature* **43** (6) 810
5. Miftachov M N, Tazmeev A Kh, Tazmeev Kh K and Fridland S V 2006 *Journal of Engineering Physics and Thermophysics* **79** (3) 532
6. Fridland S V, Tazmeev A Kh and Miftachov M N 2006 *Vestnik Kazanskogo technologicheskogo universiteta* No. 6 10
7. Tazmeev A Kh, Fridland S V and Miftachov M N 2006 *Vestnik Kazanskogo technologicheskogo universiteta* No. 6 43
8. Tazmeeva R N, Ziganshin R R and Tazmeev Kh K 2008 *Vestnik mashinostroeniya* No. 10 87
9. Tazmeev Kh K, Arslanov I M and Tazmeev G Kh 2014 *J. Phys.: Conf. Ser.* **567** 012001
10. Tazmeev A K and Tazmeeva R N 2017 *J. Phys.: Conf. Ser.* **789** 012058
11. Gibadullina G R, Tazmeev A H. and Tazmeeva R N 2015 *International Journal of Applied Engineering Research* 2015 **10** (24) 45015
12. Khafizov A A, Valiev R I, Shakirov Yu I and Valiev R A 2014 *J. Phys.: Con. Ser.* **567** 012026
13. Shakirov Yu I, Valiev R I, Khafizov A A, Valiev R A and Khakimov R G 2016 J. Phys.: Con. Ser. **669** 012064
14. Kashapov L N, Kashapov N F, Kashapov R N and Denisov D G 2016 *J. Phys.: Conf. Ser.* **669** 012029
15. Tazmeev G K, Timerkaev B A and Tazmeev K K 2017 *Plasma Physics Reports* **43** (7) 771
16. Suris A L 1985 *Termodinamika vysokotemperaturnykh protsessov. Spravochnik* (Moscow: Metallurgiya)
17. Vatolin N A, Moiseyev G K and Trusov B G 1994 *Termodinamicheskoye modelirovaniye v vysokotemperaturnykh neorganicheskikh sistemakh* (Moscow: Metallurgiya)
18. *Termodinamicheskiye svoystva individual'nykh veshchestv. Spravochnoye izdaniye v 4-kh tomakh* ed. L V Gurvich , I V Veyts, V A Medvedev i dr. 1978 T **1** Kniga 2 (Moscow: Nauka)
19. *Termodinamicheskiye svoystva individual'nykh veshchestv. Spravochnoye izdaniye v 4-kh tomakh* ed. L V Gurvich , I V Veyts, V A Medvedev i dr. 1979 T **2** Kniga 2 (Moscow: Nauka)
20. *Fizicheskiye velichiny. Spravochnik* ed. I S Grigor'yev and Ye Z Meylikhov 1991 (Moskow: ergoatomizdat)

IOP Conf. Series: Journal of Physics: Conf. Series **1058** (2018) 012037 doi:10.1088/1742-6596/1058/1/012037

About the nature of electrical conductivity a gas discharge plasma with a water-solution cathode

G K Tazmeev[1], B A Timerkaev[1] and K K Tazmeev[2]

[1]Kazan National Research Technical University named after A.N. Tupolev, 10 K. Marx str., Kazan, 420111, Russian Federation
[2]Kazan Federal University, Naberezhnye Chelny Institute, 68/19 Mira Street, Naberezhnye Chelny, 423810, Russian Federation

E-mail: tazmeevg@mail.ru

Abstract. A gas discharge between an aqueous solution of sodium chloride and a copper electrode – anode in air at atmospheric pressure was studied. It was been shown that the electrical conductivity of a gas discharge plasma with a water-solution cathode is predominantly ionic.

1. Introduction
In recent decades, studies of physicochemical processes in gas discharges of atmospheric pressure in contact with a liquid have been actively conducted [1]. The range of possibilities for practical applications of plasma of such discharges is continuously expanding. Plasma sources, in the use of liquid as electrodes, are promising for water purification, initiation of chemical reactions in solutions, and many other practical applications [2-7]. However, understanding the nature of the complex phenomena that form processes in a discharge with an intensely evaporating liquid electrode, including a water-solution cathode, at present remains extremely limited. In particular, the nature of the electrical conductivity of the plasma is not fully disclosed. As is known, a heated metallic cathode emits electrons. Thermionic emission occurs. There are no free electrons in the water. Therefore, in the case of the water-solution cathode, electron emission is practically absent. The main current carriers in aqueous solutions are cations and anions. The electrical conductivity of the aqueous solution is ionic. What is the electrical conductivity of the plasma above the water-solution cathode? What are the mechanisms of the formation of current carriers in it? This work is aimed at studying these issues.

2. Experiment
In figure 1 schematically shows a cathode assembly of a gas discharge device. It consists of a small container *1,* in which a metal rod *2* is mounted.

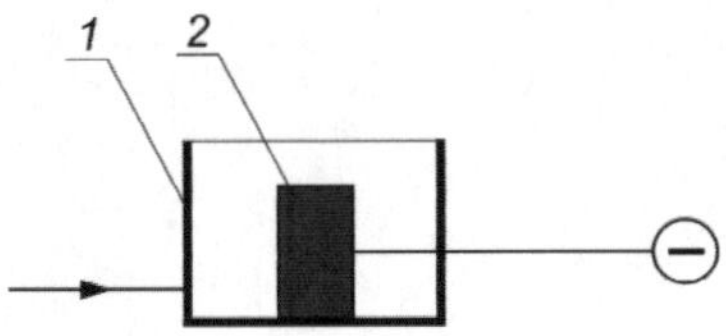

Figure 1. The cathode assembly. *1* – cylindrical container; *2* – metal rod.

After ignition of the gas discharge, an aqueous solution was supplied to container *1*. The liquid level was rose and the metal rod is immersed in an aqueous solution. Thus, a smooth transition from a metallic cathode (thermionic emission) to a water-solution cathode (without electron emission) was carried out.

The characteristics of a gas discharge with a water-solution cathode were investigated in an experimental setup, a detailed description of which is given in [8].

High-speed video shooting was made by the Photron FASTCAM SA4 camera. Its technical characteristics make it possible to study in detail the processes in systems with a liquid medium [9].

3. The experimental results and their analysis

In figure 2 shows instant photographs of two types of gas discharge. To the left is a gas discharge with thermionic cathode. It burns between two copper electrodes in the air. On the right is a gas discharge with a water-solution cathode.

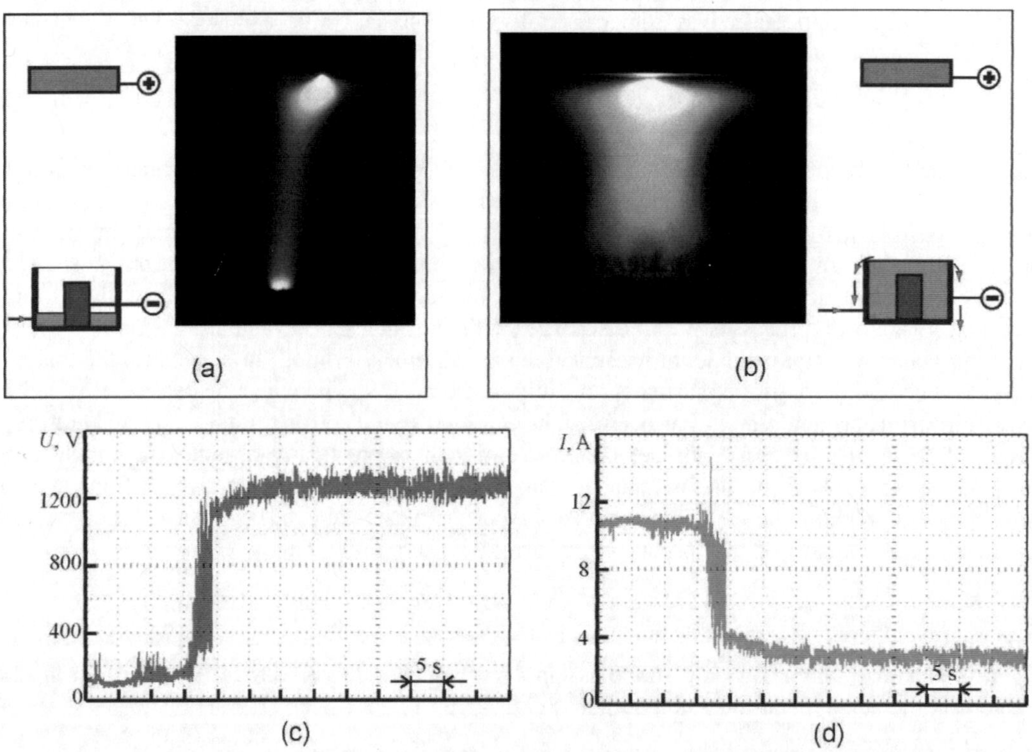

Figure 2. Transition from a thermionic cathode to a water-solution cathode. Cathode: (a) – copper rod with a diameter of 10 mm; (b) – an aqueous solution of sodium chloride with a concentration of 0.5% by mass. The interelectrode distance is 5 cm. The exposure of photos is 0.2 ms. (c) and (d) are oscillograms of voltage and current.

The transition from one type of gas discharge to another was carried out without changes in the electric power supply system. However, as can be seen from the photos, the cathode spot has disappeared. The binding of the discharge to the cathode expanded and covered a relatively large area on the surface of the aqueous solution. The plasma column above the cathode also expanded. Anode binding of the discharge remained unchanged. In both cases it pointwise.

As can be seen from the presented oscillograms, changes in the voltage and current of the discharge was abruptly occurred. The voltage increased by more than an order of magnitude: from 100 to 1200 V. It should be noted that the increase in voltage is largely due to an increase in the cathode voltage drop. The latter depends on the concentration of the aqueous solution. In this embodiment, it is about 600 V. There is a voltage drop inside the aqueous solution (approximately 100 V). When these parameters are taken into account, approximately a 5-fold increase in voltage is obtained. The discharge current decreased by approximately 3 times: from 11 to 3 A.

The most likely cause of these transformations in a gas discharge is the change in the electrical conductivity of the plasma. In the variant with a metal cathode, electronic conduction predominates. Electrons move easily under the influence of a weak electric field. Therefore, the voltage in the discharge is small. Compared to electrons, ions are less mobile. To create an ion current, a stronger electric field is necessary, and an increase in voltage is required, respectively. Consequently, ionic conductivity predominates in the variant with water-solution cathode.

In the plasma column above the water-solution cathode, many small flashes are observed. In the video frame, they are visible in the form of light circles (Figure 3a). Flashes are formed as a result of the explosion of small droplets of solution, which fall into the discharge region. They can be called secondary explosions. Primary explosions occur on the surface of a water-soluble cathode. As a result of primary explosions, cathode sputtering occurs and droplets are formed. Each droplet is a bunch of cations and anions taken out of the aqueous solution. After a secondary explosion, they fly off and become carriers of current. With each explosion, a certain number of current carriers are added. Because of this, current ripples occur. They are recorded on oscillograms (Figure 3b). The current ripple frequency is quite large. It is in the megahertz range.

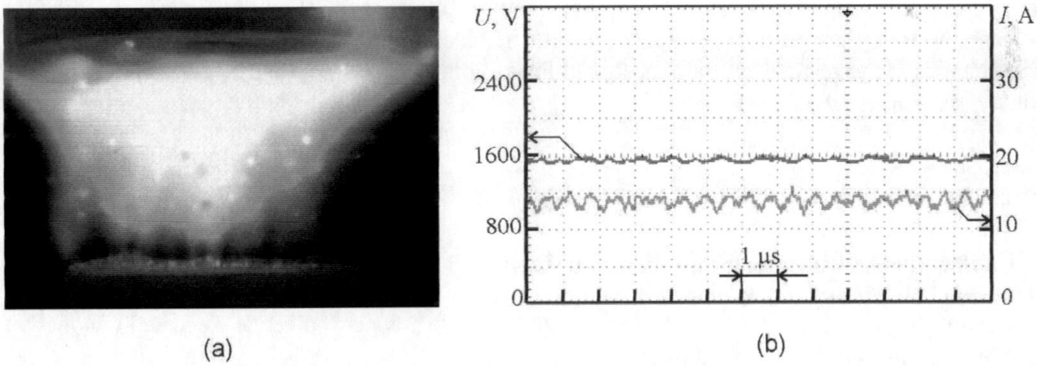

Figure 3. Secondary explosions and high-frequency current pulsations. (a) – video frame, shooting frequency 10,000 frames per second. (b) – oscillograms of current and voltage.

In the electric field of the discharge, cations and anions drift in opposite directions. The chlorine ions Cl^- are drifting towards the anode and thus they are removed, forming molecular chlorine. The amount of chlorine in the aqueous solution decreases. Sodium ions Na^+ are drifting towards the cathode and are returned in an aqueous solution. On the surface of the graphite plate with a negative potential, they are restored. Metallic sodium immediately reacts with water, forming the hydroxide NaOH, which in turn dissociates into ions Na^+ and OH^-. Thus, chlorine ions Cl^- that have departed the aqueous solution are replaced by the same amount of hydroxyl OH^- ions. The solution is alkalinized. Hydroxyl OH^- ions also participating in the creation of ion current in the discharge. In the end at the anode, in addition to chlorine, oxygen and water vapor are released. And molecular hydrogen is released on the cathode. In the discharge gap, water vapor is formed from the droplets.

Sodium ions Na^+ released from the aqueous solution in the composition of droplets, practically they all return back. The amount of sodium ions Na^+ is not changed in a solution and the amount of solvent (water) is reduced. The loss can be compensated by adding distilled water, i. e. during the burning of the discharge, the volume of the aqueous solution can be maintained unchanged. Such experiments were carried out at a fixed value of the current. Aqueous solutions with different initial concentrations C_0 were used.

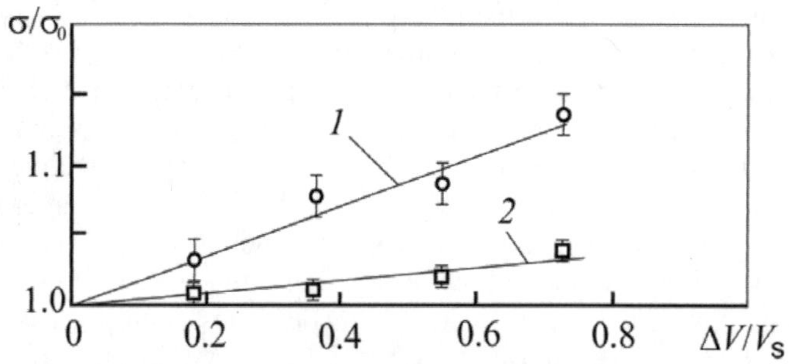

Figure 4. Change of the specific electrical conductivity of the aqueous solution. ΔV – is the volume of distilled water added to the solution. V_S – is the total working volume of the solution. *1* – C_0 = 0.02 mol / l; *2* – 0.1. The discharge current is 8 A.

Measurements performed before and after experiments showed that, the specific electrical conductivity of the solution varies insignificantly. Such results confirm the circulation of sodium ions Na^+. As a part of the drops, they leave the solution, and then return back. Their concentration practically does not change. As can be seen from the presented graphs (Figure 4), there is a tendency to increase the specific electrical conductivity with increasing duration of the discharge burning. To some extent, this is explained by the fact that chlorine ions Cl- are replaced in the solution by more mobile hydroxyl ions OH-.

4. Conclusions

Experimental facts are obtained, which are explained by the fact that the electrical conductivity of a gas discharge plasma with a water-solution cathode is predominantly ionic.

1. The exchange of metal cathode an aqueous solution causes increase burning voltage of a gas discharge under otherwise identical conditions.

2. In compliance with measures that compensate the loss of water (solvent), electrical conductivity of the aqueous solution changes slightly.

3. In gas discharge plasma the explosion of droplets of the solution are transpired, which cause the ripple current in the megahertz range.

References

1. Bruggeman P J et al. 2015 *Plasma Sources Science and Technology* **25** 053002
2. Titov V A, Stokozenko V G, Titova Y V and Moryganov A P 2015 *High Energy Chemistry* **49** (6) 459
3. Valiev R I, Gaisin A F, Gaisin F M, Gumerov A Z, Nasibullin R T, Sadriev R S, Sarimov L R and Khafizov A A 2014 *Izvestiya VUZov. Fizika* **57** No. 3-3 66
4. Tazmeev A Kh, Tazmeeva R N and Sarvarov F S 2016 *J. Phys.: Conf. Ser.* **669** 012056

5. Gibadullina G R, Tazmeev A K and Tazmeeva R N 2017 *Journal of fundamental and applied sciences* **9** (SI) 1779
6. Tazmeev A K and Tazmeeva R N 2017 *J. Phys.: Conf. Ser.* **789** 012058
7. Tazmeev G K, Timerkaev B A and Tazmeev K K 2017 *Plasma Physics Reports* **43** (7) 771
8. Tazmeev Kh K, Arslanov I M and Tazmeev G Kh 2014 *J. Phys.: Conf. Ser.* **567** 012001
9. Popov I A and Shchelchkov A V 2014 *Journal of Engineering Physics and Thermophysics* **87** (6) 1420

IOP Conf. Series: Journal of Physics: Conf. Series **1058** (2018) 012038 doi:10.1088/1742-6596/1058/1/012038

Study of a DC electric discharge with a cathode loaded in the water flow

G K Tazmeev[1], B A Timerkaev[1], K K Tazmeev[2] and I M Arslanov[2]

[1]Kazan National Research Technical University named after A.N. Tupolev, 10 K. Marx str., Kazan, 420111, Russian Federation
[2]Kazan Federal University, Naberezhnye Chelny Institute, 68/19 Mira Street, Naberezhnye Chelny, 423810, Russian Federation

E-mail: tazmeevh@mail.ru

Abstract. A gas discharge in the air between two solid-state electrodes, one of which is in a stream of water and serves as a cathode, has been studied. Two variants are considered. In the first version, the working end of the cathode protruded from the water, and in the second version it was completely immersed in water. Electric and spectral characteristics of discharges obtained under the same conditions of electric power are presented. As the cathode, aluminum, copper, titanium and graphite rods were used. The discharge current was in the range of 13-15 A.

1. Introduction

Electric discharges excited in the presence of a liquid medium are of interest as a tool that allows one to obtain a plasma with extensive possibilities of practical applications. They find application in the processes of water purification, the synthesis of metal nanoparticles, the preparation of finely dispersed oxide powders and many other technological processes [1-9].

In experiments, the metal electrode is either mounted deep inside the liquid, or above it. In this work, an intermediate variant is considered when a solid-state cathode is located near the surface of a liquid medium – water.

2. Experiment

The experiments were carried out on an experimental setup, a detailed description of which is given in [10].

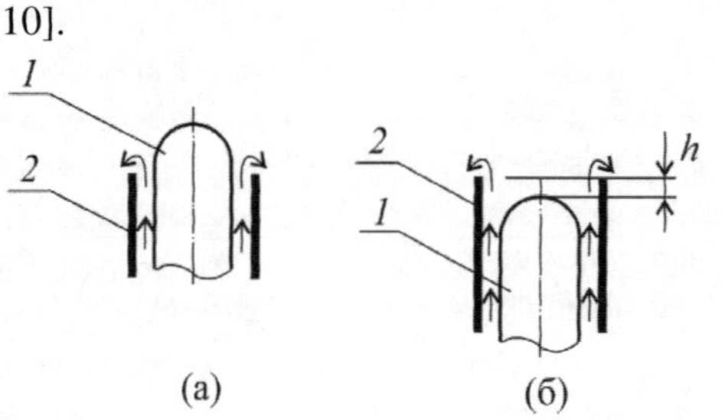

Figure 1. The cathode assembly. *1* – rod cathode; *2* – dielectric tube. Arrows indicate the direction of the water flow.

IOP Conf. Series: Journal of Physics: Conf. Series **1058** (2018) 012038 doi:10.1088/1742-6596/1058/1/012038

In the figure 1 schematically shows the cathode assembly. It consists of a rod cathode 1 inserted into the dielectric tube 2. The figure shows two options: (a) - above the water; (b) - under water. The working end of the cathode in version (a) comes out of the water and is not wetted by water, in version (b) it is under a thin layer of water. Depth of immersion h with allowance for water rise is within 2-3 mm.

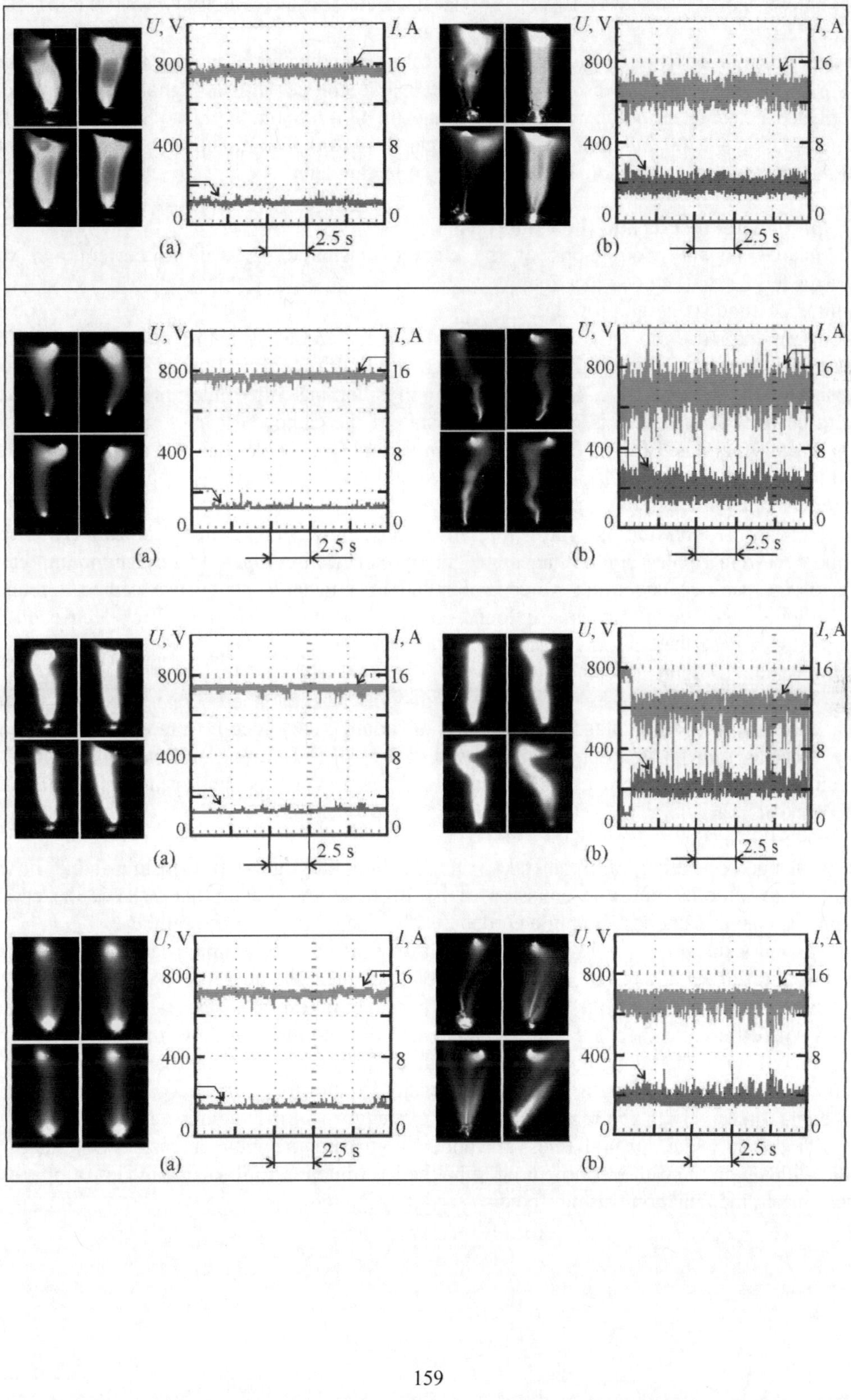

Figure 2. Instant photos and waveforms. Exposure 0.2 ms. Location of the cathode: (a) - above water; (b) - under water.

Cathodes from various materials were used in the experiments. The anode was made of copper. The mass flow rate of water was 10 g/s. The specific electrical conductivity was within 150 ± 10 μSm/cm.

The interelectrode distance (from the end of the cathode to the anode) was set to 4 cm in version (a) and 5 cm in version (b). The discharge radiation was recorded with a high-speed AvaSpec-3648 optical fiber spectrometer in the wavelength range of 484-708 nm with a resolution of 0.15 nm (diffraction grating – 1200 slits/mm, input optical gap – 10 μm). The radiation was projected onto the input of the spectrometer by a short-focus (28 mm) collecting lens.

3. The experimental results and their analysis

Figure 2 shows instant photographs of the electric discharge, as well as current and voltage oscillograms for the two cathode locations.

Aluminum cathode (figure 2, row 1).

In both variants ((a) and (b)), the cathode intensively erodes. Erosion products are solid fine particles of white color. The working surface is covered with a solid non-metallic crust, which has strong adhesion to the metal base. The plasma column is dominated by blue-blue radiation. In variant (b), a thin channel of red color is formed on the side of the cathode, characterizing the presence of radiating hydrogen atoms. Both in variant (a) and in variant (b), oscillations of current and voltage are recorded in the oscillograms, and in variant (b) the amplitude of pulsations is much larger.

Copper cathode (figure 2, row 2).

In variant (b), the emission of hydrogen atoms in the near-cathode zone becomes brighter. In this embodiment, current ripples and voltages are greatly enhanced compared to the aluminum cathode. The end of the cathode under the water is purified to a metallic state. The cathode eroded and smoothed. Fine particles of copper are found in the products of erosion. The electric discharge characteristics are described in more detail in work [9].

Titanium cathode (figure 2, row 3).

Radiation of a discharge is extremely bright in both variants. There are no specific shades. In variant a, current ripples and voltages occur rarely, in variant b they become intense and large-scale. In this case, some asymmetry is observed in the oscillograms of the current, and the voltage. At the end of the cathode a nonmetallic crust of black color is formed, which is destroyed intensely during the burning of the discharge.

Graphite cathode (figure 2, row 4).

With prolonged operation, unlike metallic cathodes, graphite lights up to form a flame. In variant (b), in the near-cathode zone, the emission of hydrogen atoms predominates, as is the case with aluminum and copper cathodes. Graphite eroded evenly along the end of the cathode.

Figure 3 shows spectra of discharges recorded at the time of the maximum radiation in the options of cathodes location under water. All spectra were obtained in the same operating mode of the spectrometer. Therefore, from the spectra it is possible to estimate the radiation intensities of discharges with different cathodes. From a comparison of the spectra it follows that the most intense is the emission of a discharge with a titanium cathode. Moreover, the discharge produces intense radiation in the entire wavelength range in the visible region. The discharge with the aluminum cathode emits intensively in the region of small wavelengths. In the radiation of a discharge with a copper cathode, only the spectral lines of copper atoms are the most intense. The radiatio of a discharge with a graphite cathode turned out to be the least intense. In the same modes of operation of the spectrometer, the radiation was not fixed.

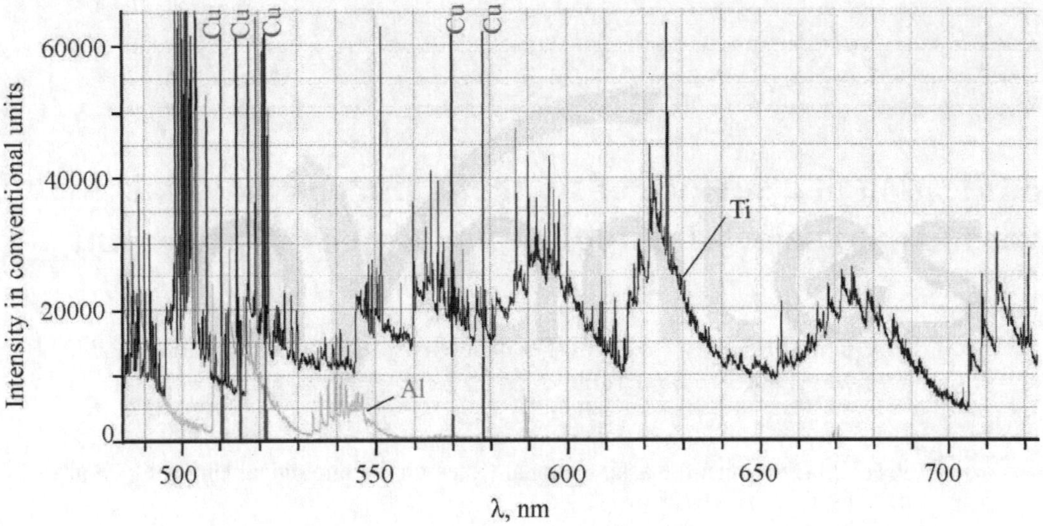

Figure 3. Spectra of radiation.

4. Conclusions

When the cathode is under a thin layer of water, its erosion increases. A large number products of erosion comes in the plasma column. Particles heated in a plasma column intensify its radiation. The most intense radiation occurs in the case of a titanium cathode.

An electric discharge with a cathode immersed in an aqueous stream can be used to produce a heterogeneous plasma and to study its properties.

References

1. Bruggeman P J et al. 2015 *Plasma Sources Science and Technology* **25** 053002
2. Titov V A, Stokozenko V G, Titova Y V and Moryganov A P 2015 *High Energy Chemistry* **49** (6) 459
3. Valiev R I, Gaisin A F, Gaisin F M, Gumerov A Z, Nasibullin R T, Sadriev R S, Sarimov L R and Khafizov A A 2014 *Izvestiya VUZov. Fizika* **57** No. 3-3 66
4. Gibadullina G R, Tazmeev A K and Tazmeeva R N 2017 *Journal of fundamental and applied sciences* **9** (SI) 1779
5. Tazmeev A K and Tazmeeva R N 2017 *J. Phys.: Conf. Ser.* **789** 012058
6. Kashapov L N, Kashapov N F, Kashapov R N and Denisov D G 2016 *J. Phys.: Conf. Ser.* **669** 012029
7. Akhmetov M N, Akhmetov N D, Gimadeev M M and Krivosheev V.A. 2017 *Plasma Physics Reports* **43** (3) 393
8. Gabdrakhmanov A T, Israphilov I H, Galiakbarov AT and Gabdrakhmanov A T 2017 *J. Phys.: Conf. Ser.* **789** 012009
9. Timerkaev B A, Sofronitskiy A O and Andreeva A A 2016 *J. Phys.: Conf. Ser.* **669** 012062
10. Tazmeev G K, Timerkaev B A and Tazmeev K K 2017 *Plasma Physics Reports* **43** (7) 771

IOP Conf. Series: Journal of Physics: Conf. Series **1058** (2018) 012039 doi:10.1088/1742-6596/1058/1/012039

Modification of a surface of synthetic fibrous materials by silver nanoparticles with application of plasma processing

Y A Timoshina, E S Tskhay, E F Voznesensky, G R Rakhmatullina, V P Tikhonova and F S Sharifullin

PNTVM Department, Kazan National Research Technological University, Kazan, 420015, Russia

ybuki@mail.ru

Abstract. Results of experimental researches on modification of textile polypropylene fibrous materials by a solution of silver nanoparticles with application of plasma processing are presented. Processing of fibrous materials by a current of nonequilibrium low-temperature plasma allows to apply effectively on materials antibacterial preparation and to fix nanoparticles of silver in a superficial layer of a material.

1. The Introduction

The use of impregnation of woven and non-woven synthetic materials by silver nanoparticles makes it possible to receive textile materials with antibacterial properties. High antibacterial activity of silver nanoparticles, a full range of antimicrobial action, as well as the suppression of pathogenic microflora make irreplaceable data modified textile materials for the production of textiles for medical purposes: medical wear and linens, both reusable and disposable use; Disposable surgical clothes and underwear; As well as medical disposable personal protective equipment. Thus, the need for production of high-quality textile materials of special purpose with high hygienic characteristics, as well as antimicrobial and antistatic properties, determines the urgency of development this topic.

Modification of textile materials by silver nanoparticles with application of plasma processing, unlike other technologies, gives an opportunity to receive textile materials, possessing not only various bactericidal properties, but also high hydrophilic, in particular the ability to absorb sweat, which is an important hygienic indicator and especially relevant for materials from synthetic fibers, which are hydrophobic. Plasma processing gives an opportunity to receive nanomodified textiles of a full spectrum of antimicrobial action, while there is no destroy influence on materials of processed products, and technology of plasma processing is environmentally safe. Plasma processing of high-frequency capacitive discharge, gives the possibility of obtaining textile materials with antibacterial properties resistant in time (prevents the particle entrainment from the surface of the material during wash and operation).

2. Materials, methods and equipment

For modification of textile polypropylene materials the experimental high-frequency plasma installation of the capacitive discharge of low pressure was used.

To evaluate the results of modification of materials by nanoparticles of silver with the use of low-temperature plasma processing is important quantitative analysis of changes in their properties. Property parameters were defined under normal conditions. The research used standard and special methods. The contact angle, characterizing the degree of wetting material, was determined on the device Kruss Easy drop DSA 20E of the sitting drop's method. The time of drop spreading on the surface of the material is determined by means of a stopwatch and is counted from falling drop on the surface of the material until it is completely absorbed by the material. Influence of efficiency of processing of materials impregnated with a solution of nanoparticles of silver, low-temperature plasma on durability of fastening of silver nanoparticles in a superficial layer of materials was investigated by determination of quantity of the silver contained in the sample before and after cycle of gas sterilization of specimens. The amount of silver contained in each sample was determined by the atomic emission analysis method on the atomic emission spectrometer ICAP 6300 DUO.

3. Results

In order to establish the patterns of impact of low-temperature plasma on samples of materials, their processing was carried out when changing the input parameters of the installation in the following limits: the discharge power - from 0.4 to 2.2 kW, processing time - from 60 to 600 seconds, the flow of plasma gas - from 0 to 0.06 gram - second (g/s) and pressure in the working chamber- from 13.3 to 30 Pa. From previous researches [1] it follows that for low-temperature plasma processing of capillary-porous and fibrous materials the optimum value of pressure in the working chamber is 26.6 Pa, consumption of plasma gas – 0.04, g/s. At these parameters of plasma the necessary energy of ions is provided, which allows the ions of plasma gas to be introduced into the surface area of the modified samples, which leads to a change of surface properties of materials, which allows to provide the maximum energy and intensity of the ion flow required for modification, but not leading to destructive processes.

The experimental estimation of change of values of an angle of capillarity of fibrous polypropylene fiber processed in optimum modes, depending on structure of plasma gas is carried out. The obtained experimental data are presented in the table.

Table 1. Hydrophilic properties of non-woven cloth depending on the composition of plasma gas (P = 26.6 Pa; G = 0.04 g/s)

Plasma treatment mode	Plasma forming environment	Angle of wetting (θ), degree	The time of the spreading, second
No processing	–	118	-
W = 1.6 kW; τ = 180 sec.	Argon	35	10
W = 1.6 kW; τ = 180 sec.	Argon - Air 70:30	36	12
W = 1.4 kW; τ = 180 sec.	Argon - nitrogen 70:30	0	0
W = 1.4 kW; τ = 180 sec.	Air	36	10

The greatest increase of wetability for polypropylene fibrous material is observed in the treatment of argon - nitrogen in the mixture of gases in the ratio of 70:30, respectively. After low-temperature plasma modification there is an instant absorption of a drop on the surface of the material.

To research the influence of plasma modification on the efficiency of fastening of silver nanoparticles in the surface layer of fibrous materials determined the amount of elemental silver on the surface of samples before and after the cycle of gas sterilization. Analysis of previous researches [1] has shown that for fastening of silver nanoparticles in a superficial layer of textile fibers the optimum plasma forming environment for plasma modification is a mixture of gases argon propane-butane in the ratio of 70:30, respectively. Repeated processing by plasma leads as to additional "pounding" of nanoparticles into the surface of materials due to ion bombardment, which promotes physical adsorption of silver nanoparticles in the surface layers, and to formation of superficial grids, due to the presence of propane - butane, capable in a ionized state to interact with formed by ion bombardment free radicals in the surface layer of fiber polymers.

The search for optimal parameters of processing of knitted and non-woven fibrous materials by high-frequency capacitive discharge plasma of low pressure was carried out by varying the power of discharge and the duration of processing plasma modification. The optimum mode of processing of materials modified by nanoparticles of silver, which leads to steady fixing of nanoparticles of silver in a surface layer of these materials is experimentally established: $W_p = 1.4$ kW, $t = 180$ s, $P = 26{,}6$ pa, $G = 0.04$ g/s, Plasma forming environment - a mixture of gases argon propane-butane in the ratio of 70:30, respectively. Processing of polypropylene materials modified by nanoparticles low-temperature plasma prevents of silver nanoparticles washing out from the surface of fibers in the course of their exploitation of the preoperational gas sterilization, whereas with the unprocessed plasma Samples after the gas sterilization cycle, the amount of silver on the material surface decreases by 18%.

4. Conclusions

The results of the research showed that the effect of plasma modification depends on the composition of plasma-supporting gas, as well as the technological modes of plasma modification. Plasma modification of textile fibrous materials occurs mainly due to the kinetic impact of the flow of ions, due to the formation of the plasma layer of positive charge around the area sample [2], recombination of ions on the surface of the fibers and the effects of the flow of heat.

Application of silver nanoparticles on a surface of polypropylene fibrous materials and subsequent processing of materials in low - temperature plasma in plasma forming mixture of gases argon-propane-butane leads to steady consolidation of silver nanoparticles on a surface of fibers , as evidenced of atomic emission analysis of samples. In this case, the functional purpose of plasma is to activate nanoparticles of silver, the removal of poorly adsorbed and the strengthening of well adsorbed nanoparticles on the surface of materials. The interaction of free radicals on the surface of fibrous materials with plasma gas monomer unit (argon-propane-butane) leads to the formation of surface mesh and smoothing surface of specimens.

References

[1] Sergeeva E A 2011 *Modificatsia sinteticheskih voloknistih materialov i izdeliy neravnovesnoy nizkotemperaturnoy plazmoy* [Modification of synthetic fibrous materials and products with low-temperature plasma]. Ch. 1. Teoria, modeli, metodi [Theory, models, methods] / E A Sergeeva, V S Zheltukhin, I S Abdullin. – Kazan *(Izdatel'stvo Kazanskogo tekhnologicheskogo universiteta)*, 252 p. (In Russian).

[2] Abdullin I S 2007 *Modificatsia nanosloyov v visokochastotnoy plazme ponijennogo davleniya: monografiya* [Modification of nanolayers in high-frequency plasma of low pressure: monograph] / I S Abdullin, V S Zheltukhin, I R Sagbiev, M F Shaekhov. - Kazan *(Izdatel'stvo Kazanskogo tekhnologicheskogo universiteta)*, 356 p. (In Russian).

Calculation of a plasma composition and its thermophysical properties in cases of maintaining or quenching of electric arcs

V Ya Frolov, D V Ivanov

Department of Electrical Power Engineering and Equipment,
Peter the Great St.Petersburg Polytechnic University,
Polytechnicheskaya, 29, 195251, Saint-Petersburg, Russian Federation

eie@spbstu.ru

Abstract. The paper presents a technique for calculating a composition as well as thermodynamic and transport properties of plasma which is used both in plasma technologies and in arc chambers of electrical apparatuses. Examples of calculations are given.

1. Introduction

Modern plasma technologies are implemented with using of various gases depending on the technological task. Typical plasma-forming gases are argon, helium, nitrogen, oxygen, hydrogen, carbon dioxide etc. as well as mixtures of these gases (air, argon-hydrogen etc.). A quenching of electric arcs in electrical apparatuses is carried out in the medium of SF6 gas, air etc. In this case plasma processes are accompanied by erosion of electrodes and ablation of the chamber walls that greatly influences the thermophysical properties of the plasma.

Currently not only experimental methods but also methods of mathematical modeling are used during the development of modern plasma processes. To perform mathematical modeling of plasma processes it is necessary to know thermophysical properties of plasma which depend on the composition of plasma, its temperature and pressure. Dependences of those properties on temperature are given in the scientific literature [1, 2] for some gases used in plasma technologies. However, in a number of cases (for example, in case of using a mixture of gases or in case when a pressure is different from atmospheric pressure) the data available in the literature are insufficient, and the properties of plasma need to be calculated. The procedure for calculating the thermophysical properties of plasma is given below.

2. Calculation of plasma composition

Calculation of the plasma thermophysical properties starts with the calculation of the plasma composition as a result of which the number densities of plasma species are determined.

If the number of individual species considered in the calculation (including electrons) is m and the number of chemical elements forming the system is n then to determine the equilibrium composition it is possible to write (m–n–1) equations of the mass action law. For neutral components it is convenient to write these equations for the dissociation of molecules into atoms [3-5]:

$$K_j(T,p)=\prod_{i=1}^{n}\frac{p_i^{a_{i,j}}}{p_j}, \qquad (1)$$

where $K_j(T,p)$ is the constant of the reaction; p_i is partial pressure of elements (atoms); p_j is partial pressure of complex components (molecules); $a_{i,j}$ are stoichiometric coefficients of reaction of dissociation of molecules onto atoms.

For electrically charged components of plasma (ions) the mass action law is expressed by the well-known Saha equation:

$$\frac{n_i \cdot n_e}{n_a} = \frac{Z_i(T) \cdot g_e}{Z_a(T)} \cdot \left(\frac{2\pi \cdot m_e \cdot k \cdot T}{h^2} \right)^{3/2} \cdot \exp\left(-\frac{E_{ioniz}}{k \cdot T} \right), \tag{2}$$

where n_i, n_e, n_a are number densities of ions, electrons, atoms respectively; $Z_i(T)$, $Z_a(T)$ are partition functions of the ion and atom; $g_e = 2$ is the statistical weight of the electron; m_e is the electron mass; k is the Boltzmann constant; h is Planck's constant; E_{ioniz} is the ionization energy.

Dalton's law on the sum of the partial pressures also needs to be included to the system of equations of thermodynamic equilibrium:

$$\sum_{j=1}^{m} p_j = p. \tag{3}$$

Another equation that is necessary for calculating the plasma composition is the quasi-neutrality equation:

$$\sum_i a_i \cdot n_{i+} = \sum_j a_j \cdot n_{j-}, \tag{4}$$

where a_i and n_{i+} are the charge and the number density of positively charged ions; a_j and n_{j-} are the charge and the number density of negatively charged ions (and electrons).

To close the system of equations it is necessary to include (n–1) equations of material balances:

$$\sum_{i=1}^{m} a_{g_i} p_i \Big/ \sum_{i=1}^{m} a_{k_i} p_i = G/K, \tag{5}$$

where a_{g_i} and a_{k_i} are the quantities of atoms of the kind g and k respectively in the molecule of the i-th specie, G and K are the total numbers of atoms of the kind g and k in the plasma.

It is convenient to solve the obtained system of equations of thermodynamic equilibrium of the form (1) – (5) by the modified Newton method for increments of logarithms of unknown values [4].

As an example let us consider the plasma that appears in the discharge chamber of a multi-chamber arrester used for lightning protection of power lines [6]. The walls of the chamber are made of silicone rubber; the electrodes are made of copper. The following relationship between chemical elements is considered:

Si: O: C: H = 1: 1: 2: 6, Cu: O = 1: 10.

The composition of such a plasma at a pressure of 2 atm is shown in Fig. 1.

a) b) c)

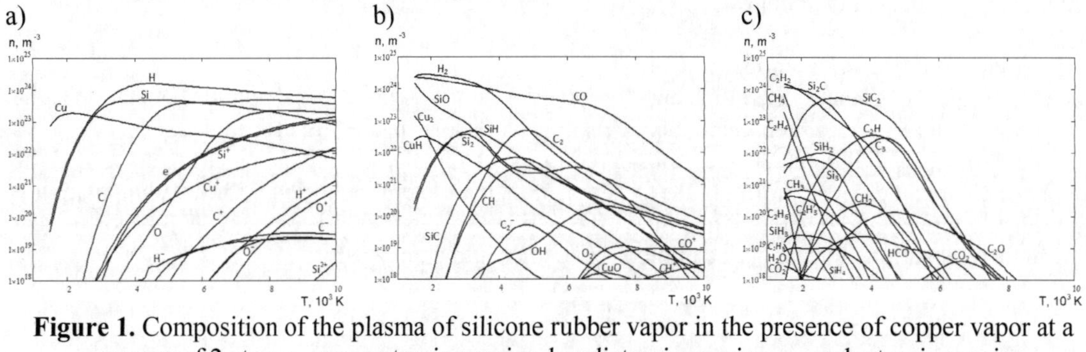

Figure 1. Composition of the plasma of silicone rubber vapor in the presence of copper vapor at a pressure of 2 atm: a – monoatomic species, b – diatomic species, c – polyatomic species

3. Calculation of thermodynamic properties

Knowing the plasma composition one can calculate its thermodynamic properties [4]:

1) Plasma mass density ρ:

$$\rho = \sum_i m_i n_i \,, \tag{6}$$

where m_i is the mass of the atom (ion, molecule) of the i-th specie of plasma, n_i is its number density.

2) The total enthalpy of the plasma (measured in J/kg):

$$h = \frac{\sum_i n_i H_i}{M_{\text{mix}} \cdot n_\Sigma} \,, \tag{7}$$

where n_i is the number density of the i-th specie; H_i is the enthalpy of the i-th specie (measured in J/mol, the temperature dependence can be found in the reference books [7, 8]); M_{mix} is the molar mass of the gas mixture (plasma), $M_{\text{mix}} = \rho \cdot N_A / n_\Sigma$; N_A is Avogadro's number; n_Σ is the total number density of all plasma species.

3) Specific heat of the plasma at constant pressure:

$$c_p = \frac{dh}{dT} \,. \tag{8}$$

An example of the calculated thermodynamic properties is shown in Fig. 2.

a) b)

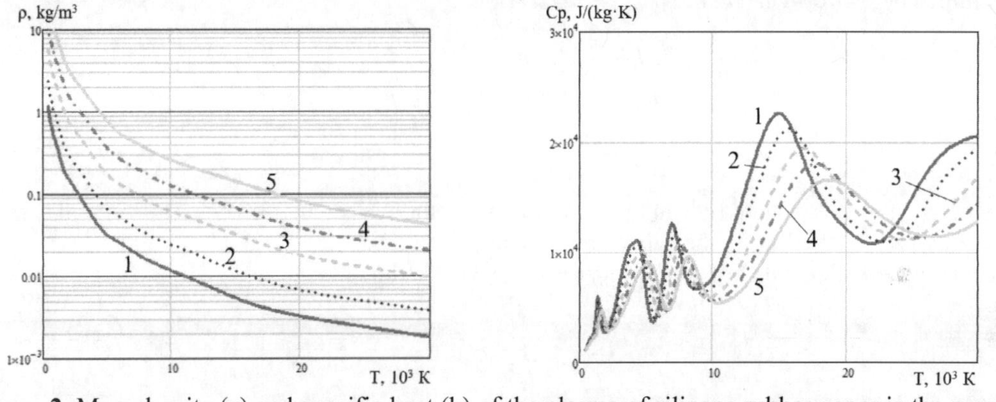

Figure 2. Mass density (a) and specific heat (b) of the plasma of silicone rubber vapor in the presence of copper vapor at different pressures: 1 – 1 atm; 2 – 2 atm; 3 – 5 atm; 4 – 10 atm; 5 – 20 atm

4. Calculation of transport properties

The main difficulty in the calculation of transport properties is their dependence on the cross sections of collisions. Knowing the reduced collision integrals for each pair of plasma species one can calculate the transport properties of the plasma:

1) The electrical conductivity of plasma can be calculated in the fourth approximation of the Chapman-Enskog theory by the following formula [9]:

$$\sigma = 3e^2 n_e^2 \left(\frac{\pi}{2kTm_e} \right)^{\frac{1}{2}} \cdot \begin{vmatrix} 1 & q_1^{01} & q_1^{02} \\ 0 & q_1^{11} & q_1^{12} \\ 0 & q_1^{21} & q_1^{22} \end{vmatrix} \Bigg/ \begin{vmatrix} q_1^{00} & q_1^{01} & q_1^{02} \\ q_1^{10} & q_1^{11} & q_1^{12} \\ q_1^{20} & q_1^{21} & q_1^{22} \end{vmatrix} \,, \tag{9}$$

where the coefficients q depend on the number densities and reduced collision integrals; formulas for calculating q can be found in [5].

2) The viscosity of plasma can be calculated in the first approximation of the Chapman-Enskog theory by the formula [10, 11]:

IOP Conf. Series: Journal of Physics: Conf. Series **1058** (2018) 012040 doi:10.1088/1742-6596/1058/1/012040

$$\mu = -\frac{5}{2}(2\pi kT)^{\frac{1}{2}} \cdot \begin{vmatrix} \hat{q}_{ij}^{00} & n_i (m_i)^{\frac{1}{2}} \\ n_j & 0 \end{vmatrix} \Big/ \left| q_{ij}^{00} \right|, \qquad (10)$$

where $\hat{q}_{ij}^{00}$ is a square matrix consisting of elements q_{ij}^{00} that depend on number densities and reduced collision integrals, formulas for calculation can be found in [5].

3) A total thermal conductivity λ of the plasma is composed of four parts: a thermal conductivity λ_h of heavy species, a thermal conductivity λ_e of electrons, a reaction heat conductivity λ_R and an internal thermal conductivity λ_{int} :

$$\lambda = \lambda_h + \lambda_e + \lambda_R + \lambda_{\text{int}} . \qquad (11)$$

In the second approximation of the Chapman-Enskog theory the thermal conductivity of heavy species is determined by the formula [10, 11]:

$$\lambda_h = -\frac{75k}{8}(2\pi kT)^{\frac{1}{2}} \cdot \begin{vmatrix} \hat{q}_{ij}^{00} & n_i \\ n_j \Big/ (m_j)^{\frac{1}{2}} & 0 \end{vmatrix} \Big/ \left| q_{ij}^{11} \right|, \qquad (12)$$

where $\hat{q}_{ij}^{00}$ is a square matrix consisting of elements q_{ij}^{11} that depend on number densities and reduced collision integrals, formulas for calculation can be found in [5].

In the second approximation of the Chapman-Enskog theory the thermal conductivity of electrons is determined by the formula [12, 13]:

$$\lambda_e = \frac{75}{8} n_e^2 k \left(\frac{2\pi kT}{m_e} \right)^{\frac{1}{2}} \cdot \begin{vmatrix} 1 & q_1^{12} \\ 0 & q_1^{22} \end{vmatrix} \Big/ \begin{vmatrix} q_1^{11} & q_1^{12} \\ q_1^{21} & q_1^{22} \end{vmatrix}, \qquad (13)$$

where the values of q_1^{11}, q_1^{12} etc. are determined by the same formulas as for the electrical conductivity.

The reaction heat conductivity can be found by the method given in [14, 15]. The internal thermal conductivity is calculated using the Eiken approximation [16].

An example of calculated transport properties is shown in Fig. 3.

a) b)

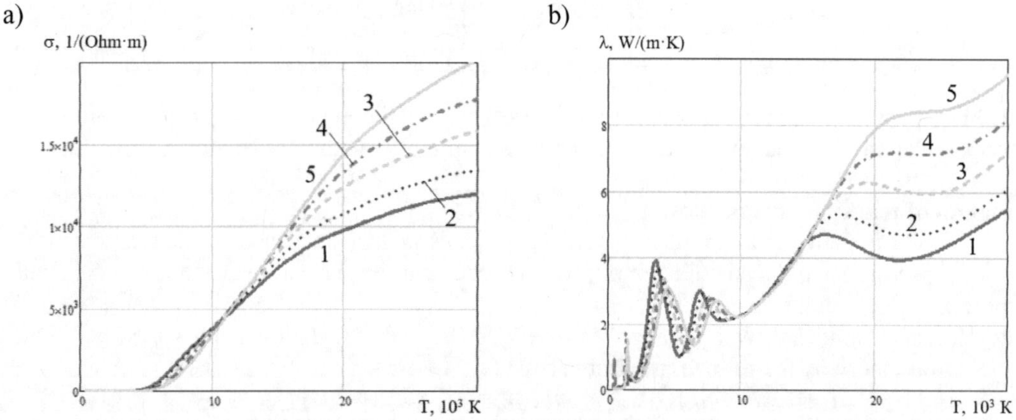

Figure 3. Electrical conductivity (a) and thermal conductivity (b) of the plasma of silicone rubber vapor in the presence of copper vapor at different pressures: 1 – 1 atm; 2 – 2 atm; 3 – 5 atm; 4 – 10 atm; 5 – 20 atm

5. Conclusions

The above described technique allows to calculate the thermophysical properties of both simple gases and their mixtures which can be used in cases of maintaining as well quenching of electric arcs taking into account erosion of electrodes and ablation of the chamber walls. The obtained distributions are necessary for the modeling of plasma processes. They can also be used to compare the properties of various gases in the development of plasma technologies.

References

[1] Engel'sht V S, Gurovich V T, Desyatkov G A 1990 *Theory of the electric arc column* (Novosibirsk: Nauka, in Russian).
[2] Boulos M I, Fauchais P, Pfender E 1994 *Thermal Plasmas: Fundamentals and Applications, Vol. 1* (NY: Plenum Press).
[3] V.P. Glushko et al., 1971 *Thermodynamic and thermophysical properties of combustion products. In 5 volumes. Vol. 1: Methods of calculation* (Moscow: VINITI, in Russian).
[4] Suris A L 1985 *Thermodynamics of high-temperature processes* (Moscow: Metallurgy, in Russian).
[5] Dresvin S V, Ivanov D V 2013 *Plasma physics* (SPb Polytechn. Univ. Publ. house, in Russian).
[6] Frolov V Y, Ivanov D V, Murashov Y V, Sivaev A D 2015 *Tech. Phys. Letters* **41(4)** 310.
[7] V.P. Glushko et al., 1978-1982 *Thermodynamic Properties of Individual Substances, Vol. I-IV,* (Moscow: National Academy of Sciences of the USSR, in Russian).
[8] M. W. Chase, Jr., Editor. 1998 *J. Phys. Chem. Ref. Data. Monograph* **9 (Part I and II)**. 1963 p.
[9] Devoto R S 1966 *Phys. Fluids* **9 (6)** 1230.
[10] Rat V, Murphy A B, Aubreton J, Elchinger M F, Faushais P 2008 *J. Phys. D: Appl. Phys.* **41 (18)** 183001.
[11] Devoto R S 1967 *Phys. Fluids* **10 (10)** 2105.
[12] Aubreton J, Elchinger M F, Fauchais P 1998 *Plasma Chem. Plasma Process* **18 (1)** 1.
[13] Devoto R S 1966 *Phys. Fluids* **9(6)** 1230.
[14] Butler J N, Brokaw R S 1957 *J. Chem. Phys.* **26 (6)** 1636.
[15] Brokaw R S 1960 *J. Chem. Phys.* **32 (4)** 1005.
[16] Hirschfelder J O, Curtiss C F, Bird R B 1964 *Molecular theory of gases and liquids* (NY: Wiley).

IOP Conf. Series: Journal of Physics: Conf. Series **1058** (2018) 012041 doi:10.1088/1742-6596/1058/1/012041

Step-Down dc-dc Converter for Low Temperature Plasma

A Hakki, N Kashapov, I Maiorova, K Sadikov
Engineering Institute, Kazan (Volga-Region) Federal University

E-mail: hakki.adnan@gmail.com

Abstract. This paper presents a step down converter with a high speed rectifier and filter for Argon plasma torches. Buck circuit with a high frequency pulse width modulation control and very fast MOSFET transistor were used for switching. For Argon (Ar) gas Plasma Torches, and by increasing the pulse width, the average output voltage has modified from 80 V up to 100 V. The frequency of the designed step down converter was steady at 20 KHz.

Keywords. Step Down Converters, Buck Circuit, MOSFET, Filters, Argon Plasma

1. Introduction

The step down converter (Buck Converter) is used in SMPS circuits where the DC output voltage needs to be lower than the DC input voltage. The DC input can be derived from rectified AC or from any DC supply. It is useful where electrical isolation is not needed between the switching circuit and the output, but where the input is from a rectified AC source, isolation between the AC source and the rectifier could be provided by a mains isolating transformer. Fig. 1. shows the schematic of a step down converter [1-5].

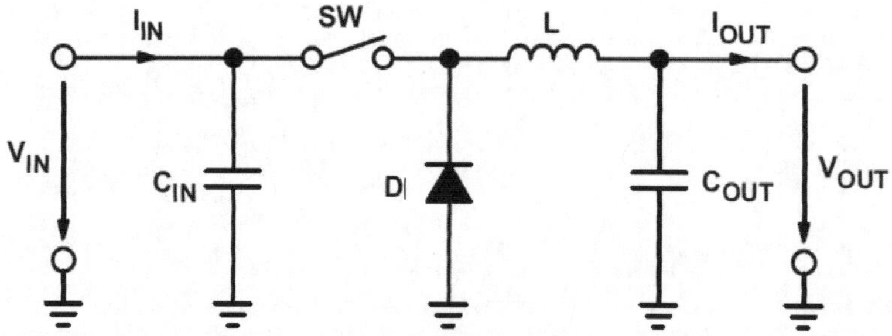

Figure 1. Schematic of a step down converter

2. Experimental Setup

Fig. 2. shows the schematic of Argon plasma power supply. The switching transistor between the input and output of the Buck Converter continually switches on and off at high frequency. To maintain a continuous output, the circuit uses the energy stored in the inductor L, during the on periods of the switching transistor, to continue supplying the load during the off periods. The circuit operation depends on what is sometimes also called a Flywheel Circuit (L, C_{out}). This is because the circuit acts rather like a mechanical flywheel that, given regularly spaced pulses of energy keeps spinning smoothly (outputting energy) at a steady rate.

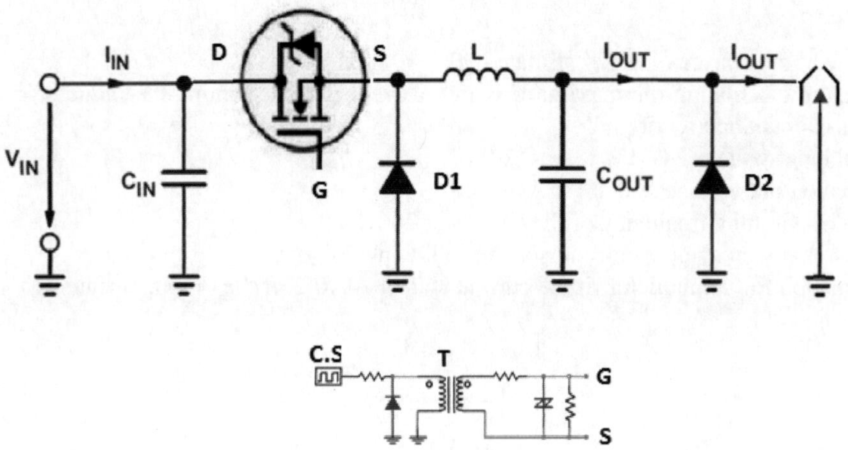

Figure 2. Schematic of Argon plasma power supply

Fig. 3. shows the typical plasma torches Volt-Ampere characteristics, The first region is the forward leakage, in this region the current is very little and no plasma emission. The second region is the reverse, in this region the current increases and the voltage decreases (negative load), for few gases there is a plasma emission. The third region is the forward conduction, in this region the current increases and the voltage increases too (positive load) ,this the case of argon gas which will be studied in this paper [6-23].

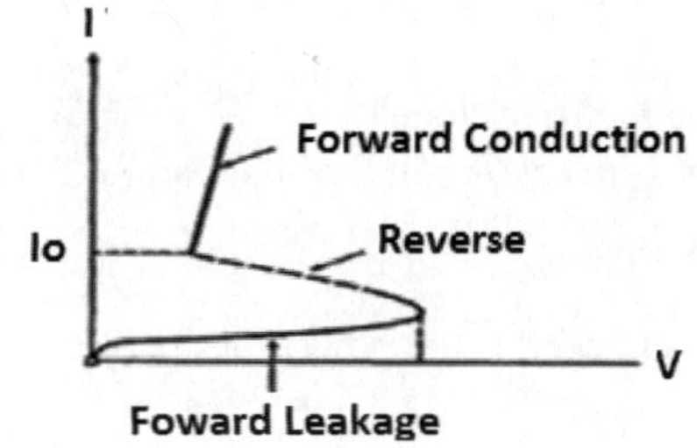

Figure 3. Typical plasma torches Volt-Ampere characteristics

3. Results and discussion

The following three parameters are needed to calculate the power stage:

1. Input voltage range: $V_{IN(min)}$ and $V_{IN(max)}$
2. Nominal output voltage: $V_{out} = 100\ V$

3. Maximum output current: $I_{out(max)}$
If these parameters are known, the power stage can be calculated.
.

Calculate the Maximum Switch Current
The first step to calculate the switch current is to determine the duty cycle D, for the maximum input voltage. The maximum input voltage is used because this leads to the maximum switch current.
$V_{IN(max)}$= maximum input voltage 350 V.
V_{Out}= output voltage 100 V.
Duty Cycle:

$$D = \frac{V_{Out}}{V_{IN(max)} \cdot \eta} = 0{,}32$$

η = efficiency of the converter, e.g., estimated 90%
The next step in calculating the maximum switch current is to determine the inductor ripple current. section of this application report.
V_{IN} = typical input voltage 310 V
V_{out} = desired output voltage 100 V
f_s = minimum switching frequency of the converter 20 KHz
ΔI_L = estimated inductor ripple current, see the following:
A good estimation for the inductor ripple current is 20% to 40% of the output current.

$$L = \frac{(V_{IN} - V_{out}).V_{out}}{\Delta I_L . f_S . V_{IN}} = 226\ \mu H$$

$$\Delta I_L = 0{,}3.I_{out(max)}$$

$I_{out(max)}$ = maximum output current 50 A.

$$\Delta I_L = 15\ A$$

Rectifier Diode Selection
To reduce losses, use Schottky diodes. The forward current rating needed is equal to the maximum output current:

$$I_F = I_{out(max)} \cdot (1 - D) = 34\ A$$

I_F = average forward current of the rectifier diode
$I_{out(max)}$ = maximum output current necessary in the application. Schottky diodes have a much higher peak current rating than average rating. Therefore the higher peak current in the system is not a problem.
The other parameter that has to be checked is the power dissipation of the diode.

$$P_D = I_F . V_F = 24\ W$$

I_F = average forward current of the receiver diode
V_F = forward voltage of the rectified diode.

Output Capacitor Selection
The best practice is to use low-ESR capacitors to minimize the ripple on the output voltage. Ceramic capacitors are a good choice if the dielectric material is X5R or better.

If the converter has external compensation, any capacitor value above the recommended minimum in the data sheet can be used, but the compensation has to be adjusted for the used output capacitance.

With internally compensated converters, the recommended inductor and capacitor values must be used, or the recommendations in the data sheet for adjusting the output capacitors to the application in the datasheet must be followed for the ratio of $L \times C$.

With external compensation, the following equations can be used to adjust the output capacitor values for a desired output voltage ripple:

$$C_{out(min)} = \frac{\Delta I_L}{8. f_s. \Delta V_{out}} = 19\ \mu F$$

$C_{out(min)}$ = minimum output capacitance
ΔI_L = estimated inductor ripple current
f_s = minimum switching frequency of the converter
ΔV_{out} = desired output voltage ripple 5 V

Table .1 shows all the parameters of this converter.

Parameters	**The value**
V_{IN}	310 V
$V_{IN(\max)}$	350 V
$V_{IN(\min)}$	250 V
V_{out}	100 V
$I_{out(max)}$	50 A
f_s	20 KHz
η	90%
D_{max}	0.32
L	226 μH
$C_{out(min)}$	19 μF
L	226 μH
I_F	34 A

4. Conclusion

In short, The working region of the step down converter of the argon plasma torches can be only in the forward conduction region (Positive resistor). For argon gas plasma step down converter (Buck), and by increasing the Duty Cycle (D) from 25 % up to 32 % the average output voltage $V_{out(ave)}$ has modified from 70 V up to 100 V, and the average output current $I_{out(ave)}$ has modified from 40 A up to 50 A . The switching frequency of the step down converter f_s was constant at the value 20 KHz.

References

[1] P. Julián, A. Oliva, P. Mandolesi, and H. Chiacchiarini, "Output discrete feedback control of a DC-DC Buck converter," in Proceedings of the IEEE International Symposium on Industrial Electronics (ISIE'97), Guimaraes, Portugal, 7-11Julio 1997, pp. 925–930.

[2] Hakki A, Sadikov K, Kashapov N 2017 *J. Phys.: Conf. Ser.* 789 012021

[3] H. Chiacchiarini, P. Mandolesi, A. Oliva, and P. Julián, "Nonlinear analog controller for a buck converter: Theory and experimental results", Proceedings of the IEEE International Symposium on Industrial Electronics (ISIE'99), Bled, Slovenia, 12–16 July 1999, pp. 601–606.

[4] Hakki A, Fayrushin I, Kashapov N 2016 *J. Phys.: Conf. Ser.* **669** 012022
[5] M. B. D'Amico, A. Oliva, E. E. Paolini y N. Guerin, "Bifurcation control of a buck converter in discontinuous conduction mode", Proceedings of the 1st IFAC Conference on Analysis and Control of Chaotic Systems (CHAOS'06), pp. 399–404, Reims (Francia), 28 al 30 de junio de 2006.
[6] Oliva, A.R., H. Chiacchiarini y G. Bortolotto "Developing of a state feedback controller for the synchronous buck converter", Latin American Applied Research, Volumen 35, Nro 2, Abril 2005, pp. 83–88.
[7] Hakki A, Kashapov N and Sadikov K 2017 *J. Phys.: Conf. Ser.* **927** 012022
[8] Kashapov L N, Kashapov N F and Kashapov R N 2014 *J. Phys.: Conf. Ser.* **567** 012025
[9] Dautov G, Fayrushin I and Kashapov N 2014 *J. Phys.: Conf. Ser.* **567** 012006
[10] Fayrushin I and Dautov G 2013 *J. Phys.: Conf. Ser.* **479** 012013
[11] Saifutdinov A I, Fairushin I I and Kashapov N F 2016 *JETP Lett.* **104** 180–185
[12] Kornienko E E, Lapushkina E J, Kuzmin V I, Vaschenko S P, Gulyaev I P, Kartaev E V, Sergachev D S, Kashapov N, Sharifullin S, Fayrushin I 2014 *J. Phys.: Conf. Ser.* **567** 012010
[13] Denisov D G, Kashapov N F and Kashapov R N 2015 *IOP Conf. Ser.: Mater. Sci. Eng.* **86** 012005
[14] Kashapov N F and Sharifullin S N 2015 *IOP Conf. Ser.: Mater. Sci. Eng.* **86** 012021
[15] Kashapov L N, Kashapov N F and Kashapov R N 2013 *J. Phys.: Conf. Ser.* **479** 012003
[16] Zaripov R G, Kashapov N F, Tkachenko L A, Shaydullin, L R 2016 *J. Phys.: Conf. Ser.* **669** 012053
[17] Galyautdinov R T, Kasparov N F and Luchkin G S 2002 *Inzhenerno-Fizicheskii Zhurnal* **75** 170–173
[18] Fayrushin I I, Dautov I G and Kashapov N F 2017 *International Journal of Environmental Science and Technology* **14** (12) 2555–2560
[19] Fadeev S A and Saifutdinov A I 2017 *Plasma Physics Reports* **43** (11) 1080–1088
[20] Shaydullin L R 2016 *IOP Conf. Ser.: Mater. Sci. Eng.* **134** 012022
[21] Fairushin I I, Dautov I G, Kashapov N F and Shamsutdinov A R 2017 *Technical Physics Letters* **43** (1) 27–30
[22] Gubaidullin D A, Kashapov N F, Zaripov R G, Tkachenko L A and Shaydullin L R 2017 *J. Phys.: Conf. Ser.* **789** 012017
[23] Luchkin A G, Hakki A, Rahimov N F, Sadikov K G and Luchkin G S 2017 *J. Phys.: Conf. Ser.* **789** 012074
[24] Hakki A and Kashapov N 2015 *IOP Conf. Ser.: Mater. Sci. Eng.* **86** 012002

Influence of electric arc treatment of inner surfaces of ceramic molds on the formation of the surface of titanium casting

E Yu Petrov, A I Khatmullina and A M Lyakhovich

Kazan Federal University, ul. Kremlevskaya, 18, Kazan, 420008, Russia

Ezio19@mail.ru

Abstract. Influence of the effect of electric arc titanium spraying on the inner surfaces of ceramic molds used for casting titanium was observed. Casting surface and cross-section were examined by electron microscopy with a probe microanalysis system and X-ray photoelectron spectroscopy. It was observed there is an "alpha case" on the casting surface. Its thickness and elemental composition varies through the casting surface.

1. Introduction

When obtaining castings from titanium alloys, a gas-saturated diffusion layer is formed in their surface layers consisting of titanium oxides and compounds containing elements of the mold called "alpha case". It is characterized by increased hardness, reduced plasticity and tendency to crack formation. It's an undesirable component of the alloy. One way to reduce the formation of the "alpha case" is the application of technological coatings to the inner surfaces of the molds by an arc method.

Objective: to investigate the effect of electric arc titanium spraying on the inner surfaces of ceramic molds used for casting titanium on the formation of casting surface.

2. Experiments

Arc spraying is a cold process of thermal spraying of metal on a product where the heating of its surface usually does not exceed 150 °C. This excludes changes in the structure of the product and its deformation. This method is used to deposit a coating of titanium on the inner surfaces of a ceramic mold to reduce the thickness of the alpha case. The arc spraying pistol circuit is shown in figure 1. Two wires are fed into the nozzle of the pistol, one of which serves as anode and the other is cathode. An electric arc lights up between them and wire begins to melt. Spraying occurs by the supply of compressed air. The process proceeds at a direct current.

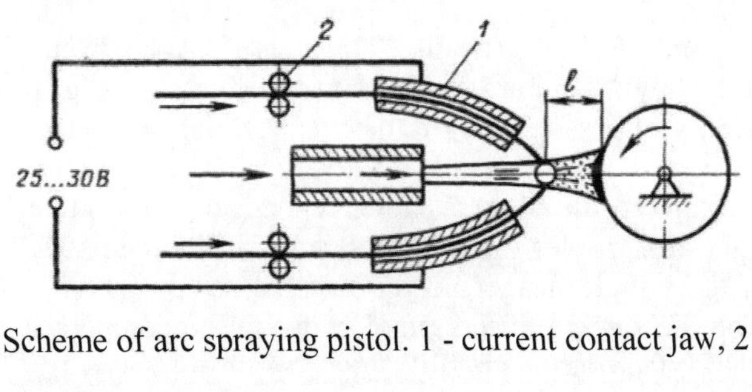

Figure 1. Scheme of arc spraying pistol. 1 - current contact jaw, 2 - Ti wire.

This method has the following advantages:

a) high productivity (up to 40 kg / h of sprayed metal);

b) stronger coatings with high adhesion in comparison with the flame plating method;

c) wires from different metals makes it possible to obtain a coating - a "pseudo-alloy";

d) low operating costs.

The disadvantages of electric arc metallization are:

a) low speed of wire supply terminates to overheating and oxidation of the sputtered materials;

b) burning out of alloying elements of the sprayed materials.

Volt – ampere characteristic of an arc discharge is falling. Working regime: U=25 V, I=200 A, the feed rate of air ~150 m/s.

Ceramic mold consisted of oxides of Mg, Ca, Si, as a binder it was used liquid glass [1]. In the study a casting sample (figure 2) the separation surface of the mold and the cross-section of the boundary layers were considered.

Figure 2. Photo of a casting sample.

Samples were examined by electron microscopy with a probe microanalysis system (analysis depth 1-3 μm), the chemical composition of sample's surface was studied by X-ray photoelectron spectroscopy (XPS) (depth of analysis is 5-10 nm).

3. Results and discussions

Electron microscopy researches showed that the sample surface is heterogeneous in structure. Light areas of the surface (Fig. 3, section 1) are characterized by an increased content of Fe, presence of Cr and Mn (Table 1). Since Cr and Mn are not inherent in a titanium alloy, and Fe is present only as an impurity (<0.25%), it can be concluded that these elements are introduced from outside. Most likely, they are the result of destruction of the cutting tool used to make the sample.

Despite the fact that before the test the samples were washed in an alcohol solution with an ultrasonic bath, this did not lead to the removal of oil-soluble impurities introduced into the sample under production conditions, as evidenced by the significant carbon content in all areas of the surface (table 1).

The areas of surface 2 and 3 characterize the "alpha case". Apart from Ti and Al alloy elements (4% Al is included in the alloy), there are elements of the form: Si, Mg, Fe, and O. The elemental composition of considered sections (table 1) is different, which indicates the heterogeneity of the surface chemical composition.

To study the concentration profile of the samples' chemical composition, slices were prepared in depth (figure 4). The dark area in electronic images refers to the compound where the sample was pressed. An analysis of the sample image showed that the sample is heterogeneous over the cross section. Light particles on the surface of the sample refer to the components of the grinding wheel, which were brought to the surface during grinding.

IOP Conf. Series: Journal of Physics: Conf. Series **1058** (2018) 012042 doi:10.1088/1742-6596/1058/1/012042

The outer layer of the sample is an alpha layer. Based on the electronic images (figure 4) and the map of the distribution of elements along the thickness of the section (figure 5), it is established that the thickness of the alpha layer varies from 2 μm to 14 μm.

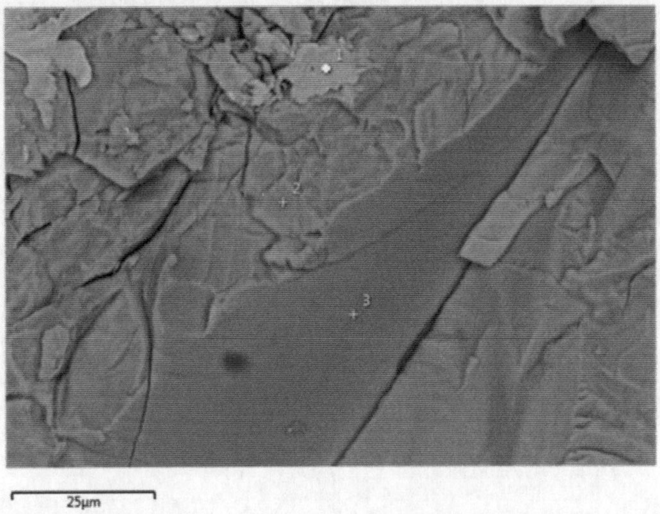

Figure 3. Image of investigated sections of the casting surface.

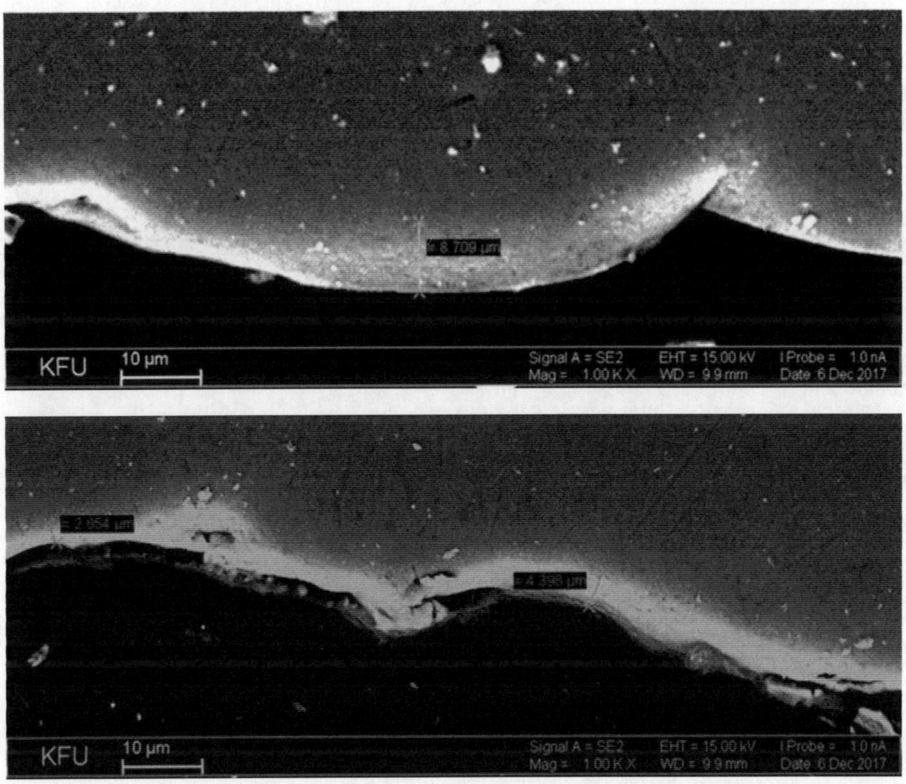

Figure 4. Image of the cross section of the casting slice.

Table 1. The elemental composition of the surface portions of the sample (at.%).

Regions of the surface	Elements								
	C	O	Mg	Al	Si	Ti	Cr	Mn	Fe
1	51.53	7.16	0.15	0.22	0.74	3.10	0.15	0.23	36.73
2	28.46	9.28	-	3.19	0.94	57.94	-	-	0.18
3	26.27	21.03	-	3.72	1.07	47.92	-	-	-

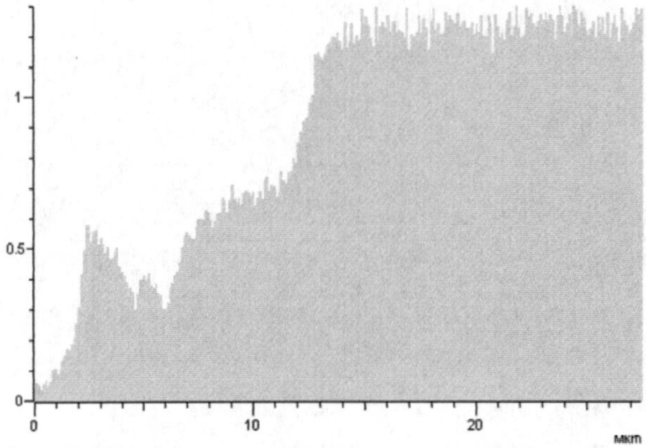

Figure 5. The profile of the Ti concentration in the depth of the "alpha case".

The elemental composition varies through the depth of alpha case (table 2). Thus, an area 1 μm from the surface is characterized by an elevated content of the mold elements and carbon. When moving into the interior of the sample (a point at a depth of 3 μm from the surface), a decrease in the content of the shape elements and an increase in the content of the alloy elements (titanium and aluminum) is observed.

Table 2. The elemental composition of the surface portions of the sample (at.%).

Distance from surface, μm	Elements							
	C	O	Mg	Al	Si	Ti	Ca	Fe
1	50.87	26.27	0.14	0.96	0.95	18.89	0.36	0.23
3	19.98	28.91	-	2.61	1.05	47.21	0.17	0.06

4. Conclusion

The thickness of the "alpha case" for the TL-3 alloy, which was contacting a 25 mm thick casting, varies from 2 to 13 μm.

The "alpha case" has a variable composition in depth, boundary layers that close to the mold contain chemical elements of the form, and deep layers of the casting mainly contain oxidized titanium.

5. References

[1] A I Khatmullina, E Yu Petrov and A M Lyakhovich Research of boundary layers of molds, used in titanium casting 2017 *8th Int. Science and Technical Conf.(Kazan)* vol 1 (Kazan: OOO Folianth) pp 357-361

Acknowledgements
The results were obtained as part of the state task of the Ministry of Education and Science of Russia 11.1300.2017/ПЧ.

IOP Conf. Series: Journal of Physics: Conf. Series **1058** (2018) 012043 doi:10.1088/1742-6596/1058/1/012043

Efficiency of ENO-schemes for computation of linear waves in an elastic body

N A Khismatullina

Institute of Mechanics and Engineering - Subdivision of the Federal State Budgetary Institution of Science "Kazan Scientific Center of the Russian Academy of Sciences" (IME - Subdivision of FIC KazanSC of RAS) 2/31, Lobachevsky str., Kazan 420111 Russia

E-mail: nailya_hism@mail.ru

Abstract. The efficiency of the UNO- and TVD-modifications of the Godunov method in computing one and two-dimensional linear waves in an elastic body is numerically studied. The body is simulated by an isotropic linear-elastic semi-space. Riemann problem of cylindrical discontinuity in hydrostatic pressure in a body and a problem of propagation of waves caused by the impulsive action on the free surface of a metal body are considered. It is shown that in all the cases both modifications resolve the waves in the body significantly better than the classic Godunov method. At that, in the case of presence of the pronounced extrema in the solution, the UNO-modification appears more preferable because the TVD- modification "cuts" the extrema to implement the TVD-property.

1. Introduction

Computations of elastic waves in bodies are necessary, in particular, when studying deformation and destruction of the body surfaces under the action of different factors. For example, if a multiphase gas-liquid medium with variable pressure contacts with a body, the surface of the body can be exposed to the action of the pressure pulses arising at cavitation bubble collapse. That is the reason the surfaces of structural elements contacting with liquid are subjected to cavitation erosion. Cavitation can destroy any material. Therefore, it is important to investigate the waves in a body in order to understand its internal processes to predict them and to determine the optimal operating conditions for the corresponding machines and mechanisms. To this end, one needs appropriate mathematical models and efficient methods of computation.

The classic Godunov method is widely used to solve many problems of the dynamics of continuous media [1, 2] due to its stability and monotonicity. Its disadvantages are low (first-order) accuracy and, as a result, strong smearing of jumps and contact discontinuities. In this paper, the efficiency of two modifications of the Godunov method in computing linear waves in an elastic body is examined. Those modifications belong to the class of ENO-schemes (ENO - Essentially Non Oscillatory) comprising UNO - and TVD-schemes of the second order (UNO - Uniformly Non Oscillatory, TVD - Total Variation Diminishing). They are similar to those used in [3] to compute nonlinear waves in gas and liquid. When constructing such schemes the time derivatives of the unknown functions are expressed in terms of their spatial derivatives. With the help of those expressions the values of the unknown functions at the next temporal semi-layer in the center of the grid cells and on both sides of each of their faces are computed. The values on the faces are found by solving Riemann problem. Limiters are usually used to confine the values of spatial derivatives in a cell. The modifications considered in the present work differ in the form of the limiters. The TVD-limiter provides the diminishing of total variation of solution of the difference scheme (TVD-property). In extreme points the order of accuracy reduces to the first one. The UNO-scheme limiter includes approximations of derivatives of not only the first order but also the second one. The main advantage of the UNO-scheme is its second order of accuracy throughout the areas of smooth solutions. The efficiency of the considered UNO- and TVD-modifications is demonstrated using one- and two-dimensional problems on the propagation of linear waves in an elastic metal body.

2. Problem statement and computational technique

The efficiency of computing linear waves in an elastic body using the UNO- and TVD-modifications of the Godunov method is studied. The body is simulated by a semi-space. Its dynamics is described by a system of seven linear equations [4] in the two components of the velocity vector, the four deviator components of the stress tensor in Cartesian or cylindrical coordinates, and the all-round (hydrostatic) pressure. The waves in the body result from its initial state or the action on its surface. In the case of one-dimensional problems, the computational domain is a segment $[0, R]$ the left end of which is assumed to be the surface of a body, the right end being an artificial boundary. In the case of two-dimensional problems, the computational domain is a square the upper and left sides of which are assumed to be the body surface and the symmetry axis, respectively, the right and the lower ones being artificial boundaries. In the problems of the present paper, the surface of the body is a free boundary with a specified pressure distribution. Artificial boundaries are usually introduced to limit the computational domain to the body area most interesting from a particular point of view. Non-reflecting conditions are posed on the artificial boundaries [5]. The computation domain is covered by the uniform grid. The time step is determined from the Courant condition.

The first of the three steps of the UNO- and TVD-algorithms computes the values of the unknowns at a semi-integer time layer. In this case, the values of the derivatives with respect to time are expressed in terms of the spatial derivatives. While computing the spatial derivatives, some limiters are used. The UNO- and TVD-schemes differ in the form of those limiters. The limiter of the UNO-scheme includes the approximations to both the first and second derivatives, while that of the TVD-scheme just utilizes the approximation to the first derivative.

The second step is the computation of the numerical fluxes across the cell sides using the solution of the Riemann problem.

On the third step, the unknown values of the new time layer are computed applying the explicit finite-difference formula.

A more detailed description of the present UNO- and TVD-algorithms can be found in [4].

3. Results

3.1 Cylindrical Riemann problem

At the initial time $t = t_0$, there is a discontinuity in the all-round pressure P in the computational domain of radius R: $P = 0$ in the area $r < 0.5R$ and $P = p^*$ in the area $r > 0.5R$ where R and p^* are known quantities.

Figure 1 demonstrates the solution of this problem on a uniform grid of 200 cells at some time $t > t_0$. It is seen that both schemes of the second order describe the fronts and extrema of the waves much better than the Godunov scheme. At that, the UNO scheme resolves the extrema better than the TVD scheme. For example, the deviations of the extremum value and its position computed by the TVD-scheme in the vicinity of $r \approx 0.2R$ from those found by the UNO scheme are respectively 2% and 3%. The similar deviations of the results computed by the Godunov scheme are 7% and 22%.

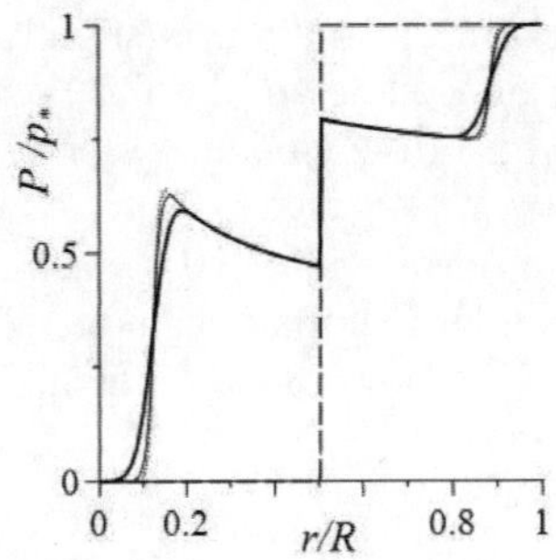

Figure 1. Riemann problem on the cylindrical discontinuity in the all-round pressure P. The dashed curve is the initial distribution, the thick-solid, fine-solid and dotted curves are respectively the results by the Godunov method and its TVD- and UNO-modifications

3.2. Dynamics of a body under the action on its surface

Two problems of propagation of waves inside an unperturbed body are considered. The waves are excited by the action on the free surface of the body located at $y = 0$. In the first case, the wave in the body results from the impulse action onto the whole surface of the body (the problem is one-dimensional), while in the second case the wave arises due to the action just onto a part of the bode surface (the problem is two-dimensional). The time dependences of the pressure defining the corresponding impulse actions onto the body surface are shown in the insets of figure 2 and figure 3.

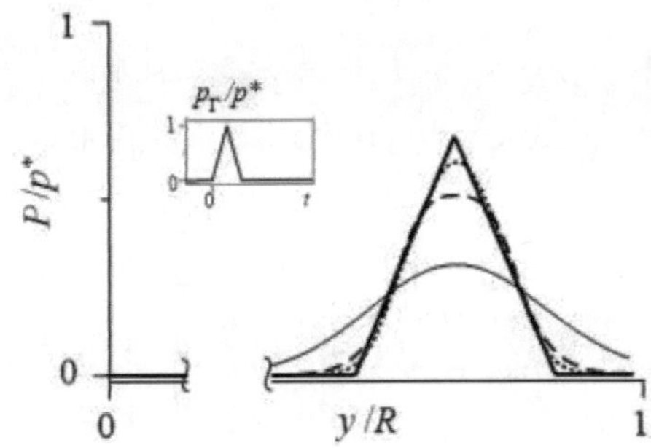

Figure 2. The distribution of the all-round pressure P in the body at some instant of time in the case the wave is excited by the pressure pulse on the whole free surface $y = 0$ of the body (the boundary pulse is shown in the insert). The thick solid line is the exact analytical solution, the fine-solid, dashed and dotted lines correspond to the Godunov, TVD and UNO schemes, respectively.

Some results of computations of the first problem on a grid with a step $h = R / 100$ are shown in figure 2. It can be seen that the UNO scheme solution is the closest to the exact solution. Its maximum

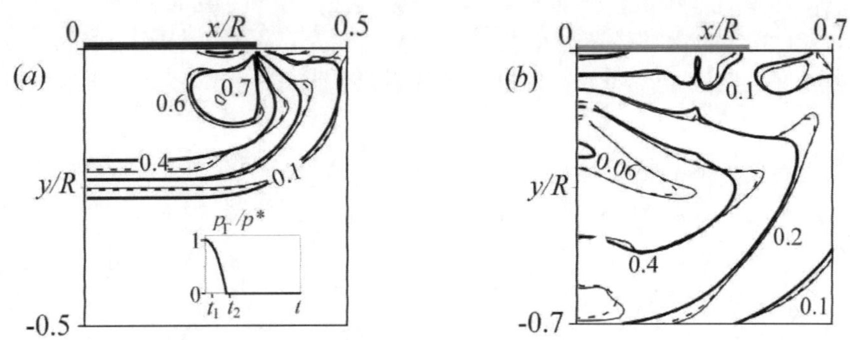

Figure 3. The dynamics of the body (in the case of impact onto a part of its free surface $y = 0$) in the vicinity of the impact spot: the stress intensity isolines (presented in fractions of p^*) in a fragment of the computational domain at two consecutive time moments t_1 and t_2 (marked in the inset where the boundary pulse is shown) computed by the Godunov (solid thick curves), TVD (dashed curves) and UNO (solid thin curves) schemes; the black stripes at the top are the impact spot.

deviation in the amplitude of the wave is 10%. The TVD scheme resolution is slightly worse. Its amplitude error of 24% is due to a decrease of the second order of accuracy of the TVD scheme to the first one at the extreme points of the solution. The amplitude error of the Godunov scheme is as large as 54%.

Some results of computations of the second problem on a uniform grid of 100 × 100 cells at two consecutive time moments t_1 and t_2 are shown in figure 3. It can be seen that again the second-order schemes give much more accurate numerical solutions than that by the Godunov scheme. And again, the UNO scheme resolves the extremes better than the TVD scheme.

4. Conclusion

Some results of a numerical investigation of the efficiency of the second-order accurate UNO- and TVD-modifications of the Godunov method in computing one- and two-dimensional linear waves in an elastic body are presented. Riemann problem of a cylindrical discontinuity in the all-round pressure inside a body and that of propagation of waves caused by the impulse action on the free surface of a body are considered. The exact solution and that computed by the Godunov method are used for comparison in the first and second cases, respectively. It is shown that in all the cases considered both modifications of the Godunov method describe the dynamics of the body considerably better than the Godunov method. Moreover, the UNO-modification is more preferable since the TVD-modification "cuts" the extremes to implement the TVD-property.

Acknowledgments

The work was supported by the Russian Foundation for Basic Research (project No. 16-01-00433).

References

[1] Godunov S K, Zabrodin AV, Ivanov M Y, Krayko A N and Prokopov G P 1976 *Numerical solving of gas dynamics multidimensional problem* (Moscow: Nauka) p 400

[2] Cheban V G, Naval I k, Sabodash P f and Cherednichenko R A 1976 *Numerical methods for solving problems in the dynamic theory of elasticity* (Kishinev: Shtiints publishing house) p 226

[3] Aganin A A, Khalitova T F and Khismatullina N A 2008 *Computational Technologies* **13** No 6 54

[4] Aganin A A and Khismatullina N A 2016 *Proceedings of the Mavlyutov Institute of Mechanics* **11** 119

[5] Ilgamov M A and Gilmanov A N 2003 *Nonreflecting conditions on the boundary of computational domain* (Moscow: FIZMATLIT) p 240

IOP Conf. Series: Journal of Physics: Conf. Series **1058** (2018) 012044 doi:10.1088/1742-6596/1058/1/012044

Simulation of electron moving in RF capacitively coupled discharge

R M Askhatov[1], I B Badriev[1], V Yu Chebakova[1] and V S Zheltukhin[2]

[1]Kazan Federal University, 18 Kremlevskaya Street, Kazan, 420008, Russia
[2]Kazan State Technological University, 68 Karl Marx Street, Kazan, 420015, Russia

E-mail: vchebakova@mail.ru

Abstract. An algorithm of simulation of kinetic movement of electrons in alternating-current field of radio-frequency capacitively coupled discharge is presented in the paper. The results of solving of modelling problems are presented also.

1. Introduction

Mathematical models of capacitively coupled radio-frequency discharge (CCRF discharge) at both "local" and "non-local" approximating are described in papers [1, 2]. The results of numerical calculations showed that large gradients of the electric field appear in near-electrode areas, which can result in strong anisotropy of the electron energy distribution function (EEDF) and inability of using of binomial splitting for solving Boltzmann equation. For greater certainty of the distribution function of charged particles in the areas of large gradients of an electric field under the conditions, characteristic to RFCC discharge, statistic simulation by Monte Carlo method is supposed to be used. Statistical simulation is widely used for describing discharges. A new spectral method for quick evaluation by Fokker-Planck-Landau (FPL) is presented in the paper [4]. The method of direct statistic simulation of Monte Carlo for kinetic equation with non-linear operator of Coulomb collisions by FPL is discussed in the paper [5]. This method is based on approximating of Landau collision integral by Boltzmann collision integral. A general approach towards Monte Carlo methods for Coulomb collisions is proposed in the paper [6]. The key idea is approximation of equations of FPL by Boltzmann equations of quasi-Maxwell type. It means that total frequency of collisions for respective Boltzmann equation does not depend on particle speeds. Main features of ion-atomic collisions, based on Monte Carlo modeling are reviewed and discussed in the paper [7]. Numerical modeling of EEDF of high voltage impulse discharge of beam type by Monte Carlo method is conducted in paper [8]. Modifications of Monte Carlo method for using in application tasks can also be viewed in the papers [8-12]. It is expected, that the results of calculations of the functions of distribution of charged particles by the energies, received by Monte Carlo method, can be used in modeling low-temperature plasma, for instance, the one described in the papers [13-17].

2. Numerical method for solving the problem

An algorithm of modeling of movements of an electron in an alternating electric field, based on the works [18,19,20] is presented in this paper. The realization of Monte Carlo method in this paper is combined with methods of computing the model of continuum, proposed in the article [20]. It is assumed that the Oz axis is directed perpendicular to the surface of the electrodes. Computational grid is introduced by the following way: spatial uniform grid within the interval [0, *b*] is $\omega_h = \{z_l = lh, l = 0,1,\ldots,M\}$, $h = (b-0)/M$, then temporal grid is $\omega_T = \{T_s = st, s = 0,1,\ldots\}$, where *h* is a space-wise step of the grid, *t* is a temporal step, and *b* is inter-electrode distance. Grid functions are defined on the 2D grid $\omega_h \times \omega_t$. Consequently, the model of elementary cell is considered as rectangular parallelepiped which size along the *Oz* axis is *h,* sizes along the *Ox, Oy* axes are arbitrary (fig.1). Changes in the field within the cell are considered linear and directed along the *Oz* axis (fig.2).

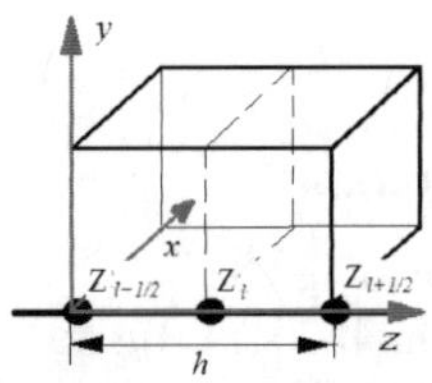

Figure 1. Elementary computational cell.

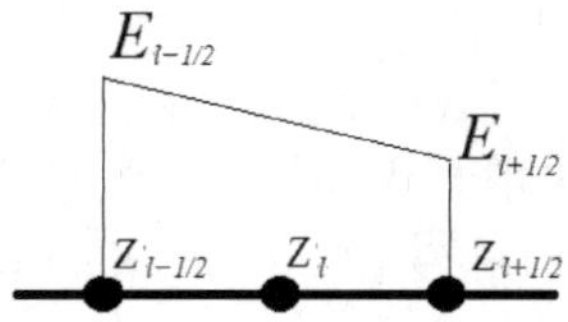

Figure 2. Approximation of an electric field in the cell

The direction of the field in the CCRF discharge is changed periodically. Consequently, the electrons in the reviewed model start from the plane which is parallel to electrodes surfaces and goes through the middle point of the cell on the *Oz* axis (fig.1, 2). The direction of the electron moving depends on the field direction. The movement of electrons is observed either before they leave the cell borders, or within time interval *t*, which is entered when solving non-stationary problems [1]. The movement of an electron is considered as a discrete random process, that is the electron trajectory is divided into separate intervals, in which the properties of the movement do not change. The length of these intervals depends on the time of non-collision movement, which is defined at random. Between these intervals, the electron participates in the process of collision, the type of which is generated using a geometrical probability. The spatial measure of all elementary outcomes is estimated as a sum of maximum measures of elementary outcomes, whereas the measure of the collision process is considered to be equal to its frequency. Since the frequencies of collisions depend on the speed of an electron, which is not constant, the measures of elementary outcomes at every generating at the end of the movement interval also change, and to keep the spatial measure of all elementary outcomes constant, the "empty" type collisions are introduced.

- The time interval of non-collision movement is computed. Considering the probability of a collision equal to $P(\tau) = \exp(-\int_0^{\tau} \nu(\upsilon)dt)$, where $\nu(\upsilon)$ is the frequency of collisions, which depends of the speed υ, and equating $P(\tau)$ to a random number *S*, uniformly distributed in (0,1], the value of $\tau = -\frac{\ln S}{\nu}$ is calculated.
- Equations of movement are solved at time interval τ, projection of the speed is re-calculated along the *Oz* axis at the end of the movement, where the field is equal $E_z = (E_{l+1/2} - E_{l-1/2}) \times (z - z_{l-1/2})/h + E_{l-1/2}$.
- If an electron has moved of the computational area, then the next electron is taken into consideration. Otherwise, generating the process of mutual collision is considered. The probability of the late process *P* is generated at random, equally distributed in (0,1]. The number of the process is defined by the *P* value range in the inequation $\sum_{k=1}^{n-1} \nu_k/\nu < P < \sum_{k=1}^{n} \nu_k/\nu$, and then the *n* process is implemented.
- In case of recombination of an electron and an ion, a scenario for the next electron takes place. When implementing the process of ionization to calculate the speed of a new electron, we solve a system of equations, consisting of the equation of energy conservation and the law of conservation of an impulse by Newton's method, considering the angle of spread is

uniformly distributed on $[0,\pi]$. When “empty” processes take place, one returns to entry 1. In other instances, the speed of the electron is re-calculated, according to the implemented process. The technique of re-calculating the speed is described in the paper [18]. Returning to entry 1

After counting frequency of processes of collisions when the speed of movement is changed, their sum is observed: if the sum is greater than ν = const, then the movement of an electron is watched again from the starting point of the movement, considering that ν equals the calculated sum.

3. The results of numerical experiments

While solving model tasks, it was supposed, that the field goes along the Oz axis in the negative direction and the module of electric field strength decreases along the positive direction of the Oz axis. The results of simulation show, that at different pressures (for which the concentration of heavy particles is calculated), and the same the field strength, temperature, initial speed, the electron path at a lower pressure is greater, and the amount of mutual collisions is lesser (fig. 3a) then at greater pressure. When the pressure is increased, the electron path is decreased, the amount of mutual collisions is increased, whereas the ionization process is concentrated in the middle of the area (fig. 3b). It happened because the electric field strength was larger at the beginning the area and decrease towards the end along the Oz axis (fig. 2). An electron starting at the left edge of the grid, influenced by the field while moving, gained energy, strong enough for ionization, after which it lost the energy needed for ionization. However, after inelastic mutual collisions the speed of the electrons increased at a rate slower than in the beginning, due to the decreasing of the field magnitude (fig.3).

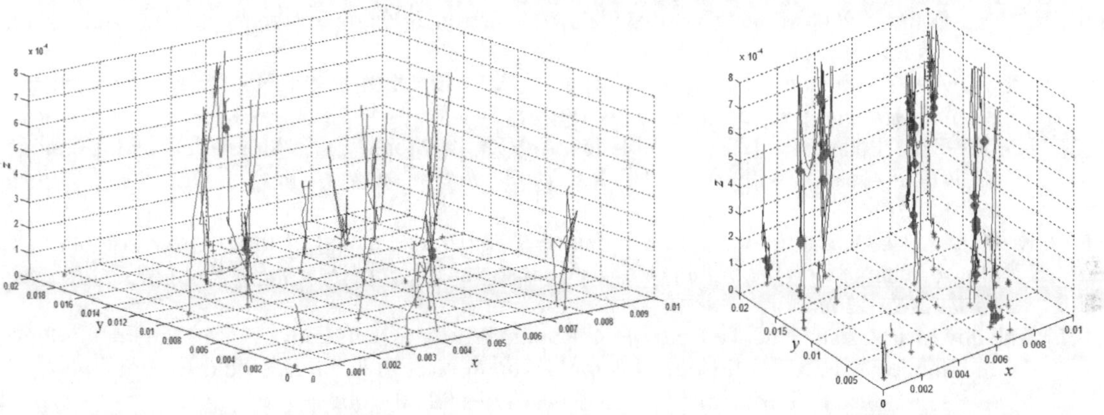

a) *b)*

Figure 3. Trajectories of movements of electrons at pressures P=13.3 Pa (a) and P=1330 Pa (b). Trajectory of movement of an electron is shown in blue. Spots, where ionization process was detected, are marked red. Beginning of movement of primary electrons is shown in green.

4. Conclusion

The obtained consistent patterns correspond with general notions of correlation of movement of electrons in an electric field under different pressures. This allows one to conclude, that the elaborated algorithm of procedure of describing movements of electrons is correct.

Acknowledgments

The work is performed according to the Russian Government Program of Competitive Growth of Kazan Federal University and supported by Russian Science Foundation, project no. 16-11-10299.

References

[1] Chebakova V Yu 2017 Modeling of Radio-Frequency Capacitive Discharge under Atmospheric

Pressure in Argon *Lobachevskii Journal of Mathematics* **38** (6)1165-78

[2] Askhatov R M, Chebakova V Yu and Zheltukhin V S 2016 Capacitive coupled RF discharge: modelling at the local and not local statement of the problem *IOP Conf. Series: Materials Science and Engineering* **158** 012009

[3] Hagelaar G J M and Pitchford L C 2005 Solving the Boltzmann equation to obtain electron transport coefficients and rate coefficients for fluid models *Plasma Sources Sci. Techn.* **14** 722-33

[4] Pareschi L, Russo G and Toscani G 2000 Fast spectral methods for the Fokker–Planck–Landau collision operator *Journal of Computational Physics* **165**, 216–36

[5] Karpova S A, Potapenkova I F and Bobylev A V 2017 On the accuracy of the direct discrete simulation of the Landau collision integral by the Boltzmann integral *Mathematical Models and Computer Simulations* **9** (2) 206-20.

[6] Bobylev A V and Potapenko I F 2013 Monte Carlo methods and their analysis for Coulomb collisions in multicomponent plasmas *Journal of Computational Physics* **246** 123-44

[7] Sysun V, Sysun A, Ignakhin V, Titov V, and Tikhomirov A 2014 Comparison of Simulation Methods of Ion-Atomic Collisions in PIC-MC *Journal of Applied Mathematics and Physics* **2** 1233-41

[8] Demkin V P, Mel'nichuk S V, Borisov A V and Bardin S S 2013 Monte-Carlo calculation of the electron energy distribution function of a He–N_2 plasma of a high voltage pulsed discharge *Russian Physics Journal* **56**(4) 486-8

[9] Ponomarev A A and Aleksandrov N L 2017 The kinetics of energetic O^- ions in discharge H_2O plasma *IOP Conf. Series: Journal of Physics: Conf. Series* **927** 012044

[10] Savino Longo 2000 Monte Carlo models of electron and ion transport in non-equilibrium plasmas *Plasma Sources Sci. Technol.* **9** 468–76.

[11] Bobylev A V., Potapenko I F, and Karpov S A 2012 DSMC Methods for Multicomponent Plasmas *AIP Conference Proceedings* **1501**(1) 541-8

[12] Karpov S A, Potapenko I F, and Bobylev A V 2014 Monte Carlo simulation of the kinetic collisional equation with external fields. *Mathematical Models and Computer Simulations* **6** (6) 598–611

[13] Kashapov N F, Saifutdinov A I and Fadeev S A 2014 The model of the positive column of a glow discharge with the influence of the acoustic oscillations *Journal of Physics: Conference Series* **567** 012004

[14] Akhatov M F, Gaisin F M, Iskhakov F R, Kayumov R R, Kuputdinova A I, Mukhametov R A, and Shpilev A I 2017 Influence of composition of electrolyte on electric discharge in Liquid *IOP Conf. Series: Journal of Physics: Conf. Series* **927** 012002

[15] Gaisin A F 2006 A multichannel discharge between a stream electrolytic cathode and a metal anode at atmospheric pressure *High Temperature* **44** (3) 336–41.

[16] Saifutdinov A I, Saifutdinova A A, Kashapov N F and Fadeev, S A 2016 Dynamics contraction of DC glow discharge in argon *Journal of Physics: Conference Series* **669** 012045

[17] Fadeev S A, and Kashapov N F 2013 The influence of transverse acoustic oscillations on contraction of the glow discharge *J. Phys.: Conf. Ser.* **479** 012009

[18] Yousfi M, Hennad A and Alkaf A 1994 Monte Carlo simulation of electron swarms at low reduced electric fields *Physical Review E* **49**(4) 3264-73

[19] Hans Rau 2000 Monte Carlo simulation of a microwave plasma in hydrogen *J. Phys. D: Appl. Phys.***33** 3214–22

[20] Gabbasov F G, Dubrovin V T and Chebakova V J 2016 Vector random fields in mathematical modelling of electron motion *IOP Conf. Series: Materials Science and Engineering* **158** 012032

[21] Badriev I B, Zheltukhin V S and Chebakova V Ju 2017 Numerical solution of the initial boundary value problems of radio - frequency capacitive coupled discharge. *IOP Conf. Series: Journal of Physics: Conf.Series* **927** 012008

IOP Conf. Series: Journal of Physics: Conf. Series **1058** (2018) 012045 doi:10.1088/1742-6596/1058/1/012045

Experimental studies of resonant nonlinear oscillations of a gas in an open and closed tube

R G Zaripov[1], L A Tkachenko[1,2] and L R Shaidullin[1,2]

[1]Institute Institute of Mechanics and Engineering, Federal Research Center KazSC, Russian Academy of Sciences, 2/31 Lobachevsky str., Kazan 420111, Russia
[2] Kazan (Volga Region) Federal University, 18 Kremlyovskaya str., Kazan 420111, Russia

E-mail: liqn@mail.ru

Abstract. Nonlinear oscillations of gas in an open and closed tube were researched experimentally. Dependencies of the of the pressure oscillation amplitude were obtained at the resonance frequencies of the gas excitation at a fixed amplitude of the tube displacement.

1. Introduction.

The development of modern technology creates new challenges, among which one of the most urgent is the study of resonant vibrations of the environment in various apparatuses of power engineering and industry. Of special interest is the tube-resonator system, where one end of the tube is closed or communicates with the environment. Excitation of the gas produces highly nonlinear pressure waves and nonlinear effects are observed [1] near the resonances in these systems, for example, flow turbulence, secondary currents and acoustothermic processes. The results of these media were analyzed in a review [2, 3] in theoretical and experimental studies and considered transient regimes of gas flow at intense excitation amplitudes [4]. It seems necessary to experimentally studies and compares the nonlinear oscillations of the gas in an open and closed tube near its first basic frequency.

2. Experimental installation.

The studies were carried out in an installation with a TIRAvib S 5220 / LS vibration generator 1 (Figure 1), which set in motion a flat tube 3. A tube with a diameter $d = 0.1$ m oscillated harmonically with a predetermined frequency ν in Hz in a glass tube 4 of length $L = 0.918$ m for the open and $L = 0.938$ m for the closed pipe. The tube was opened or closed tightly with a lid 5.

The vibration table was controlled via a computer 7 by means of a piezoelectric IEPE accelerometer 2 of 4513 model by Bruel & Kjaer firm and a controller of VR9500 using specialized VibrationVIEW software. The pressure sensor 6 of the Bruel & Kjaer 6 model 8530C-15 was located next to the tube, the signal from which was fed through a three-channel bridge voltage amplifier 9 ENDEVCO model 136 of Bruel & Kjaer firm to a digital oscilloscope 8 of DSO 3062A model by Agilent Technologies firm, and then via an RS-232 interface to the computer, where the data for the gas pressure p in bar can be observed and saved using a specialized DSO3000 software.

IOP Conf. Series: Journal of Physics: Conf. Series **1058** (2018) 012045 doi:10.1088/1742-6596/1058/1/012045

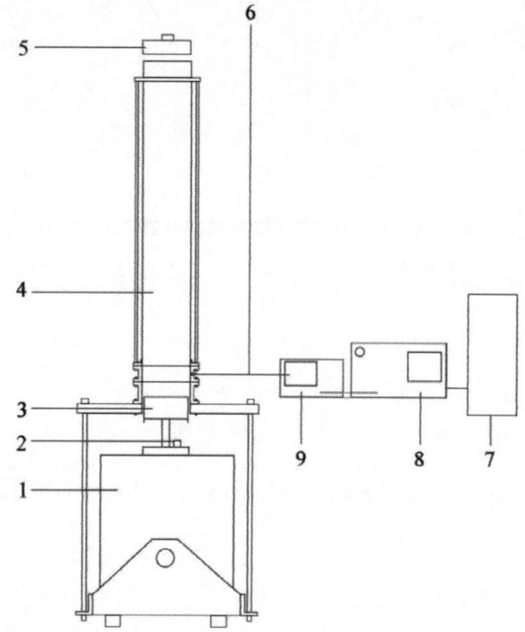

Figure 1. Basic diagram of the experimental installation.

3. Experimental results.

The dependencies are shown in Figure 2 for the range of gas pressure variations Δp from the frequency ν at the amplitude of the tube stroke l = 0.375 mm. As can be seen, the pressure value exceeds in the closed pipe almost 2 times the pressure for the open pipe [7]. This is due to the

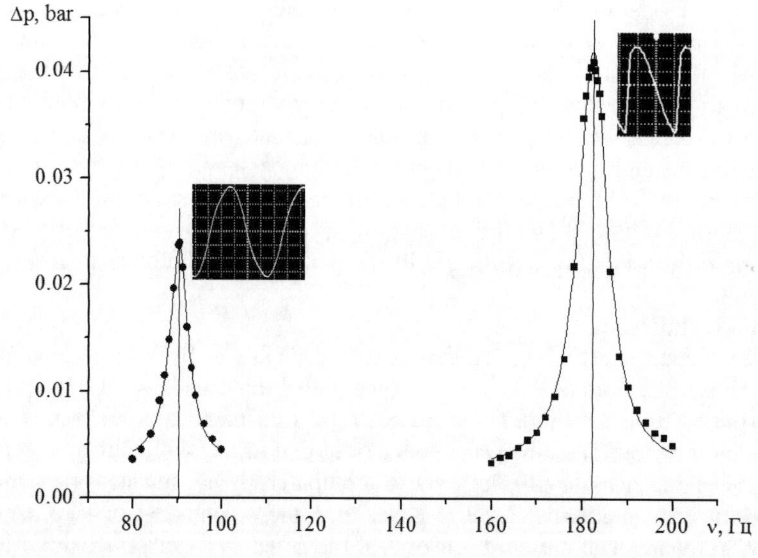

Figure 2. Dependency of the span of the gas pressure oscillation Δp on the excitation frequency ν at the amplitude of the tube stroke l = 0,375 mm, where ■ corresponds to an open (ν = 90.3 Hz), and ● is closed tube (ν = 182 Hz). Dots mark the experimental results, and the solid line is Lorentz approximation.

boundary conditions at the end of the pipe. In a pipe with an open end, a continuous interaction between the gas in the pipe and air from adjacent to the open end of the external environment, as a result of which energy losses to acoustic radiation and vortex formation are observed [5, 8]. The front of the pressure wave has a strong distortion in the closed pipe, which leads to an increase in the average axial velocity [6] and the oscillations are close to discontinuous, in the open tube - the waveform is sinusoidal.

4. Conclusion.

The obtained data show that at a resonance frequency of gas oscillations at an equal amplitude of the tube stroke, the gas pressure variations in an open and closed pipe are distinguished, which is associated with losses at the open end. The waveform of the pressure wave has an asymmetrical leading and trailing edge, more pronounced for a closed pipe, as a result of which the pressure is higher.

References

[1] Ilgamov M A, Zaripov R G, Galiullin R G, Repin V B 1996 *Appl. Mechanics Reviews* 49 137
[2] Zaripov R G, Galiullin R G and Galiullina E R 2001 *Actual problems of continuum mechanics. On the 10th anniversary of IME KazSC RAS* 19-35 (Kazan: IME KazSC RAS)
[3] Tkachenko L A and Zaripov R G 2013 *Russian Aeronautics* (1) 28-32
[4] Zaripov R G, Kashapov N F, Tkachenko L A, Shaydullin L R 2016 *J. Phys.: Conf. Series* **669** 012053
[5] Van Wijngaarden L. and Van Wormgoor L. 1974 *Finite-Amplitude Wave Eff. Fluid proc. Symp. Copenhagen* 75-80
[6] Zaripov R G, Tkachenko L A, Shaydullin L R 2014 *J. Phys.: Conf. Series* **567** 012042
[7] Shaydullin L R 2016 *IOP Conf. Ser.: Mater. Sci. Eng*. **134** 012022
[8] Gubaidullin D A, Kashapov N F, Zaripov R G, Tkachenko L A, Shaydullin L R 2017 *J. Phys.: Conf. Series* **789** 012017

Application of nanostructured coatings by plasma spraying

I.N. Kravchenko[1], A.V. Kolomeichenko[2], S.N. Sharifullin[3,4], Yu.A. Kuznetsov[2], Yu.N. Baranov[5], M.A. Glinsky[1], A.A. Kolomeichenko[1]

[1] Russian State Agrarian University-Moscow Agricultural Academy named after K.A. Timiryazev, Russia
[2] Orel State Agrarian University named after N.V. Parakhin, Russia
[3]Kazan Federal University, Russia
[4]Kazan state agrarian University, Russia
[5] Orel State University named after I.S. Turgenev, Russia

Email: Saidchist@mail.ru

Abstract. Nanostructured coatings demonstrate lower porousness by an order of magnitude, a far higher tenacity to wear, corrosion and erosion. The article presents the investigations of new method of nanostructured coatings application consisting in plasma spraying of material supplied into plasma jet and nanoparticles in suspension or solution followed by coating surface melting by air plasma jet.

Key words: coating, plasma, wear, nanoparticles, plasma jet, wear resistance, carbides, oxides, additive, microstructure, substrate

1. Introduction.

Gas thermal spraying technologies are developing intensively during the last years [1-16]. Such coatings are characterized with multilayer structure, porousness, limited values of adhesion and cohesion. Reduction of the particles forming coating considerably improves their quality. Coatings from finely divided powders and nanostructured materials demonstrate lower porousness by an order of magnitude, a far higher tenacity to wear, conditioned with abrasive particles presence, and to corrosion and erosion. In modern times of science and technology development there is a possibility to obtain new nanostructured coatings with grains sized less than 100 nm. These coatings possess improved mechanical features [17, 18].

At the correctly selected parameters nanoparticles of the sprayed material will form coating layer oncoming by integrity to the coating material in compact state. As a result of formation of complex of mechanical and technological characteristics it is possible to expect principal improvement of performance property of parts and assembly components of equipment. Over the long term it will allow to replace expensive materials with cheaper ones. One of these characteristics is wear resistance.

2. The work objective is to create nanostructured coatings based on nickel-chrome matrix, which are more resistible to wear in comparison with well-known and widely used coatings.

3. Methods.

To study the possibility to increase wear resistance of coatings based on Ni-Cr-B-Si-alloy several variants of nanostructured coatings with wear resistant additive different content of wear resistant additive in the form of tungsten carbide and aluminum oxide nanoparticles plasma sprayed with subsequent melting were obtained.

Investigation of element composition and microstructure of coatings was carried out by X-ray microanalysis (XRMA) and scanning electron microscopy (SEM) methods.

Nanostructured coatings being applied on low-carbon steel samples were tested on wear resistance according to GOST 23.208-79 (method of testing materials on wear resistance at friction on not rigidly fixed abrasive particles) [19]. The essence of the method is in the fact that at equal terms we produce the friction of the tested and reference material on the abrasive particles supplied in the friction area and pressed against the sample by means of rotary rubber roller. The wear of the samples of the tested and reference material is measured. Wear resistance of the tested material is estimated by comparing its wear with the reference material wear.

4. Investigation results.

Sprayed coatings microstructure is presented in figure 1. Basic phases of coating – γ – Ni (matrix) and carbides of different types. After melting close to base carbides acicular shape is unchanged. (Fig. 1, *b*). It is typical for coating deposition using plasma spraying method, carbides at the surface take a form of polygons and globes (Fig. 1, *a*). Tungsten carbide particles are distributed close to the coating surface. Their dimensions are distributed in the range from several hundreds of nm to 10 μm.

For the samples being obtained by plasma spraying method with the following melting. It is typical that there is an intermediate layer on the boundary «coating-base» (Fig. 2). Ni content in the intermediate layer is more than in the base material.

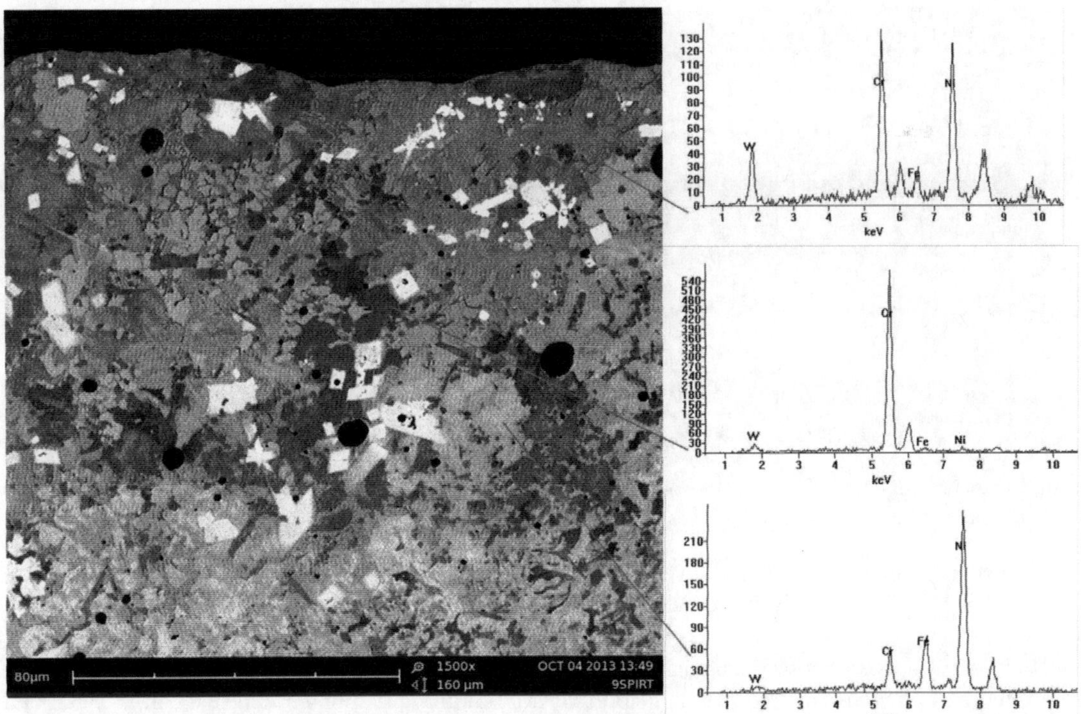

a)

IOP Conf. Series: Journal of Physics: Conf. Series **1058** (2018) 012046 doi:10.1088/1742-6596/1058/1/012046

b)

Figure 1. Coating microstructure: *a* – external surface of coating; b – boundary «coating-base»

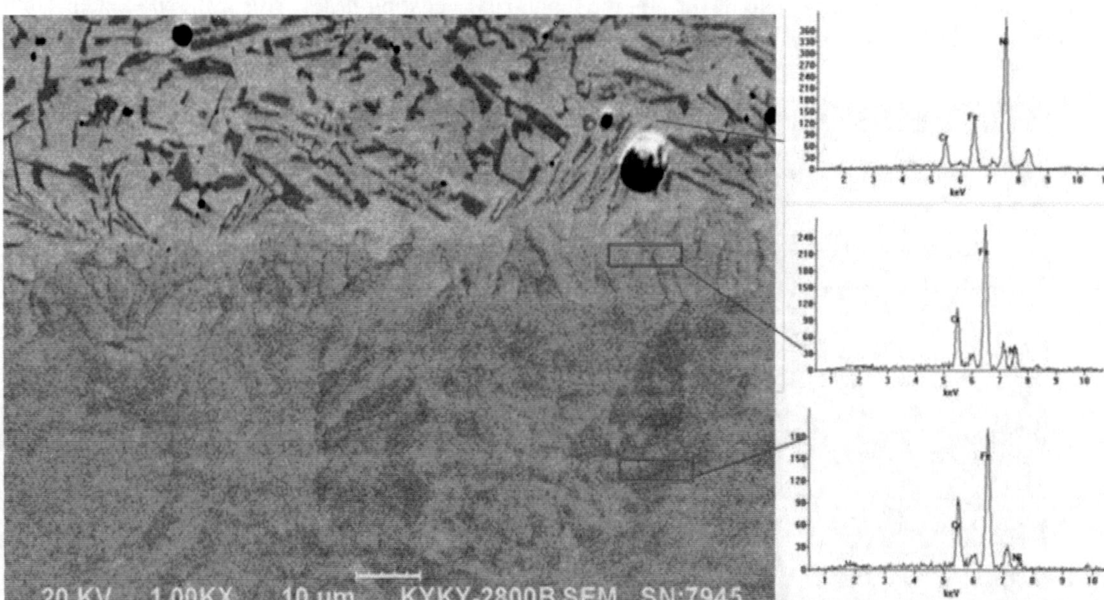

Figure 2. Formation of the interlayer and distribution of Ni on boundary «coating-base» (X-ray microanalysis (XRMA) and scanning electron microscopy (SEM)

Under test nanostructured coatings of granulated silicium boron powder (GSBP). granulated silicium boron powder (GSBP) with nanoadditives of aluminum oxide and tungsten carbide, generally acquire the characteristics, presented in table 1.

Nanostructured coatings of granulated silicium boron powder (GSBP) with Al_2O_3 and WC demonstrated the best tenacity to abrasive wear. Application of nanosized additives of tungsten carbide allows reducing residual stresses in coatings, which terminate in failure on application of conventional powder mixtures.

Table 1. Results of sample testing on wear

Coating material	Sample weight before test, g	Test Time, s	Sample weight after test, r	Weight loss of sample, g		Relative wear resistance
Steel 45	93,44 92,61 90,01	480	92,81 91,93 89,35	0,63 0,68 0,66	0,65	1,00
GSBP	176,60 176,18 178,24	480	176,17 175,73 177,80	0,43 0,45 0,44	0,44	2,76
GSBP + WC	165,72 168,07 173,50	480	165,34 167,71 173,10	0,38 0,36 0,40	0,38	3,89
GSBP + 15% Al_2O_3	94,011 94,61 95,72	480	93,531 94,11 95,25	0,48 0,50 0,47	0,48	3,25

The obtained results can be explained that inadequate penetration of particles, forming the coating, results in its brittleness and porousness increase. Hence, micro-hardness and wear resistance of coating are decreased. In contrast, increase of carbide phase mass fraction in coating composition, results in its wear resistance increase. This is because particles of tungsten carbide have high kinetic energy that is transformed into heat energy at particle impact on the surface at spraying. It provides more sound coating at the account of increase of coating particle cohesion among themselves.

5. Conclusion.

Plasma sprayed with powder material supplied into plasma jet and with nanoparticles in the suspension or solution form with the following melting by air plasma jet, coatings demonstrate high resistance to abrasive wear. Probably nanostructured coating properties are connected with saving nanostructure in coating during spraying. As follows from the analysis of wear resistance of sprayed nanostructured coatings the importance of qualitative characteristics of materials being selected to achieve the required level of coatings properties has become obvious.

References.

[1] Baldaev, L.Kh. Gas thermal spraying: study book /L.Kh. Baldaev, V.N. Borisov, V.A. Vakhalin; under the general editorship of L.Kh. Baldaev. – M.: Market DS, 2007. – 344 p.
[2] Puzryakov, A.F. Theoretical foundations of plasma spraying technology. – 2-nd ed., updated and revised /A.F. Puzryakov. – M.: Publishing house Moscow State Technical University named after N.E. Bauman, 2008. – 360 p.

[3] Kolomeichenko, A.V. Technolpgy of machine repair. Laboratory course: study book. Part II. /A.V. Kolomeichenko, V.N. Logachev, N.V. Titov [et al.]. - Orel: Publishing house Orel State Agrarian University, 2013. - 156 p.
[4] Kravchenko, I.N. Investigation of process of self-fluxing alloys plasma spraying by external arc with coating melting /I.N. Kravchenko, A.V. Kolomeichenko, A.A. Kolomeichenko, D.M. Butenko //Welding engineering. - 2016. - No. 10. - Pp. 22-25.
[5] Kuznetsov, Yu.A. Innovative methods of gas thermal spraying of coatings: monograph / Yu.A. Kuznetsov, V.V. Goncharenko, K.V. Kulakov. – Orel: Orel State Agrarian University, 2011. – 124 p.
[6] Kuzmin, V.I. Plasma spraying of wear resistant coatings from self-fluxing alloys powders /V.I. Kuzmin, S.P. Vashhenko, I.P. Gulyaev [et al.]. //Bulletin of Ugra State University. – 2015. – Issue 2(37). – Pp. 45-52.
[7] Sosnin, N.A. Plasma technologies /N.A. Sosnin, S.A. Ermakov, P.A. Topolyansky. – SPb.: Publishing house St. Petersburg Polytechnic University, 2008. – 406 p.
[8] Science and Engineering of Thermal Spray Coatings, 2nd ed., Pawlowski L., John Wiley & Sons, Ltd. – 2008. – 656 p.
[9] Temperature measurements for Ni-Al and Ti-Al phase control in SHS Synthesis and plasma spray processes /P. Gulyaev, H. Cui, I. Gulyaev, I. Milyukova //High Temperatures-High Pressures. – 2015. – V. 44. – No. 2. – Pp. 83–92.
[10] Kuzmin, V.I. The technique of formation of the axisymmetric heterogeneous flow for thermal spraying of powder materials /V.I. Kuzmin, A.A. Mikhalchenko, O.B. Kovalev [et al.] //J. of Thermal Spray Technology. – 2012. – V. 21. – No. 1. – Pp. 159–168.
[11] Kornienko E E, Lapushkina E J, Kuzmin V I, Vaschenko S P, Gulyaev I P, Kartaev E V, Sergachev D S, Kashapov N, Sharifullin S, Fayrushin I 2014 *J. Phys.: Conf. Ser.* **567** 012010
[12] Kashapov N F and Sharifullin S N 2015 *IOP Conf. Ser.: Mater. Sci. Eng.* **86** 012021
[13] Kolomeichenko, A.V. Improvement of equipment and technology at thermal spraying /A.V. Kolomeichenko, V.N. Logachev, N.N. Litovchenko /Education, science and industry. - 2015. - No. 4 (13). - Pp. 27-32.
[14] Kashapov L N, Kashapov N F and Kashapov R N 2014 *J. Phys.: Conf. Ser.* **567** 012025
[15] Saifutdinov A I, Fairushin I I and Kashapov N F 2016 *JETP Lett.* **104** 180–185
[16] Puzryakov, A.F. New approaches to increase the resource of engineering parts by methods of nanostructured material gas thermal spraying /A.F. Puzryakov, I.N. Kravchenko, A.V. Kolomeichenko [et al.] //Repair. Reconditioning. Modernization. – No. 6. – 2014. – Pp. 32-35.
[17] Kravchenko, I.N. Method of management of residual stresses in plasma coatings /I.N. Kravchenko, A.F. Puzryakov, E.V. Pankratova // Repair. Reconditioning. Modernization. – 2011. – No. 10. – Pp. 6–11.
[18] Kravchenko, I.N. Thermophysical model of determination of residual stresses in plasma coatings /I.N. Kravchenko, E.V. Pankratova, O.Ya. Moskal' // Repair. Reconditioning. Modernization. – 2012. – No.11. – Pp. 34 – 40.
[19] Kravchenko, I.N. Evaluation of machine and equipment reliability. Theory and practice: study book /I.N. Kravchenko, E.A. Puchin, A.V. Cheputin [et al.]. – M.: Alfa-M., INFRA-M, 2012. – 336 p.

Calculation of parameters of particles in a plasma jet and modeling of kinematic modes of spraying of wear resistant material

S N Sharifullin[1,2], G I Trifonov[3,4], I N Vyachina[5]

[1]Kazan Federal University, Russia
[2]Kazan state agrarian University, Russia
[3]Voronezh state technical University, Russia
[4]Air force Academy named after Professor N. E. Zhukovsky and Y. A. Gagarin, Russia
[5]Naberezhnochelninsky Institute, Kazan (Volga) Federal University, Russia

Email: Saidchist@mail.ru

Abstract. In this article the calculations on the strength of the coupling halves of the CV joint by the finite element method in the system of APM FEM. This calculation method is a method of creating a mathematical model and method of investigation. The assigned technological parameters of plasma spraying, namely, the estimated capacity of the plasma generator is 50 kW, and reach air temperature at the output of 3000 K, and the corresponding enthalpy of 3.8·106 j/kgK. Based on the analysis of the wear of the contact surfaces of the coupling halves, and also carried out strength calculations, was chosen as the material for plasma spraying – tungsten carbide. Developed mathematical model to determine the temperature of the powder particles and their velocities in a plasma jet. Based on the earlier dependence of the kinematic regimes of plasma spraying, the resulting system of equations for the kinematic modes that enable you to organize control of the kinematics of the deposition process details with the internal spherical surface.

Key words: speed of the plasma jet, the temperature of the sprayed particles, kinematic modes, plasma spraying, mathematical modeling

1. Introduction

Today, there are a large number of chemical, galvanic and physical methods for applying wear-resistant, heat-resistant, corrosion-resistant and other coatings that enhance the performance properties of the surfaces. A detailed analysis of a particular method or method in a scientific article is impossible, so it was decided to choose a promising area to establish reliable, durable and wear-resistant coatings, namely, plasma spraying [1, 2].

The aim of this work is the simulation of mathematical model calculate the velocity and temperature of the composite powder in the stream of the air plasma. The output of the system of equations for the kinematic modes of plasma spraying to the inner spherical surface of the coupling half.

2. Materials and methods of research

As a parts representative for coating was chosen as the spherical coupling of the hinge of equal angular speeds, the 3D model of which is shown in Fig. 1. To confirm and verify the appropriateness of applying a plasma spray coating on the inner spherical surface of the half couplings, were calculated by the finite element method in the system stress analysis APM FEM program KOMPAS-3D V15.1. This calculation method is a method of creating a mathematical model and method of investigation.

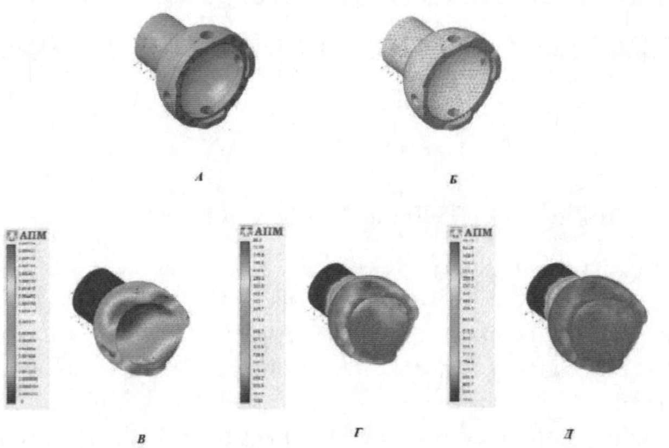

a – model coupling half; b – finite element mesh; v – the total linear displacement; g – factor of safety for strength; d – safety factor for yield strength

Figure 1 – Results of strength calculation of the APM FEM program KOMPAS-3D V15.1

To solve this problem it is necessary to determine the change similar parameters for the plasma jet created in a plasma torch. The estimated capacity of the plasma generator is 50 kW. Achieved temperature at the output of 3000 K. The Corresponding enthalpy of 3.8·106 j/kgK [3]. The thermal efficiency of the plasmatron 70 %. Air flow of plasma-forming G = 9 g/s.

The choice of material of the deposition surface of parts produced with the help of table 1, which presents a selection of materials functional coatings [4].

Table 1 –Ffunctionality of the coating

The coating material	Characteristic of the surface layer, improving a coating
Al_2O_3, Al-Ni, Ni-Ti, Mo, Cr_2O_3, Cr, **WC**, $MgAl_2O_4$	Durability
Al, Al_2O_3, Cr, Ti, Zn, Cr_2O_3, Al_2O_3-TiO_2	Corrosion resistance
Al_2O_3, ZrO_2, Al-Ni, Mo, ZnB, $MgAl_2O_4$	Temperature Resistance
ZrO_2, WC-CO, TiC, Cr_2O_3, Cr-B-Ni	Erosion resistance
W, Mo, **WC**, Cr-Ni, $MoSi_2$, $MgAl_2O_4$	The heat resistance
Al-Ni, Ti-Ni, Al_2O_3, ZrO_2	Intihuatana mobile nodes
Al_2O_3, $NiSi_2$, Al-Ni, ZrO_2	Heat insulation
Al_2O_3, $BaTiO_3$, SiO_2, MgO- Al_2O_3	Electrical isolation

Based on the analysis of the main types of wear undergone by the inner spherical surface of the coupling hub, was selected as the sprayed material is tungsten carbide.

3. The study and results

The speed of the sprayed particles is the subject of experimental study. Therefore, to compare data obtained from different researchers in this field, it is not possible because the experiments were carried out on different equipment and under certain laboratory conditions. In this regard, we propose a mathematical approach to determination of the velocity of sprayed particles.

Determine the speed of the powder [5]:

$$W_p = W\sqrt{\frac{3\rho_p C_x}{2\rho d_p}x} \quad (1)$$

where ρ – the density of the air at the appropriate temperature, kg/m3; ρ_p – the density of tungsten carbide (15,63 g/cm3); d_p – size of powder (μm); x – distance from nozzle exit (m); C_x - the drag coefficient.

Modeling the motion of particles of tungsten carbide powder in a plasma jet is taken that the tungsten carbide is spherical in shape.

Then $C_x = \frac{24}{\mathrm{Re}}$, where Re – Reynolds number. In this case $\mathrm{Re} = \frac{Wd_p}{\nu}$, where ν – the kinematic viscosity coefficient. Given these equations, equation (1) becomes:

$$W_p = 6\sqrt{\frac{W\rho_p \nu}{\rho d_p^{\,2}}x} \quad (2)$$

Determining the temperature of the sprayed particles in plasma spraying should be carried out with the radius of the particles and changes in material properties along the radius.

Heating of the particles of the powder material of tungsten carbide was seen in the assumption of their spherical shape, which is justified by the micron size of particles and their rotation due to asymmetrical entry of the powder into the plasma jet.

The temperature of the sprayed particles was found from the equation of heat transfer of powder particles with the gas. After mathematical modelling of process of heating the tungsten carbide powder, was obtained the equation for the temperature of sprayed particles:

$$T_p = T - e^{-\frac{6Nu\lambda}{\rho_p d_p^2 C_p}\tau + const} \quad (3)$$

where λ – coefficient of thermal conductivity, W/mK; C_p – the heat capacity of tungsten carbide; τ – the movement of the particles. It is worth considering that $\tau = \frac{2x}{W_p}$, a N – the Nusselt number in the flow of the ball, in this case $Nu = 2 + 0,03\mathrm{Pr}^{0,33}\,\mathrm{Re}^{0,51}\,0,35\mathrm{Pr}^{0,35}\,\mathrm{Re}^{0,58}$ [6], где $\mathrm{Pr} = \frac{\nu Cp}{\lambda}$ (Prandtl number).

Knowing that reach the temperature output of 3000K, and the melting point of tungsten carbide 3143K, then deal with the case when the particles are two-phase state, that is composed of two layers: the inner (solid) and outer (liquid). As further heat supply boundary melting is shifted inwards. In the end, during the deposition of tungsten carbide on the surface of the solid layer in the form of tungsten particles settle to the bottom border of the sprayed layer due to its density and phase state, and the carbon and all related impurities and gases rise to the upper level of the sprayed layer (carbon glass).

Kinematic regimes of plasma spraying, namely, the speed of rotation of the coupling halves $\frac{d\varphi}{dt}$, the longitudinal moving speed of the plasma torch $\frac{dz}{dt}$, the radial speed of the plasma torch $\frac{d\rho}{dt}$ and the contour speed of the plasma torch s_k , in this case, the plasma spraying of the inner spherical surface after the process simulation and mathematical transformations, drawing on the work of [6], will take the form of:

$$\begin{aligned} \frac{d\varphi}{dt} &= \frac{v}{\sqrt{(\mathrm{tg}\,\psi z)^2 + \beta^2}} \\ \frac{d\rho}{dt} &= \frac{ns_k}{\sqrt{(1+tg\psi^2)}} \\ \frac{dz}{dt} &= \frac{ns}{\sqrt{(1+tg\psi^2)}} \\ s_k &= sn \end{aligned} \tag{4}$$

where v – the speed of the sprayer relative to the surface, ψ – the angle between the tangent and forming a body of revolution at the current point and the axis z; β – the pitch of the spiral path of the spot deposition on the surface; n – the number of revolutions of the main drive; s – the reverse flow. It should be noted that the cylindrical z coordinate for the spherical surface of the half-coupling is expressed by the equation [7]:

$$z = \frac{\varphi\beta}{\sqrt{1+tg\psi^2}} \; . \tag{5}$$

The number of revolutions of the main drive is given by:

$$n = \frac{60v}{\sqrt{(2\pi kz)^2 + s/(1+tg\psi^2)}} \tag{6}$$

4. Conclusions

1. The developed mathematical model allows to determine the temperature of the powder particles, for example, tungsten carbide, and their speed in a plasma jet.
2. Based on the earlier dependence of the kinematic regimes of plasma spraying, the resulting system of equations for the kinematic modes that will allow to systematize the management of the kinematics of the deposition process details with internal spherical surface.

References

[1] Lomuhin VB Sharifullin S N and Lapteva I V 2016 *J. Phys.: Conf. Series* **669** 012051.
[2] Kornienko E E, Lapushkina E J, Kuzmin V I, Vaschenko S P, Gulyaev I P, Kartaev E V, Sergachev D S, Kashapov N, Sharifullin S, Fayrushin I 2014 *J. Phys.: Conf. Ser.* **567** 012010
[3] Vargaftik, N. B. Handbook on thermophysical properties of gases and liquids. Second edition, revised and supplemented /N. B. Vargaftik. – M.: "Nauka". – 1972. – 720 p.
[4] Trifonov, G. I. Influence of plasma spraying of composite powder materials on the wear resistance of machine parts [Electronic resource] /G. I. Trifonov, S. Y. Jackin //MASTER'S JOURNAL, Perm national research Polytechnic University, 2017. – Pp. 30-36.
[5] Ilyushchenko, A. F., the Processes of forming thermal spray coating and modeling /A. F. Ilyushchenko, A. I. Shevtsov, V. A. Okovity, G. F. Gromyko. – Minsk: Belarus. Science, 2011. – 357 p.
[6] Kutateladze, S. S. Fundamentals of the theory of heat transfer. Fifth edition, revised and enlarged /S. S. Kutateladze. – M.: Atomizdat. – 1979. – 416 p.
[7] Puzriakov, A. F. Theoretical bases of technology of plasma spraying. Proc. manual for the course "Technology of structures made of metal composites". Second edition, revised and enlarged /A. F. Puzriakov. – M.: Publishing house of Moscow state technical University named after N. E. Bauman. – 2008. – 360 p.

IOP Conf. Series: Journal of Physics: Conf. Series **1058** (2018) 012048 doi:10.1088/1742-6596/1058/1/012048

Magnetron sputtering deposition of ultra-thin metal coatings for the visualization of protein-containing objects of nanometer size by electron microscopy

I D Shumov[1], S L Kanashenko[1], V S Ziborov[2], Yu D Ivanov[1], A I Archakov[1], T O Pleshakova[1]

[1]Institute of Biomedical Chemistry, Moscow, Russia.
[2]Joint Institute for High Temperatures RAS, Moscow, Russia.

E-mail: shum230988@mail.ru

Abstract. Electron microscopy is widely used in biological and medical research, enabling to obtain images of the investigated objects with nanometer resolution. Exposure of biological objects under electron beam during the visualization can, however, cause their damage. Destructive effect of the electron beam can be significantly decreased by formation of metal coating on the surface of the objects to be studied. In the present study, an approach for the formation of ultra-thin metal coatings on the surface of protein-containing structures is proposed. Magnetron sputtering in argon plasma in DC mode was employed to form 8 Å-thick tungsten films on the surface of model protein-containing objects— tobacco mosaic virus particles — for their subsequent visualization by electron microscopy.

1. Introduction

Electron microscopy (EM) is widely used in bio medical research [1], enabling to obtain images of single protein-containing particles (viruses [2, 3] and proteins [4]) with nanometer resolution. Upon visualization by EM, the biological objects under study can be damaged by the electron beam, what causes significant loss of quality of the obtained images [3, 5, 6]. Due to this circumstance, to enhance clearness of the EM images obtained during visualization of biological objects, techniques employing preliminary treatment of the studied sample with heavy metals (Os, Ta, Au, W, Pt, Pd) or their compounds (osmium tetraoxide [1], phosphotungstic acid [2], potassium silicotungstate [4] etc.) are employed. Advantages and disadvantages of these techniques are discussed in more detail in our previous study [3]. Destructive effect of the electron beam can be significantly decreased by formation of metal coating on the surface of the sample to be studied [3]. In the present study, we have experimentally demonstrated that magnetron sputtering of ultra-thin metal coatings (whose thickness is less than 1 nm) for the visualization of nanometer-sized protein containing objects by EM.

Magnetron sputtering allows one to obtain coatings with good adhesion to the treated surface and low porosity, even at low thickness (<10 nm [7]). The advantage of magnetron sputtering consists in that it allows one to vary the deposition rate in a broad range by changing a number of parameters (power applied to the magnetron, sputtering mode — DC or RF, working gas pressure etc.). This provides flexibility of the method and enables one to obtain coatings with different characteristics, including ultra-thin ones with ~1 nm thickness [3, 7, 8].

IOP Conf. Series: Journal of Physics: Conf. Series **1058** (2018) 012048 doi:10.1088/1742-6596/1058/1/012048

In the present study, magnetron sputtering has been employed to obtain ultra-thin (of 8 Å thickness) layers of tungsten on the surface of biological protein-containing structures. As model object, tobacco mosaic virus (TMV) particles, which were extensively studied and characterized in much detail in the literature — have been used. In comparison with our previous report [3], the thickness of the formed coating has been significantly (from 13 Å to 8 Å) decreased. The thermal effect on the treated sample during the coating deposition with regard to the coating thickness is discussed.

2. Methods

The techniques used in this study were similar to those described in our previous paper [3]. The tungsten deposition procedure was modified to provide thinner coating.

2.1. Sample preparation

Tobacco mosaic virus sample (21.1 mg/mL in standard Dulbecco modified phosphate buffered saline (PBS-D) buffer) used in the study was kindly provided by Dr. V. Makarov (Belozerskii Institute Physicochemical B., Moscow State University). The samples were prepared as follows. An aliquot of the initial TMV suspension was diluted with PBS-D buffer (in the ratio of 1:99) and then incubated in a shaker (Eppendorf Thermomixer Comfort (Germany), 600 rpm, 20◦C) for 30 min. After that, 5 μL of the resulting diluted suspension was carefully dispensed onto an amorphous carbon film adhered to copper grid (Ted Pella, Inc., USA; Prod # 01843-F), and incubated for 6 min, and then rinsed off with 1 mL of ultrapure water. The samples prepared in such a way were finally dried in air at room temperature.

2.2. Ultra-thin tungsten film deposition

To form ultra-thin metal coatings on the surface of the samples, Orion-3 magnetron sputtering system (AJA Inc., USA) with installed tungsten target (99.95%, thickness 0.25'', diameter 2'', Girmet, Russia) was used. Base pressure in the system was not higher than 6×10^{-7} Torr. The sputtering was carried out using argon plasma in DC mode at constant power 70 W and sputtering gas pressure 50 mTorr during 16 s. The distance between the target and the specimen was 15 cm. During sputtering, rotation of the substrate to be coated with constant angle velocity 40 rpm was provided. At that, the sample plane was tilted relative to the plane of the disc by 45° (analogous to [3, 9]). The target was pre-sputtered for six minutes with the closed shutter prior to the deposition onto the substrate. The thickness of the tungsten layer was determined by QCM (according to the mass of tungsten deposited onto the quartz sensor surface during fixed time period) using MCM-160 thickness monitor (McVac Manufacturing, USA), installed into Orion-3 system. Under the specified conditions, the deposition rate was 0.5 Å/s; accordingly, the thickness of the coating deposited in 16 s was 8 Å.

2.3. Electron microscopy

EM images of TMV particles were obtained with Hitachi S5500 electron microscope in both SE and STEM modes at 25 kV accelerating voltage.

3. Results obtained

Figure 1 displays EM images of TMV particles coated with 8 Å tungsten film.

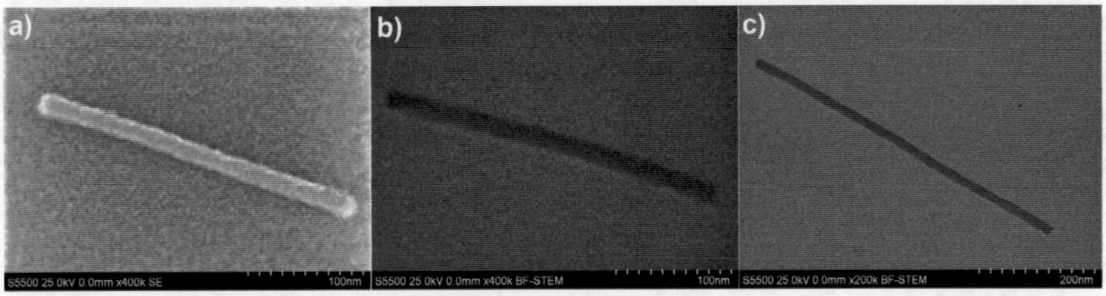

Figure 1. EM images of TMV particle adsorbed onto amorphous carbon film and coated with 8 Å tungsten film by magnetron sputtering. The images were obtained in SE (a) and STEM (b, c) modes. Magnification 400 000^X (a, b), 200 000^X (c). Panels (a) and (b) display the images of a single TMV particle obtained in different modes. Panel (c) displays «end-to-end» dimeric TMV aggregate.

The images presented in Figure 1 (a, b) indicate that the length of single TMV particles makes up 295-300 nm. This is in a good agreement with the data obtained by AFM [10]. Dimers of these particles (in the form of «end-to-end» aggregates) have a characteristic length of 600 nm (Figure 1 (c)). The data obtained confirm our previously obtained results [3].

It is known that the image quality depends on its resolution, i.e. on the relative number of small details distinguishable in this image [11]. In this connection, it has to be noted that, with regard to the study of biological objects by electron microscopy, decreasing the thickness of the formed metal coatings down to sub-nanometer values allows one to improve quality and resolution of images of the surface of the studied objects in comparison with thicker coatings.

When employing magnetron sputtering for coating biological objects it is necessary to take into account thermal effect on the treated sample, i.e., its heating during the deposition process. This heating is caused by the influence of a number of factors: the heat, released during the condensation of a sputtered metal on the sample surface; kinetic energy of the sputtered metal atoms and of sputtering gas ions; thermal radiation from the target etc. [12]. In our case, short sputtering time (16 s) and low thickness (8 Å) of the formed coating have allowed us to minimize the impact of the above-listed factors on the objects to be studied.

4. Conclusions

The use of DC magnetron sputtering deposition of ultra-thin (8 Å) tungsten coatings onto biological protein-containing structures of nanometer size has been demonstrated with the example of tobacco mosaic virus particles. These coatings allow one to minimize the destructive impact of the electron beam on these structures during their imaging by electron microscopy. Short sputtering time (16 s) and low thickness (8 Å) of the formed coatings have allowed us to minimize the impact of thermal effect on the treated sample during the deposition process.

5. Acknowledgements

The present study was performed in the framework of the Program for Basic Research of State Academies of Sciences for 2013-2020. Shumov I.D. is a recipient of Russian Federation President scholarship for young scientists for 2016-2018 (project identificator SP-4280.2016.4). Electron microscopy measurements were performed using the equipment of "Human Proteome" Core Facility of the Orekhovich Institute of Biomedical Chemistry (Russia) which is supported by Ministry of Education and Science of the Russian Federation (agreement 14.621.21.0017, unique project ID RFMEFI62117X0017).

References

[1] Kuo, J. 2014 Methods in Molecular Biology. Electron microscopy. N.Y.: Humana Press.
[2] Brenner S, Horne R W. 1959 *Biochim. Biophys. Acta.* **34** 103-10

[3] Shumov I D, Kanashenko S L, Archakov A I, Ivanov Yu D, Pleshakova T O. 2017 *Prikladnaya Fizika* **4** 10-5
[4] Tsuprun V L, Myasoedova K N, Berndt P, Sograf O N, Orlova E V, Chernyak V Ya, Archakov A I, Skulachev V P. 1986 *FEBS Lett* **205** 35-40
[5] Ogura T. 2012 *PLOS ONE* **7** e46904
[6] Adrian M, Dubochet J, Lepault J, McDowall A W. 1984 *Nature* **308** 32-6
[7] Petroff P, Sheng T T, Sinha A K, Rozgonyi G A, Alexander F B. 1973 *J. Appl. Phys.* **44** 2545-54
[8] Salamon K, Milat O, Radic N, Dubcek P, Jercinovic M, Bernstorff S. 2013 *J. Phys. D: Appl. Phys.* **46** 095304
[9] Hart R G. 1961 *J.Mol. Biol.* **3** 701-2
[10] Dubrovin E V, Yaminsky I V, Kirikova M N, Novikov V K, Drygin Y F. 2004 *Colloid J.* **66(6)** 673-8
[11] Sai S V, Sorokin N Yu. 2012 *Dokl. TUSUR* **2 (1)** 78-82
[12] Andritscky M, Guimaraes F, Teixeira V. 1993 *Vacuum* **44(8)** 809-13

IOP Conf. Series: Journal of Physics: Conf. Series **1058** (2018) 012049 doi:10.1088/1742-6596/1058/1/012049

The role of molecular ions in cryogenic dusty plasmas of dc discharge in neon

V V Shumova[1,2], D N Polyakov[1] and L M Vasilyak[1]

[1]Joint Institute for High Temperatures of Russian Academy of Sciences
[2]Semenov Institute of Chemical Physics of Russian Academy of Sciences

shumova@ihed.ras.ru

Abstract. The diffusion-drift model of positive column of glow discharge in neon have been used to study the effect of molecular ions in the interaction of plasma with dust particles of micron size. Simulations have been carried out at a pressure of 20 – 100 Pa and gas temperature of 77 K. The electric field of the discharge, the concentrations of discharge components, including metastable atoms and molecular ions, and the charge of dust particles have been simulated.

1. Introduction

Plasma of electric discharges of noble gases is widely used in engineering at low temperatures [1, 2]. Plasma with dust particles (dusty plasmas) is used in various technical devices for plasma coating, surface modification, in plasma-chemical reactors with rf or dc discharges for production of fine powders and novel materials [3-5] and in other processes of plasma-cryogenic synthesis [6]. The dusty plasma simulates the properties of non-ideal systems observed in the micro-world in a wide range of gas temperature including cryogenic [7-10].

The composition of the plasma (the partial concentrations of atomic ions, molecular ions and metastable atoms) is determined by the parameters of the discharge and the gas density. Usually, a large number of metastable atoms (A^*) is contained in the noble gas plasma. When the gas is cooled at constant pressure, the gas density increases and the concentration of molecular ions (A_2^+) also increases. The presence of excited atoms and molecular ions in the plasma can influence the interaction of the plasma with dust particles, which determines the conditions for the existence and charging of dust formations in the plasma.

The glow discharge in neon, taking into account the influence of metastable atoms on its properties at pressures up to 187 Pa at cryogenic and room temperature, was studied in [2]. The gas pressure in the discharge tube was maintained constant in the course of cooling. The gas density in a discharge corresponded to the measured temperature at given pressure, namely, was proportional to the relation of 295 K to its measured value. It was noted that an increase in the concentration of molecular ions that may be observed with an increase of the gas density in the result of discharge cooling, can affect the longitudinal electric field of the discharge. With an increase of concentration of molecular ions the role of associative ionization increases. In positive-column plasma of rare-gases, the molecular ions appear for the most part in the reaction of termolecular association (conversion) of atomic ions $A^+ + 2A \rightarrow A_2^+ + A$, as well as in reaction of associative ionization, while their destruction and loss proceed in the process of dissociative recombination in collisions with electrons $A_2^+ + e^- \rightarrow A + A$ [11-13] and diffusion to the wall of the discharge tube. It was noted in [2] that associative ionization can be a source of charged particles in neon plasma at $P > 600$ Pa at room temperature ($T = 295$ K), while at $P < 190$ Pa and $T = 77$ K, the partial input of associative ionization was supposed to be low and was not included in the model. In neon, associative ionization occurs in reactions of the interaction of excited atoms (Ne^*) with the formation of molecular ions (Ne_2^+) and electrons (e^-) as $Ne^* + Ne^* \rightarrow Ne_2^+ + e^-$ [14] and leads to a decrease in the electric field.

The role of metastable atoms in the interaction of plasma with dust particles was studied in [15-19] at room temperature for various dust particle distributions in dust structures, but no molecular ions were considered. This paper is devoted to the investigation of neon plasma with dust particles at cryogenic temperature. In this paper, we numerically study the influence of molecular ions on discharge parameters and the charge of dust particles in a dc glow discharge in neon at $T = 77$ K and a pressure up to 100 Pa (measured at 295 K).

2. Numerical model

Numerical study of the effect of molecular ions on the parameters of glow discharge plasma in neon and the charge of dust particles was carried out on the basis of the diffusion-drift model of positive column of a glow discharge in neon with microparticles [15, 16]. The model takes into account the formation, drift, diffusion and losses of electrons, atomic and molecular ions and metastable neon atoms in the plasma volume, on the walls of the discharge tube and on dust particles. In addition to the previously considered processes, the formation of molecular ions in the reaction of the termolecular association $Ne^+ + 2Ne \rightarrow Ne_2^+ + Ne$, with the rate constant taken from [11], as well as their losses in the process of dissociative recombination of $Ne_2^+ + e^- \rightarrow Ne + Ne$ in collisions with electrons with the rate constant taken from [13]. As in [13], it was assumed that the rate of formation of molecular ions in the process of associative ionization can be neglected in comparison with the rate of termolecular association (ion conversion).

The average electron energy and their transport coefficients were calculated using the BOLSIG+ package [20]. The mobility of molecular ions was taken from [21].

The charge of dust particles was calculated taking into account the ion-atom collisions in the CEC approximation (collision enhanced collection) [22], and the fluxes of metastable atoms to dust particles were simulated in the gas kinetic approximation. The possible heating of ions [23] was not taken into account, and the temperature of the heavy components of plasma was assumed to be equal to gas temperature.

3. Results

The simulation of plasma-dust systems has been carried out in a discharge with a radius of $R = 8$ mm, at neon pressure $P = 20 - 100$ Pa and a discharge current $I = 0.6$ mA. To simulate the dust structures formed by spherical dust particles with a diameter of 4.14 μm, the dust cloud radius r_d with a density distribution of dust particles $n_d(r)$, described by a step blurred function, have been chosen:

$$n_d(r) = n_{d,0},\ r \leq r_d \quad (1)$$

$$n_d(r) = n_{d,0}\exp[10(r_d - r)/R],\ r > r_d, \quad (2)$$

where $n_{d,0}$ is the concentration of dust particles on the discharge axis. The two different sets of the parameters r_d and n_{d0} of dust particle distribution were studied: $r_d = 2$ mm, $n_{d,0} = 10^{11}$ m^{-3} and $r_d = 0.5$ mm, $n_{d,0} = 10^{12}$ m^{-3}, that were observed in experiments [9, 10]. To analyze the effect of the plasma composition on the charge of dust particles, simulations also have been made of the charge of a solitary test dust particle introduced in the discharge. Examples of calculation are presented in figures 1 and 2 for a pressure of 84 Pa and discharge current of 0.6 mA.

Figure 1 shows the radial profiles of the concentrations of plasma particles in a discharge without dust particles (a) and in a discharge with a dust structures (b, c). It was discovered that the concentrations of electrons (1) and atomic ions (2) in discharge without dust particles (figure 1(a)) differ little from each other, since the concentration of molecular ions (4) is small in comparison with

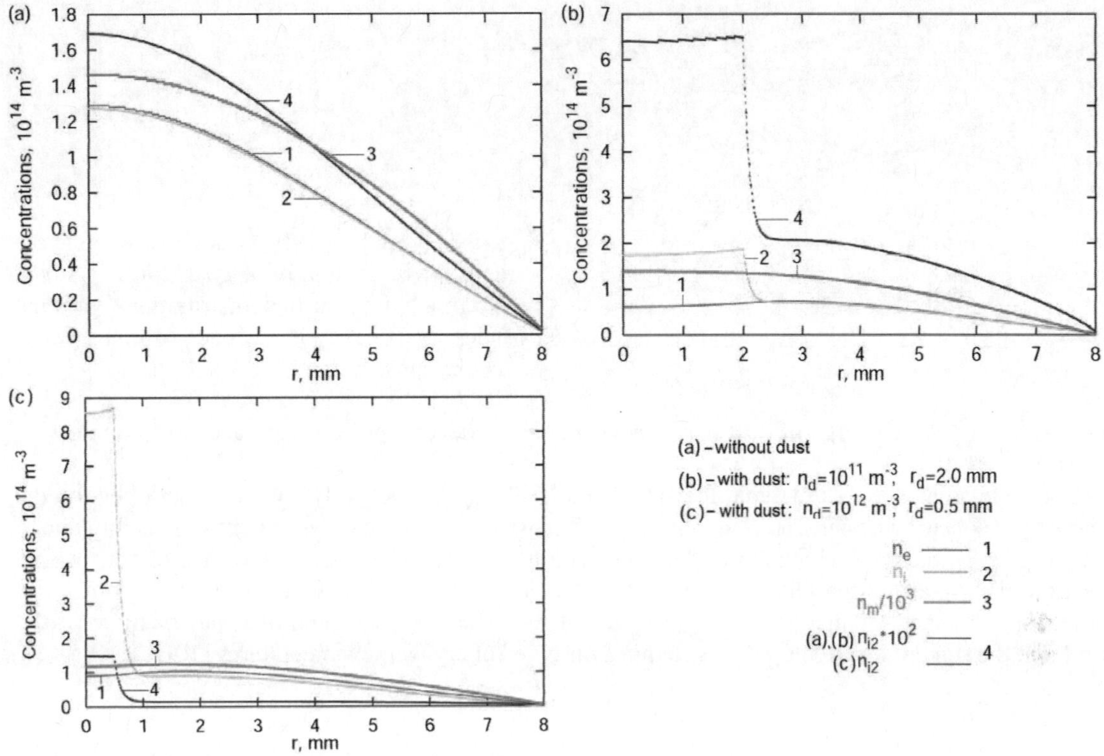

Figure 1. Radial profiles of electron concentration n_e (1), atomic ions n_i (2), metastable atoms n_m (3) and molecular ions n_{i2} (4) in the positive column of discharge at $P = 84$ Pa, $I = 0.6$ mA in a discharge without dust particles (a) and in presence of dust structures with various distributions of dust particles (b) and (c).

the concentration of atomic ions (2). In presence of dust structures, the ion and electron profiles principally differ. One can see in figure 1(b, c) that in the region of localization of the dust structure the total concentration of ions is much higher than the concentration of electrons due to the effect of a plasma-dust trapping [24, 25].

Simulations show that in the discharge with dust particles the rate of termolecular conversion increases inside the dust structure. This leads to the fact that within the dust structure, the fraction of molecular ions (n_{i2}/n_i) is significantly higher than in a discharge without dust particles increasing with the increase of n_d.

The corresponding values of the longitudinal electric field were calculated, which amounted to 11.89 V/cm in the discharge without dust particles, 15.64 V/cm for $r_d = 2$ mm and $n_{d,0} = 10^{11}$ m^{-3}, and 12.00 V/cm for $r_d = 0.5$ mm, $n_{d,0} = 10^{12}$ m^{-3}. Without taking into account the molecular ions, the corresponding values were 11.94, 15.4, and 11.96 V/cm, respectively. Apparently, the main factor responsible for the influence of molecular ions on the plasma parameters is the higher mobility of molecular ions compared to atomic ions. The concentration of molecular ions in a discharge without dust particles increased with increasing gas pressure and, when the pressure converted to the value at room temperature, was close to the data given in [13] obtained at a temperature of 300 K.

One can analyze the change of the of dust particle charge Z_d in presence the molecular ions. Figure 2 represents the charge of a dust particle in a discharge with the same values of neon pressure and discharge current, but various distributions of dust particles. At $n_{d,0} = 10^{11}$ m^{-3}, the change in the charge of dust particles when the molecular ions were taken into account, was negligible. However, at a higher concentration of dust particles $n_{d,0} = 10^{12}$ m^{-3} observed in dust structures formed at cryogenic

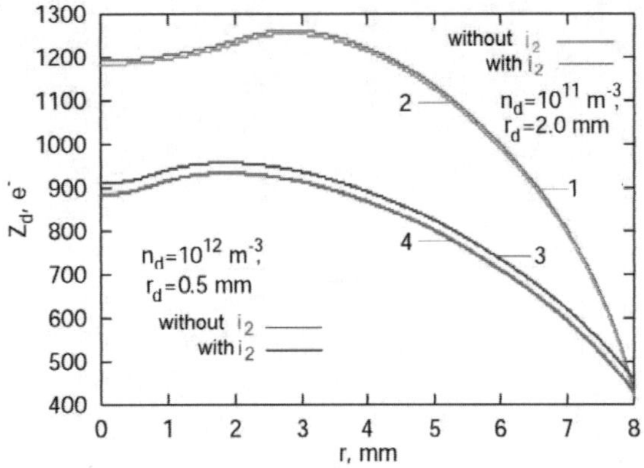

Figure 2. Radial profiles of dust particle charge in the positive column of discharge at P = 84 Pa, I = 0.6 mA with various distributions of dust particles and with (1, 3) and without (3, 4) molecular ions.

temperature [10], the charge of dust particles decreased by several percent with account of molecular ions.

The simulations have shown that the flux of molecular ions practically had no direct effect on the charge of a solitary dust particle in the investigated range of parameters, their effect was mainly due to a change in an electron temperature. Though, within a dense dust structures with a deficit of free electrons in a plasma, the charge of a dust particles may decrease due to the higher mobility of molecular ions than atomic ones. The account of molecular ions can improve the accuracy of the results in the simulation of dense dust structures observed at cryogenic temperatures [10].

4. Conclusion

The effect of molecular ions on the parameters of dusty plasmas at cryogenic temperature in dc discharge in neon have been studied. It have been found that the role of molecular ions in dusty plasmas is higher than in a discharge without dust particles at the same neon pressure and discharge current. In the presence of dust particles in the discharge, the fraction of molecular ions increases in the region of localization of dust structure. At dust particle concentrations up to 10^{11} m $^{-3}$, molecular ions do not significantly affect the electron temperature, and there is no significant effect of their flux to the dust particle surface, on the dust particle charge. When the concentration of dust particles increases by an order of magnitude, the fraction of molecular ions increases, and their effect on the charge of dust particles becomes noticeable. Thus, taking into account molecular ions one can improve the accuracy of simulation with dense dust structures with dust particle concentration exceeding 10^{12} m $^{-3}$, which were observed at cryogenic temperatures.

Acknowledgments

The authors are grateful to the Russian Foundation for Basic Research for the financial support (Grant No. 16-02-00991).

References

[1] Adamovich I *et al* 2017 The 2017 Plasma Roadmap: Low temperature plasma science and technology *J. Phys. D: Appl. Phys.* **50** 323001
[2] Shumova V V, Polyakov D N and Vasilyak L M 2017 Neon dc glow discharge at cryogenic cooling: experiment and simulation *J. Phys. D: Appl. Phys.* **50** 405202
[3] Vasilyak L M, Vasil'ev M N, Vetchinin S P, Polyakov D N, and Fortov V E 2005 Vapor Phase Deposition of Coatings onto Dust Particles in Combined Plasma *Tech. Phys. Let.* **31** 827–9
[4] Polyakov D N, Vasilyak L M and Shumova V V 2013 Positive Column of Glow Discharge with Dust Particles *Surf. Eng. Appl. Electrochem.* **49** 114–24
[5] Polyakov D N, Vasilyak L M and Shumova V V 2015 Synergetics of Dusty Plasma and Technological Aspects of the Application of Cryogenic Dusty Plasma *Surf. Eng. Appl. Electrochem.* **51** 143–51

[6] Huttel Y 2017 *Gas-Phase Synthesis of Nano-Particles* (Weinheim: Wiley VCH)

[7] Fortov V E, Vasilyak L M, Vetchinin S P, Zimnukhov V S, Nefedov A P and Polyakov D N 2002 Plasma-Dust Structures at Cryogenic Temperatures *Dokl. Phys.* **47** 21–4

[8] Polyakov D N , Shumova V V and Vasilyak L M 2015 Electric field on the diagram of phase transitions in cryogenic dusty plasmas *Rom. Rep. Phys.* **67** 1040–8

[9] Polyakov D N, Shumova V V and Vasilyak L M 2016 Formation of structures from dusty clusters in dc discharge at plasma cooling *J. Phys.: Conf. Ser.* **774** 012181

[10] Polyakov D N, Shumova V V and Vasilyak L M 2017 Transformations of dust structures in glow DC discharge in neon: effect of gas temperature and discharge current *Plasma Sources Sci. Technol.* **26** 08LT01

[11] Dielis J W H, de Hoog F J, Schram D C 1980 Formation and destruction of molecular ions in a Townsend discharge in neon *J. Appl. Phys.* **51** 5708–17

[12] Ivanov V A 1992 Dissociative recombination of molecular ions in noble-gas plasmas *Sov. Phys. Usp.* **35** 17–36

[13] Ichikawa Y, Teii S 1980 Molecuar ion and metastable atom formations and their effect on the electron temperature in medium-pressure rare-gas positive-column plasmas *J. Phys. D: Appl. Phys.* **13** 2031–43

[14] Rutscher A and Pfau S 1967 Zum Mechanismus der positiven Niederdrucksäule im Grenzfall kleiner Elektronendichten. III. Die Diffusionstheorie der homogenen positiven Säule *Beiträge aus der Plasmaphysik.* **7** 187–201

[15] Shumova V V, Polyakov D N and Vasilyak L M 2014 Effect of metastable neon atoms in positive column of glow discharge with dust particles *Plasma Sources Sci. Technol.* **23** 065008

[16] Shumova V V, Polyakov D N and Vasilyak L M 2015 Influence of dust particles on ionization and excitation in neon dc discharge *J. Phys.: Conf. Ser.* **653** 012132

[17] Shumova V V, Polyakov D N and Vasilyak L M 2017 Influence of dust void on neon DC discharge *Plasma Sources Sci. Technol.* **26** 035011

[18] Polyakov D N, Shumova V V, Vasilyak L M 2017 Positive Column of a Glow Discharge in Neon with Charged Dust Grains (a Review) *Plasma Phys. Rep.* **43** 397–404

[19] Vasilyak L M, Polyakov D N, Shumova V V 2013 Glow Discharge Positive Column with Dust Particles in Neon *Contrib. Plasma Phys.* **53** 432-5

[20] Hagelaar G J M, Pitchford L C 2005 Solving the Boltzmann equation to obtain electron transport coefficients and rate coefficients for fluid models *Plasma Sources Sci. Technol.* **14** 722–33

[21] Brown S C 1959 *Basic Data of Plasma Physics* (New York: Wiley)

[22] D'yachkov L G, Khrapak A G, Khrapak S A, Morfill G E 2007 Model of grain charging in collisional plasmas accounting for collision less layer *Phys. Plasmas.* **14** 042102

[23] Golyatina R I and Maiorov S A 2016 On Ion Temperature in Cryogenic Discharge *Bulletin of the Lebedev Physics Institute.* **43** 237–41

[24] Shumova V V, Polyakov D N and Vasilyak L M 2016 Direct current discharge with microparticles: The electrical characteristics and the plasma trap parameters *J. Phys.: Conf. Ser.* **774** 012182

[25] Shumova V V, Polyakov D N and Vasilyak L M 2017 Dust particle trapping of ions in a dc glow discharge in neon *J. Phys.: Conf. Ser.* **946** 012159

IOP Conf. Series: Journal of Physics: Conf. Series **1058** (2018) 012050 doi:10.1088/1742-6596/1058/1/012050

Transients in a longitudinal Glow Discharge

R F Yunusov[1], M M Garipov[1] and E R Yunusova[2]

[1]Kazan National Research Technical University named by A.N. Tupolev,420111, K. Marx str., 10, Kazan, Russian Federation
[2]Kazan Medical College, 420055, Professor Kamai str., 34, Kazan, Russian Federation

E-mail: optanir@mail.ru

Abstract The article is devoted to an experimental study of transient processes in a glow discharge (GD). The longitudinal GD was ignited in a stream of air in a cylindrical discharge chamber (DC). Copper hollow electrodes were located at a distance of a = (2 – 8) cm from each other along the axis of the DC. The gas pressure P, its flow rate G, discharge current I and its voltage U varied accordingly in the ranges: P = (4.7 – 26) kPa, G = (0 – 0.05) g / s, I = (20 – 100) mA, U = (1. 5 – 4. 2) kV. The current-voltage characteristics of the discharge were measured. The voltage decreased with increasing current, which is typical for a given type of discharge. The oscillations of the discharge voltage were registered by an oscilloscope; the current-voltage characteristic of the discharge was recorded by a two-coordinate recorder. Low-frequency (f ~ 300 Hz) and high-frequency (f ~ 100 kHz) oscillations of the discharge voltage were observed. The influence of the gas flow on the current-voltage characteristics of the discharge, its shape and the frequency of the detected oscillations is shown.

1. Introduction

Currently, GD is widely used not only as a source of light, but also as an active medium for lasers, in technological processes for surface treatment, production of nanomaterials and for medical purposes [1]. For various practical and technological problems, it is necessary to know the distributions of electrical and thermal parameters of a given type of discharge [2-12]. Knowledge of the physics of the processes of interaction of the discharge with the gas flow, the instabilities arising in it, will allow controlling the discharge parameters for various technological operations [13-17]. Therefore, the study and research of GD plasma parameters is of great scientific and practical interest [18-21]. This is due to the needs of practice, since for specific technological tasks various DC and ways of organizing the flow of matter in them are required. Obtaining a stable discharge with the parameters necessary for practice is one of the problems in this area. The GD, are generally no stationary [22]. The purpose of this article is an experimental study of transient processes in a longitudinal GD.

2. Experiment

The scheme of the experimental setup is shown in Fig. 1. The air through the needle cone **1** and the flow meter **3** (rheometer RDS-10) was supplied to the discharge chamber DC **4**. The DC is a quartz tube with an internal diameter of 10 mm (Fig.1.b). The discharge was ignited in the longitudinal flow between the copper hollow cylindrical electrodes: the cathode **8** and the anode **9**.

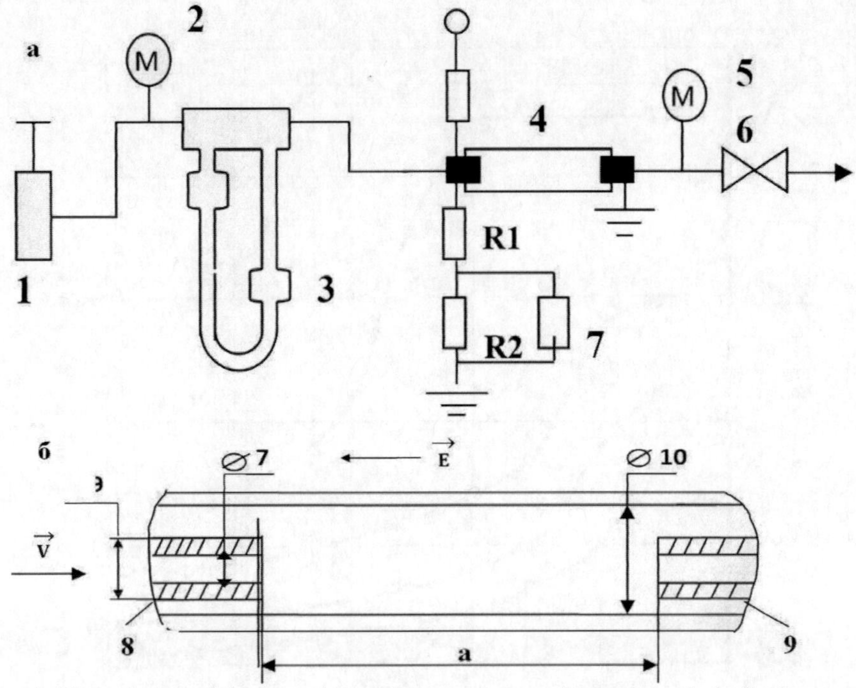

Figure1. The scheme of the experimental setup: a - gas-dynamic path; b-discharge chamber.

The gas moved in a direction from the cathode to the anode. The distance between the electrodes could be changed in the range from 2cm to 8cm. The spent medium from the DC entered the capacity and was removed into the atmosphere by a vacuum pump. The static pressure in the RC was measured by U-shaped mercury manometers **2** and **5**. The flow rate and the pressure value were controlled by the valve **6**. The oscillations of the discharge voltage were registered by an oscilloscope **7**. The current-voltage characteristic of the discharge was recorded by a two-coordinate recorder of the type H-306.

3. Results

Figure 2 shows the current-voltage characteristics of an electrical discharge in an air stream. The experiment showed three characteristic regions of current-voltage characteristics. After the ignition of the discharge, the starting point falls on the first region I of the current-voltage characteristics. Therefore, for the full description, the inverse branch of the U-I characteristic is considered. Figure 2 shows two volt-ampere characteristics: 1- for a discharge with an air flow, 2 - for a discharge in still air. For the first curve, with a decrease in current strength from 100 to 70 mA (curve 1, region I), the value of the discharge voltage slowly increases. The positive discharge column reaches almost to the cathode. In the current range I = 70 ... 40 mA there is a smooth transition (region II) of the luminous discharge to the discharge without visible glow. With a further decrease in the current intensity (region III), the discharge exists almost at a constant voltage up to quenching. The glow in this case is observed only from the ends of the hollow electrodes. In the quiescent gas (curve 2), there are also all three regions of the current-voltage characteristic. However, region I shifts towards lower currents, and regions II and III are respectively narrowed. The current-voltage characteristics practically did not change when the polarity of the electrodes changed.

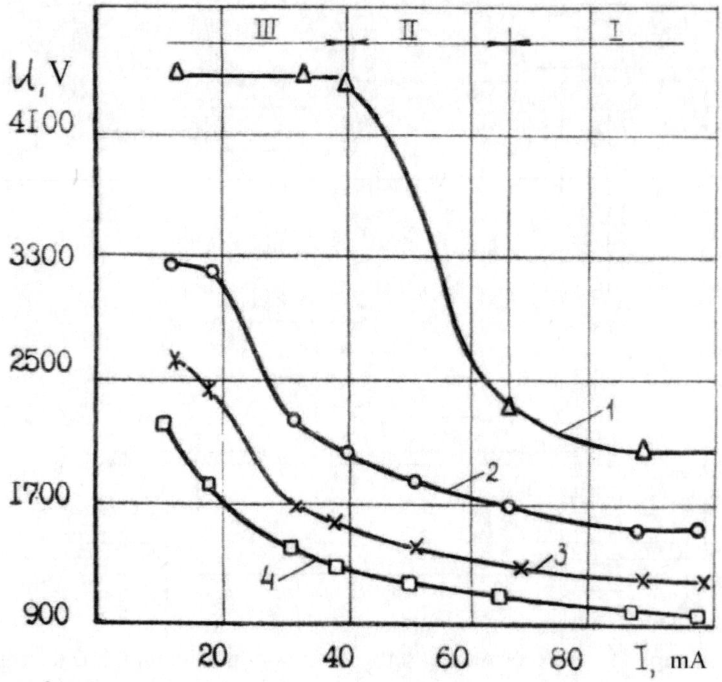

Figure2. Current-voltage characteristic of an electric discharge in the longitudinal air flow: a = 0.04 m; P = 26 kPa, 1 – G = 0.04 g/s; 2 – G =0; P = 6. 25 kPa, 3 – G=0. 01 g/s, 4 – G=0.

Photos of the discharge are shown in Figure 3. As can be seen from the photographs (Fig. 3), the discharge consists of a brightly shining region extending up to the anode and a dark space near the cathode. Directly at the cathode there is the luminous layer. It was found that as the current increases, the length of the dark space decreases, and the surface of the cathode, covered with a luminous layer, increases. In a discharge with a gas flow (Fig. 3), the positive column of the discharge contracts and loses its axial symmetry.

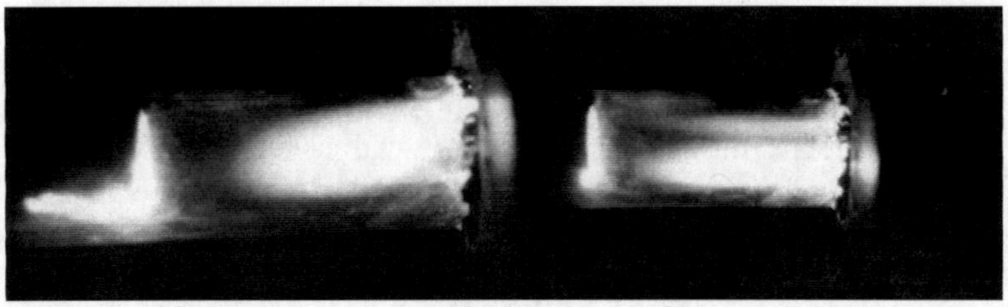

Figure3. Photos of the discharge with a gas flow (right) G = 0. 02 g/s and without it (left), a = 0. 03 m, P= 3. 6 kPa, I = 50 mA.

With increasing air flow, the discharge voltage first rises, reaches a maximum, then decreases and remains constant. With increasing pressure and decreasing current strength, discharge stability deteriorates, and voltage pulsations appear. Low-frequency periodic and aperiodic damped voltage fluctuations occur in a discharge without an air flow. These voltage pulsations occur due to low frequency oscillations of the power supply voltage. The frequency of the variable component is 300 Hz. These low-frequency ripple voltages correspond to the discharge without visible glow of the PC.

With further reduction of the current intensity, the discharge is extinguished. In the air flow, the nature of the oscillations varies. High-frequency voltage oscillations (Fig. 4) with a frequency f = (5 – 100) kHz appear in a discharge with an air flow. This feature is observed throughout the investigated range of currents. High-frequency voltage pulsations have the shape of a sinusoid, and their frequency decreases with increasing airflow. The region in which high-frequency voltage oscillations occurred was determined by analyzing the oscillograms. High-frequency voltages oscillations are absent at currents greater than 20 mA.

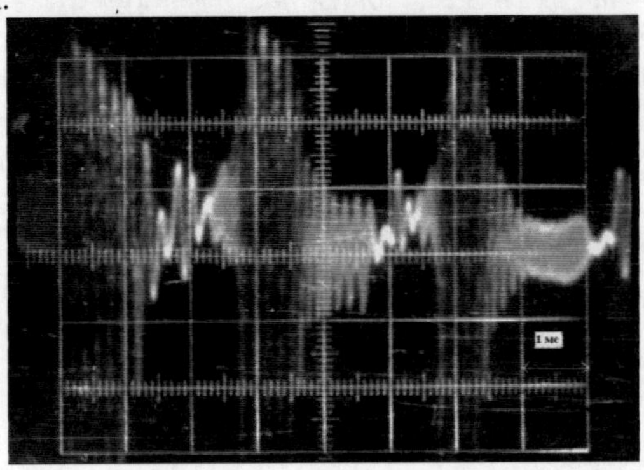

Figure4. The typical oscillogram of discharge voltage pulsations: G = 0.03 g / s, I = 8 mA, U = 1380 V

4. Conclusions

A longitudinal glow discharge in the air flow was studied in this paper. The effect of air flow on the current-voltage characteristics of the discharge was obtained. An increase in the air flow in the interval G = (0-0.03) g / s causes an increase in the discharge voltage. With a further increase in the flow rate to the value G = 0.06 g / s, the discharge voltage practically does not change. Simultaneously with this increase in the discharge voltage, its structure changed: the discharge lost its axial symmetry ancontracted. With a decrease in the discharge current in the interval I = (70-40) mA, the luminous region of the discharge disappears, and its voltage increases sharply. In this region of transition to the dark form of the discharge, high-frequency and low-frequency oscillations of the discharge voltage are observed. An increase in the air flow rate leads to an increase in the discharge current, at which high-frequency voltage pulsations occur.

References

[1] Raizer Y P 1992 *Gas Discharge Physics* (Moscow: “Science”) p 536
[2] Yunusov R F 1982 *Journal of Engineering Physics and Thermophysics* **43** 1100-03
[3] Yunusov R F and Mukhametzyanov R B 1982 *Izvestiya Vysshikh Uchebnykh Zavedenii Aviatsionaya Tekhnika* **2** 40-44
[4] Yunusov R F 1983 *Inzhenerno-Fizicheskii Zhurnal* **43** 1100-03
[5] Yunusov R F 1985 *Izvestiya Vysshikh Uchebnykh Zavedenii Aviatsionaya Tekhnika* **1** 68-72
[6] Yunusov R F 1985 *Journal of Engineering Physics and Thermophysics* **48** 591-592
[7] Yunusov R F 1985 *Journal of Engineering Physics and Thermophysics* **48** 214-219
[8] Yunusov R F 1985 Soviet Aeronautics **28** 78-82
[9] Yunusov R F 1988 *Journal of Engineering Physics and Thermophysics* **54** 76 – 80
[10] Dautov G Y and Yunusov R F 1988 *Izvestiya Vysshikh Uchebnykh Zavedenii Aviatsionaya Tekhnika* **2** 21-25
[11] Yunusov R F 1990 *Journal of Engineering Physics and Thermophysics* **59** 990-994
[12] Yunusov R F 1991 *Inzhenerno-Fizicheskii Zhurnal* **59** 990-994
[13] Israphilov I H, Zvezdin V V, Israphilov D I and Chernova M A 2014 *J. Phys.: Conf. Ser.* **567** 012030

[14] Yunusov F S, Khisamutdinov R M and Yunusov R F 2008 *Russian Engineering Research* **28** 965-973
[15] Yunusov F S and Yunusov R F 2011 *Russian Engineering Research* **31** 15-21
[16] Yunusov F S and Yunusov R F 2011 *Russian Engineering Research* **31** 660-665
[17] Yunusov F S and Yunusov R F 2011 *Russian Engineering Research* **31** 951-959
[18] Israphilov D I 2017 *Journal of Physics: Conference Series* **789** 012023
[19] Yunusov R F 2017 *Journal of Physics: Conference Series* **789** 012069
[20] Yunusov R F and Garipov M M 2017 *Journal of Physics: Conference Series* **927** 012076
[21] Yunusov R F and Garipov M M 2017 *Journal of Physics: Conference Series* **927** 012077
[22] Tazmeev Kh K, Timerkaev B A and Tasmeev G Kh 2017 *Journal of Physics: Conference Series* **789** 012061

Heat Balance in the Positive Column of a Glow Discharge

R F Yunusov[1], M M Garipov[1] and E R Yunusova[2]

[1]Kazan National Research Technical University named by A.N. Tupolev,420111, K. Marx str., 10, Kazan, Russian Federation
[2]Kazan Medical College, 420055, Professor Kamai str., 34, Kazan, Russian Federation

E-mail: optanir@mail.ru

Abstract. According to classical concepts, a cylindrical positive column (PC) of the glow discharge (GD) is a luminous volume of ionized gas that fills up the whole space in the discharge chamber (DC) and which parameters are indiscrete, i.e. they do not vary along the axis of the chamber. In this case, the heat balance of the PC is that the electric power released in the discharge is expended on the heat flux to the wall of the discharge chamber (DC) due to heat transfer; also, a certain fraction of energy is carried away due to heat transfer by radiation. In case of an open system when gas flows through the area of the PC, it is possible to change the discharge parameters along its axis, and it is necessary to take into account heat transfer by convection in the energy balance. It is important to have knowledge of heat balance and its components both for gas discharge physics and for any DC designed for various applications. This paper is devoted to a comprehensive experimental study of the distribution of potential, electric field intensity, and gas temperature in the longitudinal GD. The gas pressure P, its flow rate G, the discharge current I and its voltage U varied accordingly in the following ranges: P = (2 – 9) kPa, G = (0 – 0.05) g/s, I = (30 – 80) mA, U= (1 – 2) kV. The obtained distribution of potential, electric field intensity, and gas temperature make it possible to calculate all the heat fluxes in the plasma of the discharge PC.

1. Introduction

The non-equilibrium nature of GD when the electron temperature is much higher than that of the neutral component makes this type of discharge attractive for use in a variety of technologies to produce new types of materials, apply coatings by cathode sputtering, to generate ozone that can be used to sterilize medical products, etc.[1-6]. Another feature of GD is a significant cathode potential drop that ensures knockout of electrons from the cathode due to the kinetic energy of incident ions. This allows to make devices with a cold cathode and to create uniformly luminous surfaces of the required configuration. A new round of interest in GD has emerged with the development of gas lasers where it is used as an active medium. To use effectively all the advantages of GD, it is necessary to understand the regularities associated with GD in the gas flow, in DCs with different geometries. Analytical formulas based on the equations of charge continuity and Ohm's law in integral form, the energy equation was obtained in [7-10]. The temperature fields of neutral particles, the heat balance were calculated in [11-17]. Experimental studies of the gas temperature were carried out in [18-22]. The objective of this work is to further study the heat balance in the PC of GD.

2. Experiment
The scheme of the experimental situation is shown in Fig. 1.

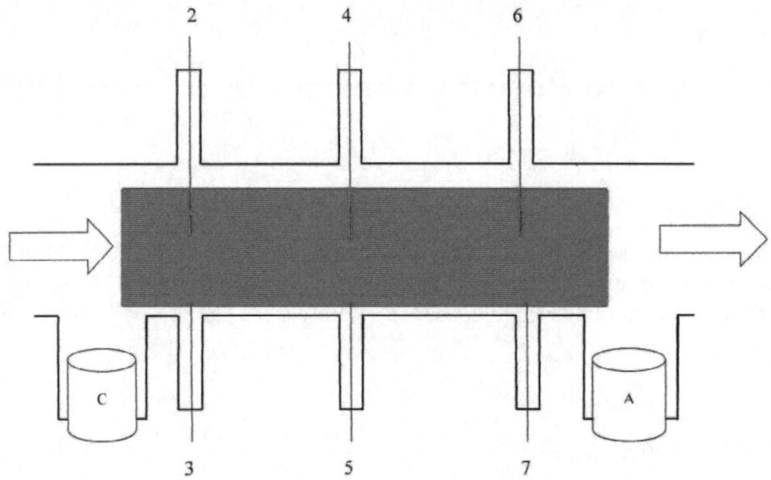

Figure1. The scheme of the experimental setup: C, A - cathode, anode; 2 ... .7 – thermocouples or probes.

Copper electrodes located at a distance of l = (0.08 – 0.1) m, are located in the side branches of the glass tube with a diameter of 0.01m. The thermocouples are arranged in such a way that from the measured values of the gas temperature it is possible to calculate the heat flux to the wall due to thermal conductivity, as well as convective energy transfer from the region of the PC of the discharge. The arrows show the direction of the gas flow. Also, to study the distribution of the potential, the electric field strength along the discharge chamber, probes were inserted into the guiding capillaries instead of thermocouples 2.4 and 6. Cylindrical probes made of platinum with a diameter of 0.1 mm and working lengths of 2 mm were used. The non-working part of the probes was isolated by a quartz capillary with a diameter of 0.2 mm. To measure the potential, static voltmeters of the C50 type of accuracy class 1.0 with measurement limits of 300, 1000, 1500 and 3000 V were used. All measurements of the potential of the electric field were made with respect to the grounded anode. The procedure for measuring gas temperatures by chromel-alumel thermocouples and other discharge parameters is described in [19].

3. Results
Fig.2 shows the distribution of the potential φ along the axis of the discharge chamber in the air flow. In these graphs, the origin of coordinates Z = 0 corresponds to the right edge of the cathode, whose potential is assumed to be zero.

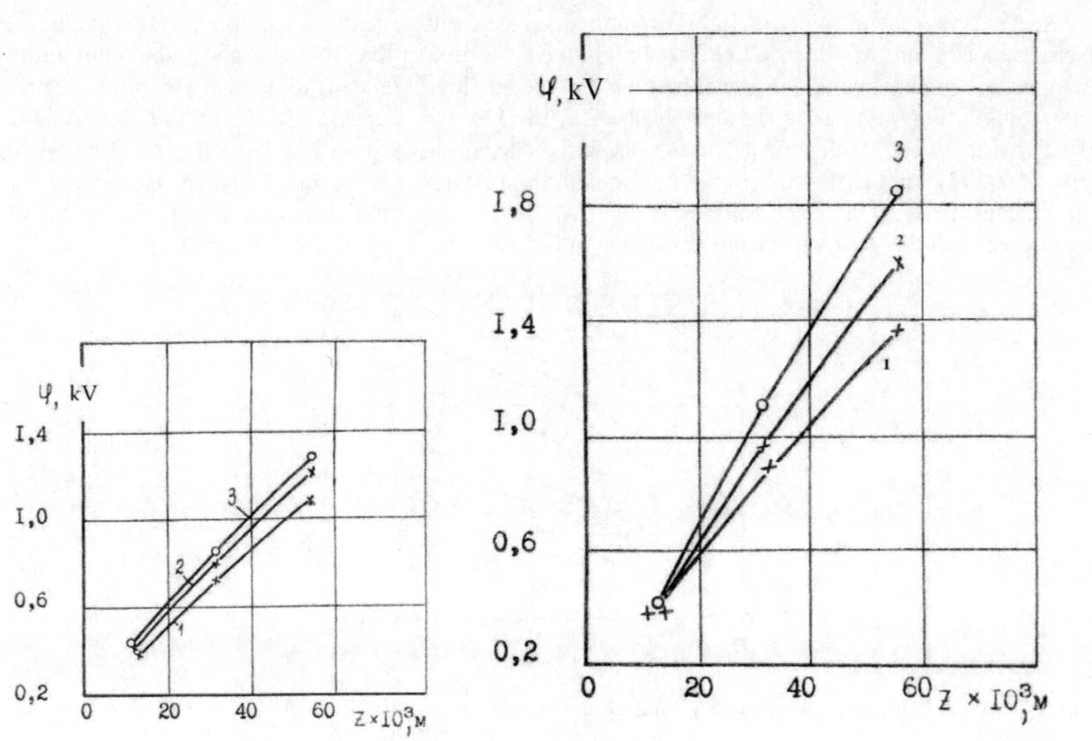

Figure2. The potential distribution in the glow discharge: P = 4.6 kPa, 1- I = 50 mA, 2- I = 30 mA, 3-I = 20 mA, to the left-G = 0, to the right-G = 0.04 g / s.

The magnitude of the potential from the cathode to the anode increases approximately linearly and therefore the average electric field strength remains almost constant. With increasing current, the magnitude of the potential decreases, this corresponds to the dropping current-voltage characteristic of the discharge. For a more accurate verification of the variation or constancy of the electric field strength along the DC axis, it is necessary to place more probes along the axis or use a mobile probe, which introduces additional errors. The average electric field strength was calculated from the gradient of the potential. Dependences of the electric field strength on the discharge current are shown in Fig. 3 (left).

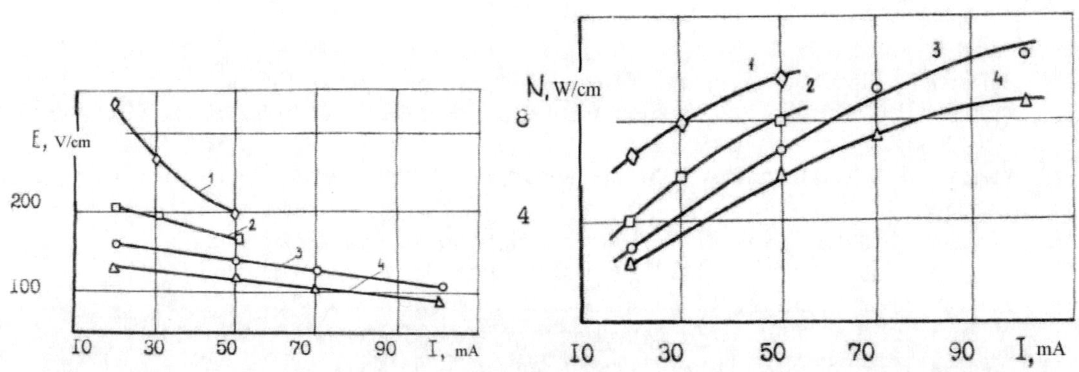

Figure3. Dependences of the electric field strength E and power density N on the discharge current I: 1,2,3 - P = 4.7 kPa, 4 - P = 6.1 kPa; 1, 4-G = 0.04 g / s; 2-G = 0; 3- G = 0.05 g / s; 1,2-air; 3-argon; 4-helium.

As can be seen, the electric field strength decreases with increasing current. Under identical discharge conditions, the electric field strength in the air is greater than that in argon and helium. From a comparison of curves 1 and 2, it can be seen that the air flow rate leads to an increase in the electric field strength. For calculating the heat balance of the PC, it is necessary to know the value of the power density, which is dependent on the discharge current, as is shown in Fig.3 (right). With equal flow rates and currents, the discharge power density for air is approximately twice as large as for helium. This is due to the large value of the electric field strength in the molecular gas. The results of measurements of the gas temperature are shown in Fig.4.

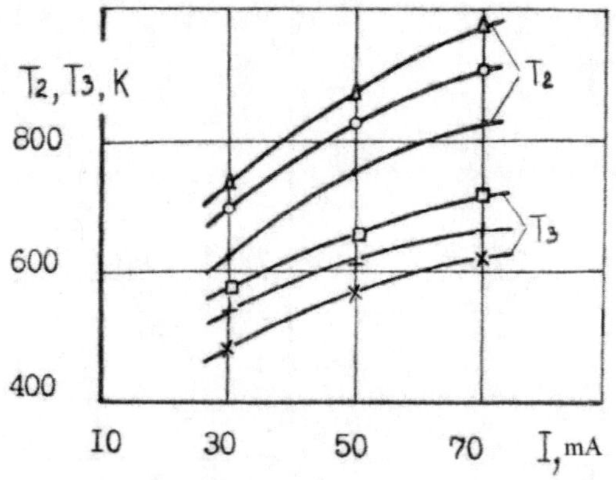

Figure4. Dependences of temperatures 2 and 3 on discharge current, G = 0.007 g / s; •, x - P = 2.5kPa; ◦, + - P = 3.9 kPa; Δ, □ -P = 5.5 kPa.

The gas temperature decreases along the radius of the DC and increases in the direction of gas flow. The temperature at the axis of the DC increases with increasing current, discharge power and gas pressure. The effect of air consumption on the temperature distribution and the shape of the positive column of the discharge are also revealed.

4. Conclusions

This paper presents the results of the comprehensive experimental research on energy properties of a longitudinal GD. The gas potential and temperature distributions in the longitudinal and transverse directions are obtained. The dependencies of the electrical field strength and power density on the discharge current for air, argon and helium are obtained. The obtained distributions and dependencies allow you to calculate the heat balance in the PC of GD.

References

[1] Raizer Y P 1992 *Gas Discharge Physics* (Moscow: "Science") p 536
[2] Israphilov I H, Zvezdin V V, Israphilov D I and Chernova M A 2014 *J. Phys.: Conf. Ser.* **567** 012030
[3] Yunusov F S, Khisamutdinov R M and Yunusov R F 2008 *Russian Engineering Research* **28** 965-973
[4] Yunusov F S and Yunusov R F 2011 *Russian Engineering Research* **31** 15-21
[5] Yunusov F S and Yunusov R F 2011 *Russian Engineering Research* **31** 660-665
[6] Yunusov F S and Yunusov R F 2011 *Russian Engineering Research* **31** 951-959
[7] Yunusov R F 1982 *Journal of Engineering Physics and Thermophysics* **43** 1100-03
[8] Yunusov R F 1983 *Inzhenerno-Fizicheskii Zhurnal* **43** 1100-03
[9] Yunusov R F 1985 *Journal of Engineering Physics and Thermophysics* **48** 591-592
[10] Yunusov R F 1985 *Journal of Engineering Physics and Thermophysics* **48** 214-219

[11] Yunusov R F and Mukhametzyanov R B 1982 *Izvestiya Vysshikh Uchebnykh Zavedenii Aviatsionaya Tekhnika* **2** 40-44
[12] Yunusov R F 1985 *Izvestiya Vysshikh Uchebnykh Zavedenii Aviatsionaya Tekhnika* **1** 68-72
[13] Yunusov R F 1985 Soviet Aeronautics **28** 78-82
[14] Yunusov R F 1988 *Journal of Engineering Physics and Thermophysics* **54** 76 – 80
[15] Dautov G Y and Yunusov R F 1988 *Izvestiya Vysshikh Uchebnykh Zavedenii Aviatsionaya Tekhnika* **2** 21-25
[16] Yunusov R F 1990 *Journal of Engineering Physics and Thermophysics* **59** 990-994
[17] Yunusov R F 1991 *Inzhenerno-Fizicheskii Zhurnal* **59** 990-994
[18] Israphilov D I 2017 *Journal of Physics: Conference Series* **789** 012023
[19] Yunusov R F 2017 *Journal of Physics: Conference Series* **789** 012069
[20] Yunusov R F and Garipov M M 2017 *Journal of Physics: Conference Series* **927** 012076
[21] Yunusov R F and Garipov M M 2017 *Journal of Physics: Conference Series* **927** 012077
[22] Timerkaev B A, Amirzyanov D R and Israphilov D I 2016 *Journal of Physics: Conference Series* **789** 012063

IOP Conf. Series: Journal of Physics: Conf. Series **1058** (2018) 012052 doi:10.1088/1742-6596/1058/1/012052

The fluid flow, containing flexible particles, in the channel of microfluidic devices

Tazyukov F.Kh., Kutuzova E.R.*, Tazyukova A.F.
Russia, Kazan, 420012, Karl Marx street,68

Elvira.kutuzova@list.ru

Abstract. In this paper we investigated the flow of an elastic fluid, represented by the rheological constitutive relation FENE-P in a typical planar T-shaped channel as a part of microfluidic device. This model succeeds to predict necessary rheological properties of many real fluids. Distributions of the main flow characteristics for different sections were obtained for constant values of the simulation parameters (Reynolds number Re = 0.01, Weissenberg number We = 0.6, retardation coefficient β = 0.1 and degree of unraveling of the flexible particle L^2 = 50). The isolines and main flow quantities distributions are drawn, comparison of the results of the elastic and Newtonian fluids is provided, areas of increased stresses are considered.

1. Introduction

Construction of microfluidic devices contains a system of microchannels of various shapes, where different types of flow can be realized [1,2]. Peculiarity of above-mentioned fluids is the presence of flexible particles which can change their configuration in the flow. It leads to the appearance of nonlinear effects in the flow, even for small values of the Reynolds number (Re << 1). Such behavior is attributed to non-Newtonian fluids and caused by large stresses in comparison to Newtonian fluids where such kind of effects are absent [3].

The aim of the present paper is to investigate the viscoelastic fluid flow in a planar T-shaped microchannel under perpendicularly located input parts at small value of the Reynolds number (Re << 1).

2. Problem statement

The isothermal flows of viscoelastic fluids are described by the mass conversation equation and the momentum equations:

$$\rho\left(\frac{\partial \vec{v}}{\partial t}+\vec{v}\cdot\vec{\nabla}\vec{v}\right)=-\vec{\nabla}p+\vec{\nabla}\cdot\tilde{\tau}, \tag{1}$$

$$\vec{\nabla}\cdot\vec{v}=0. \tag{2}$$

$$\tilde{\tau}=\tilde{\tau}^{p}+\tilde{\tau}^{s}, \qquad \tilde{\tau}^{p}=\frac{\eta^{p}}{\lambda}\left[\frac{\tilde{A}}{1-\frac{tr(\tilde{A})}{3L^2}}-\frac{\tilde{I}}{1-\frac{1}{L^2}}\right], \qquad \tilde{\tau}^{S}=2\eta^{S}\tilde{D}. \tag{3}$$

$$\frac{\tilde{A}}{1-(tr\tilde{A})/(3L^2)}+We\overset{\nabla}{\tilde{A}}=\frac{\tilde{I}}{1-1/L^2},\quad \overset{\nabla}{\tilde{A}}=\frac{\partial\tilde{A}}{\partial t}+\vec{v}\cdot\vec{\nabla}\tilde{A}-\vec{\nabla}\vec{v}\cdot\tilde{A}-\tilde{A}\cdot(\vec{\nabla}\vec{v})^{T}\cdot \tag{4}$$

The modeling parameters are We = 1.5, Re = 0.01, ß = 1/9, L2 = 50. Governing equations are solved by the finite volume method (FVM) using the OpenFoam source - CFD (Computational Fluid Dynamics) software package [3].

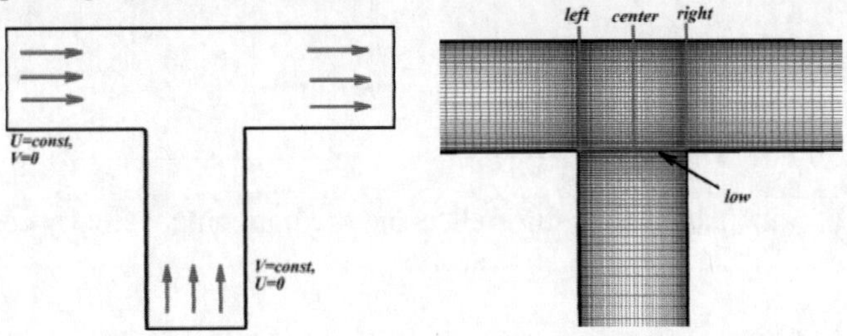

Fig. 1 – Schematic representation of the channel and refined mesh

The length of the channels was set to 10 channel's width for the formation of the velocity profile at the inlet flow and to establish the outlet flow.

3. Results

At first glance, streamline patterns, obtained for Newtonian and viscoelastic fluids have minor differences (Fig. 2). The stagnant area emerges near the right corner for a viscoelastic fluid. Such formation is associated with the specific properties of the fluid (viscosity anomaly, finite stress relaxation time, longitudinal viscosity's dependence on longitudinal deformation rate) and channel shape.

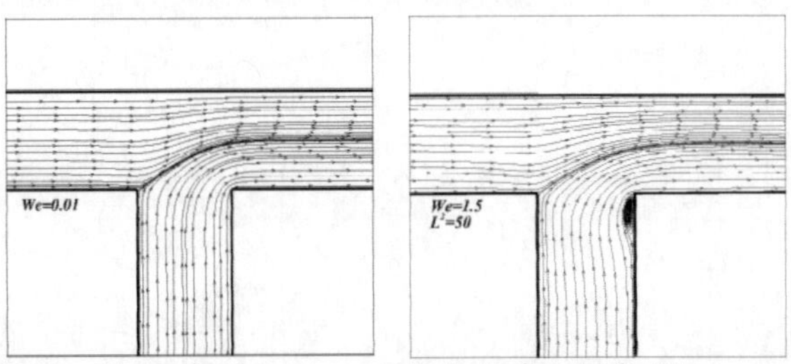

Fig. 2 Streamlines for Newton and viscoelastic fluids, respectively

Comparing the distributions of principal stresses difference in the channel for Newtonian and viscoelastic fluids, it might be seeing that the distribution of these quantities has both a qualitative and a quantitative difference (Fig. 3).

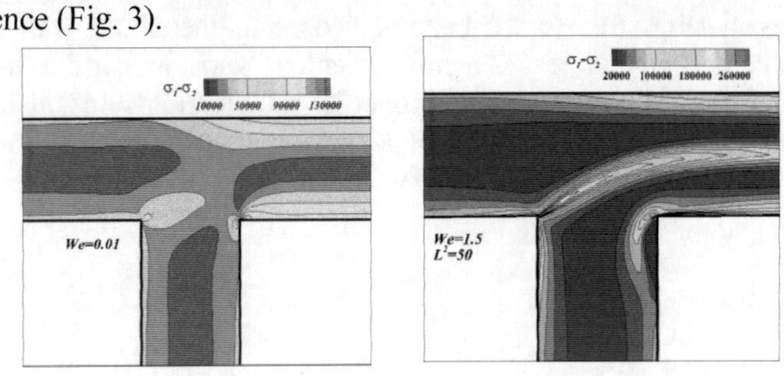

Fig. 3. Isolines of principal stresses difference in the channel

In the left section, the most significant area is $0.8 < y < 0.84$. For a viscoelastic fluid, a flow from the left sleeve will "bump" the flow region, containing partially oriented flexible particles, and flow around it (Fig. 4).

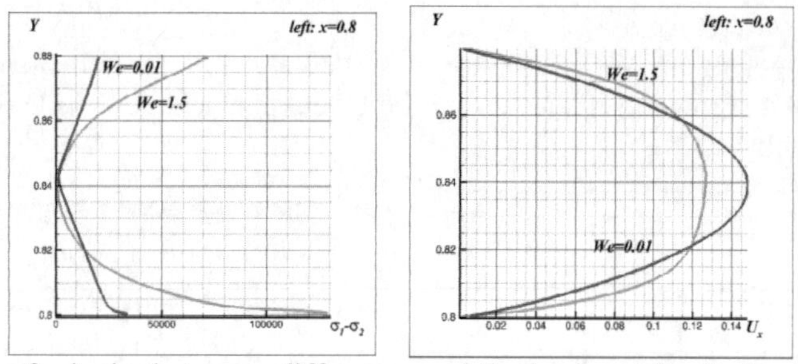

Fig. 4. Distributions of principal stresses difference and the horizontal velocity component in the left section

Significant increase of principal stresses difference's value occurred due to the formation of a long-range order in the arrangement of the flexible particles near the left corner point.

For the center section near the central point of the channel (y = 0.84), the influence of the walls and angular points of the channel is not so significant, so the velocities in this part are greater, and the flexible particles are close to their equilibrium state (Fig. 5). In the upper region, the principal stresses difference decreased due to the fact that at this point a part of the flow is "clamped" between the oriented flexible particles of below and the upper wall of the channel. The maximum value of the principal stresses difference at the point y = 0.823 is explained by the "collision" of two flows. The zigzag change in the velocity of viscoelastic fluid in the interval $0.81<y<0.83$ corresponds to the width of the oriented flexible particles.

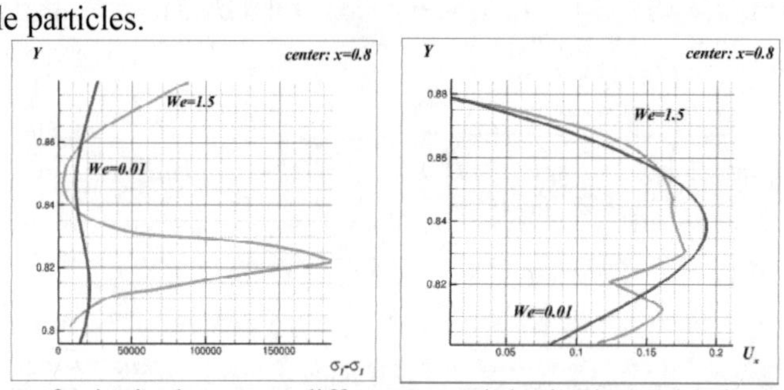

Fig. 5. Distributions of principal stresses difference and the horizontal velocity component in the center section

For the right section in the area of the point y = 0.85, we can suppose that the orientation of the flexible particles is absent (Fig. 6). As we become closer to the upper wall of the channel, the orientation of the flexible particles increases again. Oriented bands appear due to the peculiarities of the channel's shape, the type of flow and the properties of the fluid, and also due to the mutual influence of oriented macromolecules segments. The velocity distribution shows that the maximum value of the velocity for a viscoelastic fluid is observed near the right corner point.

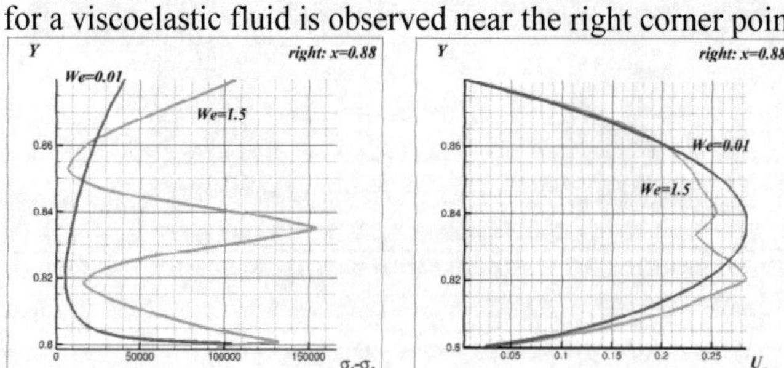

Fig. 6. Distributions of the principal stresses difference and the horizontal velocity component in the right section

In the low section, same as for the left section, the increased orientation of the flexible particles (Fig. 7) is observed near the corner points for a viscoelastic fluid. This is a consequence of the flow from below to the right rotation, oriented bands of flexible particles in the central and outlet parts of the channel, symmetry line of the channel in the turning direction.

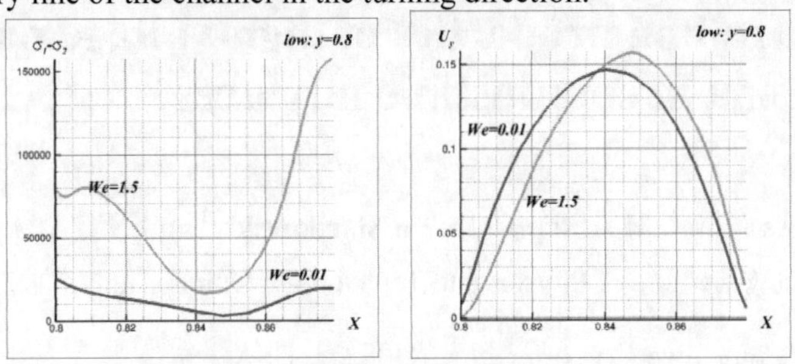

Fig. 7. Distributions of principal stresses difference and the horizontal velocity component in the low section

4. Conclusions

The peculiarity of this problem is the mutual perpendicular arrangement of the input parts of the channel. Such model configuration cases the following 2 features for the viscoelastic fluid:
1. A band of oriented flexible particles appears in the central part of the channel, along the separation line of the input flows.
2. A stagnant region near the right corner point.

The possible degree of extension of a flexible particle for this case exceeds its length in the equilibrium state by no more than 50 times. Even having such relatively small values, the picture of the stress state in a fluid differs significantly from the Newtonian one, for which the presence of normal stresses in the central part of the channel occures due to the channel's shape and the flow's type. Large stress values in microfluidic devices can adversely affect the components, causing their deformation and, possibly, breaking.

References

[1] Tabeling P. Introduction to mivrofluidics. Oxford: Oxford University Press, 2005. 301 p.
[2] Karnidakis G., Beskok A., Aluru N. Microflows and nanoflows // Interdisciplinary Applied Math. Springer Science + Business Media, Inc., 2005. Vol. 29. 817 p.
[3] F. Kh. Tazyukov The influence of fluid-flexible particle interaction on fluid flow optical non-Homogeneity in channel bifurcation / Tazyukov F.Kh., E.R. Kutuzova, F.A. Garifullin // Conf. Series: Journal of Physics 789 - 2017 – 012062. DOI: 10.1088/1742-6596/789/1/012062
[4] OpenFoamUserGuide: [*Electronic Materials*]. URL: http://openfoam.com (01.09.17)

IOP Conf. Series: Journal of Physics: Conf. Series **1058** (2018) 012053 doi:10.1088/1742-6596/1058/1/012053

Processes controlling properties of high-voltage nanosecond discharge plasma in combustible mixtures

E M Anokhin[1], M A Popov[1], A Yu Starikovskiy[2] and N L Aleksandrov[1]

[1]Moscow Institute of Physics and Technology, Dolgoprudny, 141700, Russia

[2]Princeton University, Princeton, NJ08544, USA

E-mail: nick_aleksandrov@mail.ru

Abstract. We present the results of the experimental study of high-voltage nanosecond repetitively pulsed discharge and of its afterglow in ethane, dimethyl ether and their mixtures with oxygen. The measurements were made for room temperature and pressures from 2 to 6 Torr. The measured specific deposited energy and mean discharge current varied with the number of voltage pulses monotonously in ethane and dimethyl ether and nonmonotonously in the hydrocarbon:oxygen mixtures. A microwave interferometer was used for time-resolved electron density measurements in the discharge afterglow. The effective recombination coefficients were obtained from the analysis of the measured data. These coefficients varied with the number of voltage pulses monotonously in ethane and peaked in pure dimethyl ether and its mixture with oxygen. Possible mechanisms of the nonmonotonous behavior of the discharge characteristics were discussed.

1. Introduction

Repetitive, nanosecond-pulse discharges are widely used for plasma generation. In such discharges, ionization, dissociation and excitation of molecules efficiently occur because of high reduced electric fields created in the discharge plasmas. Therefore, nanosecond discharges are promising for many applications including aerospace applications. Over the last decades, considerable progress has been made in studies of ignition and combustion by nanosecond discharge plasmas [1-7]. It was shown that, using non-equilibrium nanosecond discharges, it is possible to reduce ignition delay time and ignition temperature, to stabilize flame and to increase flammability limits in various combustible mixtures.

The effect of non-equilibrium discharge plasmas on oxidation and ignition processes are hard to study at temperatures below the self-ignition threshold when fuel is not ignited without plasma. For low temperatures, oxidation processes have been studied in fuel-containing mixtures activated by nanosecond repetitively pulsed discharges [8-14]. Plasma-assisted oxidation was considered for H_2 [8], alkanes [9, 10, 13, 14] and ethylene [11, 12]. In these studies, emission and absorption spectroscopy and gas chromatography were used to measure species density and gas temperature.

Previous studies of fuel oxidation under the action of nanosecond discharge plasmas considered the effect of discharge on oxidation processes. There is also the influence of oxidation on the properties of discharge plasmas. It was shown in [15] that fuel oxidation influences discharge characteristics and the rate of plasma decay in discharge afterglow. The aim of this work was to further study the effect of oxidation on discharge development and the decay of plasma generated by a nanosecond repetitively pulsed discharge. Energy deposition, discharge current and electron density history in the discharge afterglow were measured to obtain information about plasma properties and lifetime for various numbers of discharge pulses leading to partial or complete fuel oxidation. The study was extended to oxygenated hydrocarbons (dimethyl ether) to compare the effect of oxidation in these hydrocarbons with that in saturated hydrocarbons considered in [15]. For the comparison purposes, measurements were also made in pure hydrocarbons.

2. Experiment

The experimental setup and the methods used to measure the characteristics of a high-voltage nanosecond repetitively pulsed discharge and electron density in its afterglow are the same as in our previous work [9, 16]. The discharge developed in a quartz tube with an inner diameter of 47 mm. The high-voltage electrode and the low-voltage electrode were a cone-shaped one and a grounded ring, respectively. The distance between the electrodes was 20 cm. The plasmas were generated at gas room temperature and pressures in the range 2 - 6 Torr. The amplitude of the pulses applied to the high-voltage electrode was up to 24 kV. The pulse duration was 25 ns and the pulse rise time was 5 ns. The pulse frequency was 10 or 20 Hz. A back-current shunt was used to measure the voltage amplitude, the shape of the pulses, the mean discharge current, and the deposited energy.

We also measured the temporal evolution of electron density in the discharge afterglow by a microwave interferometer with a reference wave frequency of 94 GHz. An incident microwave signal crossed the plasma volume in the middle of the discharge gap perpendicularly, reflected from the angle reflector, crossed the plasma volume again and returned to the waveguide. Electron density was determined using the measured phase shift of the reflected wave. The uncertainty of the electron density measurements was ≈ 30 %.

3. Measured results and discussion

The properties of discharge plasmas were studied in pure C_2H_6 (ethane), the stoichiometric C_2H_6:O_2 mixture, CH_3OCH_3 (dimethyl ether) and the stoichiometric CH_3OCH_3:O_2 mixture. Figure 1 shows the measured average discharge current, I, in pure C_2H_6 and in the C_2H_6:O_2 mixture versus the number of voltage pulses. Similar results were obtained in pure CH_3OCH_3 and in the CH_3OCH_3:O_2 mixture. In pure C_2H_6, the mean current increases monotonously with increasing number of pulses, N_p, and I(2 Torr) > I(4 Torr). This behavior changes when O_2 is added to C_2H_6. In the C_2H_6:O_2 mixture, the mean current varies nonmonotonously with N_p and shows minimum at $N_p \approx$ 1100 for 2 Torr and at $N_p \approx$ 2000 for 4 Torr. The minimum is pronounced for 2 Torr and less marked for 4 Torr. As a result, we have I(2 Torr) < I(4 Torr) at $N_p < 6000$ and I(2 Torr) > I(4 Torr) for higher values of N_p.

The specific deposited energy, W, in pure C_2H_6 and in the C_2H_6:O_2 mixture versus the number of voltage pulses is shown in figure 2. The specific energy input per single voltage pulse is between 0.002 and 0.009 eV per one neutral particle. This quantity, as well as I, depend on gas pressure, gas composition and the number of voltage pulses. In pure C_2H_6 and CH_3OCH_3, the specific deposited energy is inversely proportional to the gas pressure p ($W \sim 1/p$). This energy is independent of N_p in pure C_2H_6 and decreases with increasing N_p in CH_3OCH_3. In the O_2-containing mixtures, the value of W varies nonmonotonously when N_p is increased. This energy shows profound minimum for 2 Torr and less marked minimum for 4 Torr at the values of N_p at which the minimum is observed for I in these mixtures.

According to our estimates, rapid gas heating during one voltage pulse and discharge afterglow and gas heating due to energy release in the oxidation processes were less than 15K. Therefore, each new discharge developed in a room temperature gas. The absence of gas heating was controlled by measuring gas pressure in the discharge tube.

IOP Conf. Series: Journal of Physics: Conf. Series **1058** (2018) 012053 doi:10.1088/1742-6596/1058/1/012053

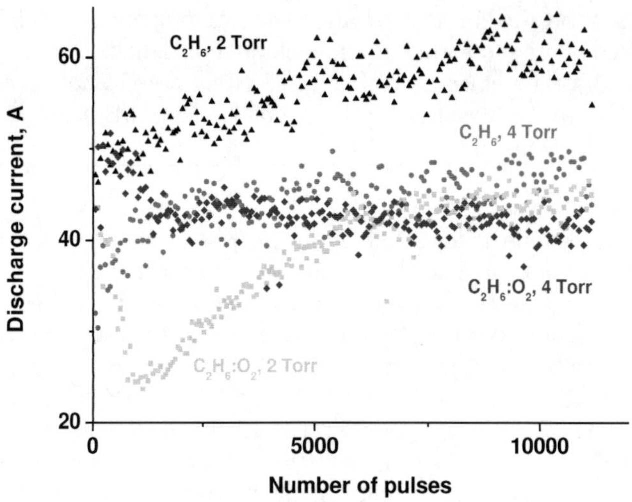

Figure 1. Mean discharge current in C_2H_6 and stoichiometric C_2H_6:O_2 mixture for various pressures as a function of the number of voltage pulses.

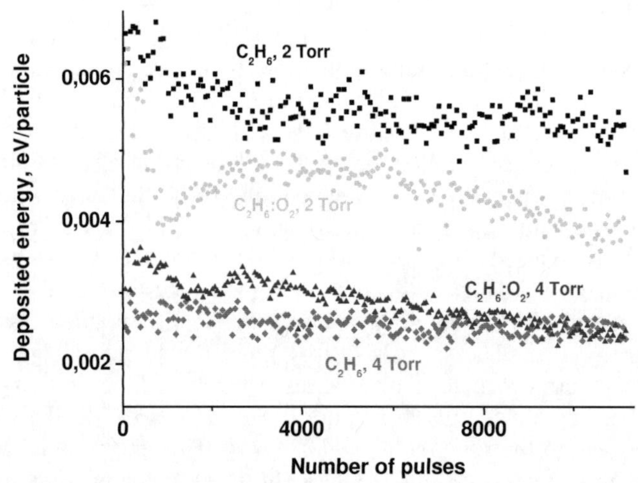

Figure 2. Specific deposited energy in the nanosecond discharge in C_2H_6 and stoichiometric C_2H_6:O_2 mixture for various pressures as a function of the number of voltage pulses.

Using a microwave interferometer we studied how the plasma decay in hydrocarbons and combustible mixtures depends on gas mixture parameters. In our case, recombination with molecular ions is the dominant mechanism of electron loss. From the measurement of electron density history, we obtained the effective recombination coefficient defined as $\alpha_{eff} = (n_{e0}\tau_{1/2})^{-1}$, where $\tau_{1/2}$ is the time required to halve the electron density, n_{e0}. We took $n_{e0} = 2\times10^{11}$ cm^{-3}; this value is an order of magnitude lower than the value of n_e in the beginning of the afterglow and corresponds to $t \sim 2$ μs.

It is worth noting that the electron temperature T_e in the discharge afterglow in hydrocarbons and hydrocarbon-containing mixtures is close to the gas temperature T = 300 K. The calculation showed that the electron temperature is thermalized for 100 ns during plasma decay in ethane and propane at 2 Torr [17]. According to our estimates, the same is also expected in CH_3OCH_3 and in all mixtures considered in this work.

The effective recombination coefficient in pure CH_3OCH_3 and in the stoichiometric CH_3OCH_3:O_2 mixture versus N_p is shown in figure 3. In pure CH_3OCH_3 the recombination coefficient peaks at $N_p \approx$ 2300 for 2 Torr and at $N_p \approx$ 4300 for 4 Torr. The recombination coefficient in the CH_3OCH_3:O_2 mixture shows much more profound maximum with increasing number of pulses and this peak is observed at much lower number of pulses. Here, this coefficient peaks at $N_p \approx$ 400 for 2 Torr and at $N_p \approx$ 600 for 4 Torr. The dependence of α_{eff} on N_p in the CH_3OCH_3:O_2 mixture is similar to that obtained in the mixtures of saturated hydrocarbons with O_2 [15]. However, a nonmonotonous dependence of α_{eff} on N_p was not observed in pure saturated hydrocarbons, whereas this was obtained in pure CH_3OCH_3.

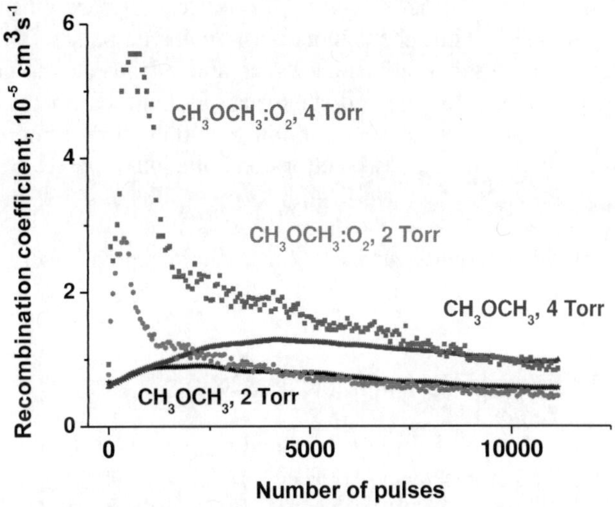

Figure 3. The effective recombination coefficient in CH_3OCH_3 and stoichiometric CH_3OCH_3:O_2 mixture afterglows as a function of the number of voltage pulses prior to these afterglows.

It was suggested in [15] that the nonmonotonous behaviors of discharge characteristics and α_{eff} with increasing N_p in hydrocarbon:oxygen mixtures can be associated with the production of some neutral intermediates that favor plasma decay. A number of neutral hydrocarbon intermediate species have been detected during oxidation of C_1 – C_5 alkanes after a high-voltage nanosecond discharge. These intermediates include CH_2O, CH_3OH, CH_3CHO, and CH_3CH_2CHO (see, for instance [13, 14]). Some hydrocarbon intermediate species have high dipole moments. The production of such intermediates is followed by the formation of cluster ions for which the recombination coefficients with electrons are anomalously high [19]. The observed increase of the effective recombination coefficient with increasing gas pressure (see figure 3) can be associated with the production of more complicated cluster ions in three-body reactions and with the increased contribution of three-body electron-ion recombination to the total rate of plasma decay.

In [15], it was suggested that the intermediates favoring plasma decay are formed in saturated hydrocarbon:oxygen mixtures and are absent in pure saturated hydrocarbons. It was shown in this work that the nonmonotonous behavior of the effective recombination coefficient with increasing N_p also can be observed in pure oxygenated hydrocarbons. We suggest that the intermediates leading to the increase in the recombination rate can be also produced in these hydrocarbons without the presence of oxygen in the initial gases.

4. Conclusions

The discharge characteristics of a high-voltage nanosecond repetitively pulsed discharge in ethane, dimethyl ether and their mixtures with oxygen were measured at room gas temperature and at pressures from 2 to 6 Torr. In addition, time-resolved electron density was measured in the afterglow of this discharge. Observations were made after various number of voltage pulses when the neutral composition of the gaseous mixtures was changed due to partial fuel oxidation. Observations showed that the mean discharge current and specific deposited energy varied with the number of voltage pulses monotonously in pure hydrocarbons and nonmonotonously in the hydrocarbon:oxygen mixtures.

The effective recombination coefficients were obtained from the analysis of the measured electron density histories. These coefficients peaked with the number of voltage pulses in the ethane:oxygen mixtures, in pure dimethyl ether and in its mixture with oxygen. In pure ethane, the recombination coefficient varied monotonously with the number of voltage pulses. We suggest that the nonmonotonous behaviors of the discharge characteristics and of the recombination coefficients are associated with the formation of dipole intermediate species in hydrocarbon:oxygen mixtures and in pure oxygenated hydrocarbons. As a result, electron-ion recombination is more efficient due to the formation of cluster ions and due to three-body electron-ion collisions.

Acknowledgments
This work was supported by the Russian Foundation for Basic Research under the project No. 17-02-00481.

References

[1] Starikovskaia S M 2006 *J. Phys.D: Appl. Phys.* **39** R265
[2] Popov N A 2007 *High Temp.* **45** 261
[3] Adamovich I V, Choi I, Jiang N, Kim J-H, Keshav S, Lempert W R, Mintusov E, Nishihara M, Samimy M and Uddi M 2009 *Plasma Sources Sci. Technol.* **18** 034018
[4] Starikovskiy A and Aleksandrov N 2013 *Progr. Energy Combust. Sci.* **39** 61
[5] Starikovskaia S M 2014 *J. Phys. D: Appl. Phys.* **47** 353001
[6] Ju Y and Sun W 2015 *Progr. Energy Combust. Sci.* **48** 21
[7] Adamovich I V and Lempert W R 2015 *Plasma Phys. Control. Fusion* **57** 014001
[8] Zatsepin D V, Starikovskaia S M and Starikovskii A Yu 2001 *Combust. Theory Modelling* **5** 97
[9] Anikin N B, Starikovskaia S M and Starikovskii A Yu 2004 *Plasma Phys. Rep.* **30** 1028
[10] Anikin N B, Starikovskaia S M and Starikovskii A Yu 2006 *J. Phys. D: Appl. Phys.* **39** 3244
[11] Mintusov E, Serdyuchenko A, Choi I, Lempert W R and Adamovich I V 2009 *Proc. Combust. Inst.* **32** 3181
[12] Lefkowitz J K, Uddi M, Windom B C, Lou C and Ju Y 2015 *Proc. Combust. Inst.* **35** 3505
[13] Lefkowitz J K, Guo P, Rousso A and Ju Y 2015 *Phys. Trans. R. Soc. A* **373** 201403333
[14] Tsolas N, Lee J G and Yetter R A 2015 *Phys. Trans. R. Soc. A* **373** 201403344
[15] Anokhin E M, Popov M A, Starikovskiy A Yu and Aleksandrov N L 2017 *Combust. Flame* **185** 301
[16] Aleksandrov N L, Kindysheva S V, Kirpichnikov A A, Kosarev I N, Starikovskaia S M and Starikovskii A Yu 2007 *J. Phys. D: Appl. Phys.* **40** 4493
[17] Anokhin E M, Popov M A, Kochetov I V, Starikovskiy A Yu and Aleksandrov N L 2016 *Plasma Sources Sci. Technol.* **25** 044006
[18] Anokhin E M, Popov M A, Kochetov I V, Aleksandrov N L and Starikovskii A Yu 2016 *Plasma Phys. Rep.* **42** 59
[19] Florescu-Mitchell A I and Mitchell J B A 2008 *Phys. Rep.* **430** 277

IOP Conf. Series: Journal of Physics: Conf. Series **1058** (2018) 012054 doi:10.1088/1742-6596/1058/1/012054

Reflection of acoustic waves from the boundary of contaminated fog

D A Gubaidullin, Yu V Fedorov, R R Zaripov

Institute of Mechanics and Engineering - Subdivision of the Federal State Budgetary Institution of Science "Kazan Scientific Center of the Russian Academy of Sciences", 2/31 Lobachevsky str., Kazan 420111, Russia

Kazan (Volga Region) Federal University, 18 Kremlyovskaya str., Kazan 420008, Russia

E-mail: gubaidullin@imm.knc.ru

Abstract. The problem of oblique incidence of an acoustic wave from pure air to the boundary of a vapor-gas mixture with droplets of liquid and solid particles is considered. Dependencies of the reflection coefficient of an acoustic wave on the volume content of inclusions and the angle of incidence of a wave are given. The influence of phase transitions is studied.

1. Introduction

The study of the reflection of acoustic waves from the vapor-gas medium is actual. This is mainly due to natural phenomena, such as a layer of fog, dusty gas in the atmosphere and atmospheric clouds. It is also possible to use various gas-droplet mixtures (fogs) to hide various objects from radars, for example ships. Such media are unusual, since they are characterized by the dispersion of the speed of sound. The dynamics of wave processes in such media was presented in well-known monographs [1, 2, 3]. Numerical and experimental analysis of the processes occurring in such media was considered in [4, 5]. A more interesting medium is a polydisperse medium, because such medium more accurately describes models of real medium. The influence of polydispersity on the propagation of sound in complex media was considered in [6, 7]. The problem of reflection of an acoustic wave from a fog boundary without phase transformations was considered in [8]. In this work the effect of phase transformations on the reflection of an acoustic wave from the boundary of a two-fraction vapor-gas-droplet mixture with polydisperse inclusions is studied.

2. Mathematical model.

We consider two mediums and denote them by numbers 1 and 2. Suppose that there is a planar interface between them, which coincides with the x-axis. The acoustic wave falls from the medium 1 to the interface of the medium 2. The angle at which the wave falls is measured from the normal (z axis) of the interface between the two media and denoted by θ_1 (Fig. 1). In this case, the reflection coefficient can be represented in the form [9]

$$R = \frac{Z_2 - Z_1}{Z_2 + Z_1} \tag{1}$$

$$Z_i = \frac{\rho_i C_i}{\cos\theta_i}, i = 1, 2, \theta_2 = \arcsin\left(\frac{C_2}{C_1}\sin\theta_1\right)$$

Here R - a reflection coefficient, ρ - density, C - speed of sound, subscripts 1 and 2 refer to mediums 1 and 2.

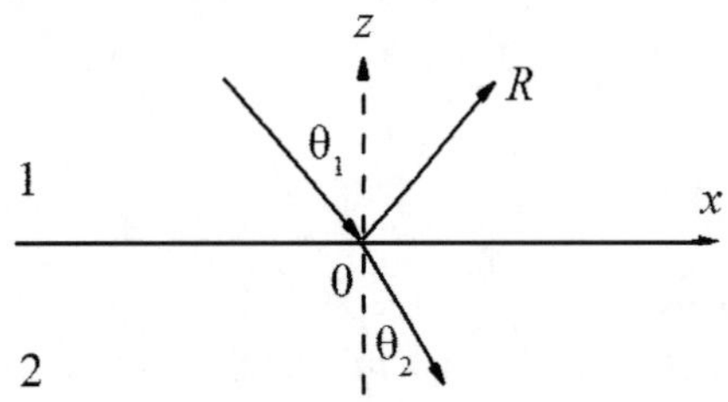

Figure 1. Scheme of the media.

It can be seen from formula (1) that to determine the reflection coefficient it is necessary to know the density and speed of sound in mediums 1 and 2. Analysis shows that the effect of phase transformations on the velocity in aerosol is more pronounced at low frequencies [2]. Thus, to study the effects of phase transformations on the reflection coefficient, we will use the equation for equilibrium velocity in the medium. The equation for the equilibrium speed of sound in a vapor-gas-droplet mixture with two fractions of inclusions has the form [10]

$$C_2 = C_1\left(\frac{\gamma_e}{m_1\gamma_1}\right)^{1/2}, m_1 = 1 + m_a + m_b \tag{2}$$

$$\gamma_e = \frac{m_2 R_V^*(1 - k_V R_V^*) + L}{(m_2 - 1)R_V^* + L - L_1 R_V^*/\gamma_1}, R_V^* = \frac{R_V}{R_1}, L = \gamma_1(\gamma_1 - 1)k_V l_0^{*2},$$

$$m_2 = 1 + \frac{m_a c_{2a}}{c_{p1}} + \frac{m_b c_{2b}}{c_{p1}}, L_1 = k_V(\gamma_1 - 1)(2l_0^*\gamma_1 - R_V^*) - 1, l_0^* = \frac{l_0}{C_1^2}$$

$$R_1 = k_V R_V + (1 - k_V)R_G, \gamma_1 = k_V\gamma_V + (1 - k_V)\gamma_G$$

Here, subscripts V refer to the vapor component of the carrier phase, G - to the gas component, a - to the first fraction, b - to the second fraction, γ - ratio of specific heat, k -the mass concentration of the vapor or gas, m - mass concentration, R - constant value of vapor or gas, c_p - specific heat, l_0 - specific heat of vaporization, C_1 - speed of sound in the carrier phase (gas).

3. The results of calculations.

Let us consider the fall of a low-frequency acoustic wave to the interface between two mediums. Let the wave fall from the side of air to the boundary of the vapor-gas mixture with droplets of water and ash particles at some angle. From the formulas (1) and (2) we obtain:

$$R = \frac{\rho_2 C_2/\cos\theta_2 - \rho_1 C_1/\cos\theta_1}{\rho_2 C_2/\cos\theta_2 + \rho_1 C_1/\cos\theta_1}, \theta_2 = \arcsin\left(\frac{C_2}{C_1}\sin\theta_1\right) \tag{3}$$

Calculations were obtained using formulas (2) and (3), the density of this mixture was determined as $\alpha_1\rho_1 + \alpha_{2a}\rho_{2a} + \alpha_{2b}\rho_{2b}$.

Fig. 2 shows the results of the reflection coefficient in dependence on the volume content of the drops of the first fraction, when acoustic wave falls at different angles. Here, the initial concentration of vapor $k_{V\,0} = 0.01$, the volume content of particles of the second fraction $\alpha_{2b} = 0.005$. Under normal incidence $\theta_1 = 0$ of an acoustic wave, the curve $|R^2(\alpha_{2a})|$ is

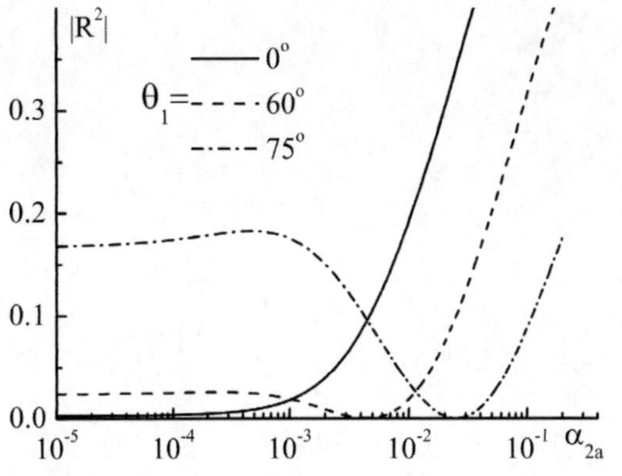

Figure 2. Dependencies of the reflection coefficient on the volume content of the inclusions.

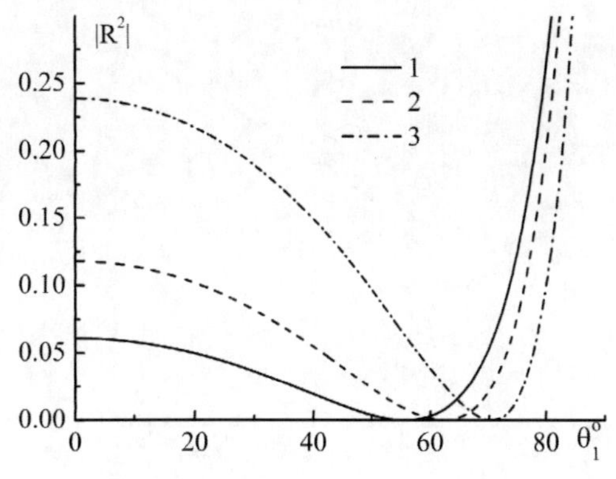

Figure 3. Dependencies of the reflection coefficient on the angle of incidence of the wave

monotonic, and if the angle of incidence of the wave grows up to 60° and to 75°, the curve $|R^2(\alpha_{2a})|$, as we can see, is not monotonic. Thus, when the wave falls at an angle θ_1 = 60° the curve $|R^2(\alpha_{2a})|$ practically does not change at volume drops $\alpha_{2a} < 10^{-3}$ and assumes a certain constant value. Then the curve decreases, which reaches a minimum at $\alpha_{2a} \approx 4.5 \cdot 10^{-3}$ and takes a zero value. In this way, an acoustic wave passes through the boundary completely for a given volume content. When the wave falls at an angle θ_1 = 75° a similar behavior of the reflection coefficient curve is observed. The only significant difference between them, is that the total passage of the wave through the boundary occurs with the volume content of the droplets $\alpha_{2a} \approx 2.5 \cdot 10^{-3}$.

Fig. 3 shows the reflection coefficient in dependence on the incidence angle of acoustic wave θ_1. The volume content of particles of the second fraction α_{2b} = 0.001, initial vapor concentration $k_{V\,0}$ = 0.01. Curves: 1 - α_{2a} = 0.0023, 2 - α_{2a} = 0.005, 3 - α_{2a} = 0.013. Here we should be noted, that the reflection coefficient does not depend monotonically on the angle of incidence of the wave. At the different value of the incidence angle the reflection coefficient reaches a minimum value, where the reflection coefficient becomes zero. The value of the angle at which the acoustic wave passes completely across the boundary depends on the volume content of the inclusions. For the case, when the volume content of the droplets α_{2a} = 0.0023 the wave is completely absorbed at θ_1 = 55°.The value of the angle, at which the wave passes without reflection across the boundary, increases with an increase in the volume content of the inclusions. To find out at which values of the incidence angles of the wave and the volume content of the inclusions, the incident wave will not be reflected from the interface of the two media, we turn to the analytical expression [11, 12]

$$\theta_1^* = \arcsin\sqrt{\frac{\rho_2^2 C_2^2 - \rho_1^2 C_1^2}{C_2^2(\rho_2^2 - \rho_1^2)}} \qquad (4)$$

where θ_1* - is the critical angle of incidence of the wave, under which the condition for the complete passage of the wave across the boundary is satisfied.

Fig. 4 shows calculations of the dependences of the critical angle on the volume content of the inclusions, in which the incident wave is not reflected from the interface between the two media. Here, the curves in the form of lines correspond to different values of the vapor concentration obtained with the help of formula (4), and the points are constructed according to theoretical data [13]. As can be seen from the figure 4, the curve obtained by formula (4) agrees well with theoretical data [13].

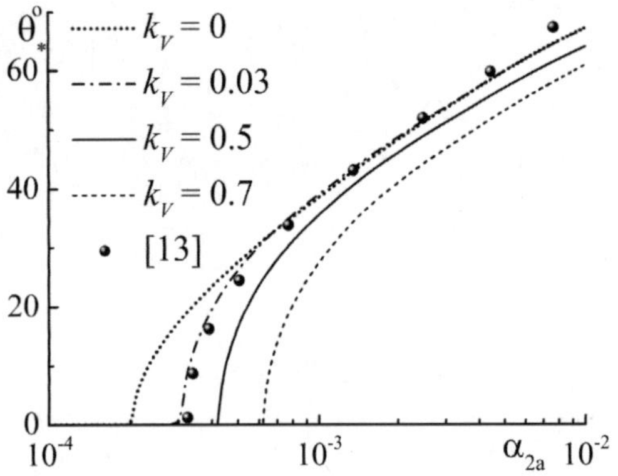

Figure 4. Dependencies of the critical angle of incidence on the volume content of the inclusions

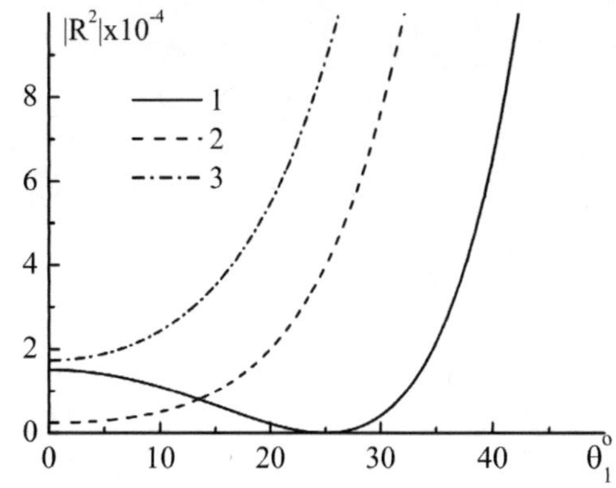

Figure 5. Dependencies of the reflection coefficient on the angle of incidence of the wave

Fig. 5 shows the reflection coefficient in dependence of the incidence angle of acoustic wave θ_1. Here $\alpha_2a = 10^{-4}$, $\alpha_{2b} = 10^{-4}$, curves 1 - $k_{V0} = 0$, 2 - $k_{V0} = 0.01$, 3 - $k_{V0} = 0.05$. As can be seen from the figure, when there are no phase transformations, the reflection coefficient depends non-monotonically on the angle of incidence of the wave. And for given volume contents of inclusions the wave passes without reflection at $\theta_1 = 25^\circ$. For curves $|R^2(\theta_1)|$ with allowance for phase transitions, a monotonic dependence is observed.

Conclusions

The fall of a low-frequency acoustic wave from pure gas at different angles to the boundary of a vapor-gas mixture with droplets of water and ash particles is considered. The effects of the initial vapor concentration, the volume content of inclusions and the angle of incidence of the wave on the reflection coefficient were analyzed. It is shown, that the dependence of the reflection coefficient on the volume content is monotonic, in the case of a normal incidence of the wave, and in the case of an oblique incidence of the wave, a nonmonotonic dependence is observed. It is illustrated that for certain volume contents of inclusions and the angle of incidence of the wave the reflection coefficient assumes zero values. Namely, the incident wave passes without reflection through the interface between two media.

References

[1] Nigmatulin R I 1989 *Dynamics of multiphase media* (Moscow: Nauka)
[2] Gubaidullin D A 1998 *Dynamics of two-phase gas-vapor-droplet media* (Kazan: Kazan mathematical society)[3] Temkin S 2005 *Suspension acoustics: An introduction to the physics of suspension* (New York: Cambridge University Press)
[4] Bezrukov N A and Chernookulsky A V 2015 *Proceedings of the Russian Academy of Sciences. Physics of the atmosphere and ocean* 52 577
[5] Baudoin M, Coulouvrat F and Thomas J L 2011 *J. Acoust. Soc. Am.* 130 1142
[6] Gumerov N A and Ivandaev A I 1988 *PMTF* 115
[7] Gubaidullin D A and Ivandaev A I 1993 *PMTF* 75
[8] Shagapov V Sh and Sarapulova V V 2014 *Izvestiya Atmospheric and Oceanic Physics* 50 602
[9] Brekhovskikh L M and Godin O A 1989 *Acoustics of Layered Media* (Moscow: Nauka)
[10] Gubaidullin D A and Fedorov Yu V 2012 *High Temperature* 50 616
[11] Gubaidullin D A and Fedorov Yu V 2018 *Acoustical Physics* 64 164
[12] Gubaidullin D A and Fedorov Yu V 2017 *Doklady Physics* 62 478
[13] Shagapov V Sh and Sarapulova V V 2015 *Applied Mechanics and Technical Physics* 56 119

IOP Conf. Series: Journal of Physics: Conf. Series **1058** (2018) 012055 doi:10.1088/1742-6596/1058/1/012055

Reflection and Transmission of Sound Waves through the Layer of Two-fractional Bubbly Liquid

D A Gubaidullin[1,2] **and R N Gafiyatov**[1]

[1] Institute of Mechanics and Engineering - Subdivision of the Federal State Budgetary Institution of Science «Kazan Scientific Center of the Russian Academy of Sciences», 2/31 Lobachevsky str., Kazan, 420111, Russia
[2] Kazan (Volga Region) Federal University, 18 Kremlyovskaya str., Kazan, 420008, Russia

E-mail: gafiyatov@mail.ru

Abstract. The mathematical model that determines reflection and transmission of sound waves through a medium containing two-fractional bubbly liquid is presented. For the model of media with three layer the wave reflection and transmission coefficients are calculated. The influence of the volume content of bubbles on the investigated coefficients is shown.

1. Introduction

The investigations of wave dynamics and acoustics of disperse media are of significant interest. Currently, the basics of mechanics and thermophysics of bubbly liquids, as well as the most significant results in the study of wave processes in these environments are presented in monographs [1-5]. Hao and Prosperetti [6] obtained the results of study of the vapor bubble dynamics in liquid. Commander and Prosperetti [7] presented a model of the propagation of plane waves with small amplitude in a mixture of liquid and of gas bubbles. This model is adequate for the frequencies far from resonance at the volume content of a disperse phase of 1-2%. Nigmatulin et al. [8] showed the need of allowing for the effect of liquid compressibility for problems of acoustics of bubbly liquids. Shagapov [9] considered a problem of propagation of small disturbances in liquids with polydisperse bubbles. Nigmatulin et al. [10] studied the small radial oscillations of vapor-gas bubbles in liquid under the influence of an acoustic field. They showed that capillary effects and phase transitions lead to the new resonant frequency of small bubbles. This frequency is different from the Minnaert frequency.

The propagation of acoustic waves in a two-fractional mixture of liquid with bubbles is investigated in [11-13]. A mathematical model of the propagation of sound waves in the two-fractional mixture of liquid with polydisperse vapor-gas and gas bubbles with phase transitions is shown in [14-16]. The propagation of acoustic waves in multifractional mixtures of liquid with gas and vapor-gas bubbles of different sizes and different compositions is studied in [17-20].

In this study the sound wave reflection and transmission through a multilayer object containing a two-fractional bubbly fluid layer is investigated.

2. Dispersion relation

Linearized equations for one-dimensional disturbances in a two-fractional mixture of liquid with bubbles are obtained from the general equations of motion for bubbly mixtures [1]. The dispersed phase consists of two fractions having various gases in bubbles and different in the bubbles radii. The total bubble volume concentration is small. Solving this system this dispersion relation is obtained:

$$\left(\frac{K_*}{\omega}\right)^2 = \frac{\alpha_{10}^2}{C_1^2} + \frac{\alpha_{20}^{\mathrm{I}}\rho_{10}}{p_0 N_R^{\mathrm{I}}}\left(1 + \frac{H_1^{\mathrm{I}} + H_1^{\mathrm{II}} + H_2^{\mathrm{II}}\chi_1}{M_3^{\mathrm{I}}\chi_2 / b_1^{\mathrm{I}}}\right) + \frac{\alpha_{20}^{\mathrm{II}}\rho_{10}}{p_0 N_R^{\mathrm{II}}}\left(1 + \frac{H_1^{\mathrm{I}} + H_1^{\mathrm{II}} + H_2^{\mathrm{I}}\chi_3}{M_3^{\mathrm{II}}\chi_4 / b_1^{\mathrm{II}}}\right) \quad (1)$$

Dispersion relation (1) (i.e. the function of the complex wave number K_* on the frequency ω) determines a propagation of sound waves in the two-fractional mixtures of liquid with vapor-gas bubbles of different gases (different initial radii, different initial volume contents and different thermal properties of fractions) with the interphase diffusion mass transfer.

3. Mathematical model

In analysing the interaction between an acoustic signal and a multilayer object (Fig. 1), the following method of calculations is used. According to [21], the result of the reflection and transmission of a plane monochromatic wave $\exp(iK_* x - \omega t)$ from a multilayer object is the plane waves $R\exp(iK_* x + \omega t)$ and $T\exp(iK_* x - \omega t)$, where R and T are the wave reflection and transmission coefficients, respectively, determined in terms of the layer impedances Z_i and the entry impedances of the layer boundaries Z_i^{in}. For a three-layer medium the coefficients R and T are as follows:

$$R = \frac{Z_2(Z_1 - Z_3) - i(Z_2^2 - Z_1 Z_3)\mathrm{tg}(K_2 d_2)}{Z_2(Z_1 + Z_3) - i(Z_2^2 + Z_1 Z_3)\mathrm{tg}(K_2 d_2)}$$

$$T = \prod_{j=1}^{2}\left(\frac{Z_j^{in} + Z_j}{Z_j^{in} + Z_{j+1}}\exp[iK_j d_j]\right) \quad (2)$$

$$Z_j = \rho_j \frac{\omega}{K_j}, \quad Z_2^{in} = \frac{Z_1 - iZ_2\mathrm{tg}(K_2 d_2)}{Z_2 - iZ_1\mathrm{tg}(K_2 d_2)} Z_2, \quad j = 1,2,3.$$

Here, d_2 is the thickness of the layer with bubbles, K is the wave number, ω is the perturbation frequency, and ρ is the layer density. For a homogeneous layer, the wave number is determined as $K_j = \omega / C_j$, where C_j is the speed of sound in the j-th layer.

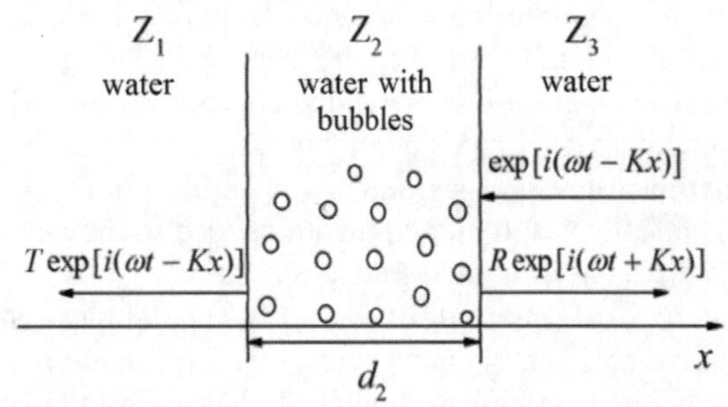

Figure 1. Scheme of the medium.

4. Results

We will consider the acoustic signal transmission through the three-layer medium, namely, water-water with bubbles-water. Disperse phase of bubbly liquid contain vapor-air bubbles and bubbles of helium with water vapor. Let the bubble layer thickness be d_2=5 mm, the bubble radii – 2mm (vapor-air bubbles), 1mm (bubbles of helium with water vapor), f_0=1630Hz (resonance frequency of 2mm bubbles) and the volume content α_{20}=0.01. The mixture pressure p_0=0.1 MPa and the temperature T_0=288 K. The calculations were performed according equations (1), (2). The dependences of the coefficients of the wave reflection and transmission through the given bubble layer on the dimensionless disturbance frequency are presented in the figure 2. Clearly that a minimum of the transmission coefficient and a maximum of the reflection coefficient are observable in the region of the resonance frequency of the bubbles. This means that at the given frequency the bubble layer almost completely reflects the incident acoustic wave. In addition, two fractions with different initial radii of the disperse phase leads to the appearance of two local minima and maxima on this coefficients. This is due to the difference in the values of the resonance frequencies of the intrinsic vibrations of bubbles of each fraction.

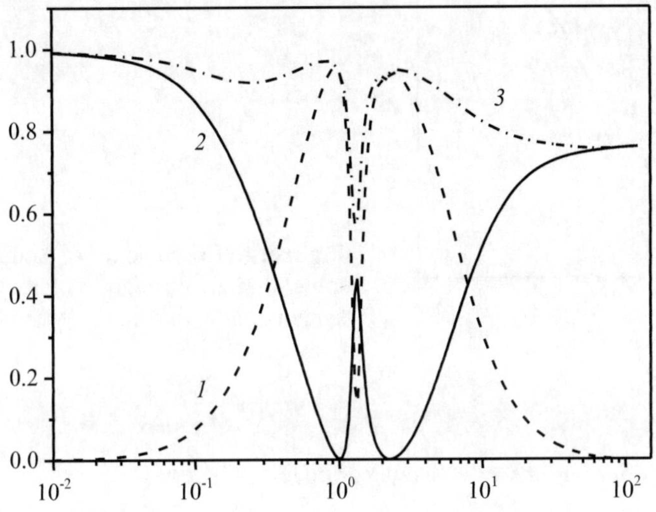

Figure 2. Coefficients of sound reflection and transmission through a bubbly screen, *1* – R^2, *2* – T^2, *3* – R^2+T^2.

On the figures 3, 4 the influence of the volume content of bubbles in bubble layer on the sound wave transmission and reflection is illustrated. Parameters of bubbly liquid are the same as on figure 2, curves *1* – α_{20}=0.0025, curves *2* – α_{20}=0.005, curves *3* – α_{20}=0.01. An increasing the volume content of bubbles leads to a decrease in the transmission coefficient and, correspondingly, to an increase in the reflection coefficient on the entire frequency range. The increasing the volume content of bubbles, the range near the resonance frequency, where the wave transmission coefficient takes near-zero values (opposite effect on the reflection coefficient), also expands.

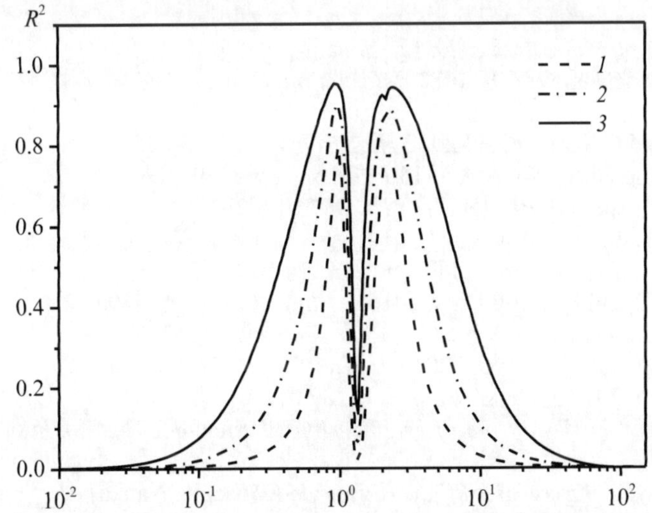

Figure 3. Coefficient of sound reflection through a bubbly screen at different volume content.

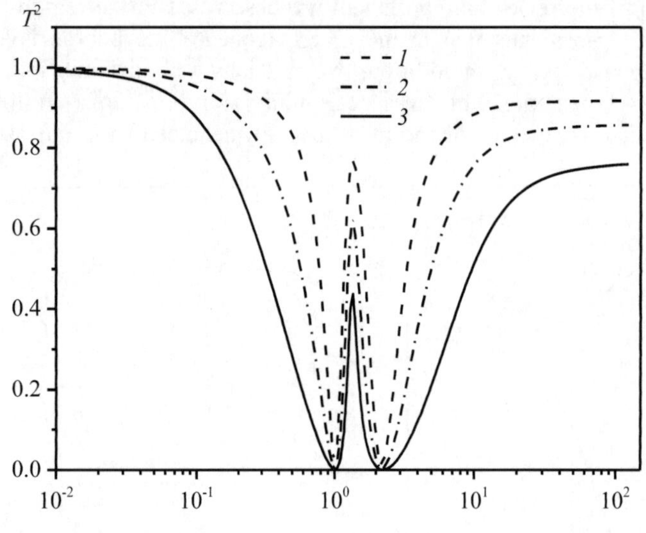

Figure 4. Coefficients of sound transmission through a bubbly screen at different volume content.

References

[1] Nigmatulin R I 1991 *Dynamics of Multiphase Media* (New York: Hemisphere)
[2] Temkin S 2005 *Suspension Acoustics: An Introduction to the Physics of Suspensions* (New York: Cambridge University Press)
[3] Leighton T G 1994 *The acoustic bubble* (London: Academic Press)
[4] Nakoryakov V E, Pokusaev B G and Shreiber I R 1993 *Wave Propagation in Gas-Liquid Media* (New York: CRC Press)
[5] Gubaidullin D A 1998 *Dynamics of two-phase gas-vapor-droplet media* (Kazan: Kazan mathematical society)
[6] Hao Y and Prosperetti A 1999 *Physics of Fluids* **11** 2008
[7] Commander K W and Prosperetti A 1989 *Journal of the Acoustical Society of America* **85** 732
[8] Nigmatulin R I, Shagapov V Sh and Vakhitova N K 1989 *USSR Academy of Sciences Doklady* **304** 1077
[9] Shagapov V Sh 1977 *Journal of Applied Mechanics and Technical Physics* **18** 77
[10] Nigmatulin R I, Khabeev N S and Nagiev F B 1981 *International Journal Heat and Mass Transfer* **24** 1033
[11] Gubaidullin D A, Nikifirov A A and Gafiyatov R N 2012 *High Temperature* **50** 250
[12] Gafiyatov R N, Gubaidullin D A and Nikiforov A A 2013 *Fluid Dynamics* **48** 366
[13] Gubaidullin D A and Gafiyatov R N 2014 *J. Phys.: Conf. Ser.* **567** 012020
[14] Gubaidullin D A, Gubaidullina D D and Fedorov Yu V 2014 *J. Phys.: Conf. Ser.* **567** 012011
[15] Gubaidullin D A and Fedorov Yu V 2016 Acoustical Physics **62** 179
[16] Gubaidullin D A, Gubaidullina D D and Fedorov Yu V 2016 *J. Phys.: Conf. Ser.* **669** 012011
[17] Gubaidullin D A, Nikiforov A A and Gafiyatov R N 2015 *High Temperature* **53** 240
[18] Gubaidullin D A, Nikiforov A A and Gafiyatov R N 2016 *J. Phys.: Conf. Ser.* **669** 012019
[19] Gubaidullin D A and Gafiyatov R N 2017 *J. Phys.: Conf. Ser.* **789** 012019
[20] Gubaidullin D A and Gafiyatov R N 2017 *Journal of Engineering Physics and Thermophysics* **90** 1433
[21] Brekhovskikh L M and Godin O A 1989 *Acoustics of layered media* (Moscow: Nauka)

IOP Conf. Series: Journal of Physics: Conf. Series **1058** (2018) 012056 doi:10.1088/1742-6596/1058/1/012056

Particle drift with account of the acoustic flow in 2D resonator

D A Gubaidullin[1], P P Osipov[1] and R R Nasyrov[1]

[1] Institute of Mechanics and Engineering – Subdivision of the Federal State Budgetary Institution of Science «Kazan Scientific Center of the Russian Academy of Sciences», 2/31, Lobachevsky str, 420111, Kazan, Russia.

E-mail: petro300@rambler.ru

Abstract. The particle drift in the standing wave taking into account the acoustic flow in the 2D resonator is investigated. Streamlines of acoustic flow and particle trajectories for different entrainment coefficients are calculated. It is shown that the entrainment coefficient has a great influence on deviation of the particle trajectory from the acoustic streamlines.

Standing wave in resonator induces an acoustic flow in form of Rayleigh and Schlichting vortices. Besides of particle eigen drift, these vortices transfer particles. Particle trajectories deviates from streamlines of acoustic flow. The degree of such deviation depends on the particle entrainment coefficient. In [1, 2] mathematical models of particle dynamics are described. Mathematical model and numerical analysis of acoustic streaming are discussed in [3]. In present paper a particle drift near the viscous boundary layer in the presence of Rayleigh and Schlichting vortices for different entrainment coefficients is investigated. This question has not been studied and represents both theoretical and practical interest.

1. Acoustic streaming model

Resonator presents a rectangular region $0 \le x \le L$, $0 \le y \le H$. 1D standing wave along x with velocity $U = U_0 \sin(k_w x)\sin(\omega t)$ is considered, where $k_w = 2\pi/\lambda$ – is wave number, $\lambda = 2L$ – is wave length, $T = 2\pi/\omega$ – is period. Near the bottom $y = 0$ boundary viscous boundary layer with thickness $\delta = (2\nu/\omega)^{0.5}$ arises. The height of the flow region is taken as $H = 8\delta$ [5].

Acoustic flow is described by Prandtl equations [4]

$$\frac{\partial u}{\partial t} + u\frac{\partial u}{\partial x} + v\frac{\partial u}{\partial y} - \eta\frac{\partial^2 u}{\partial y^2} = \frac{\partial U}{\partial t} + U\frac{\partial U}{\partial x}, \qquad \frac{\partial u}{\partial x} + \frac{\partial v}{\partial y} = 0,$$

where η is kinematic viscosity of fluid. Boundary conditions are taken from [5].

2. Particle dynamics

The particle dynamics is described by equations

$$\frac{dx_p}{dt} = u_p, \qquad \frac{du_p}{dt} = -\frac{u_p - u}{\tau_v}\left(1 + \frac{\sqrt{\mathrm{Re}_{12}}}{6} + \frac{\mathrm{Re}_{12}}{60}\right),$$

$$\frac{dy_p}{dt} = v_p, \qquad \frac{dv_p}{dt} = -\frac{v_p - v}{\tau_v}\left(1 + \frac{\sqrt{\mathrm{Re}_{12}}}{6} + \frac{\mathrm{Re}_{12}}{60}\right),$$

where (x_p, y_p) – particle position, (u_p, v_p) – particle velocity, $\mathrm{Re}_{12} = \mathrm{Re}_0 \sqrt{(u-u_p)^2 + (v-v_p)^2} / U_0$, $\mathrm{Re}_0 = 2rU_0/\eta$, $\tau_v = \frac{2}{9} r^2 \frac{\rho_p}{\mu}$, μ – dynamic viscosity.

3. Results

The trajectory of water droplets, having the same radius r, in the air for Mach number $M = 0.015$ are computed. At the initial time particles are placed evenly over the rectangular enclosure. In figs. 1, 2, 3 the streamlines of acoustic flow are depicted by thin curves (Schlichting and Rayleigh vortices), and particle trajectories by thick curves. Each figure corresponds to a different Reynolds numbers Re_0 and entrainment coefficient $\mu_p = \{1 + (\omega\tau_v)^2\}^{-0.5}$. It is seen that trajectories of some particles remain in the zone of Rayleigh vortex, and trajectories of another particles are "attracted" to the Schlichting vortex. For the entrainment coefficient $\mu_p = 0.03$ (fig. 1) particle trajectories are close to the streamlines. For $\mu_p = 0.7$ and 0.99 (fig. 2, 3) particle trajectories deviate significantly from the acoustic streamlines. For $\mu_p = 0.7$ the number of particles "attracted" to the Schlichting vortex is maximal.

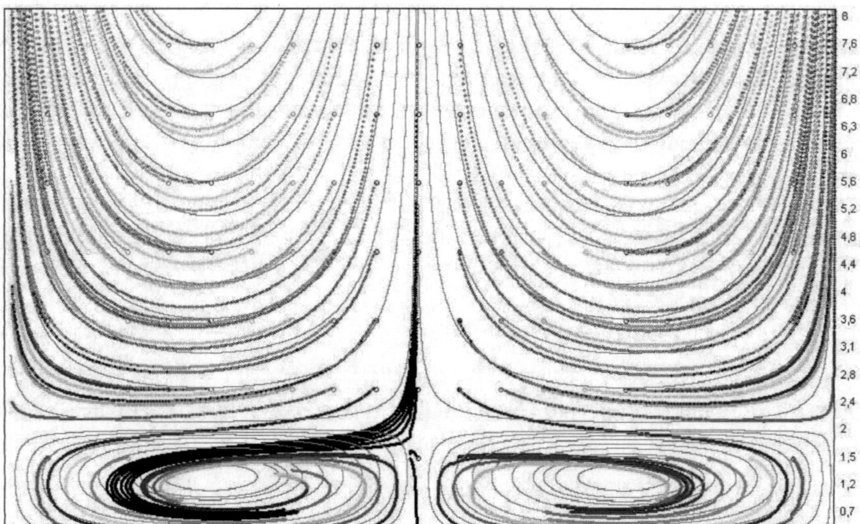

Fig. 1 Particle trajectories for $\mu_p = 0.03$, $\mathrm{Re}_0 = 33.25$

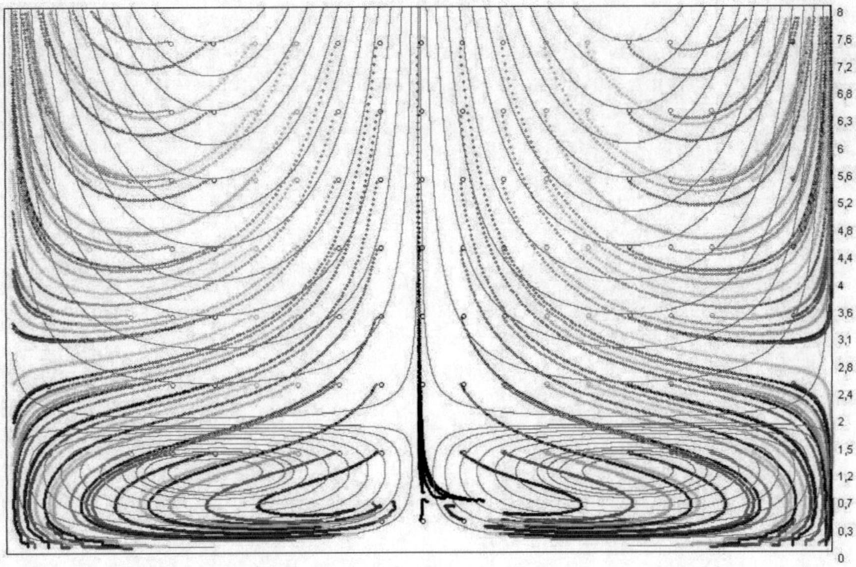

Fig. 2 Particle trajectories for $\mu_p = 0.7$, $\mathrm{Re}_0 = 6.65$

Fig. 3 Particle trajectories for $\mu_p = 0.99$, $\mathrm{Re}_0 = 0.08$

Acknowledgments
This work was supported by the Russian Science Foundation (grant No. 15-11-10016).

References

[1] Nigmatulin R I 1978 *Dynamics of multiphase media* (*Part I, II*) (Moscow: "Nauka")
[2] Ganiev R F and Ukrainky L E 2008 *Non-Linear Wave Mechanics & Technologies* (Moscow: "R&C Dynamics")
[3] Aktas M K and Farouk B 2004 Numerical simulation of acoustic streaming generated by finite-amplitude resonant oscillations in an enclosure *J.Acoust. Soc. Am.* **116** 2811-2831

[4] Zarembo L K and Krasilnikov V A 1966 *Vvedenie v nelinejnuyu akustiku* (Moscow: "Nauka")
[5] Gubaidullin D A, Osipov P P and Nasyrov R R Numerical simulation of Schlichting streaming induced by standing wave in rectangular enclosure *J. Phys: Conf. Ser*. **567** 012017

Numerical simulation of vortex structures in the spraying of electrically charged gas suspension on a plate

A L Tukmakov[1,2], N F Kashapov[3], D A Tukmakov[1,2] and M G Fazlyyyakhmatov[3]
[1]Kazan National Research Technical University named after A N Tupolev, 31 K. Marksa str., Kazan 420111, Russian Federation
[2]Institute of Mechanics and Engineering, Kazan Science Center, Russian Academy of Science, 2/31 Lobachevsky str., Kazan 420111, Russian Federation
[3]Kazan (Volga Region) Federal University, 18 Kremlyovskaya str., Kazan 420008, Russian Federation

E-mail: tukmakovda@imm.knc.ru

Abstract. Based on the numerical solution of polydisperse gas suspension dynamics system of equations, the process of powder spraying on a plane surface in an electric field is described. The model includes motion equations of the carrier medium and disperse phase fractions under aerodynamic friction and Coulomb forces action taking into account interfacial exchange of pulse and energy. The system solves by an explicit predictor-corrector method with splitting on the spatial directions and the non-linear correction scheme. The numerical model is applied to gas suspension speed and density fields receiving in interelectrode space and on target electrode surface. The system of equations was written in the generalized curvilinear coordinates: the physical region of a current in variables (x, y) was written in a canonical computational region in variables (ξ,η) and was solved by an explicit second-order Mac-Cormack method. The Poisson equation for the electric field potential was written in the generalized coordinates and was solved by finite difference method with the use of the iterative Seidel scheme on a gas-dynamic computational mesh.

1. Introduction

Movement of the charged gas mixture in the bearing environment under the influence of aerodynamic and electrical forces is the cornerstone of technology of a powder coating [1-5]. For describing of the deposited layer formation process it is required to solve the problem about the movement of the charged aerosol particles in the space between the spray nozzle, on which the negative potential is supplied, and the painted surface, which potential is the positive or zero. In case of polydisperse composition of particles of pigment powder, their movement in interelectrode space depends on value of a charge, aerodynamic resistance of particles, electrodes layout, electric intensity, speed and concentration of solid fractions.

2. The equations of the movement of the charged gas-suspension and method of their decision

The process model can be based on a system of equations of movement of the multispeed multitemperature medium [2] in which the particle, along with forces of the aerodynamic nature, is influenced by Coulomb force. The system of equations includes motion equations of the carrying medium and disperse fractions accounting of interfacial force interaction and heat exchange and also taking into account interaction of the charged gas suspension with an electric field which is created by both fixed electrodes, and the electrical charge spatially distributed which is transferred by a disperse phase [6, 7]:

$$\frac{\partial \rho}{\partial t}+\frac{\partial(\rho u)}{\partial x}+\frac{\partial(\rho v)}{\partial y}=0, \tag{1}$$

$$\begin{aligned}
&\frac{\partial(\rho u)}{\partial t}+\frac{\partial}{\partial x}\left(\rho u^2+p-\tau_{xx}\right)+\frac{\partial}{\partial y}\left(\rho uv-\tau_{xy}\right)=-\sum_{i=2,n} F_{xi}+\sum_{i=2,n}\alpha_i\frac{\partial p}{\partial x},\\
&\frac{\partial(\rho v)}{\partial t}+\frac{\partial}{\partial x}\left(\rho uv-\tau_{xy}\right)+\frac{\partial}{\partial y}\left(\rho v^2+p-\tau_{yy}\right)=-\sum_{i=2,n} F_{yi}+\sum_{i=2,n}\alpha_i\frac{\partial p}{\partial y},
\end{aligned} \tag{2}$$

$$\begin{aligned}
&\frac{\partial(e)}{\partial t}+\frac{\partial}{\partial x}\left(\left[e+p-\tau_{xx}\right]u-\tau_{xy}v+\lambda\frac{\partial T}{\partial x}\right)+\frac{\partial}{\partial y}\left(\left[e+p-\tau_{yy}\right]v-\tau_{xy}u+\lambda\frac{\partial T}{\partial y}\right)=\\
&=\sum_{i=2,n} Q_i-\sum_{i=2,n}\left(\left|F_{xi}\right|(u-u_i)-\left|F_{yi}\right|(v-v_i)\right)+\sum_{i=2,n}\alpha_i\left(\frac{\partial(pu)}{\partial x}+\frac{\partial(pv)}{\partial y}\right),
\end{aligned} \tag{3}$$

$$\tau_{xx}=\mu\left(2\frac{\partial u}{\partial x}-\frac{2}{3}D\right),\ \tau_{yy}=\mu\left(2\frac{\partial v}{\partial y}-\frac{2}{3}D\right),\ \tau_{xy}=\mu\left(\frac{\partial u}{\partial y}+\frac{\partial v}{\partial x}\right),\ D=\frac{\partial u}{\partial x}+\frac{\partial v}{\partial y}. \tag{4}$$

The movement of disperse phase fractions is described by the fraction's average density conservation equation, the impulse components conservation equations and the equation of the energy conservation, which are written taking into account heat exchange, impulse exchange with the carrying medium and Coulomb force acting on particles of fraction i of the disperse phase:

$$\frac{\partial \rho_i}{\partial t}+\frac{\partial(\rho_i u_i)}{\partial x}+\frac{\partial(\rho_i v_i)}{\partial y}=0, \tag{5}$$

$$\begin{aligned}
&\frac{\partial(\rho_i u_i)}{\partial t}+\frac{\partial}{\partial x}\left(\rho_i u_i^2\right)+\frac{\partial}{\partial y}\left(\rho_i u_i v_i\right)=F_{xi}+F_{ex\,i}-\alpha_i\frac{\partial p}{\partial x},\\
&\frac{\partial(\rho_i v_i)}{\partial t}+\frac{\partial}{\partial x}\left(\rho_i u_i v_i\right)+\frac{\partial}{\partial y}\left(\rho_i v_i^2\right)=F_{yi}+F_{ey\,i}-\alpha_i\frac{\partial p}{\partial y},
\end{aligned} \tag{6}$$

$$\frac{\partial(e_i)}{\partial t}+\frac{\partial}{\partial x}\left(e_i u_i\right)+\frac{\partial}{\partial y}\left(e_i v_i\right)=-Q_i, \tag{7}$$

$$\rho_i=\alpha_i\,\rho_{i0},\ e_i=\rho_i C_{vi} T_i, \tag{8}$$

$$\frac{\partial^2\varphi}{\partial x^2}+\frac{\partial^2\varphi}{\partial y^2}=\rho q \tag{9}$$

$$\begin{aligned}
F_{xi}=&\frac{3}{4}\frac{\alpha}{(2r)}C_d\,\rho\sqrt{(u-u_i)^2+(v-v_i)^2}\,(u-u_i)+\alpha_i\rho\left(\frac{\partial u}{\partial t}+u\frac{\partial u}{\partial x}+\right.\\
&\left.+v\frac{\partial u}{\partial y}\right)+0.5\alpha_i\rho\left(\frac{\partial u}{\partial t}+u\frac{\partial u}{\partial x}+v\frac{\partial u}{\partial y}-\frac{\partial u_i}{\partial t}-u_i\frac{\partial u_i}{\partial x}-v_i\frac{\partial u_i}{\partial y}\right),
\end{aligned} \tag{10}$$

$$F_{yi} = \frac{3}{4}\frac{\alpha_i}{(2r)} C_d \rho \sqrt{(u-u_i)^2 + (v-v_i)^2}\,(v-v_i) + \alpha_i \rho\left(\frac{\partial v}{\partial t} + u\frac{\partial v}{\partial x} + v\frac{\partial v}{\partial y}\right) + 0.5\alpha_i \rho\left(\frac{\partial v}{\partial t} + u\frac{\partial v}{\partial x} + v\frac{\partial v}{\partial y} - \frac{\partial v_i}{\partial t} - u_i\frac{\partial v_i}{\partial x} - v_i\frac{\partial v_i}{\partial y}\right), \tag{11}$$

$$C_{di} = \frac{24}{Re_{i1}} + \frac{4}{Re_{i1}^{0.5}} + 0.4,\ M_{i0} = |\bar{V} - \bar{V}_i|/c\ ,\ Re_{i1} = \rho_1 |\bar{V}_1 - \bar{V}_i| 2r/\mu, \tag{12}$$

$$Pr = \gamma\, C_p \mu/\lambda\ ,Q_i = 6\alpha_i Nu\lambda(T_1 - T_i)/(2r)^2\ \ i=2,..n, \tag{13}$$

$$Nu_i = 2\exp(-M_{i0}) + 0.459 Re_{i0}^{0.55} Pr^{0.33},\ \Pr = \gamma\, C_p \mu/\lambda\ , 0 \le M_{i0} \le 2\ , 0 \le Re_{i0} < 2\cdot 10^5 \tag{14}$$

Thermal energy of the solid phase fractions suspended in gas is defined as $e_i = \rho_i C_{pi} T_i$. The system of equations was complemented with the corresponding initial and boundary conditions. On input and output computational region borders, excepting the spray nozzle inlet section, for all gas-dynamic functions, including the carrying medium and disperse phase velocity components Neumann boundary conditions were set. Coulomb force components per unit of gas-suspension volume were defined by its specific charge, solid phase volume density and electric intensity: $F_{ex\,i} = -q_0\rho_i \,{}^{\partial\varphi}/_{\partial x}$, $F_{ey\,i} = -q_0\rho_i \,{}^{\partial\varphi}/_{\partial y}$, where q_0 is the specific charge of solid fraction mass unit, φ – electric field potential. Electric field potential in computational region was found from the Poisson equation solution with second type boundary conditions. In the right member of Poisson equation the gas-suspension charge density divided by carrying medium absolute dielectric constant [8]:

$$div\,\boldsymbol{E} = \frac{\rho_{el}}{\varepsilon\varepsilon_0}, \boldsymbol{E} = -\Delta\,\varphi, \rho_{el} = \rho_{sum}\cdot q, \rho_{sum} = \textstyle\sum_i \rho_i\,, \varepsilon_0 = \frac{10^{-9}}{36\pi}\ F/m. \tag{15}$$

System of equations was written in the generalized curvilinear coordinates: physical region of a current in variables (x, y, t) was mapped on canonical computational region in variables (ξ, η, t) [9-12] and was solved by an explicit second order Mac-Cormack method [8] with the subsequent application of the solution non-linear correction scheme. The Poisson equation for the electric field potential was written in the generalized coordinates and was solved by finite difference method by means of the iterative Seidel scheme on a gas-dynamic computational mesh.

Acknowledgments
This research is executed with support of the RFBR and Tatarstan Republic government (project No. 17-48-160359) on the state instructions of the Ministry of Education and Science of the Russian Federation to Higher Education Institutions (project No. 9.5542.2017/6.7).

References

[1] Salyanov F A 1997 *Theory of the low-temperature plasma physics, plasma devices and technologies* (Moscow: Nauka) p 240
[2] Anoshin O A, Belogolovsky A A et al 2000 *High-voltage electrotechnics* ed I P Vereshchagin (Moscow: MPEI) p 204
[3] Fazlyyyakhmatov M and Kashapov N 2014 *High Temperature Material Processes* **18(4)** 273-279
[4] Gavrilova V A, Fazlyyyakhmatov M G and Kashapov N F 2014 *J. Phys.: Conf. Series* **567** 012023
[5] Fazlyyyakhmatov M 2017 *J. Phys.: Conf. Series* **789** 012005
[6] Kutushev A G 2003 *Mathematical modelling of wave processes in aerodisperse and powder media (*St. Petersburg: Nedra) p 283

[7] Gavrilova V A, Kashapov N F, Tukmakov A L, Tukmakov D A 2012 *Bulletin of the Kazan state technical university named after A N Tupolev* **2** 177-182
[8] Tukmakov A L 2015 *Journal of Applied Mechanics and Technical Physics* **56(4)** 636-643
[9] Tukmakov A L, Tukmakov D A 2017 *High Temperature* **55(4)** 491-495
[10] Falkovsky O I 2009 *Technical electrodynamics* (St. Petersburg: Lan') p 432
[11] Fletcher K 1991 *Computing methods in dynamics liquids* Vol. 2 (Moscow: Mir) p 551
[12] Steger J L 1978 *AIAA Journal* **16** 679-686

Method of assessing the thermal state of the heat transfer surfaces

E R Saifullin[1], Yu V Vankov[2], E V Izmaylova[2], E R Bazukova[2]
[1]Kazan Federal University, 18 Kremlyovskaya str., Kazan 420008, Russian Federation
[2]Kazan state power engineering university, 51 Krasnoselskaya St., Kazan , 420066, Russia

yvankov@mail.ru, evgeniya-izmailova@yandex.ru

Abstract. One of methods of increase in efficiency of the heat exchange equipment is timely detection of a sludge formation and scum on heat exchange surfaces. Deposits of salts and products of corrosion from technical water on the surface of heating of heat exchange devices reduce coefficient of a heat transfer and efficiency of heat exchange that leads to a considerable overexpenditure of energy carriers, overheating of surfaces of heating of coppers, decrease in service life, increase in costs of service and repair of the heat exchange equipment. Reliable methods and control devices of intensity of formation of deposits in the heat exchange equipment don't exist now. In work the method the express of control of thickness of deposits on heat exchange surfaces allowing to make control of a thermal condition of heat exchange surfaces is offered and in due time to carry out washing of devices.

1. Introduction

Deposits of salts and products of corrosion from technical water on the surface of heating of heat exchange devices reduce heat transfer coefficient, efficiency of heat exchange and as a result there are essential losses of energy [1, 2].

At power stations at the growth rate of deposits on the heat exchange surfaces of turbines condensers from technical water from 0,5 to 3 mm/year there is a decrease in power production from 120 to 250 million kWh or from 1,7 to 3,47% respectively for 1000 MW of rated capacity [3].

The cost of replacement of heat exchangers and coppers tubes insignificantly is less than the cost of the new equipment. Existence of deposits on the surface of heat exchange in coppers considerably reduces their efficiency (see Table 1). So, existence of 1 mm of deposits increases fuel consumption approximately by 12% [4].

Table 1. Losses from a scum in the systems of heating depending on layer thickness

Thickness of a scum layer (mm)	0.4	0.8	1.6	3.2	4.8	6.4	9.6	12.7	15.9	19.1
Excessive consumption of liquid fuel on 1000 l	40	70	110	180	270	380	480	600	740	900
Excessive consumption of coal on 1 t	80	140	220	360	540	760	960	1,200	1,400	1,800

IOP Conf. Series: Journal of Physics: Conf. Series **1058** (2018) 012058 doi:10.1088/1742-6596/1058/1/012058

Excessive consumption of gas on 1000 m3	40	70	110	180	270	380	480	600	740	900
Energy carrier overexpenditure, in %	4	7	11	18	27	38	48	60	74	90

In housing and communal services and the industry cases of operation of heat exchange devices and package boilers without special preparation of water aren't rare. It leads to formation on the internal surface of devices and units (drums, tubes) of the scum worsening a heat transfer and in whole reducing the efficiency of installations. Heat conductivity of a scum of metals more than 40 times lower than heat conductivity. Therefore even the thin layer of scale leads to sharp decrease in heat conductivity and temperature increase of metal surfaces of heating which reaches dangerous sizes and reduces the mechanical durability of metal. Metal damages are a consequence of it, namely - emergence it is made bulge out, fistulas, and it is frequent also ruptures of pipes of boilers.

At high content in water the scum-forming salts and products of corrosion the settlement operating mode of heat exchangers is quickly broken. Most heat exchangers of tap water are subject to pollution. The polluted heat exchangers have an increase in an expense of the heat carrier and growth of temperature in the return pipeline that, in turn, leads to electricity consumption growth by pumps and thermal losses in pipelines. Restoration of a settlement operating mode requires an eduction of heat exchangers from operation and cleaning of the polluted surfaces.

The serious problem of fight against deposits arises in heat exchangers of hot water supply when the section of pipes through passage and the surfaces of heat exchange almost completely grows with deposits (Figure 1).

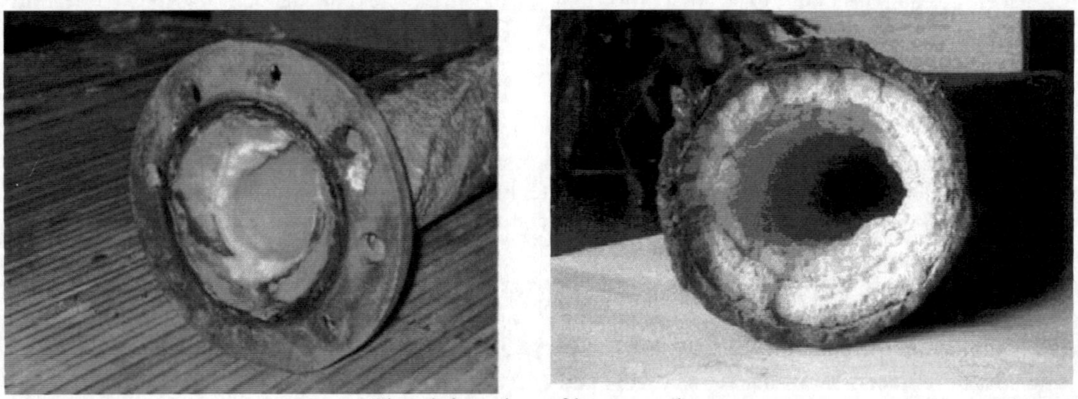

The giving pipe of heat supply system

Element of the lamellar heat exchanger with a scum

Figure 1. Reduction of throughput section because of existence of deposits

Overgrowing by deposits of pipelines of thermal systems including reverse water supply, leads to significant increase in their hydraulic resistance, maladjustment of heating systems and big power losses by pumping of system.

There is similar situation with power losses from deposits in air conditioning systems (Table 2).

Table 2. Losses from deposits in air conditioning systems depending on layer thickness

Thickness of a scum layer (mm)	0.5	1	2	4	6	8	10	12.7
Decrease of the conditioner power	5%	9%	17%	23%	29%	34%	50%	56%
Increase of temperature at the exit from the evaporator (°C)	0.4	0.8	1.6	3.2	4.8	6.4	8	10
Electric power over expenditure	5.8%	10.6%	20.2%	29.4%	35.6%	46.8%	66%	76%

It is possible to reduce power losses by timely cleaning of internal surfaces of a scum and deposits. For this purpose it is necessary to have a reliable, available technique and the equipment for control of thickness of deposits.

Research objective: Decrease in power losses in the heat exchange equipment by control of thickness of deposits on the surfaces of heat exchange.

2. Methods

Reliable methods and control devices of intensity of deposits formation in the heat exchange equipment don't exist now.

Now acoustic methods of a research and tests of materials of products and constructions widely develop [5-7].

Low-frequency acoustic methods of nondestructive testing significantly differ from the ultrasonic methods which are also using elastic fluctuations and waves. These distinctions consist in other physical principles, scopes, range of working frequencies, designs of converters, etc.

The correlation dependence between elastic constants of material and physicomechanical properties of products is the basis of low-frequency control methods. There are following ways of excitation of oscillations in a product:

1) shock excitation;

2) continuous excitation.

The block diagram of the devices realizing low-frequency control methods is provided on Figure 2 [8,9].

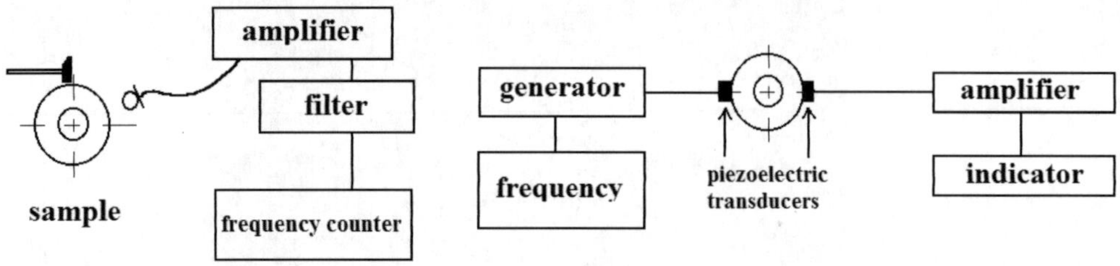

a) Method of free fluctuations b) Method of the compelled fluctuations

Figure 2. The block diagram of the devices realizing low-frequency control methods

In the first case (Figure 2, a) at blow to a surface of a controlled product arise mechanical oscillations which will be transformed to an electric signal by the sensor, amplify, pass through the filter allocating from a range of the frequencies of own fluctuations (FOF) that, corresponding to the type of fluctuations chosen for registration and are registered a frequency counter. At a method of the compelled fluctuations (Figure 2, b), the electric fluctuations created by the generator with the changing frequency will be transformed by means of the converter to mechanical and transferred to a controlled product. Fluctuations of a product are accepted by the sensor, amplify and move on the indicator of a resonance which fixes the moment of coincidence of frequency of the generator with measured by the device of FOF frequency counter, a controlled product, and on previously received dependences, physicomechanical properties of a controlled product are defined [10].

Two absolutely identical bodies have identical own frequencies. Existence of deposits changes thickness, mass of surfaces of heat exchange, therefore, and own frequencies of fluctuations. Having "the acoustic passport" of a new product, knowing the current frequencies of fluctuations of controlled surfaces it is possible to determine thickness of deposits. For definition of a frequency range of fluctuations of products it is offered to use a finite element method.

3. Accounting of interaction of a controlled design with the device equipment

For increase in sensitivity of a technique of control it is desirable to force to fluctuate only the site of a design explored at present. It can achieve by means of the damping frame pressed to a design. After a frame clip to a design on area, the limited frame, strikes blow and in the same place the sensor fluctuations are registered.

Let's write down the equation of the movement taking into account interaction of a design with a frame:

$$[M]\left\{\frac{\partial^2 q}{\partial t^2}\right\}+[C]\left\{\frac{\partial q}{\partial t}\right\}+[K]\{q\}+[M_r]\left\{\frac{\partial^2 q_r}{\partial t^2}\right\}+[C_r]\left\{\frac{\partial q_r}{\partial t}\right\}+[K_r]\{q_r\}=P_0(t) \qquad (2)$$

where M_r, C_r, K_r - matrixes of masses, damping and rigidity of a frame of the device; $\frac{\partial^2 q_r}{\partial t^2}$, $\frac{\partial q_r}{\partial t}$, q_r - vectors of accelerations, speeds and amplitudes of a frame knots, M, C, K - matrixes of masses, damping and rigidity of a construction; $\frac{\partial^2 q}{\partial t^2}$, $\frac{\partial q}{\partial t}$, q - vectors of accelerations, speeds and amplitudes of constructions knots.

Let's consider a subvector from $\frac{\partial^2 q}{\partial t^2}$, $\frac{\partial q}{\partial t}$, q, corresponding to contact knots. Let's designate them $\frac{\partial^2 q^r}{\partial t^2}$, $\frac{\partial q^r}{\partial t}$ and q^r. At small value of force of a clip P_{cm} of a frame to a construction there is $\frac{\partial^2 q^r}{\partial t^2} \neq \frac{\partial^2 q_r}{\partial t^2}$, $\frac{\partial q^r}{\partial t} \neq \frac{\partial q_r}{\partial t}$, $q^r \neq q_r$, then the equation (2) breaks up to two equations:

$$[M_r]\left\{\frac{\partial^2 q_r}{\partial t^2}\right\} + [C_r]\left\{\frac{\partial q_r}{\partial t}\right\} + [K_r]\{q_r\} = P_{cm} + P_r(t) \qquad (3)$$

and

$$[M]\left\{\frac{\partial^2 q}{\partial t^2}\right\} + [C]\left\{\frac{\partial q}{\partial t}\right\} + [K]\{q\} = P_0(t) - P_{cm} - P_r(t), \qquad (4)$$

the size P_{cm} of influence doesn't render on dynamic amplitudes, and $P_r(t)$ - dynamic loading is caused by change in time of a contact zone of a frame and a construction. In other words, there is "beating" or "jingle". It is difficult to define concrete character of this loading, however it is possible to estimate her influence on design amplitudes. For this purpose let's notice that action of force $P_r(t)$ is directed to damping of knots fluctuations in a contact zone. And increase in P_{cm} leads to increase in damping of these knots. In this regard, it is necessary to conduct a research of influence of increase in coefficients of damping of separate knots (local) of model on coefficients of damping of forms of own fluctuations. The ratio can be used for this purpose:

$$2 \times \xi_n \times \omega_n = q_n^t \times C \times q_n, \qquad (5)$$

where ξ_n - coefficients of modal attenuation; numerical values of these coefficients directly depend on a force, pressing a frame.

With the certain force of a clip P_{cm}, we receive:

$$\frac{\partial^2 q^r}{\partial t^2} = \frac{\partial^2 q_r}{\partial t^2}, \ \frac{\partial q^r}{\partial t} = \frac{\partial q_r}{\partial t}, \ q^r = q_r. \qquad (6)$$

Then the equation (2) is given to standard record:

$$[M + M_r]\left\{\frac{\partial^2 q}{\partial t^2}\right\} + [C + C_r]\left\{\frac{\partial q}{\partial t}\right\} + [K + K_r]\{q\} = P_0(t). \qquad (7)$$

I.e., there is "gluing" of a frame to a design that involves change of the rigidity and weight making fluctuation, constructions and, as a result, change of frequencies and forms of fluctuations. At a certain ratio of a frame and a construction rigidity happens effect of hinged fixing of a construction in contact knots that allows reducing borders of the studied range of fluctuations significantly.

4. Conclusion

Preliminary calculations have shown that the offered method allows to find thickness of deposits in 0,5 mm. Carrying out further researches will allow to define such parameters as thickness, structure, porosity and heat conductivity of deposits.

Realization of the offered control method of thickness of deposits on heat exchange surfaces will give essential energy saving effect.

Acknowledgements

The work is performed according to the Russian Government Program of Competitive Growth of Kazan Federal University. The reported research was funded by Russian Foundation for Basic Research and the government of the region of the Russian Federation, grant № 18-48-160051\18.

References

[1] Influence of deposits in the heatexchange equipment on thermal resistance and power expenses on subjects to trade preparation of oil. A.I. Antipov, L.G. Golubev, Ya.S. Mukhtarov//Power problems, 2003, No. 5-6, pp. 28-35.

[2] Development of methods of diagnosing of the heatexchange equipment of nuclear power plants on existence of deposits in him. P.A. Budarin, abstract of the dissertation of the candidate of technical sciences:05.14.03 Nuclear power stations, including design, operation and an eduction from operation, 2007, 22 pages.
[3] Structural features and heatphysical properties of intra pipe deposits on heatexchange surfaces in the systems of technical water. I.A. Bublikov //Power system. 1998. No. 2, pp. 30-34.
[4] I.A. Bublikov Pollution of the heatexchange equipment of TPP and NPP. Rostov N / D.: SKNTs VSh publishing house, 2003, 232 pages.
[5] Control of technical condition of pipelines by acoustic methods / Sh.G. Ziganshin, Yu.V. Vankov, E.V. Izmaylova. – Kazan: Publishing house of Kazan University, 2015. – 160 pages.
[6] Larionov, V.M., Saifullin, E.R., Semenova, E.V. Self-exited gas oscillations in Helmholtz resonator type combustor (2016) Journal of Physics: Conference Series, 669 (1), 012047.
[7] Bykovets, A.P., Larionov, V.M., Marchukov, E.Yu. The influence of a steam injection on vibrational burning in a modelled combustion chamber (1992) Izvestiya Vysshikh Uchebnykh Zavedenij. Aviatsionnaya Tekhnika, (3), pp. 71-74.
[8] The effect of the size and position of transverse cracks in the natural frequencies of the rods Yu.V. Vankov, Sh.G. Ziganshin, E.V. Izmailova, A.R. Zagretdinov IOP Conference Series: Materials Science and Engineering 2015.
[9] Acoustic spectral defectoscope for defects detection in composite materials (part 2) Yu.V. Vankov, 2004, Instruments and control systems.
[10] Determination of the oscillation frequencies of corrosion defects finite element methods in order to develop methods of acoustic monitoring of pipelines Yu.V. Vankov, Sh.G. Ziganshin, E.V. Izmailova, V.V. Serov, 2015, IOP Conference Series: Materials Science and Engineering.

The algorithm diagram of combustion optimizing of a hydrocarbon fuels variable composition in thermal power plants

E R Saifullin[1], S A Nazarychev[1], A O Malahov[1], E Y Misoedova[2], I V Larionova[3]

[1]Kazan Federal University, 18 Kremlyovskaya str., Kazan 420008, Russian Federation
[2]Kazan State Power Engineering University, 51Krasnoselskaya str., Kazan 420066, Russian Federation
[3]Kazan National Research Technical University named after A. N. Tupolev , 10 K.Marx St., Kazan 420111, Russian Federation

E-mail: nazarichev.sa@gmail.com

Abstract. In modern power engineering, there is a problem of rational and efficient use of hydrocarbon fuels variable composition. The composition of the fuel can vary in time and from different sources. These changes lead to combustion optimum regime shift in thermal power plants. Previously, an algorithm of combustion optimizing of a hydrocarbon fuel variable composition in thermal power plants was developed. According to the algorithm, the fuel and air flows are regulated depending on the outlet temperature of the heat carrier. In this paper, a diagram of the implementation of this algorithm is presented.

1. Introduction

Due to reduction of fuels reserves and the electricity consumption increasing the problem of energy efficiency and energy saving in combustion processes is topical. The composition of any fuel can vary depending on the time and source. The change in the fuel composition leads to a changes of the fuel characteristics: fuel calorific value, burning temperature, burning velocity, heat release rate, composition of exhaust gases et al. As a result, the stability of the heat power plant is disturbed, that leads to negative effects: blowout, flashback, autoignition, underburning of fuel, change in temperature and composition of combustion products et al. [1].

According to the Government Decree of the Russian Federation since 2009 [2] 95% of produced associated petroleum gases (APG) should be rationally used, for example, as fuel to produce thermal energy. At the same time, the USA Environmental Protection Agency, since 2008, considers any gases generated at the petroleum refinery as fuel gases [3], which also obliges to use these wastes to generate energy.

At this time, special expert organizations are engaged to regulating the operation of thermal power plants in the Russia. However, during the calculations, the expert organization takes the fuel composition as a constant. Thus, regulation of the boiler according to the calculated parameters occurs with an error, when the fuel composition changes.

There are several methods to optimize the combustion process in heat power plants [4-7]. However, the available methods do not allow reacting quickly and efficiently on fuel compositions changes or requiring a continuous analysis of fuel calorific value or exhausting gas composition. To implement these methods, it is necessary to develop complex automated control systems.

The authors developed a mathematical model of the composition changing effect of the alkane fuels (C_nH_{2n+2}) on the heat release of the combustion process [8]. Conditions for stabilizing the thermal effect of combustion during short-term and long-term changes in the composition of the alkane fuels were obtained. Also, a numerical simulation of the combustion processes of APG various composition in a steam boiler was carried out [9]. As a result of mathematical and numerical modeling an algorithm for stabilizing the boiler's heat release rate was developed [10, 11]. The algorithm includes interrelated changes in fuel and air flows, which ensure established temperature of the heat carrier at the boiler outlet that corresponds to an established heat release rate.

The purpose of this work is the development of the diagram of the combustion optimization algorithm of hydrocarbon fuels variable composition in thermal power plants. The diagram will allow develop the software and the automated control system of fuel and air flow rates in the thermal power plants. Using of the automated system in thermal power plants will improve the efficiency of traditional fuels combustion and reduce their consumption through the use of alternative fuels, such as APG. Optimization of the combustion process will reduce the amount of harmful emissions into the atmosphere and will allow using alternative fuels for their utilization and generation of thermal energy.

2. Diagram of the combustion optimization algorithm.

Earlier, an algorithm for stabilizing the boiler's heat release rate under optimal combustion conditions for fuels variable composition was developed. The stability of the boiler's heat release rate was evaluated by the outlet temperature of the heat carrier. It is established that in case of deviation of outlet temperature from the set value, it is necessary to change fuel and air consumption in order to restore the initial operating mode of the boiler. Two cases were considered: a decreasing and increasing of the fuel calorific value, associated with a change in the fuel composition. In accordance with this, the proposed diagram will have two independent blocks (Figure 1).

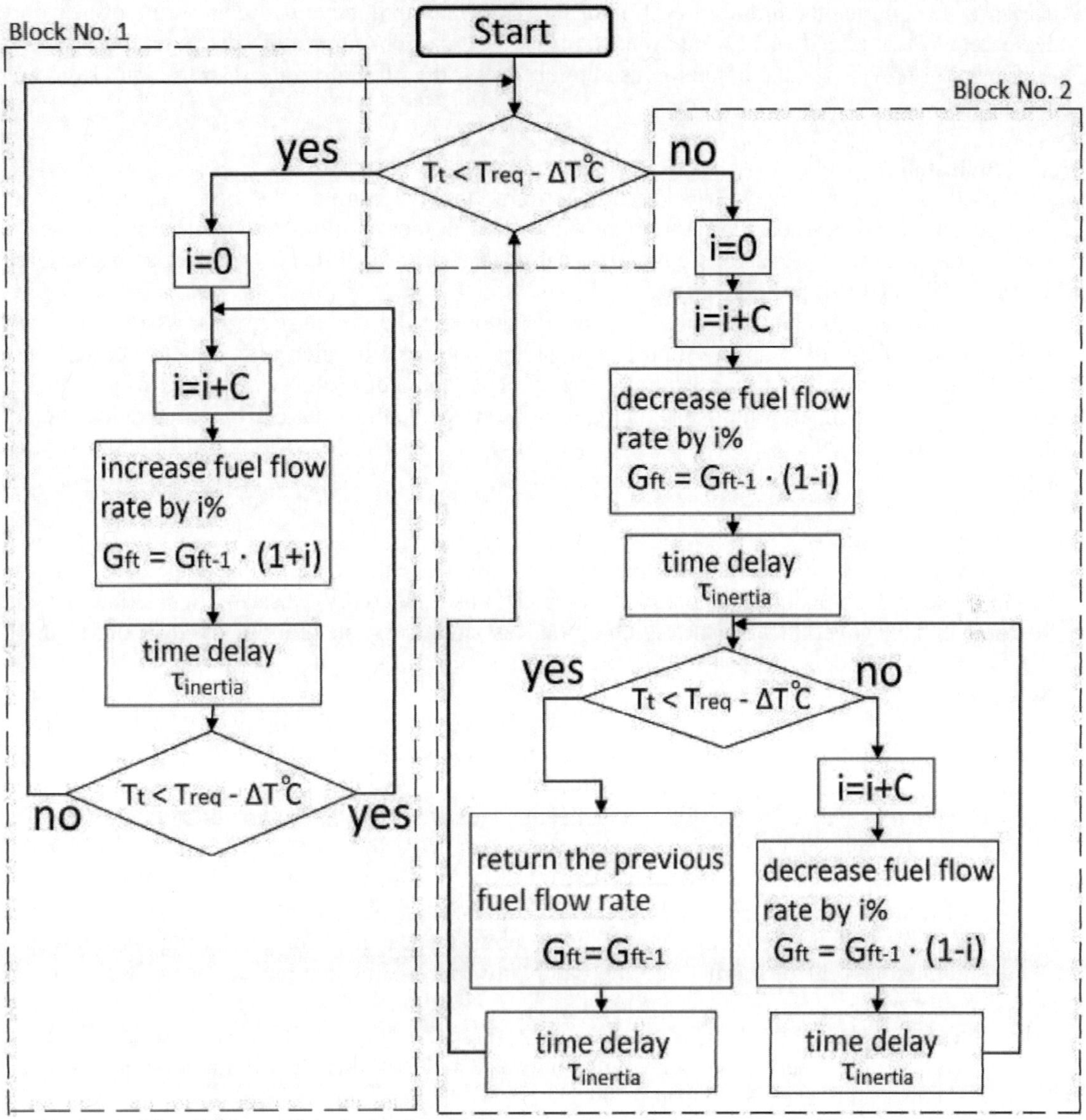

Figure 1. The diagram of optimization. T_t–heat carrier outlet temperature, T_{req} – required heat carrier outlet temperature, ΔT – variation value of heat carrier outlet temperature changing, i – cycle variable, C – variation value of fuel flow rate changing, G_{ft} – fuel flow rate, $\tau_{inertia}$ - boiler inertia time.

Block No. 1 begins to work if the outlet temperature of the heat carrier decreases, which indicates a reduction of fuel calorific value. In this case is necessary to start a gradual increase in the fuel flow rate until the outlet temperature of the heat carrier becomes to the required. After that, the program returns to the initial stage of the work.

Block No. 2 is a test block and it works when the outlet temperature of the heat carrier does not change, which indicates an increasing or stability of fuel calorific value. The test is carried out by reducing the fuel flow rate. After reducing the fuel flow, the program operates based on a change in the outlet temperature of the heat carrier. Decreasing the outlet temperature of the heat carrier means the stability of the fuel calorific value and it is necessary to return to the previous value of fuel flow rate. If the temperature does not change, this means that the fuel calorific value has increased and it is necessary to continue the further reduction of fuel flow rate until the outlet temperature of heat carrier decreases. When, after fuel flow rate reduction, the heat carrier temperature decreases, it is necessary to return the previous value of fuel consumption. Next, the program returns to the initial mode and checks the work of the algorithm again.

3. Conclusions

So, based on the previously developed algorithm, the diagram of the optimization process of combustion of hydrocarbon fuels variable composition in thermal power plants has been developed. The diagram shows how to stabilize boiler's heat release rate, when fuel calorific value is changing in the case of the fuel composition change.

Based on the algorithm diagram, software to automate the combustion process of fuel variable composition will be developed. Approbation of the optimization algorithm and the corresponding software will be carried out on a laboratory setup. The laboratory setup will consist of a combustion chamber, heat exchangers, automatic fuel and air flow regulators, heat carrier temperature meters at the inlet and outlet of the heat exchanger and an automated system to control digital and analog signals.

Acknowledgements

The work is performed according to the Russian Government Program of Competitive Growth of Kazan Federal University. The reported research was funded by Russian Foundation for Basic Research and the government of the region of the Russian Federation, grant № 18-48-160051\18.

References

[1] Lieuwen, T., McDonell, V., Petersen, E., Santavicca, D. Fuel flexibility influences on premixed combustor blowout, flashback, autoignition, and stability (2008) Journal of Engineering for Gas Turbines and Power, 130 (1), 011506.

[2] Government Decree of the Russian Federation since 08.01.2009 (edition of 28.02.2017).

[3] U.S. Environmental Protection Agency, Standards of Performance for Petroleum Refineries, 40 CFR 60, Subpart J, Section 60.101(d), U.S. Government Printing Office, Washington, DC, Electronic Code of Federal Regulations, current as of June 12, 2008.

[4] Ismail, O.S., Umukoro, G.E. Modelling combustion reactions for gas flaring and its resulting emissions (2016) Journal of King Saud University - Engineering Sciences, 28 (2), pp. 130-140.

[5] Lee, C.-L., Hou, S.-S., Lee, W.-J., Jou, C.-J.G. Improving cost-effectiveness for the furnace in a full-scale refinery plant with reuse of waste tail gas fuel (2010) International Journal of Hydrogen Energy, 35 (4), pp. 1797-1802.

[6] Bykovets, A.P., Larionov, V.M., Marchukov, E.Yu. The influence of a steam injection on vibrational burning in a modelled combustion chamber (1992) Izvestiya Vysshikh Uchebnykh Zavedenij. Aviatsionnaya Tekhnika, (3), pp. 71-74.

[7] Larionov, V.M., Saifullin, E.R., Semenova, E.V. Self-exited gas oscillations in Helmholtz resonator type combustor (2016) Journal of Physics: Conference Series, 669 (1), 012047.

[8] Saifullin, E.R., Vankov, Yu.V. Optimization of burning process of hydrocarbon fuels with varying specific heat of combustion (2015) IOP Conference Series: Materials Science and Engineering, 86 (1), 012006/

[9] Saifullin, E.R., Nazarychev, S.A., Malahov, A.O., Larionov, V.M., Iovleva, O.V. The heat effect of combustion process depending on fuel composition fluctuations (2017) Journal of Physics: Conference Series, 789 (1), 012045.

[10] Saifullin, E.R., Larionov, V.M., Busarov, A.V., Busarov, V.V. Optimization of hydrocarbon fuels combustion variable composition in thermal power plants (2016) Journal of Physics: Conference Series, 669 (1), 012037.

[11] Saifullin, E.R., Larionov, V.M., Busarov, A.V., Busarov, V.V. Thermal effect of hydrocarbon

fuels combustion after a sudden change in the specific calorific value (2016) Journal of Physics: Conference Series, 669 (1), 012043.

IOP Conf. Series: Journal of Physics: Conf. Series **1058** (2018) 012060 doi:10.1088/1742-6596/1058/1/012060

Development of the setup for study of the gas ionization in the pulsating mode of combustion

K G Sadikov[1], A O Malahov[1], V M Larionov[1], E R Saifullin[1], I V Larionova[2]

[1]Kazan Federal University, 18 Kremlyovskaya str., Kazan 420008, Russian Federation
[2]Kazan National Research Technical University named after A. N. Tupolev, 10 K.Marx St., Kazan 420111, Russian Federation

E-mail: aleks19bass@yandex.ru

Abstract. At present, known methods of direct conversion of the chemical energy of gaseous fuels to electrical energy are not so effective. This work is directed to developing the setup for experimental study of the gas ionization processes in the pulsating mode of combustion. This problem being solve for the first time. As result, schematic diagram and methodological recommendations will be developed which will be subsequently used for experimental studies.

1. Introduction

It is known that ionization of the gas begins at sufficiently high temperatures, for example, at the combustion temperature of the gaseous fuel. In a magnetic field, moving electric charges are separated and can be concentrated at electrode surfaces located in the combustion zone. This principle used in magnetohydrodynamic (MHD) generators - power plants in which working medium energy, moving in a magnetic field, is converted directly into electrical energy. In such devices, liquid or gaseous electrically conductive media can be used as a working fluid.

At the present time, development of methods for intensification of the ionization processes in burning gases is being carried out with the aim of increasing an efficiency of direct conversion of the organic fuels energy into electric energy. Many developments in the MHD energy generation field used technologies with pulsating conductive medium in a channel, creating regions of spatial distribution of various characteristics of working fluid, synchronizing the flow oscillations with alternating magnetic field. As example, there is an MHD generator in which a high-temperature gas injected into a closed toroidal channel in such way, that the gas flow moves along a circular channel and form of clear number of alternating high and low-pressure regions [1]. There is directly converting method of pulsating detonation combustion fuel energy into electrical energy by feeding portions of the fuel mixture to two counter directed detonation combustion chambers. The ignition and detonation combustion of the fuel creates a shock wave that moves back and forth in the working channel. The wave carries a positive space charge, which is created by excluding electrons from the combustion chambers by assistance of a catalytic electrode and electromagnetic induction. Electromagnetic field is induced by windings of magnetic contour that enfold the working channel. The useful alternating current voltage is consumed from the catalytic electrode [2]. When vortex burners operate in acoustic resonators, the energy of combustion of the fuel is converted into the vibrational energy of the exhaust gases. There are theoretical studies of conducting fluid oscillations in converters of acoustic energy into electrical energy, which is applicable to pulsating MHD generators [3].

IOP Conf. Series: Journal of Physics: Conf. Series **1058** (2018) 012060 doi:10.1088/1742-6596/1058/1/012060

The above examples allow us to consider that the development of effective methods for direct conversion the combustion energy of fuel into electric energy is an actual problem. In this project it is proposed to use the pulsed combustion mode, in which self-excitation of acoustic oscillations intensifies the process of ionization of burnt gases [4, 5]. This is due with the burning regime, in which the flame periodically changes its shape and size. Simultaneously with the same frequency, all process parameters in a combustion chamber and at different points of the path oscillate with the same frequency: pressure, gas velocity, mixture composition, temperature, etc. [6]. It is proposed to impact to an ionized gas with constant and alternating magnetic fields. At present the processes of the gas ionization in the pulse combustors, where acoustic oscillations are synchronized with alternating magnetic field, are investigated insufficiently.

The purpose of this work is the rationale and development of the schematic diagram of the experimental setup for experimental study of gas ionization processes in pulsating mode combustion of the fuels under influence a constant and alternating magnetic fields.

2. Prehistory of the research.

In study of interactions between electric discharges and acoustic oscillations, under turbulent combustion mode of gaseous fuels, it was found the space-time distribution of ions in acoustic field occurs [7,8]. In this work were used two types of burners with tangential feeding of a gas mixture into a combustion chamber (Figure 1). First burner was a tube with one nozzle and an electrode placed in the center. In the operation of single nozzle burner, acoustic oscillations didn't excited, but nevertheless, an electric potential of direct current was induced on the probe, which varied depending on stoichiometry of the fuel mixture composition. A direction of the electric potential depended on the stoichiometry of premixed fuel supply and varied from -800 to +990 mV. When a mixture were lean, with an excess air ratio of $\alpha = 1.37$, the voltage was 990 mV. With a rich mixture, with an excess air ratio $\alpha = 0.8$, the voltage at the probe was minus 800 mV (Figure 2).

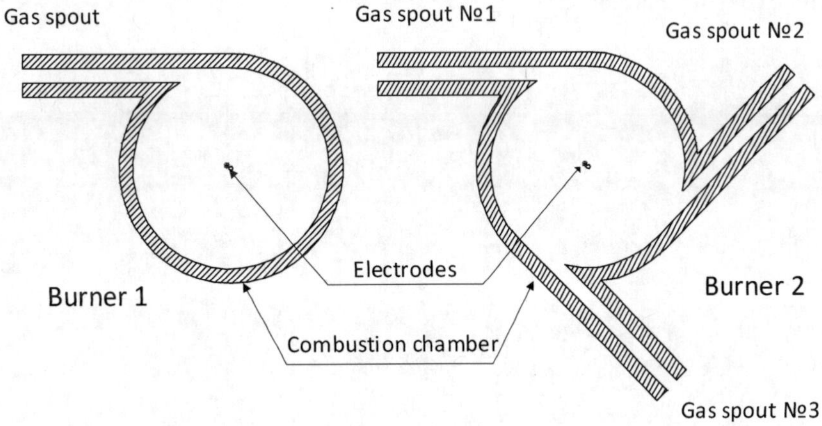

Figure 1. Scheme of burners.

Burner with a single nozzle was excited acoustic oscillations not so well. Therefore a second burner with three nozzles was made. In this burner, acoustic vibrations were excited in a wide stoichiometric range. As in the first burner, three nozzles burner has similar central electrode. The second electrode was the combustion chamber body itself. During steady-state burning on probe electrode inducted constant voltage. During transition to the pulsating combustion mode, a voltage had been becoming alternate and also depends on stoichiometry.

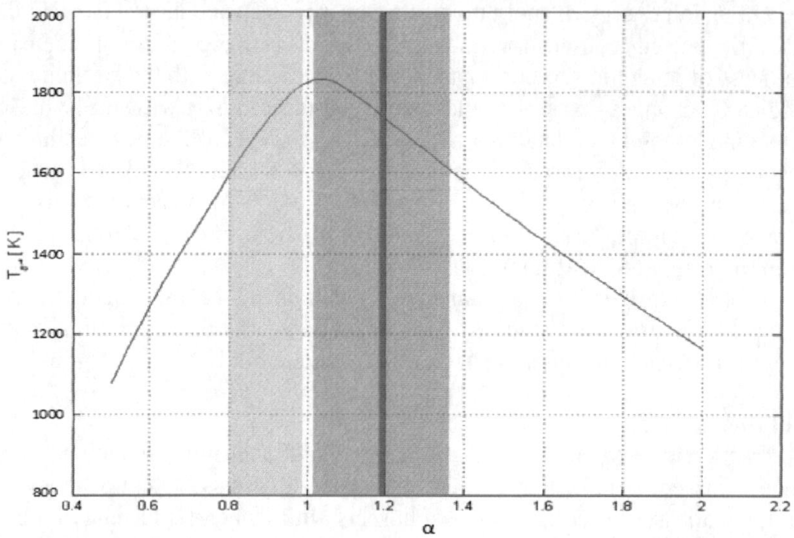

Figure 2. Dependence of temperature on the coefficient of excess air.

The three nozzles burner combusted in pulsating mode in the wide stoichiometric range. At appropriate temperature, the oscillation frequency was equal to 400 Hz, as like as the resonance frequency of the combustion chamber. An amplitude of oscillations potential at the central electrode also depended on stoichiometry. For $\alpha = 1.2$, the amplitude of the oscillations was about 1000 mV (marked as red line, Figure 2). Figure 3 shows the oscillogram of signal from the probe at a stoichiometry for $\alpha = 1.4$. The frequency of oscillations in this mode was equal to main acoustic resonance of the tube (the first harmonic).

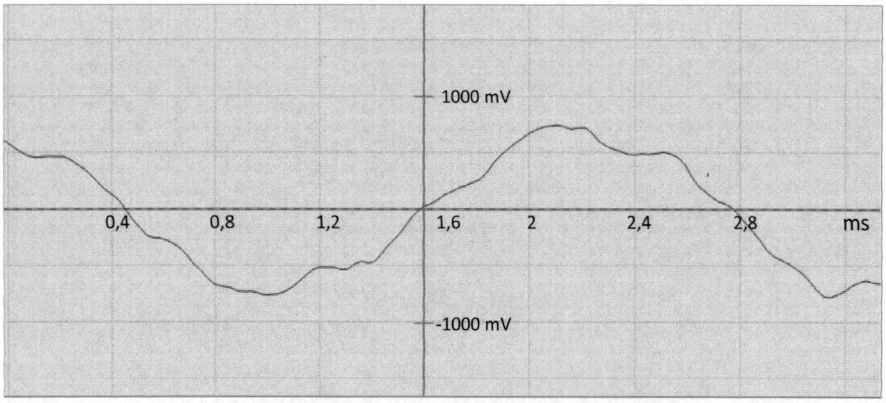

Figure 3. Oscillogram of the signal on probe.

These experiments prove that during operation of swirl burners in turbulent and pulsating modes, the space-time distribution of ions take place. Moreover, during operation of swirl burners in the pulsating mode powerful acoustic waves arise with a power level up to 140 dB, and the processes of ionization of burnt gases become intensely.

IOP Conf. Series: Journal of Physics: Conf. Series **1058** (2018) 012060 doi:10.1088/1742-6596/1058/1/012060

3. Experimental setup.

The observed phenomena prompted us to create an experimental setup to study hydrocarbon fuel combustion in pulsating mode for space-time distribution of ionization. Schematic diagram of experimental setup is shown in Figure 4.

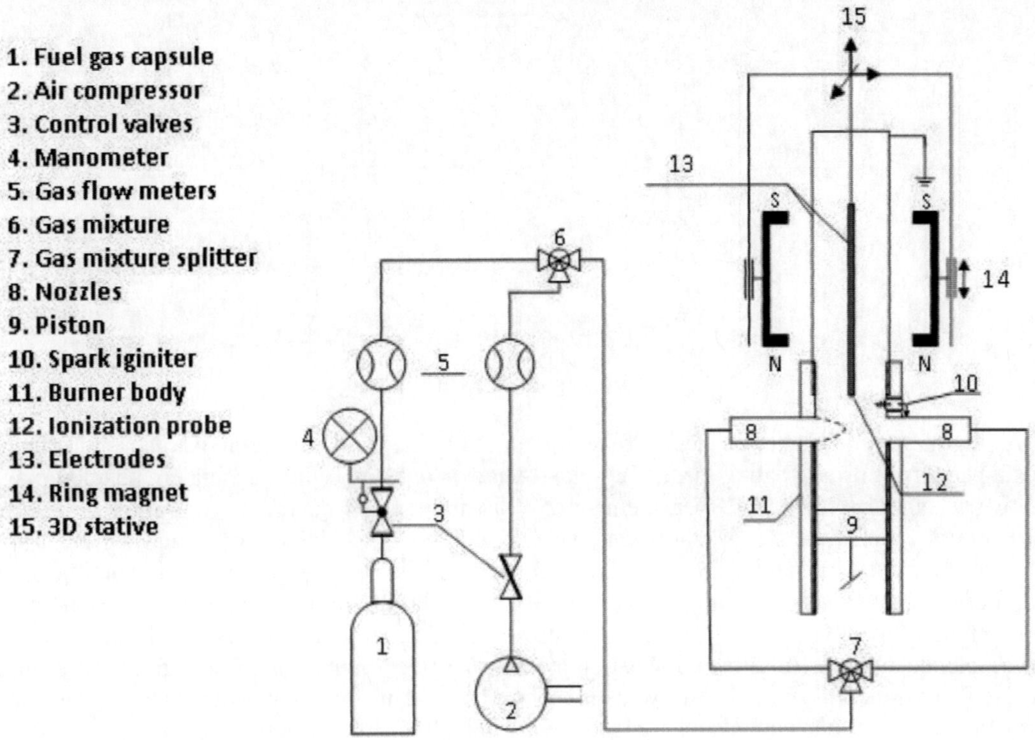

Figure 4. Schematic diagram of the experimental setup.

Laboratory setup is based on a swirl burner (Figure 5). The burner body has tangential nozzles through which the prepared combustible mixture was supplied. Gas-air mixture ignition plug was installed in the swirling area. At the lower part of burner body was closely fitted piston. The piston was needed to fine retune pulsation combustion mode. In addition to the piston, it is possible adjusting tube length to change resonant frequencies. Additional advantages of the solution are an ability using other materials or designs of tubes.

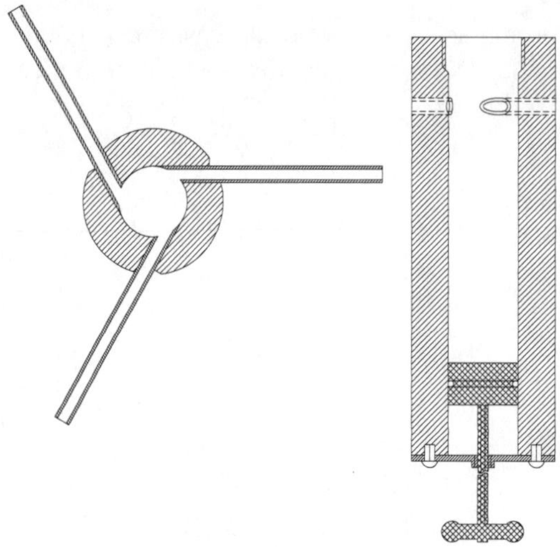

Figure 5. Swirl burner.

A ring magnet or electromagnet is placed along the burner body on the movable stand. They will create permanent or alternating magnetic fields. A movable stand will to accurately position magnets relative to an ionization zones. Measurement of a distribution and maximum ionization zones will be carried out by ionization sensor that place inside the burner. A special sensor's stand will be designed for allowing placed it in all three dimensions (3D stative). It is necessary to study distribution zones of light or heavy ions, not only along the vertical axe of the burner in the combustion zone, but also for determining of the concentration zones of the positive and negative ions in near-wall layers. Standing waves excited during a pulsating combustion mode, can effect on plasma quasineutrality. Presumably, there will be zones of quasineutrality disturbance observed in the combustion tube. As result, ions formation depends on not only charge potential, but also on distribution nodes and antinodes at standing waves.

4. Conclusion

Thus, a schematic diagram of the pulsating combustor has been developed to study gas ionization processes in a magnetic fields. This scheme and recommendations will be used for further research.

Acknowledgements

The work is performed according to the Russian Government Program of Competitive Growth of Kazan Federal University. The reported research was funded by Russian Foundation for Basic Research and the government of the region of the Russian Federation, grant № 18-48-160051\18.

References

[1] Patent RU № 2109393, 20.04.1998. Sposob polucheniya elektricheskoy energii i rezonansnyiy MGD - generator dlya ego realizatsii /Danilin A. V.

[2] Patent RU № 2418968, 20.05.2011. Sposob pryamogo preobrazovaniya energii impulsnogo detonatsionnogo sgoraniya topliva v elektricheskuyu energiyu i generator peremennogo toka dlya ego realizatsii /Danilin A. V.

[3] Vogin, C., Alemany, A. Analysis of the flow in a thermo-acoustic MHD generator with conducting walls (2007) European Journal of Mechanics, B/Fluids, 26 (4), pp. 479-493.

[4] Ermakov, R.A., Galiullin, R.G., Larionov, V.M., Nikolaev, A.N. A gas compressor based on a piston-type acoustic resonator (2008) Chemical and Petroleum Engineering, 44 (7-8), pp. 387-392.

[5] Ermakov, R.A., Sadykov, A.F., Galiullin, R.G., Larionov, V.M. Assessment of energy

efficiency of gas supercharger based on acoustic resonator (2011) Chemical and Petroleum Engineering, 46 (11-12), pp. 749-751.

[6] Bykovets, A.P., Larionov, V.M., Marchukov, E.Yu. The influence of a steam injection on vibrational burning in a modelled combustion chamber (1992) Izvestiya Vysshikh Uchebnykh Zavedenij. Aviatsionnaya Tekhnika, (3), pp. 71-74.

[7] Larionov, V.M., Sadikov, K.G., Mitrofanov, G.A. Interaction of electric and acoustic vibrations in combustion / (2016) Journal of Physics: Conference Series, 669 (1), 012042.

[8] Larionov, V.M., Saifullin, E.R., Semenova, E.V. Self-exited gas oscillations in Helmholtz resonator type combustor (2016) Journal of Physics: Conference Series, 669 (1).

Evaluation of heat transporting losses due to changes of insulation properties during operation

E R Saifullin[1], Yu V Vankov[2], E R Bazukova[2], A V Maryashev[3]

[1]Kazan Federal University, 18 Kremlyovskaya str., Kazan 420008, Russian Federation
[2]Kazan state power engineering university, 51 Krasnoselskaya St., Kazan , 420066, Russia
[3]Volga State University of Technology, b.3Lenin Sq., Yoshkar-Ola, Russia, 424000

yvankov@mail.ru, ell17@mail.ru

Abstract. The need for measuring the thermal conductivity of various materials exists in different fields of science and industry. First of all, it relate to constructing and power engineering. Thermal control allows evaluating the thermal insulation properties of the structure, identifying problem areas and assessing the energy efficiency of the pipeline. The paper assesses the transporting heat losses taking into account the actual operating conditions of the pipelines and the heterogeneity of the thermal insulation properties. A number of experiments were carried out. Experiments have shown that the destruction of insulation properties depends on such factors as the surface temperature of the pipe, the humidity of the environment, the service life.

1. Introduction

Currently, Russia produces about 1/2 billion tons - this is about 10% of the world's total production. The explored reserves of "light oils" in the world are decreasing, and the reserves of high-viscosity oils and natural bitumen are increasing [1]. Some experts consider that in the next 20-25 years the technology of extraction of high-viscosity oils will be the main ones in the development of new fields.

In the Russian Federation, the Republic of Tatarstan is in the first place in the extraction of highly viscous oils (Figure 1).

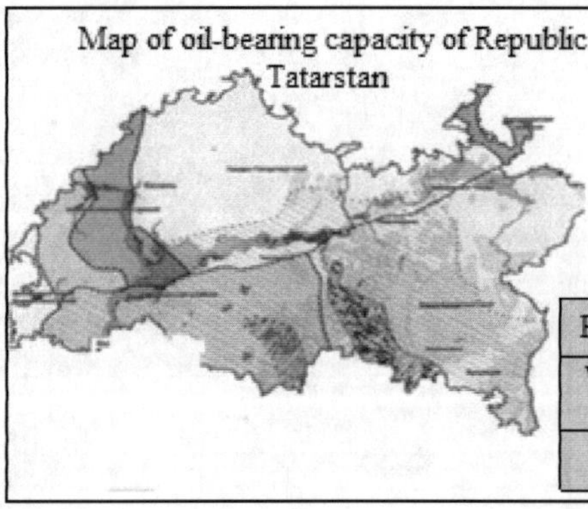

Depth of productive layer, m	70-250
Thickness of productive layer, m	10-38 (average-20)
Porosity, %	31.6
Transmittivity, darcy	2.69

Formation temperature	8 °C	281.2
Viscosity in formation conditions	27350 MPa·s	28342cP
Density	965 kg/m^3	15 API

IOP Conf. Series: Journal of Physics: Conf. Series **1058** (2018) 012061 doi:10.1088/1742-6596/1058/1/012061

Figure 1. The scheme of location of deposits of ultra-viscous oil on the territory of the Republic of Tatarstan, the properties of oil and reservoir layer.

Development of energy-saving technologies for the production of high-viscosity oil (HVO), ultra-high viscosity oil (UHVO) and bitumen (B) is one of the most important tasks of oil production.

To extract HVO, UHVO, B the methods of thermal impact on the stratum are promising. Due to the effect of heat, the viscosity of the oil decreases, the mobility of oil and water in the formation increases. The methods are characterized by high energy costs for steam production [2].

To reduce heat losses during the transportation of the heat carrier, the next types of thermal insulation can be used:

1) stitched mineral wool plates in galvanized cover;
2) half-cylinders of basalt fiber in galvanized cover;
3) two-layer insulation (basalt + polyurethane foam) in galvanized and polyethylene cover;

During the operation of steam pipelines and bitumen pipelines, the temperature on the inner surface of the thermal insulation may exceed 120°C, that leads to a change in the thermal insulation properties of the materials (Figure 2).

Figure 2. The end face of double-layer thermal insulation (basalt + polyurethane foam).

Manufacturers often guarantee that the polyurethane foam operates at a temperature of 150°C. However, in fact, even at a temperature of 120°C, its destruction occurs.

From the experience of operating steam pipelines transporting a heat carrier with a temperature of 190-200 ° C, it can be concluded that a two-layer thermal insulation in which the first layer consists of basalt materials, the second layer consists of polyurethane foam, does not meet the technical requirements. [3, 4].

It is established that the specific thermal losses with increasing thermal conductivity coefficient can increase to 35-40% [5].

In works of A G Dementiev it's found that the durability of PPU insulation depends on the temperature environment in which it operates. The graphs of the dependence of the thermal conductivity on prolonged thermal aging are given [6].

According to the fact that to determine the actual heat losses taking into account the technical condition of thermal insulation, there are no specialized instruments and measurement methods, we need a research, which will have as the result the method for diagnosing thermal insulation with the help of a device-complex, developed on the basis of the meter of a heat flux density and temperature.

2. Methods

The heat flux meter is an important element in controlling the heat consumption. Thermal control makes it possible to evaluate thermal insulation properties, identify problem areas and assess the overall energy efficiency of the pipeline. The meter of density of heat flows and temperature allows conducting thermal control of constructions in the maintenance process, which contributes to the timely detection and elimination of energy losses.

The density of heat flow, as an indicator of resistance to heat transfer, characterizes the quality of thermal insulation of external enclosing structures. The heat flux density can be determined by measuring the temperature outside and on the inner surface of the testing construction. To make the measurement results more reliable, it is necessary to take measurements for a long time. Electronic meter of heat flux density allows to simplify and automate this process considerably. It is a measuring device that incorporates heat flux density sensors that convert the heat load into an electrical signal and an electronic unit that performs the necessary calculations, displays the results and forms a report.

In addition to the energy-saving aspect, heat flow meters allow to influence the flow of technological processes, providing the necessary data package on the state of the objects of the heat and power complex.

Inspecting the condition of insulation of pipelines under operating conditions is proposed to be carried out using a device that based on the method of an additional wall [7]. It consists in the fact that an additional wall with a known thermal resistance is densely pressed against the surface of the pipeline λ_0/δ_0 (Figure 3).

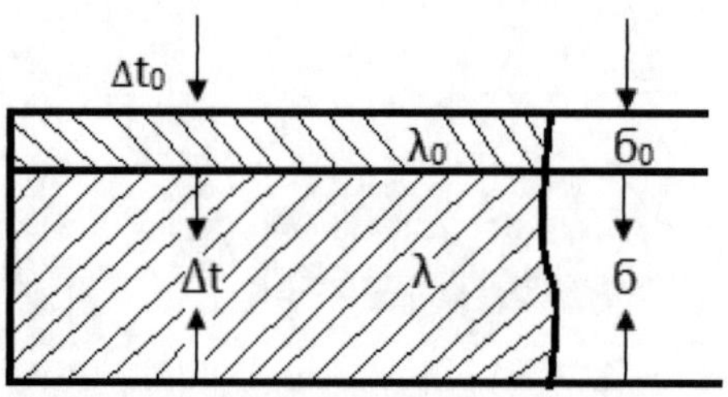

Figure 3. The illustration of the «additional» wall method.

By measuring the temperature difference Δt_0 in the additional wall, we can find the heat flux (W/m^2) passing through it, using the equation

$$q = \frac{\lambda_0}{\delta_0} \Delta t_0 \quad (1)$$

If the thermal resistance of the additional wall is small compared to the thermal resistance of the main insulation, then in the case of a steady-state thermal state, the same heat flow will pass through the main wall. This flow is the heat loss of the pipeline.

The system for controlling the thermal insulation properties is implemented in the LabVIEW environment. The scheme of the developed data acquisition system in LabVIEW is shown in Figure 4.

IOP Conf. Series: Journal of Physics: Conf. Series **1058** (2018) 012061 doi:10.1088/1742-6596/1058/1/012061

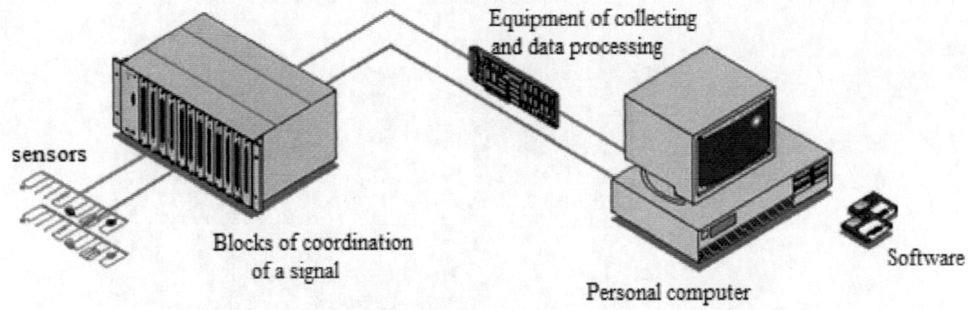

Figure 4. The data collection system implemented in LabVIEW.

A block-diagram of the developed program for registration and analyzing data is presented in Figure 5.

The program is a control system that automatically reads the signal from the sensors and converts it from analog to digital, using the SCXI-1000 and displays the measurement results in the diagram, that shows the dependence of the amplitude from time. The measurement result is automatically saved.

To conduct the research, an installation was assembled including a data acquisition device (PC, multichannel ADC); laboratory adjusting autotransformer type RNO-250-2 (220V, 9A); a fragment of a steel pipe with PPU insulation, a heater; heat flow sensors.

Prior to conducting the tests, the measuring system was calibrated. As a model device, a mass-produced heat flow meter ITP-MG4.03 / X (Y) «Potok» was used.

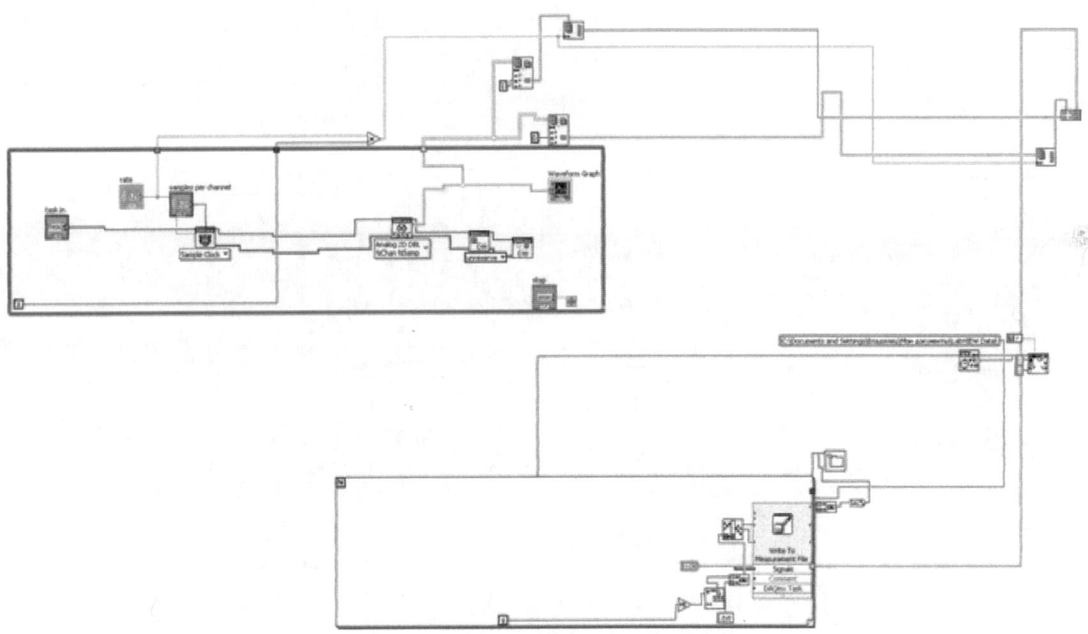

Figure 5. Block-diagram of the developed program.

When assessing heat losses with aging of insulation the measurements were carried out with samples of insulation of different degrees of destruction.

The change in the heat flux density q, depending on the aging of the insulation, can be observed in the diagram (Figure 6).

IOP Conf. Series: Journal of Physics: Conf. Series **1058** (2018) 012061 doi:10.1088/1742-6596/1058/1/012061

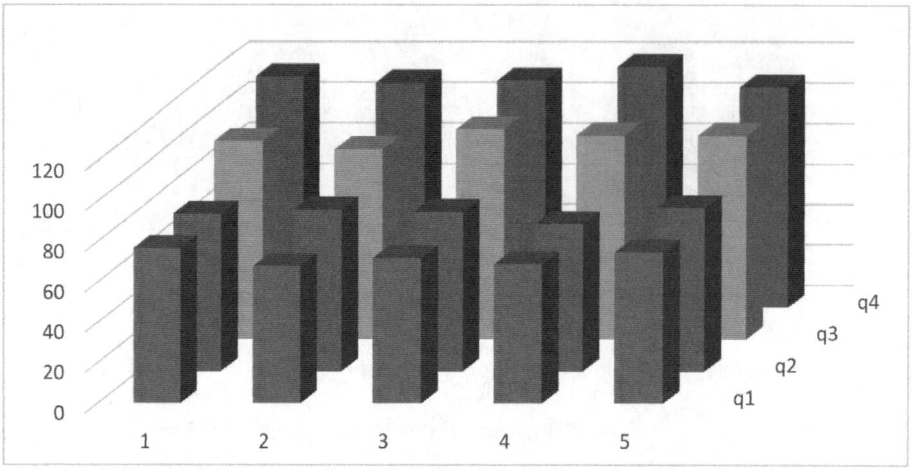

Figure 6. Dependence of the heat flux density on the aging intensity of insulation.

3. Conclusion

According to the methods given in SNaR 41-03-2003 «Thermal insulation of equipment and pipelines» (updated version of SR 61.13330.2012), calculation of specific heat losses of steam pipelines of the above-ground gasket d-76 mm и d-159 mm was carried out, through which steam with a temperature of 194 °C was transported. The results of the calculation are shown in the diagram (Figure 7). Thus, it can be stated that the difference in thermal losses, taking into account thermal destruction and without taking it into account, is significant. In the case of thermal destruction of heat insulation, losses during the transportation of steam are increased by 39%. This circumstance acquires a special urgency in connection with the increase in prices for energy carriers.

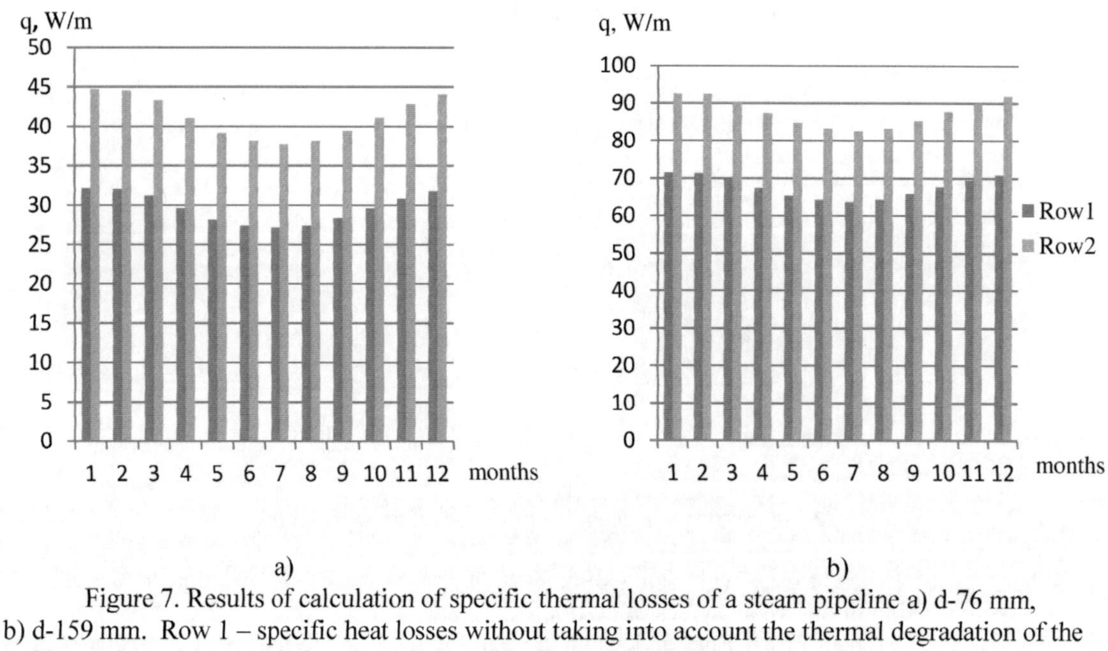

Figure 7. Results of calculation of specific thermal losses of a steam pipeline a) d-76 mm, b) d-159 mm. Row 1 – specific heat losses without taking into account the thermal degradation of the foam polyurethane insulation. Row 2 – specific heat losses with increasing coefficient of thermal conductivity due to thermal degradation by 20%.

It can be seen from Figure 7 that operation of the heat network in conditions different from projected leads to a significant increase in heat losses. The received results testify to the need to take into account the actual operating conditions when estimating heat losses. Timely detection of «problematic» places with the help of the developed system will allow ensuring the implementation of energy saving programs.

Acknowledgements
The work is performed according to the Russian Government Program of Competitive Growth of Kazan Federal University. The reported research was funded by Russian Foundation for Basic Research and the government of the region of the Russian Federation, grant № 18-48-160051\18.

References

[1] https://neftegaz.ru/science/view/1028-Vysokovyazkaya-neft-novye-tehnologii-razrabotki.-Chast-2.
[2] Ibatullin R.R. 2011 *Technological processes of development of oil fields* (Moscow: A-RRIOEOGI) p 304
[3] Dementiev A.G. 2000 *the Penopoliuretan Magazine* №2 Assessment of endurance of polyurethane foams for thermal insulation pp. 38-41
[4] Dementiev A.G., Dementiev M.A., Zinger P.A., Metlyakova I.R. 1999 Effect of the cellular structure on the thermal conductivity of rigid closed-cell foam polymers during long-term aging *Mechanics of composite materials* New York, Vol. 35, № 2 pp. 129-138.
[5] Bazukova E.R. and Vankov Y.V. 2015 *Engineering bulletin of Don* № 3 *Thermal losses at deterioration in properties of isolation in use*. ivdon/ru/ru/magazine/archive/n3y2015/3249.
[6] Dementiev A.G. 1997 Structure and properties of gas-filled polymers: thesis of the Dr.Sci.Tech.: 05.17.06. Moscow p 409
[7] Gerashchenko O.A. 1971 *Teplometriya bases* (Kiev: Publishing house «Naukova dumka») p 192

Calculation of the distribution of temperature in the form based on magnesium oxide for the casting of titanium products

N F Kashapov[1], I G Dautov[1,2], L A Tkachenko[1,3], S A Fadeev[1] and L R Shaidullin[1]

[1] Kazan Federal University, 18 Kremlyovskaya str., Kazan 420008, Russia
[2] Kazan National Research Technical University named by A.N.Tupolev, 420111, K.Marx str., 10, Kazan, Russia
[3] Institute of Mechanics and Engineering, Subdivision of the Federal State Budgetary Institution of Science «Kazan Scientific Center of the Russian Academy of Sciences», 2/31, Lobachevsky str, 420111, Kazan, Russia

E-mail: sergey_fadeev@kpfu.ru

Abstract. Calculations of the distribution of temperature in the mold based on magnesite oxide, when casting products from titanium alloys, are presented. It is shown that in the mold based on magnesite oxide it cools down exponentially to ambient temperature.

1. Introduction

In the process of interaction of molten titanium with a mold based on magnesite oxide, a large number of chemical reactions occur, both reversible and irreversible. It is necessary to single out those of them, whose products, if they fall on the "mold-casting" boundary, can lead to the formation of TiO_2. The amount of oxygen forming the alpha case layer depends on the time of thermal exposure in a certain temperature range. Thus, knowing the coefficients of thermodiffusion of molecules of compounds containing oxygen and participating in chemical reactions in a titanium melt, it is possible to determine the main causes of the formation of an alpha case layer [1-5].

2. Model and discussion

The mold cools down in the oven at an ambient temperature of 600 ° C, the initial temperature T_0 can be set. Heat removal through the gate system is rather small, since the process of melting and cooling of the casting is carried out in a vacuum. In this case, the integral energy luminosity is proportional to the outer area of the object, it can be assumed that in a sufficient distance from the gating system, the heat is removed from the melt through the mold. Assuming that the mold-casting interface is flat, setting the temperature of the molten titanium T_1 = 1961 K, for the case of an unsteady heat flux in a solid without internal sources, we consider the equation of the nonstationary temperature field (the Fourier heat equation)

$$\frac{\partial T}{\partial t} = a\nabla^2 T \tag{1}$$

At the initial time, the temperature at the "mold-casting" boundary T_1. The ambient temperature is T_0. We introduce the dimensionless temperature

$$\theta = \frac{T - T_0}{T_1 - T_0}. \tag{2}$$

As a result of the transformation of equation (1), we obtain

$$\frac{\partial \theta}{\partial t} = a\nabla^2\theta \text{ или } \frac{\partial \theta}{\partial Fo} = \nabla^2\theta \tag{3}$$

where is the thermal diffusivity.
We seek the solution of equation (3) in the form of a product of two functions, one of which is related only to the coordinate, the other only with time

$$\theta = \varphi(Fo)\psi(\xi), \tag{4}$$

where $Fo = \frac{at}{\delta^2}$ is the function of the Fourier criterion, $\xi = \frac{x}{\delta}$ is the coordinate function, and δ is the thickness of the plate. Substituting (4) into (3), we obtain

$$\varphi'(Fo)\psi(\xi) = \varphi(Fo)\nabla^2\psi(\xi) \tag{5}$$

In order for expression (4) to satisfy equation (3), it is necessary that the functions φ and ψ satisfy the condition:

$$\frac{\varphi'(Fo)}{\varphi(Fo)} = \frac{\nabla^2\psi(\xi)}{\psi(\xi)} = -\beta^2 \tag{6}$$

Also, in order for function (4) to be a solution of equation (1), equality (5) must be satisfied identically, that is, for all values of independent variables $0<x<\delta$, $t>0$. Therefore, fixing, for example, some value of x and changing t (or vice versa), we find that the right and left parts of (5) retain a constant value $-\beta^2$ when changing their arguments (the values of such a parameter are called eigenvalues).

We introduce the initial conditions

$$\theta = 1, \; Fo = 0. \tag{7}$$

The boundary conditions have the form

$$\left(\frac{d\theta}{d\xi}\right)_{\xi=0} = 0, \; \xi = 0, \tag{8}$$

$$-\lambda_T\left(\frac{\partial \theta}{\partial x}\right)_{x=\delta} = \alpha\left(T - T_\infty\right)_{x=\delta}. \tag{9}$$

A boundary condition of the third kind in the dimensionless form is written as

$$-\left(\frac{d\theta}{d\xi}\right)_{\xi=0} = Bi(\theta)_{\xi=1}, \tag{10}$$

Where $Bi = \frac{\alpha\delta}{\lambda_T}$ - Bio criterion, α - heat transfer coefficient, λ_T - thermal conductivity coefficient.

We substitute $\theta = \psi\varphi$ into the boundary conditions (8) and (10), we obtain

$$\left(\frac{d\psi}{d\xi}\right)_{\xi=0} = 0\,,\ \xi = 0, \tag{11}$$

$$-\left(\frac{d\psi}{d\xi}\right)_{\xi=0} = Bi(\psi)_{\xi=1}, \tag{12}$$

In this case, the function ψ depends only on one variable ξ, then from equation (6) we obtain

$$\frac{d^2\psi}{d\xi^2} + \beta^2\psi = 0. \tag{13}$$

The solution of equation (13) is

$$\psi = C_1 \cos\beta\xi + C_2 \sin\beta\xi. \tag{14}$$

The condition (11) is satisfied only by the even function, therefore we set

$$C_2 = 0. \tag{15}$$

We rewrite (14) in the form

$$\psi = C_1 \cos\beta\xi, \tag{16}$$

we substitute (16) into (12) and obtain

$$\beta \sin\beta = Bi\cos\beta. \tag{17}$$

The transcendental equation (17) does not have an analytical solution and is solved by the graphical method [2]. It follows from equation (17) that the boundary conditions determine the values of the constants β_i, while the constant C_1 is found by means of boundary conditions.

The function φ depends only on time, then from equation (6) we obtain

$$\varphi^{'}(Fo) + \varphi(Fo)\beta^2 = 0. \tag{18}$$

A solution of equation (18) is a temperature versus time

$$\varphi(Fo) = Ce^{-\beta^2 Fo}. \tag{19}$$

Thus, the general solution of equation (3) has the form of a series

$$\theta = \sum_{i=1}^{\infty} C_i \cos \beta_i \xi e^{-\beta_i^2 F_0}. \tag{20}$$

The coefficients β_i increase rapidly with the number of the term of the series, so the more the criterion *Fo*, the smaller the value of the highest term of the series in comparison with the previous ones. As a result, after some value of *Fo*, all the terms of the series become negligibly small in comparison with the first term of the series. The technical process of cooling the casting is up to several hours, so the temperature with a higher degree of accuracy can be expressed by a single-term formula

$$\theta = C_1 \cos \beta_1 \xi e^{-\beta_1^2 F_0}. \tag{21}$$

The constant C_1 is calculated by substituting (21) into the initial conditions (7), then the dimensionless temperature has the form

$$\theta = 2\frac{\sin(\beta_1)\cos(\beta_1 \xi)}{\beta_1 + \sin(\beta_1)\cos(\beta_1)} e^{(-\beta_1^2 F_0)}. \tag{22}$$

We reduce the expression (22) to the dimensional form

$$T = 2\frac{\sin(\beta_1)\cos(\beta_1 \frac{x}{\delta})}{\beta_1 + \sin(\beta_1)\cos(\beta_1)} e^{(-\beta_1^2 F_0)} (T_1 - T_0) + T_0. \tag{23}$$

Expression (23) is a solution of the heat equation (1). In the case of a mold based on magnesite oxide $Bi \approx 860$, which corresponds to $\beta_1 = \pi/2$ [2]. Fig. 1 shows the calculation of the temperature distribution in mold based on magnesite oxide when casting a titanium product.

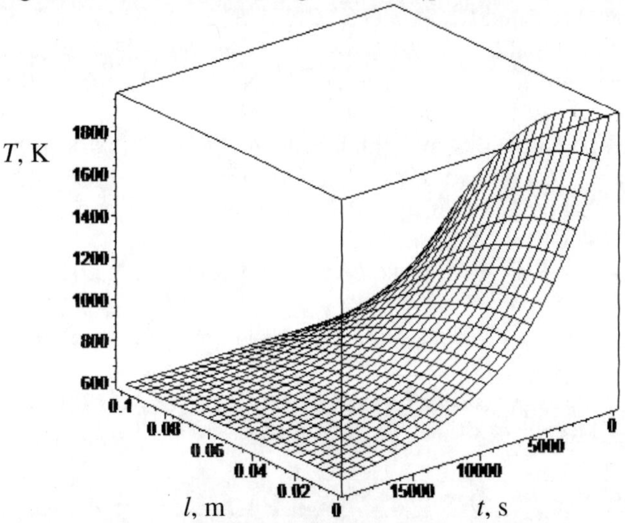

Figure 1. Calculation of temperature distribution in a mold based on magnesite oxide form when casting a titanium product.

IOP Conf. Series: Journal of Physics: Conf. Series **1058** (2018) 012062 doi:10.1088/1742-6596/1058/1/012062

When casting products from titanium, a mold based on magnesite oxide heated up to 1961 K (melting temperature of Ti) and, subsequently, it cools down exponentially to the temperature in the furnace. Knowing the temperature distribution in the mold, at each instant of time, and having the thermal diffusion coefficients of the molecules of compounds containing oxygen, it is possible to determine the amount of oxygen diffusing into the casting and participating in the formation of the alpha case layer. This will allow to develop new methods of casting titanium products in a mold based on magnesite oxide, aimed at preventing the formation of an alpha case layer [9-20]. This will lead to a reduction in the labor intensity and cost of the technological process of casting titanium products.

Acknowledgments
This work was funded by the subsidy allocated to Kazan Federal University for the state assignment in the sphere of scientific activities (11.1300.2017/4.6), Presidential Program for State Support of Young Candidates of Sciences (grant no. MK-539.2017.1), and the research grant of Kazan Federal University.

References
[1] Mauilov L A 1962 *Physical chemistry and chemistry of silicon* (Moscow, High school) p 65
[2] Tkachenko L A 2017 *Heat exchange theory* (Kazan, Publishing house of Kazan University) p 151
[3] Kutateladze S S 1979 *Fundamentals of heat exchange theory* (Moscow, Atomizdat) p 416
[4] Kutateladze S S 1990 *Heat Transfer and Hydrodynamic Resistance* (in Russian) (Moscow, Energoatomizdat Publishing House) p 366
[5] Schlichting H 1974 *Boundary layer theory* (in Russian) (Moscow, Science) p 711
[6] Kudinov V A, Kartashov E M and Stefanyuk E V 2015 *Technical thermodynamics and heat transfer: a textbook for academic baccalaureate* (in Russian) (Moscow, Yurayt Publishing House) p 566
[7] Gavrilova V A, Fazlyyyakhmatov M G and Kashapov N F 2013 *J. Phys.: Conf. Ser.* **479** 012010
[8] Saifutdinov A I, Saifutdinova A A, Kashapov N F and Fadeev S A 2016 *J. Phys.: Conf. Ser.* **669** 012045
[9] Kornienko E E, Lapushkina E J, Kuzmin V I, Vaschenko S P, Gulyaev I P, Kartaev E V, Sergachev D S, Kashapov N, Sharifullin S, Fayrushin I 2014 *J. Phys.: Conf. Ser.* **567** 012010
[10] Saifutdinov A I, Fairushin I I and Kashapov N F 2016 *JETP Lett.* **104** 180–185
[11] Luchkin A G, Kashapov N F and Luchkin G S 2013 *J. Phys.: Conf. Ser.* **479** 012019
[12] Hakki A, Kashapov N and Sadikov K 2017 *J. Phys.: Conf. Ser.* **927** 012022
[13] Fayrushin I I, Dautov I G and Kashapov N F 2017 *International Journal of Environmental Science and Technology* **14** (12) 2555–2560
[14] Zaripov R G, Tkachenko L A and Shaydullin L R 2014 *J. Phys.: Conf. Ser.* **567** 012042
[15] Shaydullin L R 2016 *IOP Conf. Ser.: Mater. Sci. Eng.* **134** 012022
[16] Gubaidullin D A, Kashapov N F, Zaripov R G, Tkachenko L A and Shaydullin L R 2017 *J. Phys.: Conf. Ser.* **789** 012017
[17] Gubaidullin D A, Zaripov R G, Tkachenko L A and Shaidullin L R 2017 *High Temperature* **55** (3) 469–471
[18] Zaripov R G, Tkachenko L A, Shaidullin L R 2017 *Journal of Engineering Physics and Thermophysics* **90** (6) 1463–1468
[19] Fadeev S A and Saifutdinov A I 2017 *Plasma Physics Reports* **43** (11) 1080–1088
[20] Tkachenko L A and Fadeev S A 2017 *Acoustical Physics* **63** (1) 7–13

Processes occurring during the manufacture of form based on magnesite oxide and casting titanium products

N F Kashapov[1], I G Dautov[1,2], R A Ariskina[1], L R Shaidullin[1], S A Fadeev[1] and M T R Zaitov[1]

[1] Kazan Federal University, 18 Kremlyovskaya str., Kazan 420008, Russia
[2] Kazan National Research Technical University named by A.N.Tupolev, 420111, K.Marx str., 10, Kazan, Russia

E-mail: sergey_fadeev@kpfu.ru

Abstract. In this work investigate the chemical processes occurring during the manufacture of magnesium oxide-based mold and casting titanium products to prevent the formation of alpha case. When molten titanium is poured into the mold due to low thermal conductivity the mold is heated unevenly. The resulting carbon dioxide and active metal compounds migrate to the more heated regions adjacent to the surface of the casting. The increased concentration on the surface of the O, Si, Ca, Na forms promotes the reactions to form active metal titanates.

1. Introduction

Titanium has a high corrosion resistance and belongs to reactive metals due to the stable passive TiO_2 film on titanium surface, which is bound up with the base metal and excludes direct contact of metal and the corrosive environment. The thickness of this film is usually 5-6 nm. Owing to the oxide film, titanium and it's alloys are not corroded in the atmosphere, in fresh and sea water, they are stable in organic acids, and are widely used in various fields of science and technology. Unfortunately, the manufacture of titanium products and it's alloys has a technological features. The high chemical activity of molten titanium leads to the alpha case formation in the casting of products. There is a sharp decrease of the technological plasticity of titanium and titanium's alloys by reason of alpha case formation. It is necessary to investigate the chemical processes occurring during the manufacture of magnesium oxide-based mold and casting titanium products to prevent the formation of alpha case. This problem can be solved with the help of plasma methods for coating forms [1-11].

2. Processes occurring during the manufacture of form based on magnesite oxide and casting titanium products

Stage 1: preparation of the starting mixture. At room temperature fine crystalline magnesium oxide is chemically active, so it reacts slowly with water:

$$MgO + H_2O \leftrightarrow Mg(OH)_2$$

(at 350 ° C magnesium hydroxide decomposes into magnesium oxide and water) magnesium oxide absorbs carbon dioxide:

$$MgO + CO_2 \leftrightarrow MgCO_3$$

(at 550-850 ° C magnesite decomposes into magnesium oxide and carbon dioxide). Calcium oxide (quicklime) interacts with water:

$$CaO + H_2O \leftrightarrow Ca(OH)_2$$

(at 520-580 ° C calcium hydroxide decomposes into calcium oxide and water)
Sodium oxide reacts with carbon dioxide:

$$NaO + CO_2 \leftrightarrow NaCO_3$$

(at 855 ° C melting of the sodium salt of carbonic acid is occurred)

Stage 2: calcination with carbon dioxide.
Purge of the mixture with carbon dioxide accelerates the process of chemical binding of water with elements of liquid glass.
Carbon dioxide chemically reacts with liquid glass with heat release according to the equation:

$$Na_2O{\cdot}nSiO_2 + mH_2O + CO_2 \rightarrow Na_2CO_3 + nSiO_2{\cdot}mH_2O$$

As a result of this reaction, the liquid glass is destroyed with the formation of soda and a gel of silicic acid, which has high adhesive properties, linking the grains of sand to each other. The grains of the charge are "dressed" in a shell of silicic acid gel or, as they say, a gel film is formed on the grains (the less water in the hydrogel, the stronger the hardening mixture, and when the portion of the attached water is lost, the gel strengthens) [12, 13].

Stage 3: drying at 110-200 ° C.
110-120 ° C - removal of hygroscopic moisture:

$$2H_2O \rightarrow 2H_2\uparrow + O_2\uparrow$$

above 200 ° C - the removal of moisture crystalline hydrate, formed in particular when the batch is moistened:

$$X{\bullet}2H_2O \rightarrow X + 2\,H_2\uparrow + O_2\uparrow$$

150-200 ° C - gel dehydration and evaporation of water - formation of silica gel:

$$H_2SiO_3 \rightarrow H_2O\uparrow + SiO_2$$

Silica gel (dried gel, T_m = 1610 ° C) is a substance that acquires a porous structure and absorbs moisture well. Silica gel has a large specific surface area consisting of -SiOH groups located at a distance of 0.5 nm from each other. These groups are active centers and the activity of a specific lot of silica gel depends on the number and activity of such centers. In the active adsorbent i. e. in which adsorbed on it's surface water is removed, many centers will be active. Such activation occurs when the gel is heated to 150-200 ° C.

IOP Conf. Series: Journal of Physics: Conf. Series **1058** (2018) 012063 doi:10.1088/1742-6596/1058/1/012063

Stage 4: calcination to 1000 ° C.
200 - above 400 ° C - the surface area of the silica gel irreversibly decreases:

$$H_2SiO3 \rightarrow H_2O\uparrow + SiO_2$$

At temperature in the range of 200-400 ° C the activity is lost as a result of the formation of Si-O bonds which occurs with the elimination of water. However this stage is reversible. When heated above 400 ° C the surface area of the silica gel irreversibly decreases.

800-900 ° C - solid phase formation of sodium silicate:

$$2Na^{+} + SiO_4^{-2} \rightarrow Na_2SiO_4$$

At temperature above 567 ° C iron oxide reacts with sodium carbonate to form sodium monoferrite and carbon dioxide:

$$Fe_2O_3 + Na_2CO_3 \rightarrow 2NaFeO_2 + CO_2$$

up to 1000 ° C - formation of magnesium metasilicate - enstatite:

$$MgO + SiO_2 \rightarrow MgO \cdot SiO_2$$

up to 1000 ° C - formation of calcium metasilicate:

$$CaO + SiO_2 \rightarrow CaO \cdot SiO_2$$

The reaction products: periclase constitute the main mass of materaial, also the material contains calcined gel, enstatite, sodium silicate, sodium monoferrite and calcium metasilicate. A porous structure forms.
Stage 5: cooling to 300 ± 25 ° C.
With slow cooling molten sodium silicates and other components are sintered with the moldable mixture. The reactions are summarized in Table 1. Here X is a chemical compound containing bound water.
Stage 6: pouring of molten titanium into the form at 1843 ° C
Formation of titanates (magnesium, sodium, calcium). For example the formation of magnesium metatitanate at 850 ° C:

$$MgO + TiO_2 \rightarrow MgTiO_3$$

The properties of certain titanates that can form during the interaction of Ti melt with magnesite (periclase) casting molds are discussed below.

Ti-Mg

In the MgO-TiO_2 system three compounds are known: Mg_2TiO_4, $MgTiO_3$ (geykilite), and $MgTi_2O_3$. The reaction between titanate dioxide and magnesium oxide in their equimolecular mixture (1: 1) proceeds with the formation of mainly magnesium metatitanate.

IOP Conf. Series: Journal of Physics: Conf. Series **1058** (2018) 012063 doi:10.1088/1742-6596/1058/1/012063

Table 1. Characteristic chemical reactions for the corresponding temperatures during cooling of magnesite forms.

№	Reaction temperature range, °C	Name of reaction	Characteristics of the reaction
1	20-25	Formation of: - gel of silicic acid as a result of blowing the form with carbon dioxide; - magnesium hydroxide; - magnesium salt carbonic acid; - Calcium hydroxide; - sodium salt of carbonic acid	$Na_2O{\bullet}nSiO_2 + mH_2O + CO_2 \rightarrow NaCO_3 + nSiO_2{\bullet}mH_2O$; $MgO + H_2O \leftrightarrow Mg(OH)_2$; $MgO + CO_2 \leftrightarrow MgCO_3$; $CaO + H_2O \leftrightarrow Ca(OH)_2$; $NaO + CO_2 \leftrightarrow NaCO_3$
2	110—120	removal of hygroscopic moisture	$2H_2O \rightarrow 2H_2\uparrow + O_2\uparrow$
3	200-275	removal of moisture crystalline hydrate formed in particular when the charge is moistened	$X{\bullet}2H_2O \rightarrow X + 2H_2O \rightarrow 2H_2\uparrow + O_2\uparrow$
4	200 - over 400	silica gel formation	$H_2SiO_3 \rightarrow H_2O + SiO_2$
5	over 567	synthesis of sodium monoferrite	$Fe_2O_3 + Na_2CO_3 \rightarrow 2NaFeO_2 + CO_2$
6	851-over 1000	melting of the sodium salt of carbonic acid	$Na_2CO_3 \rightarrow Na_2O + CO_2$
7	800—900	solid phase formation of sodium silicate	$2Na^+ + SiO_4^{-2} \rightarrow Na_2SiO_4$
8	550-850	periclase formation	$MgCO_3 \rightarrow MgO + CO_2$
9	before 1000	formation of metasilicates of magnesium and calcium	$MgO + SiO_2 \rightarrow MgO{\bullet}SiO_2$; $CaO + SiO_2 \rightarrow CaO{\bullet}SiO_2$
10	1843 and above	titanate formation	$MgO + TiO_2 \rightarrow MgTiO_3$
11	1843 and above	formation of titanium oxides	$Ti + 2H_2O \rightarrow TiO_2 + 4H$

Ti-Ca

When TiO2 interacts with $CaCO_3$ after dissociation of the carbonate, calcium metatitanate is directly formed, and only at temperatures above 1300-1350 ° C with an excess of calcium oxide a three-calcium dititanate is formed. At temperatures above 1480 ° C we can also observe the formation of a titanate of the composition $3CaOTiO_2$. Ershov showed that at temperature up to 1300 ° C $CaTiO_2$ (perovskite) is the least basic compound in the CaO-TiO_2 system [14].

At a temperature of 900 - 1200 ° C $CaTiO_3$ is formed from the anatase with higher rate than from rutile. With a content of several percent of iron oxide in the rutile (in the form of a solid solution) the rate of formation of CaTiO3 from rutile is higher than from pure anatase [15].

Ti-Al

Aluminum titanate Al_2TiO_3 as a refractory material with negative thermal expansion up to 600° C and high volume resistance is of practical interest. Aluminum titanate was obtained by double firing a mixture of fine powders (6 microns grain) of aluminum oxide and titanium dioxide initially at 1820 ° C, then at 1650 ° C. The melting point of Al2TiO3 is 1860 ° C. It is stable at temperature above 1300 ° C; at a temperature of 750 - 1300 ° C it decomposes into corundum and rutile.

Ti-K

Potassium metatitanate - K_2Ti_3 (molecular weight is 174.10) is formed in the systems K_2O-TiO_2, K_2O-TiO_2-H_2O and K_2O-TiO_2-CO_2. Practically potassium metatitanate is obtained by fusing titanium dioxide or metatitanic acid with hydroxide or potassium carbonate. In the latter case the reaction for the accumulation of carbon dioxide is reversible like the reaction with the sodium salt but occurs with a lower yield of titanate. The equilibrium is established at 900 ° C, the pressure of carbon dioxide is 1 atm. at the ratios:

$$0{,}65TiO_2 + 0{,}65K_2CO_3 = 0{,}35K_2TiO_3 + ArCO_2$$

The formation of potassium metatitanate also occurs when the elemental titanium is fused with potassium hydroxide and nitrate (or chlorate):

$$3Ti + 2KOH + 4KNO_3 = 3K_2TiO_3 + 4NO + H_2O$$

The heat of formation of K_2TiO_3 is 384.6 kcal / mol.

Anhydrous potassium metatitanate is a yellowish white crystalline mass. It crystallizes in the tetragonal system (a = 547, c = 1660 pm). Density is 3.58 g / cm^3 at 20 ° C. Melting point is 8160 ° C. The dielectric constant is 16.5; refractive index is 1,910.

Potassium ditantate KaTi2O5 (molecular weight 254.00) is formed by heating a mixture of titanium dioxide and potassium carbonate at molar ratio of 2: 1. Dititantate potassium is a brilliant needle-shaped crystals. Density is 3,485 g / cm^3 at 15 ° C. Melting point is 9800 °C The refractive index is more than 1.74. Crystals of potassium dithitanate are weathered in the air. In water dithitanate does not dissolve; decomposes with acids.

Potassium hexachitanate $K_2Ti_6O_3$ is colorless fibrous crystals with a melting point 13700 °C.

For the analysis of chemical reactions that can lead to the formation of an alpha case layer a calculation of the temperature at the "metal-form" boundary is carried out using the Fourier thermal conductivity equation [16-21]. In Fig. 1 there is showed the temperature distribution at the "metal-form" interface and the temperature intervals in which the chemical reactions listed in Table 1 are indicated.

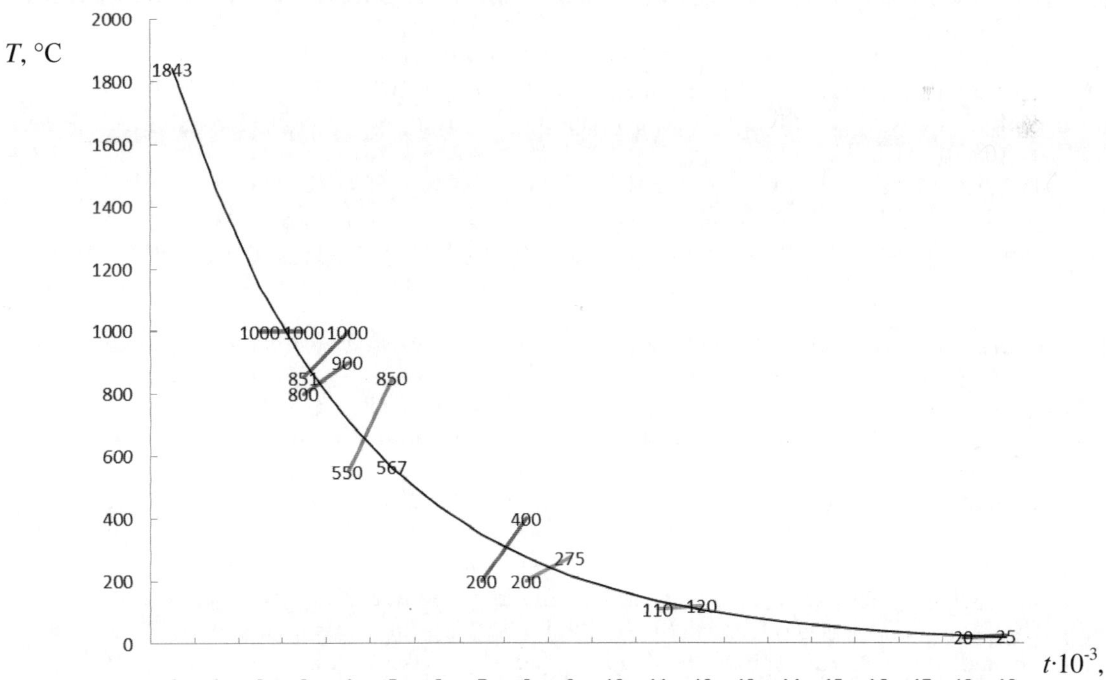

Figure 1. Temperature distribution at the "metal-form" interface and the temperature intervals in which the chemical reactions listed in Table 1.

3. Conclusions

Comparison of the obtained data allows us to consider the following model for the alpha case formation. The source of oxygen supply to the surface of the casting can be compounds of active metals. When molten titanium is poured into the mold due to low thermal conductivity the mold is heated unevenly. At temperature above 840 K, reactions 5-9 from Table 1 begin. The resulting carbon dioxide and active metal compounds migrate to the more heated regions adjacent to the surface of the casting. The increased concentration on the surface of the O, Si, Ca, Na forms promotes the reactions to form active metal titanates. In some cases due to the reversibility of the reactions this leads to the formation of rutile. At high temperature the bound water forms carbon dioxide which in turn comes into various reactions with the active metals and titanium. When temperatures favorable for the formation of rutile from various compounds on the surface of the casting are reached, phase transformations of oxygen solutions in titanium occur.

Acknowledgments
This work was funded by the subsidy allocated to Kazan Federal University for the state assignment in the sphere of scientific activities (11.1300.2017/4.6).

References

[1] Kornienko E E, Lapushkina E J, Kuzmin V I, Vaschenko S P, Gulyaev I P, Kartaev E V, Sergachev D S, Kashapov N, Sharifullin S, Fayrushin I 2014 *J. Phys.: Conf. Ser.* **567** 012010
[2] Saifutdinov A I, Fairushin I I and Kashapov N F 2016 *JETP Lett.* **104** 180–185
[3] Luchkin A G, Kashapov N F and Luchkin G S 2013 *J. Phys.: Conf. Ser.* **479** 012019
[4] Dautov G, Fayrushin I and Kashapov N 2014 *J. Phys.: Conf. Ser.* **567** 012006
[5] Hakki A, Kashapov N and Sadikov K 2017 *J. Phys.: Conf. Ser.* **927** 012022
[6] Fayrushin I I, Dautov I G and Kashapov N F 2017 *International Journal of Environmental Science and Technology* **14** (12) 2555–2560
[7] Fairushin I I, Dautov I G, Kashapov N F and Shamsutdinov A R 2017 *Technical Physics Letters* **43** (1) 27–30
[8] Fadeev S A and Saifutdinov A I 2017 *Plasma Physics Reports* **43** (11) 1080–1088
[9] Fayrushin I and Dautov G 2013 *J. Phys.: Conf. Ser.* **479** 012013
[10] Saifutdinov A I, Saifutdinova A A, Kashapov N F and Fadeev S A 2016 *J. Phys.: Conf. Ser.* **669** 012045
[11] Hakki A, Fayrushin I, Kashapov N 2016 *J. Phys.: Conf. Ser.* **669** 012022
[12] Schlichting H 1974 *Boundary layer theory* (in Russian) (Moscow, Science) p 711
[13] Kudinov V A, Kartashov E M and Stefanyuk E V 2015 *Technical thermodynamics and heat transfer: a textbook for academic baccalaureate* (in Russian) (Moscow, Yurayt Publishing House) p 566
[14] Adler Yu P et all 1971 *Planning an experiment when searching for optimal conditions* (in Russian) (Moscow, Science) p 282
[15] Sokolovskaya I Yu 2010 *Full factorial experiment* (Novosibirsk, NGAWT) p 36
[16] Gubaidullin D A, Zaripov R G, Tkachenko L A and Shaidullin L R 2017 *High Temperature* **55** (3) 469–471
[17] Zaripov R G, Tkachenko L A, Shaidullin L R 2017 *Journal of Engineering Physics and Thermophysics* **90** (6) 1463–1468
[18] Tkachenko L A and Fadeev S A 2017 *Acoustical Physics* **63** (1) 7–13
[19] Shaydullin L R 2016 *IOP Conf. Ser.: Mater. Sci. Eng.* **134** 012022
[20] Zaripov R G, Tkachenko L A and Shaydullin L R 2014 *J. Phys.: Conf. Ser.* **567** 012042
[21] Gubaidullin D A, Kashapov N F, Zaripov R G, Tkachenko L A and Shaydullin L R 2017 *J. Phys.: Conf. Ser.* **789** 012017

IOP Conf. Series: Journal of Physics: Conf. Series **1058** (2018) 012064 doi:10.1088/1742-6596/1058/1/012064

Numerical simulation of the shock wave in the closed resonator using 1D Lagrange's approach

D A Gubaidullin[1,2], P P Osipov[1,2], R R Nasyrov[1,2] and I M Almakaev[1,2]

[1] Institute of Mechanics and Engineering, Subdivision of the Federal State Budgetary Institution of Science «Kazan Scientific Center of the Russian Academy of Sciences», 2/31, Lobachevsky str, 420111, Kazan, Russia.
[2] Kazan (Volga region) Federal University, 18, Kremlin str, 420008, Kazan, Russia.

E-mail: ildar.almakayev@gmail.com

Abstract. The comparison of the results of calculating the shock wave in a two-dimensional acoustic resonator based on the 1D Lagrange approach with experimental, analytical and numerical results of other authors obtained on the basis of 2D simulation is presented. The shock wave, which arises at frequencies close to the first resonant one, is investigated. It's shown that the 1D Lagrange approach gives a satisfactory agreement with the results of 2D simulations of other authors as well as with results obtained using CFD-package Fluent in this work. It's shown that Lagrange's approach due to exact formulation of the boundary condition, high calculation speed and minimum memory requirements can be effectively used to solve a number of two-dimensional problems, as well as to verify the reliability of other numerical methods.

1. Mathematical model

The governing equations in Lagrange's coordinate are [1]:

$$\rho x_\xi = \rho_0, \quad \rho_0 u_t = -p_\xi + \sigma_\xi, \quad dp = c^2 d\rho,$$

where $x = x(\xi, t)$ is the Euler's coordinate of gas particle at time t, ξ is the Lagrange's coordinate, $u = x_t$ is the velocity, ρ, ρ_0 – are gas density and its initial value respectively, p – is the pressure, c is the local sound speed, $\sigma = (2\mu + \lambda)\left.\frac{\partial u}{\partial x}\right|_{t=const} = \left(\frac{4}{3}\mu + \zeta\right)\left.\frac{\partial u}{\partial x}\right|_{t=const} = \left(\frac{4}{3}\mu + \zeta\right)\left.\frac{\partial u}{\partial \xi}\right|_{t=const} \Big/ \left.\frac{\partial x}{\partial \xi}\right|_{t=const}$ – viscous stress, λ –Lame parameter, $\zeta = \lambda + \frac{2}{3}\mu$ – second viscosity (according to the Stokes hypothesis $\zeta = 0$).

Using the sound velocity $c^2 = \gamma p / \rho = c_0^2 / x_\xi^{\gamma-1}$ and excluding density and pressure $dp = c^2 d\rho$, the nonlinear wave equation for viscous polytropic gas can be obtained as follows:

$$x_{tt} = (c_0^2 / x_\xi^{\gamma+1}) x_{\xi\xi} + \frac{\mu}{\rho_0}\left(\frac{4}{3} + \frac{\zeta}{\mu}\right)\left(\frac{x_{t\xi}}{x_\xi}\right)_\xi \tag{1}$$

Introducing dimensionless quantities $t^* = c_0 t / L$, $x^* = x / L$, $\xi^* = \xi / L$, $\mathrm{Re}_0 = \dfrac{\rho_0 c_0 L}{\mu}$ and omitting asterisk, equations (1) can be written in the dimensionless form:

$$x_{tt} = (1 / x_\xi^{\gamma+1}) x_{\xi\xi} + \frac{1}{\mathrm{Re}_0}\left(\frac{4}{3} + \frac{\zeta}{\mu}\right)\left(\frac{x_{t\xi}}{x_\xi}\right)_\xi \tag{2}$$

1.1. Boundary value problem

One have to find the solution of (2) in the domain $\xi \in [0, 1 + l / L]$ using boundary conditions (here – the closed end on the left side):

$$\begin{aligned} &x(0,t) = 0, \\ &x(1,t) = 1 + l / L \cos \omega t, \end{aligned} \tag{3}$$

and initial conditions:

$$\begin{aligned} &x(\xi,0) = \xi, \\ &x_t(\xi,0) = 0. \end{aligned} \tag{4}$$

2. Numerical scheme

It was taken the mesh grid for spatial coordinate $\xi_i = ih$, $(i = 0,..N)$ with grid size $h = (1 + l / L) / N$ and the grid for time discretization $t^j = j\tau$ with time step $\tau (j = 0,1,2...)$.

The three-layered "cross" stencil was used in the space-time domain for calculations.

$$\frac{x_i^{j+1} - 2x_i + x_i^{j-1}}{\tau^2} = (1 / x_\xi^{\gamma+1})_i \frac{x_{i+1} - 2x_i + x_{i-1}}{h^2} + \frac{1}{\mathrm{Re}_0}\left(\frac{4}{3} + \frac{\zeta}{\mu}\right)\Lambda_i\left(\frac{\Lambda_i u}{\Lambda_i x}\right), \tag{5}$$

where $\Lambda_i \varphi = \begin{cases} -0.5(3\varphi_0 - 4\varphi_1 + \varphi_2) / h, & (i=0) \\ 0.5(\varphi_{i+1} - \varphi_{i-1})/h, & (0 < i < N), \\ 0.5(3\varphi_N - 4\varphi_{N-1} + \varphi_{N-2}) / h, & (i=N) \end{cases}$

$(x_\xi)_i = \Lambda_i x$.

The first two layers are initiated by means of the initial conditions:

$$x_i^0 = \xi_i = ih, \qquad x_i^1 = x_i^0.$$

On the boundaries we use:

$$x_0^j = 0, \qquad x_N^j = 1 + l / L \cos(\omega t^j). \tag{6}$$

The compression is calculated as follows: $(x_\xi)_i^{j+1} = \Lambda_i x^{j+1}$. The density is defined by equation $\rho_i^{j+1} / \rho_0 = 1 / (x_\xi)_i^{j+1}$. The pressure is obtained from the ratio for the polytropic gas $p_i^{j+1} / p_0 = (\rho_i^{j+1} / \rho_0)^\gamma$. The velocity is obtained via three layers of the time $u_i^{j+1} = (x_t)_i^{j+1} = 0.5(3x_i^{j+1} - 4x_i^j + x_i^{j-1}) / \tau$. After that, one have to update the quantities $x_i^{j-1} = x_i^j$, $x_i^j = x_i^{j+1}$, $u_i^j = u_i^{j+1}$.

The artificial viscosity is used to damp the oscillations that arise during numerical computation $x_i^{j+1} = x_i^{j+1} + \dfrac{4}{3}\dfrac{\alpha}{Re_0}\left(\dfrac{x_{t\xi}}{x_\xi}\right)_\xi^{j+1} \tau^2$, $(\alpha \approx 100)$. The scheme (5) is solved under Courant's condition $\tau = 0.5h$.

3. Numerical investigation of gas oscillations in the closed resonator

The theory of propagation of acoustic waves in a closed resonator is presented in [3], [4] on the basis of experimental data. In the work [2] authors numerically investigated the problem of propagation of acoustic waves in a closed resonator and compared the results obtained theoretically and experimentally.

Series of numerical experiments were performed to verify the obtained one-dimensional equation of gas motion in Lagrange coordinates (2) with the boundary conditions (3), (4).

The piston oscillates at the right boundary with an amplitude l=3.2 mm at the first resonant frequency Ω=636.76 rad/sec (f=101.3 Hz) in the pipe of length L=1.7 m.

Figure 1 shows the comparison of pressure, obtained in different works numerically and experimentally. Figure 2 shows the comparison of the pressure versus time for frequencies close to the resonant (on the left - the results [2], on the right - the results of the work). The calculations in the Lagrange coordinates are in good agreement with the known results. It can be seen from the graphs of Fig. 2 that at frequencies $\omega/\Omega<1$ for the pressure profile the predominance of the lower part is characteristic, and at the $\omega/\Omega>1$ – upper one.

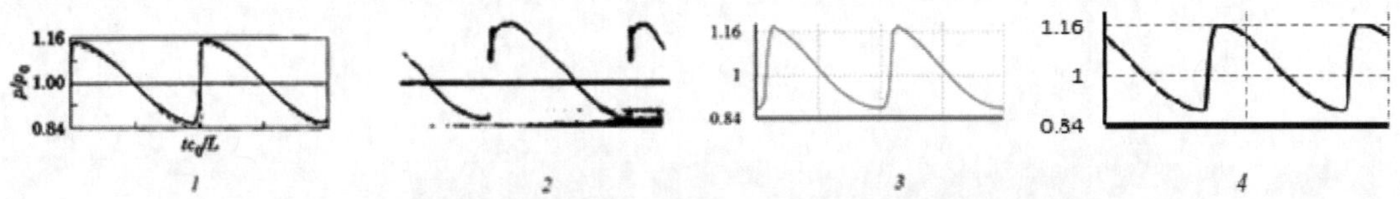

Figure 1. The pressure oscillation at the piston.
1 – solid – [2], dashed – [3]; *2* – experimental data [4];
3 – Lagrange approach; 4- CFD-package Fluent.

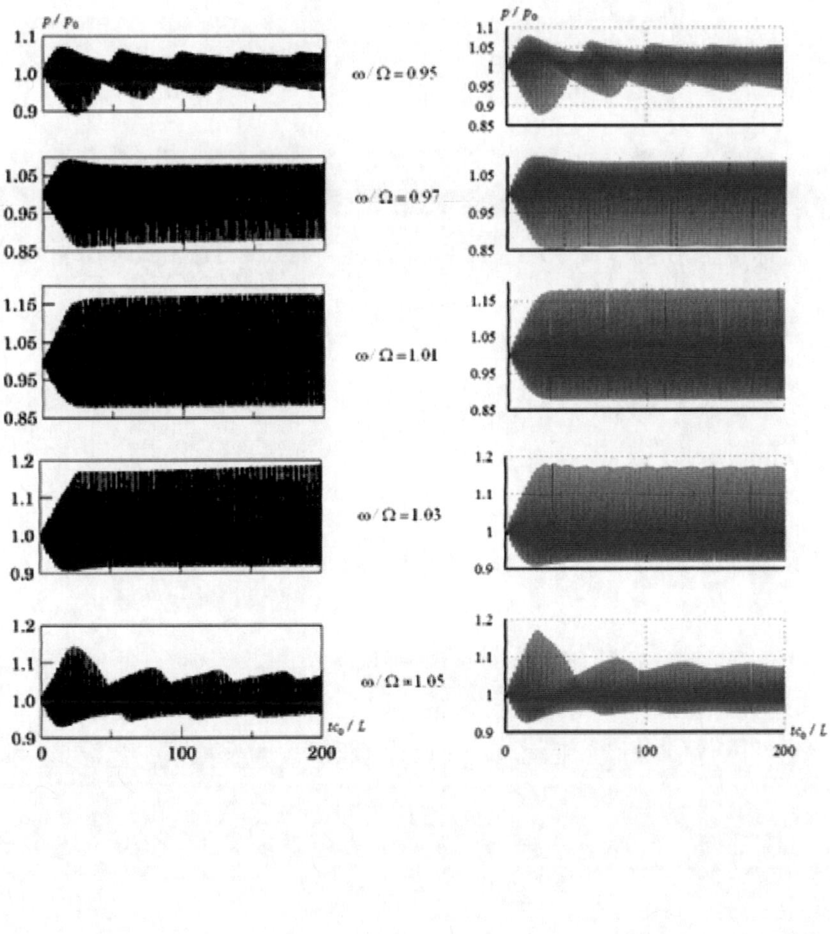

IOP Conf. Series: Journal of Physics: Conf. Series **1058** (2018) 012064 doi:10.1088/1742-6596/1058/1/012064

Figure 2. Pressure time dependencies for five frequencies in the neighborhood of resonant frequency Ω. The left-hand plots – work [2], right-hand plots – Lagrange's approach.

References

[1] Zarembo L K and Krasilnikov V A 1966 *Vvedenie v nelinejnuyu akustiku* (Moscow: "Nauka")
[2] Aganin A A, Ilgamov M A and Smirnova E T 1996 Development of longitudinal gas oscillations in a closed tube *Journal of Sound and Vibration* **195** 359-374
[3] Chester W Resonant oscillations in closed tubes 1964 *Journal of Fluid Mechanics*. **18** 44-64
[4] Saenger R A and Hudson G E 1960 Periodic shock waves in resonating gas columns *The Journal of the Acoustical Society of America* **32** 961-970

IOP Conf. Series: Journal of Physics: Conf. Series **1058** (2018) 012065 doi:10.1088/1742-6596/1058/1/012065

Investigation of coating thickness obtained by plasma electrolytic oxidation on aluminium alloys in electrolytes of type «KOH-H_3BO_3»

A Kosenko[1], A V Kolomeichenko[2], Yu A Kuznetsov[2], N V Titov[2], R Yu Solovyev[3], S N Sharifullin[4]

[1]Ariel University Center, Izrael
[2]Orel State Agrarian University named after N.V. Parakhin, Russia
[3]Federal Scientific Agroengineering Center of All-Russian Institute of Mechanization, Russia
[4]Kazan (Volga region) Federal University, Russia

E-mail: Saidchist@mail.ru

Abstract: Plasma Electrolytic Oxidation (PEO) is one of advanced methods of obtaining thin layer oxide ceramic coatings on machine elements of different functional purpose providing considerable increase their resource. The work is devoted to the investigation of coating thickness formed by Plasma Electrolytic Oxidation on different aluminium alloys in electrolyte of type «KOH-H3BO3». Plasma Electrolytic Oxidation is recommended to be implemented at different engineering and repair and technical enterprises, dealing with production, restoration and hardening of elements.

1. Introduction

The method of obtaining oxide coatings on aluminum products has been known for a long time [1]. This is chemical and electrochemical oxidation. Disadvantages of this technology: the use of such harmful acids as phosphoric, sulfuric and hydrochloric, the possibility of obtaining only decorative and slightly wear-resistant coatings of several micrometers in thickness. These and other shortcomings did not allow widespread use of this technology.

To date, existing traditional technologies do not allow the production of coatings with such physical and mechanical properties as high hardness, strength, wear resistance, corrosion resistance. It is these coatings that make it possible to increase the service life of both manufactured and restored parts and assemblies of technical products. Only highly efficient technologies based on the use of concentrated energy flows are able to realize this task. In this paper, one of the technologies for using concentrated energy flows is proposed to increase the life of parts made of aluminum alloys. This is the technology of hardening by the method of plasma electrolytic oxidation (PEO) or as it is sometimes called the micro arc oxidation (MDO) method.

Increasing the resource of parts by restoring them and hardening them is the main reserve of material saving, reducing costs for repairing machinery and one of the most important problems in conditions of rising raw material costs and energy resources. One of the promising ways, which allows to significantly increase the life of parts made of aluminum alloys (pistons of internal combustion engines, hydraulic pump casing, etc.) is plasma electrolytic oxidation (PEO). The essence of PEO is the formation of a thin-layer high-strength wear-resistant ceramic coating on the surface of the workpiece under the action of microplasma discharges, consisting primarily of solid-phase aluminum oxides [2-13].

The main advantages of PEO can be attributed [3-9, 14, 15]: the possibility of coatings on complex-shaped products, internal surfaces and hidden cavities; Obtaining coatings up to 0.3 mm thick with adhesion comparable to that of the base material; Cheapness and availability of chemical reagents and materials; Environmental safety.

At present, many aspects of the mechanism of the PEO flow remain insufficiently studied: there is no systematic information about the influence of internal and external factors on this process, many new electrolytes have not yet been tested.

Below are the results of studies of the wear resistance of coatings obtained by PEO on various aluminum alloys in the electrolyte of the "$KOH-H_3BO_3$" type.

2. Methods of conducting research

Aluminium alloys of trade marks AMg2 (1520), AD1 (1235) and D16 (2024) were considered as material for sample production. They are widely used for production of elements of different nomenclature in processing industry. Aluminium alloys chemical content is presented in Table 1. Plated samples of size 50x15x5 mm were used to carry out the investigations.

Table 1 – Mass fraction of elements in the alloy content (base – aluminium), %

	Mg	Si	Mn	Cu	Ti	Fe	Cr	Zn
AMg2 (1520)	1,7-2,4	0,4	0,1-0,5	0,15	0,15	0,5	0,05	0,15
AD1 (1235)	0,05	0,3	0,025	0,05	0,15	0,3	-	0,1
D16 (2024)	1,2-1,8	0,5	0,3-0,9	3,8-4,9	0,15	0,5	0,10	0,25

It is suggested that the aluminium alloys data application provides the realization of method of installation of the hardened extra repair element (ERE) at the restoration of elements produced from different materials (for example, made of different materials (for example, bronze, corrosion-resisting steel) and contacting with liquid food mediums (milk, fruit extracts, etc). The main task being solved in this case is obtaining of hardened oxide ceramic layers, formed inward of base of oxidized aluminium alloy.

To form coatings on above mentioned aluminium alloys electrolyte of type «$KOH-H_3BO_3$» with the following components concentration for liter of distilled water: KOH – 5 g/l, H_3BO_3 – 15-30 g/l was used.

The sample oxidation was done on the Plasma Electrolytic Oxidation installation functioning in anode-cathode mode. Current density was 10…30 A/dm^2; processing time – 120 min; electrolyte temperature was supported in the range of 30…35°C.

Coating thickness was determined on disks by means of device PMT-3.

3. Research results and discussion

The results of the researches showed that thickness of oxide ceramic coating obtained in electrolyte of type «$KOH-H_3BO_3$», depends on the modes of Plasma Electrolytic Oxidation and trade mark of aluminium alloy.

Figure 1 presents morphology and typical micro structure of coatings being formed by Plasma Electrolytic Oxidation on aluminium alloys.

Concentration KOH increase at other constant factors results in deeper penetration into alloy base, increasing thickness. Obviously it can be explained that for the sake of more intensive etching of the oxidized alloy surface, in coating formation the greater number of formed connections that had reacted with alkalis. Concentration KOH increase being more than 6 g/l does not sufficiently influence on coating thickness, because electrolyte aggressiveness increases. Obtained coatings turn out to be soft.

At KOH concentration being less than 2 g/l electrolyte has low dispersive capacity and obtaining of thick uniform coatings is difficult.

Boric acid concentration increase at other constant factors of Plasma Electrolytic Oxidation does not sufficiently influence on change of coating thickness. In our case boric acid is used as anode inhibitor, increasing break down strength. Boric acid concentration increase being more than 25 g/l is

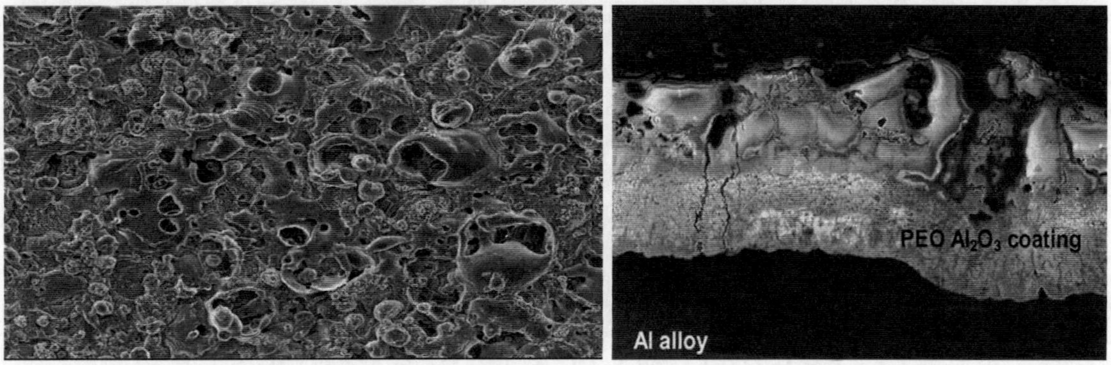

a) **b)**

Figure 1 – Morphology (a) and micro structure (b) of coatings being formed by plasma electrolytic oxidation. Magnification 600^X and 400^X

not rational from economic point of view. At boric acid concentration being less than 15 g/l to switch into Plasma Electrolytic Oxidation is not possible. The above mentioned is well coherent with the data of other scientists [3, 9].

Current density growth at other constant factors of Plasma Electrolytic Oxidation results in increase of hardened layer thickness (Figure 1). It provides more intensive process of Plasma Electrolytic Oxidation at the account of increase of microdischarges number on the oxidized surface. Nevertheless at current densities being more than 35 A/dm^2 micro plasma discharges are changed into arc ones. Deep craters and delaminations are observed on the surfaces of the oxidized samples, coating is destroyed.

Carrying out plasma electrolytic oxidation at low current densities is not rational from technological point of view, because to obtain the required thickness is necessary to have sufficient amount of time.

Coatings obtained on alloy AMg2, have maximum thickness, because magnesium in alloy with aluminium form uniform oxide connections (Figure 1). Coatings obtained on alloys AD1 and D16, do not differ sufficiently.

To maintain the main factors influence on coating thickness change and also taking into consideration their interaction the investigation was done according to the method of full factorial experiment planning 2^3.

Matrix of planning 2^3 and mean real values of optimization parameter are presented in Table 2.

Table 2 – Matrix of planning 2^3 and mean real values of optimization parameter h

Number of plan point	Code factor designations			Geometry parameters (process factors)			Mean real values of optimization parameter		
	X_1	X_2	X_3	D_t	CH_3BO_3	CKOH	h, μm АМг2	h, μm АД1	h, μm Д16
1	-	-	-	15	20	4	106,2	104,34	97,8
2	+	-	-	25	20	4	127,8	126	120,6
3	-	+	-	15	30	4	113,4	108	106,3
4	+	+	-	25	30	4	135	126	122,4
5	-	-	+	15	20	6	109,2	108	104,4

6	+	-	+	25	20	6	133,2	131,3	126
7	-	+	+	15	30	6	112,2	109,8	106,2
8	+	+	+	25	30	6	138,6	133,4	129,6

The experiment data processing, creation of mathematical model of process and check of its adequacy were done using PC by means of software product MS Excel. As the result the following regression equations were obtained:

For AMg2: $h = 121{,}95+11{,}7\ X_1+2{,}85\ X_2+1{,}75\ X_3+0{,}3\ X_1\ X_2\ X_3$; (1)

For AD1: $h = 118{,}4+10{,}81\ X_1+0{,}95\ X_2+2{,}26\ X_3+0{,}5\ X_1\ X_2\ X_3$; (2)

For D16: $h = 114{,}16+10{,}49\ X_1+1{,}96\ X_2+2{,}39\ X_3+1{,}06\ X_1\ X_2\ X_3$ (3)

Considering their ratios importance for obtaining the required thickness all three factors were controlled.

Then the experiment results were processed using the least square method by means of software product MS Excel. As the result the process mathematical models in the form of linear equations were obtained:

For AMg2: $h = 54{,}15+2{,}34\ D_t+0{,}57\ C\ H_3BO_3+1{,}35\ C\ KOH$; (4)

For AD1: $h = 59{,}04+2{,}16\ D_t+0{,}19\ C\ H_3BO_3+2{,}26\ C\ KOH$; (5)

For D16: $h = 50{,}45+2{,}09\ D_t+0{,}39\ C\ H_3BO_3+2{,}4\ C\ KOH$ (6)

For final check of the obtained models adequacy the results of coating thickness measurements obtained in the pilot experiment were used. Substituting into equation $D_t = 20\ A/dm^2$; $CH_3BO_3 = 25$ g/l; $C_{KOH} = 5$ g/l, we will get: for AMg2: h = 122 μm, for AD1: h = 118 μm, for D16: h = 114 μm. Comparison of these calculated values with experimental values points to the mathematical model adequacy.

4. Conclusions

Taking into account that the thickness of coating necessary to provide high wear resistance, corrosion resistance, low porosity must be no less than 90…120 μm, to harden surfaces of extra repair element (ERE) (for example, repair bushings) in electrolyte of type «KOH-H_3BO_3», it is possible to recommend the following:

- to produce repair bushings ERE one should use alloy AMg2;
- to use electrolyte of the following content, g/l: KOH – 4…6, H_3BO_3 – 20…25;
- the recommended electrolyte temperature 30…35 °C;
- the current density during the process one should maintain about 20…25 A/dm^2;
- the recommended duration of Plasma Electrolytic Oxidation – 110…130 min.

This method of elements surface hardening can be recommended for implementation at engineering and repair and technical enterprises, dealing with production, restoration and hardening of elements. The expected life growth of hardened elements will correspond to not less than 150-200% in relation to new elements.

References

[1] Averyanov E.E. Oxide characteristics obtained during the color protective-decorative anodizing of aluminum foil /E.E. Aver'yanov, G.F. Bakhteeva, D.Kh. Zainutdinova, R.A. Shagiskhmetov, S.N. Sharifullin //Chemical and Petroleum Engineering, 1992. – V. 27 (7-8). – Pp. 404 – 406.

[2] Kuznetsov Y. Studies on corrosion resistance of coatings formed by plasma electroltic oxidation on aluminum alloys / Y. Kuznetsov, A. Kossenko, A. Lugovskoy //Proceedings of the tenth Israeli-Russian Bi-National Workshop 2011 «The Optimization of Composition, Structure and Properties of Metals, Oxides, Composites, Nano- and Amorphous Materials». – Jerusalem, Israel, 2011. – (307). – Pp. 297 – 303.

[3] Suminov I.V. Plasma-electrolytic modification of the surface of metals and alloys. Suminov, P.N. Belkin, A.V. Epelfeld et al. - M .: Technosphere, 2011. - T. 2. - 512 p.
[4] Kolomeichenko A.V. Investigation of the corrosion resistance of articles made of aluminum alloys with protective coatings formed by plasma electrolytic oxidation. Kolomeichenko, N.S. Chernyshov, N.V. Titov, V.N. Logachev // Electronic processing of materials, 2016. - T. 52. - № 6. - Pp. 25 - 29.
[5] Minaev A.N. Composite coatings formed by plasma electrolytic oxidation. Minaev, S.V. Gnedenkov, S.L. Sinebryukhov et al. // Corrosion: materials, protection, 2011. - No. 3. - Pp. 1-10.
[6] Shatrov A.S. Protective oxide-ceramic coatings formed on aluminum alloys by the method of plasma electrolytic oxidation / AC. Shatrov // Corrosion: materials, protection, 2012. - № 4. - Pp. 28 - 37.
[7] Yegorkin VS Formation of protective coatings on aluminum and steel by the method of plasma electrolytic oxidation. Egorkin, A.V. Puz, OA Khrisanfova et al. // Bulletin of the Engineering School of the Far Eastern Federal University, 2011. - No. 1 (6). - Pp. 58 - 75.
[8] Dudareva N.Yu. Adhesion strength of a plasma electrolytic coating formed on a high-silicon alloy by microarc oxidation / N.Yu. Dudareva, R.V. Kalschikov, F.F. Musin et al. // Bulletin of the Irkutsk State Technical University. - 2014. - No. 10 (93). - Pp. 38 - 43.
[9] Korovin A.Ya. Technological bases of restoration and strengthening of details from aluminum alloys by microarc oxidation: monograph / A.Ya. Korovin, Yu.A. Kuznetsov. - The Eagle: publishing house Orel GAU, 2008. - 140 p.
[10] Kornienko E E, Lapushkina E J, Kuzmin V I, Vaschenko S P, Gulyaev I P, Kartaev E V, Sergachev D S, Kashapov N, Sharifullin S, Fayrushin I 2014 *J. Phys.: Conf. Ser.* **567** 012010
[11] Kashapov N F and Sharifullin S N 2015 *IOP Conf. Ser.: Mater. Sci. Eng.* **86** 012021
[12] Saifutdinov A I, Fairushin I I and Kashapov N F 2016 *JETP Lett.* **104** 180–185
[13] Kashapov L N, Kashapov N F and Kashapov R N 2014 *J. Phys.: Conf. Ser.* **567** 012025
[14] Vasilenko A.A. MAO coatings in electrolyte-suspension / AA. Vasilenko, D.S. Denisov, S.V. Gerasimov et al. // Trends in Science and Education in the Modern World, 2016. - No. 12-1. - Pp. 23 - 24.
[15] Kuchmin I.B. Changes in plasma formations in the process of microarc oxidation, their effect on coating characteristics. Kuchmin, G.G. Nechaev, N.D. Solovyov // Physics and Chemistry of Material Processing, 2015. - No. 4. - Pp. 45 - 49.

Numerical study of phase distribution phenomena and wall effects in bubbly two-phase flow

D A Gubaidullin[1], B A Snigerev[1]

[1]Institute of Mechanics and Engineering – Subdivision of the Federal State Budgetary Institution of Science «Kazan Scientific Center of the Russian Academy of Sciences», 2/31, Lobachevsky str, 420111, Kazan, Russia.

E-mail: snigerev@imm.knc.ru

Abstract. The results of numerical simulation of the structure of a two-phase flow of a gas–liquid bubble mixture in a vertical ascending flow in a pipe are presented. The mathematical model is based on the use of the Eulerian description of the mass and momentum conservation for the liquid and gas phases, recorded within the framework of the theory of interacting continua. To describe the bubble-size distribution, the equations of particle-number conservation for individual groups of bubbles with different constant sizes are used for each fraction, taking the processes of breakage and coalescence into account. Comparison of the results of numerical simulation with experimental data has shown that the proposed approach enables the simulation of bubble turbulent polydisperse flows in a wide range of gas concentrations.

1. Introduction

Two-phase flow refers to any fluid flow consisting of two phases, either of the same or different substances. Gas–liquid flows are by far the most complex due mainly to the fact that they combine a deformable interface and the compressibility of the gaseous phase [1-4]. This is probably the most important form of two-phase flow, and is found widely in a whole range of industrial applications. These include nuclear reactors, chemical reactors, pipeline systems for the transport of oil and gas, automotive industry, sewerage treatment plants etc. For a vertical gas–liquid two-phase flow in a pipe, the distribution of interfacial structure can take possible regimes such as bubbly flow, slug flow, churn-turbulent flow, wispy annular flow and annular flow. Thus, in order to capture and enhance the accuracy of code predictions of the transient evolution of these flow regimes, more detailed numerical methods able to model the dynamics of fluid particle interaction are necessary. In the present study, we will be concerned with isothermal bubbly flows, i.e. dispersed gas–liquid flow with the gas as the dispersed phase and liquid as the continuous phase where only momentum is exchanged between the phases. The system studied in [5] is vertical upflow of water and air in a round pipe with inner diameter D = 38.0mm. They performed experiments specifically designed to show the effects of bubble size by using a special gas injector that allows to adjust the bubble size independently of liquid and gas mass fluxes. All three parameters, average bubble size d_s, liquid mass flux $< J_l >$

and gas mass flux $< J_g >$ were varied in the investigation. Radial profiles of void-fraction α_g, bubble-size d_s, liquid velocity U_l and axial liquid turbulence intensity were measured at an axial position L/D = 60, so that fully developed flow conditions are realized. A change in the void fraction profile from wall to core peak with increasing bubble size was observed as well as turbulence suppression in the pipe center for combinations of high liquid and low gas mass flux which correspond to the smallest bubble sizes.

2. Governing equations

System the Reynolds-averaged Navier-Stokes equations to describe the dynamics of a turbulent mixture of liquid and air bubbles with regard to interfacial interaction has the form

$$\frac{\partial(\rho_k\alpha_k)}{\partial t}+\frac{\partial(\alpha_k\rho_k u_j)}{\partial x_j}=0,\quad \alpha_l+\alpha_g=1,\quad k=l,g \tag{1}$$

$$\frac{\partial(\alpha_k\rho_k u_{ki})}{\partial t}+\frac{\partial(\alpha_k\rho_k u_{ki}u_{kj})}{\partial x_j}=-\alpha_k\frac{\partial p}{\partial x_i}+\alpha_k\rho_k g_i+F_{ki}+ \tag{2}$$

$$\frac{\partial}{\partial x_j}\left[(\mu_k+\mu_{kt})\left(\frac{\partial u_{ki}}{\partial x_j}+\frac{\partial u_{kj}}{\partial x_i}-\frac{2}{3}\frac{\partial u_k}{\partial x_k}\delta_{ij}\right)\right]$$

Here x_i - cartesian coordinates,, u_i - components of velocity vector, t - time, p - pressure, ρ_l , ρ_g - density of continuous and dispersed phases respectively, α_l - is the volume fraction of liquid phase in mixture, δ_{ij} - is the Kronecker symbol, μ_l ,μ_g - dynamic coefficient of viscosity of water and air, μ_{kt} - is the coefficient of turbulent viscosity, F_{ki} - the strength of interfacial interaction forces. The turbulent viscosity for the carrier liquid phase are determined using a two-parameter turbulence model, modified for two-phase media [3,6]. In the equations for the transfer of turbulent kinetic energy and its dissipation introduces additional components to the kinetic energy caused by pulsations of the bubbles. The motion of the dispersed phase is determined by the forces of interfacial interaction. The main forces considered have the following components: Archimedes force, resistance force, joined the force, the rotational force of Magnus, the wall friction force [3]. To describe the distribution of bubble sizes in two-phase flow the equation for conservation of number of particles is written, taking into account the processes of coagulation and fragmentation. To solve the equation of conservation for the number of bubbles we use an approach based on the method of fractions [6]. Range of distribution of particle sizes is divided into a number of fractions with fixed boundaries, there can also be exchange of vesicles between different fractions as a result of coagulation and fragmentation. In this method, the distribution of bubble sizes is approximated by a piecewise uniform distribution, thus, the problem of describing the spectrum of droplet size is reduced to the solution of the equations for the volume concentrations of individual fractions.

3. Results

To verify the presented mathematical model, the experimental data of [7] were selected, where polydisperse upward bubble flow in a pipe was studied (pipe diameter D = 50.8 mm; height H = 3060 mm). Table presents the main parameters for different flow modes of a two-phase bubble medium in a vertical pipe ($<J_l>$ is the reduced velocity of the liquid phase at the pipe inlet, $<J_g>$ is the reduced velocity of the gas phase, and ϑ is the volume flow gas concentration). The Reynolds number for the liquid phase flow varies within $Re_l = \rho_l<J_l>D/\mu_l = (2.5–10.2) \times 10^4$.

The two-phase flow in a vertical pipe is assumed to be axisymmetric, so a design field consisting of a circular sector with a radius of $r_0 = D/2 = 2.54 \times 10^{-3}$ m, a length of H = 3.06 m, and a solution angle of 5° is selected for numerical simulation. Numerical calculations are performed for finite-volumes grids consisting of M_e = 32 600, 84 200, and 168 600 units. In the plane section of

ox_1x_2, the number of partitions along the axis of the pipe and along its length for different grids is $L_1 = 20 \times 200$, $L_2 = 40 \times 400$, and $L_3 = 60 \times 600$. The numerical solution for grid L_2 differs from that for L_3 by less than 2%; therefore, all the presented numerical results are obtained using a finite-volume grid with partitioning L_2. The following parameters of the components of the two-phase water–air medium

Table 1. The parameters of the experimental measurements ($d_s = 2.5$ mm)

	$<J_l>$, m/s	$<J_g>$, m/s	α_g	ϑ
L1	0.986	0.0473	0.051	0.053
L2	0.986	0.321	0.231	0.279

are set: $\rho_l = 998.2$ kg/m^3, $\rho_g = 1.2$ kg/m^3, $\mu_l = 1.1 \times 10^{-3}$ kg/(m s), and $\mu_g = 1.7 \times 10^{-5}$ kg/(m s). The particle-size distribution is presented in the form of six groups of bubbles with the minimum diameter of $d_1^{\min} = 1 \times 10^{-3}$ m and the maximum diameter of $d_6^{\max} = 1.2 \times 10^{-2}$ m. The calculated and experimental data on the distribution of the characteristics of the carrier and polydisperse flows in the developed turbulent flow in the section of $x_3/D = 53.6$ along the pipe radius are presented (r is the distance from the pipe axis). Figure 1 presents comparative data of the averaged profile of the axial component of the carrier phase velocity u_{3l} and velocity u_{3g} of the gas-liquid flow for $<J_f> = 0.986$ m/s and $<J_g> = 0.0473$ and 0.321 m/s (modes L1 and L2). The introduction of a small concentration of bubbles ($\alpha_g = 0.0473$) leads to the fact that the velocity profile of the liquid phase becomes practically flat in the flow core (Figure 1a, curve *1*). The effect of bubbles manifests itself in the fact that the velocity gradient increases near the wall. An increase in the velocity and concentration of bubbles ($\alpha_g = 0.321$) lowers the effect of the dispersed phase on the velocity profile. The gradients of the velocity

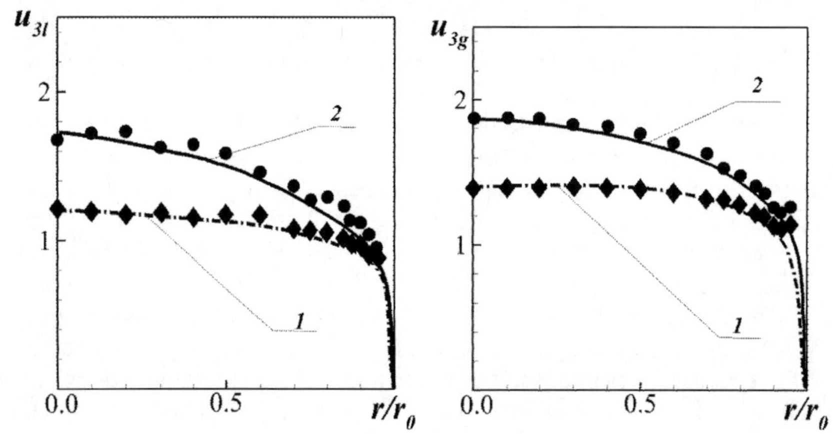

Figure 1. Comparison of (curves 1 and 2) calculated results and (symbols) experimental data obtained in the L1 and L2 modes for (a) u_{3L} ; (b) u_{3g}.

profile in the near-wall region decrease, and the profile itself becomes more parabolic, corresponding to a single-phase flow. At low flow velocities, a peak bubble concentration in the near-wall region and a uniformly low bubble concentration in the flow core at different gas concentrations are observed (Figure 2a). Similar results in the distribution diagram of α_g along the pipe radius are confirmed by the experimental data of many researchers. As an example, the saddle profiles of the gas concentration along the pipe section were found experimentally in the flow of an ascending gas-liquid flow [5].

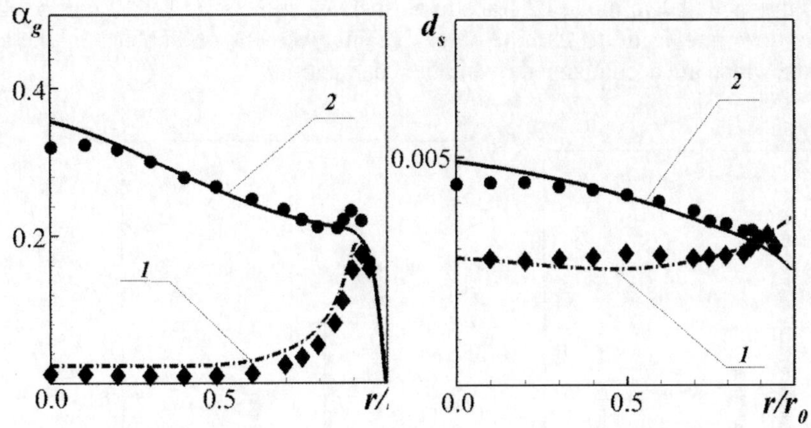

Figure 2. Comparison of (curves 1 and 2) calculated results and (symbols) experimental data obtained in the L1 and L2 modes for (a) α_g ; (b) d_s.

In these works, the peak distribution of the volume gas concentration in the near-wall region and the uniform flow in the core (called the "skin effect") were obtained in the flow of a bubble mixture in the pipe at Re = 6×10^3–6×10^4. The difference in the effect of interphase forces on bubbles of different sizes leads to the fact that the bubbles with a diameter that is smaller than the original one are displaced into the near-wall region. Note that the considered flow mode can be characterized as a flow in that the bulk of the bubbles move in the near-wall region. The distribution of the average volume–surface diameter d_s of bubbles along the pipe radius is shown in Figure 2b for two modes. At low fluid flow rates (curve *1*), a uniform distribution of the mean diameter of bubbles along the channel section is observed. In this case, there are a small number of bubbles per unit volume and their effect is small; the intensity of the coagulation and fragmentation processes is insignificant, and a certain equilibrium state takes place. The number of bubbles produced by merging is approximately the same as the number of particles that have disintegrated into smaller ones as a result of collision with other particles. The increase in the velocity and concentration of the gas phase leads to an increase of d_s up to 4 mm, which indicates the predominance of the intensity of coagulation of bubbles over the rate of their collapse. The motion of a gas--liquid flow with an initial inlet velocity of $\langle Jg \rangle = 0.321$ m/s is characterized by an intensification of coalescence and breakage processes; the action of transverse interfacial forces results in the redistribution of bubbles of different sizes along the channel section. Small bubbles move to the near-wall region and large bubbles accumulate in the center of the channel due to the action of interfacial forces. The flow structure at low concentrations and velocities of the gas phase, as characterized by the accumulation of bubbles at the walls, is reversed. At high velocities of the gas-liquid flow, the bulk of the bubbles move in the central region, while near the walls their concentration is negligible. As the velocity of the carrier and dispersed phases increases, the structure of the particle-size distribution changes. It is seen that the proportion of larger bubbles increases (with a size of > 5 mm) in the gas-liquid flow region. The distribution of d_s along the pipe radius for both modes is shown in Figure 2, indicating a change in the structure of a two-phase bubble flow for the L2 mode. The increase in the velocity of the liquid leads to the fact that the intensity of the collisions increases. It grows because of the higher concentration of bubbles per unit volume, causing a reduction in the distance between the bubbles to a size comparable to their diameters. Increasing the involvement of bubbles in turbulent pulsations enhances the intensity of coagulation, as evidenced by an increase in d_s in the core of the flow.

The particle-size distribution in section x_3/D = 53.6 is shown in Figure 3 in the form of a fraction of particles N_i of the corresponding diameter d_i from the total number of bubbles in the flow region N_s.

The transition from the mode L1 to mode L2 leads to an increase in $N_s = 0.62 \times 10^6$ up to 1.22×10^6. An increase in the flow rate leads to a more intensive involvement of bubbles in the turbulent pulsation movements, which are accompanied by bubble coagulation.

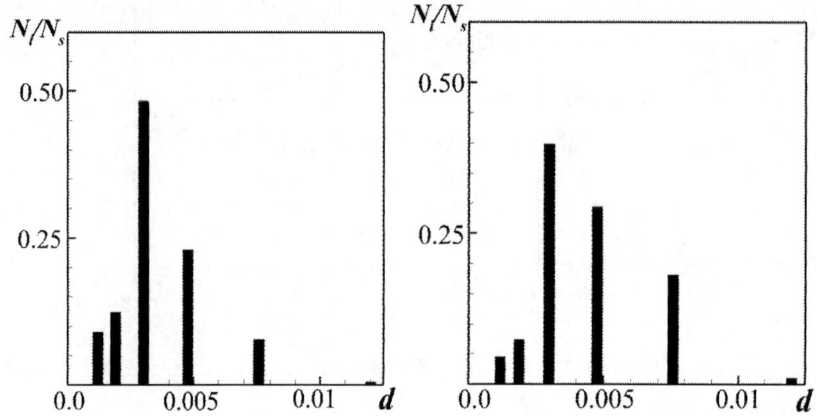

Figure 3. The particle-size distribution N_i/N_s for the different modes (a) L1; (b) L2.

4. Conclusion

An Eulerian two-fluid model is presented to describe the process of transfer of polydisperse bubbles in turbulent gas-liquid flows with significant gas concentrations of the gas phase. A modified two-parameter model of turbulence k–ε is used to simulate the turbulent flow of a liquid phase. The bubble-size distribution in a gas-liquid flow is taken into account on the basis of the partitioning of the entire particle system into separate groups (fractions) with fixed diameters, for each of which the equation of particle conservation is recorded, considering the processes of coalescence and breakage of bubbles. Based on a comparison of the results of a numerical model with the experimental data, the applicability of the proposed model for the numerical simulation of gas-liquid flows with high values of the gas concentration of the dispersed phase in a bubble upward flow in a vertical pipe is demonstrated.

References

[1] H A Jakobsen 2013 *Chemical reactor modelling. Multiphase reactive flows* (Springer)
[2] R I Nigmatulin 1987 *Dynamics of multiphase flow* (M.: Science)
[3] Ishii M, Hibiki T 2011 *Thermo-Fluid Dynamics of Two-Phase Flow* (Springer)
[4] Drew D A, Passman S L 1998 *Theory of Multicomponent Fluids* (Springer)
[5] Liu T J, Bankoff S G 1993 *Int. J. Heat Mass Transfer* **36** 1049-1079
[6] Gubaidullin D A, Snigerev B A 2018 *High Temperature* **56** 61-72
[7] Hibiki T, Ishii T, Xiao Z 2001 *Int. J. Heat Mass Transfer* **44** 1869-1882

Influence of spatial position of gas bubbles in liquid on their joint dynamics

A A Aganin and A I Davletshin

Institute of Mechanics and Engineering - Subdivision of the Federal State Budgetary Institution of Science «Kazan Scientific Center of the Russian Academy of Sciences», 2/31 Lobachevsky str., Kazan 420111, Russia

E-mail: anas.davletshin@gmail.com

Abstract. The influence of the mutual arrangement of initially identical spherical equidistant gas bubbles in liquid on their dynamics at the antinode pressure of an ultrasonic standing wave is studied. A mathematical model is used, in which small deformations of bubbles are taken into account. Two configurations of their relative position are considered: linear and planar. The first configuration is composed from three bubbles with their centers on one straight line, the second configuration is composed from five bubbles with their centers on two mutually orthogonal straight lines. It is shown that the magnitude of the deformations of the central bubbles in the linear and planar configurations is approximately the same.

1. Introduction

In the case of a relatively close arrangement of gas-vapor bubbles in liquid to each other, the hydrodynamic interaction between them can have a significant effect on their dynamics [1-4]. For example, the pressure acting on the bubbles in the center of the cluster can be much larger than at the periphery. On the other hand, the interaction between the bubbles can lead to their large deformations and destruction.

The interaction of bubbles may depend on various factors. In the present work, the influence of their mutual arrangement in the pressure antinode of an ultrasonic standing wave is studied. Two configurations of the mutual arrangement of bubbles are considered. In the first configuration, the bubble centers are on one straight line, and in the second configuration they are located on two mutually orthogonal straight lines. The dynamics of bubbles is described by a system of ordinary differential equations obtained by the method of spherical functions [3, 4] under the assumption that the liquid is weakly compressible, weakly viscous, and the bubbles are homobaric. In the equations, the possibility of strong radial oscillations of bubbles, their displacements in liquid, as well as their deformations, small in magnitude but arbitrary in shape, is taken into account.

2. Statement of the problem

We study the dependence of the dynamics of weakly-nonspherical gas bubbles on their mutual arrangement in liquid at the pressure antinode of a standing wave, where the pressure p_∞ varies according to the harmonic law

$$p_\infty = p_0 - p_a \sin \omega t .$$

Here p_0 is the static pressure of the liquid; p_a, ω are the amplitude and frequency of oscillations; t is time.

The considered configurations of the mutual arrangement of bubbles are illustrated in figure 1. In the first (linear) configuration, the bubble centers are on one straight line, whereas in the second (planar) configuration they are located on two mutually orthogonal straight lines. In the linear configuration, three equidistant bubbles are taken. The planar configuration is composed of five bubbles.

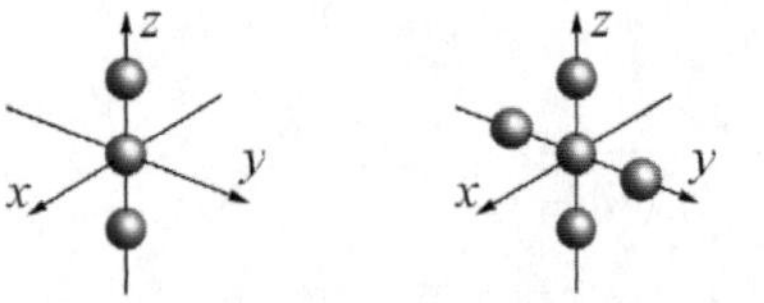

Figure 1. Linear (left) and planar (right) configurations of the mutual arrangement of bubbles at the beginning of their interaction.

Initially, all the bubbles are spherical, equal in radius, filled with air, and the peripheral equidistant from the central ones. The liquid is water at room conditions. The input parameters of the problem are the following: the sound velocity in liquid is $c_0 = 1500$ m/s, the liquid density is $\rho_0 = 998$ kg/m^3, the static liquid pressure is $p_0 = 1$ bar, the surface tension coefficient is $\sigma = 0.0725$ N/m, the dynamic viscosity of liquid is $\mu = 10^{-3}$ kg/(m s), the adiabatic index is $\kappa = 1.4$, the amplitude of pressure oscillations in the wave is $p_a = 1.2$ bar, the frequency of their oscillations is $\omega / 2\pi = 20$ kHz. At the initial time $t = 0$, the bubble radius is 3 μm, the distance between the centers of adjacent bubbles is $d_0 = 200$ μm.

3. Mathematical model

The equation of the surface of the k-th bubble ($k = 1, 2, ..., K$, K is the number of the bubbles) is

$$F_k(r_k,\theta_k,\varphi_k,t) = r_k - R_k(t) - \sum_{n=2}^{N} \mathbf{a}_{nk}(t)\cdot \mathbf{Y}_n(\theta_k,\varphi_k) = 0 .$$

Here r_k, θ_k, φ_k is the spherical coordinate system with the origin at the center of the k-th bubble; $\mathbf{Y}_n(\theta_k,\varphi_k) = (Y_{nk}^{-n}, Y_{nk}^{-n+1}, ..., Y_{nk}^{n-1}, Y_{nk}^{n})$, $Y_{nk}^{m} = P_n^{|m|}(\cos\theta_k)e^{im\varphi_k}$ is a spherical function; $P_n^{|m|}$ is the associated Legendre polynomial of degree n of order $|m|$; i is the imaginary unit; $\mathbf{a}_{nk} = \left((a_{nk}^{-n} + ia_{nk}^{n})/2, (a_{nk}^{-n+1} + ia_{nk}^{n-1})/2, ..., a_{nk}^{0}, ..., (a_{nk}^{-n+1} - ia_{nk}^{n-1})/2, (a_{nk}^{-n} - ia_{nk}^{n})/2\right)$, a_{nk}^{m} is the amplitude of the deviation of the surface of the k-th bubble from the spherical shape $r_k = R_k$ in the form of the surface harmonics with the number n and the order m; N is the maximum of the number of spherical harmonics composing the bubbles surface.

To describe the joint dynamics of the bubbles we use the following system of ordinary differential equations of second order in the bubble radii R_k, the radius vectors of their centers $\mathbf{p}_k$ and the vectors $\mathbf{a}_{nk}$, characterizing the deviation of the bubble surface from spherical,

$$R_k\ddot{R}_k + \frac{3\dot{R}_k^2}{2} - \frac{3\dot{\mathbf{p}}_k\cdot\boldsymbol{\beta}_{110}\cdot\dot{\mathbf{p}}_k}{8} + \frac{2\sigma}{\rho_0 R_k} - \frac{p_k - p_\infty}{\rho_0} + \psi_{0k} + \Delta_k = \sum_{j=1, j\neq k}^{K}\left[\frac{\dot{B}_{0j}}{d_{kj}} - \frac{B_{0j}\dot{d}_{kj}}{d_{kj}^2} - \frac{\left(R_j^3\dot{\mathbf{p}}_j\cdot\mathbf{C}_{10kj}\right)'}{2d_{kj}^2} - \right.$$
$$\left. - \frac{9B_{0j}\dot{\mathbf{p}}_k\cdot\boldsymbol{\beta}_{110}\cdot\mathbf{C}_{01kj}}{4d_{kj}^2} + \sum_{s=1, s\neq j}^{K}\frac{\left(R_j^3 B_{0s}\mathbf{C}_{01js}\cdot\mathbf{C}_{10kj}\right)'}{2d_{kj}^2 d_{js}^2} + \sum_{s=1, s\neq k}^{K}\frac{9B_{0j}B_{0s}\mathbf{C}_{01kj}\cdot\boldsymbol{\beta}_{110}\cdot\mathbf{C}_{01ks}}{8d_{kj}^2 d_{ks}^2}\right], \quad (1)$$

$$R_k\ddot{\mathbf{p}}_k + 3\dot{R}_k\dot{\mathbf{p}}_k - 2\dot{R}_k\dot{\mathbf{p}}_k\cdot(\boldsymbol{\varepsilon}_{2k}\cdot\boldsymbol{\beta}_{211}) - \frac{7R_k\ddot{\mathbf{p}}_k\cdot(\boldsymbol{\varepsilon}_{2k}\cdot\boldsymbol{\beta}_{211})}{6} - \frac{3\dot{\mathbf{p}}_k\cdot(\dot{\mathbf{a}}_{2k}\cdot\boldsymbol{\beta}_{211})}{2} + \boldsymbol{\psi}_{1k}$$
$$= \sum_{j=1, j\neq k}^{K}\left[\frac{3\left(R_k B_{0j}\mathbf{C}_{01kj}\right)'}{d_{kj}^2} - \frac{3\left(R_k R_j^3\dot{\mathbf{p}}_j\cdot\mathbf{C}_{11kj}\right)'}{2d_{kj}^3} - \frac{6R_k B_{0j}\mathbf{C}_{01kj}\dot{d}_{kj}}{d_{kj}^3} - \frac{5R_k B_{0j}\dot{\mathbf{p}}_k\cdot(\mathbf{C}_{02kj}\cdot\boldsymbol{\beta}_{211})}{d_{kj}^3} + \frac{3\left(B_{0j}\mathbf{a}_{2k}\cdot(\mathbf{C}_{01kj}\cdot\boldsymbol{\Upsilon}_{121}^2)\right)'}{2d_{kj}^2} - \frac{B_{0j}\left(\dot{\mathbf{a}}_{2k}+2\dot{R}_k\boldsymbol{\varepsilon}_{2k}\right)\cdot(\mathbf{C}_{01kj}\cdot\boldsymbol{\beta}_{121})}{d_{kj}^2}\right], \quad (2)$$

$$R_k\ddot{\mathbf{a}}_{mk} + 3\dot{R}_k\dot{\mathbf{a}}_{mk} - (m-1)\ddot{R}_k\mathbf{a}_{mk} - \delta_{2m}\frac{27\dot{\mathbf{p}}_k\cdot(\dot{\mathbf{p}}_k\cdot\boldsymbol{\beta}_{112})}{8} + \frac{3}{2}\sum_{n=m-1}^{m+1}\left(\dot{\mathbf{p}}_k\cdot(\dot{\mathbf{a}}_{nk}\cdot\boldsymbol{\Upsilon}_{n1m}^2) + \ddot{\mathbf{p}}_k\cdot(\mathbf{a}_{nk}\cdot\boldsymbol{\Upsilon}_{n1m}^{2(2-m)/3}) - \frac{(m+1)\dot{\mathbf{p}}_k\cdot\left((\dot{\mathbf{a}}_{nk}+2\dot{R}_k\boldsymbol{\varepsilon}_{nk})\cdot\boldsymbol{\beta}_{n1m}\right)}{n+1}\right) + \frac{(m^2-1)(m+2)\sigma\boldsymbol{\varepsilon}_{mk}}{\rho_0 R_k} + \boldsymbol{\psi}_{mk} = \sum_{j=1, j\neq k}^{K}\left[\delta_{2m}\left(-\frac{27B_{0j}\mathbf{C}_{01kj}\cdot(\dot{\mathbf{p}}_k\cdot\boldsymbol{\beta}_{112})}{4d_{kj}^2} + \frac{5\left(R_k^2B_{0j}\mathbf{C}_{02kj}\right)'}{d_{kj}^3} + \sum_{s=1, s\neq k}^{K}\frac{27B_{0j}B_{0s}\mathbf{C}_{01kj}\cdot(\mathbf{C}_{01ks}\cdot\boldsymbol{\beta}_{112})}{8d_{kj}^2 d_{ks}^2}\right) + \delta_{3m}\left(-\frac{10R_kB_{0j}\dot{\mathbf{p}}_k\cdot(\mathbf{C}_{02kj}\cdot\boldsymbol{\beta}_{213})}{d_{kj}^3} + \frac{7\left(R_k^3B_{0j}\mathbf{C}_{03kj}\right)'}{d_{kj}^4}\right) + \sum_{n=m-1}^{m+1}\left(\frac{3\left(B_{0j}\mathbf{a}_{nk}\cdot(\mathbf{C}_{01kj}\cdot\boldsymbol{\Upsilon}_{1nm}^2)\right)'}{2d_{kj}^2} - \frac{3(m+1)B_{0j}\left(\dot{\mathbf{a}}_{nk}+2\dot{R}_k\boldsymbol{\varepsilon}_{nk}\right)\cdot(\mathbf{C}_{01kj}\cdot\boldsymbol{\beta}_{1nm})}{2(n+1)d_{kj}^2}\right)\right]. \quad (3)$$

Here the dashes and the dots denote differentiation with respect to time, $m = 2, 3, \ldots, N$, $\mathbf{p}_k = \left((x_k + i\,y_k)/2, z_k, (x_k - i\,y_k)/2\right)$, x_k, y_k, z_k are the coordinates of the center of the k-th bubble, $\boldsymbol{\varepsilon}_{nk} = \mathbf{a}_{nk} / R_k$, d_{kj} is the distance between the centers of the k-th and j-th bubbles, p_k is the gas pressure in the k-th bubble, ρ_0 is the density of the liquid, σ is the surface tension coefficient, ψ_{0k} $\boldsymbol{\psi}_{1k}$ $\boldsymbol{\psi}_{mk}$ Δ_k are corrections for of the effects of the liquid viscosity and compressibility [4], δ_{nm} is the Kronecker symbol; $B_{0k} = -R_k^2\dot{R}_k$; $\mathbf{C}_{\gamma\varsigma kj} = \left(C_{\gamma\varsigma kj}^{\zeta\xi}\right)$, $\boldsymbol{\alpha}_{\gamma\varsigma\varrho} = \left(\alpha_{\gamma\varsigma\varrho}^{\zeta\xi\vartheta}\right)$ are the matrices of numbers, $\zeta = -\gamma, -\gamma+1, \ldots, \gamma$, $\xi = -\varsigma, -\varsigma+1, \ldots, \varsigma$, $\vartheta = \varrho, \varrho-1, \ldots, -\varrho$, $C_{\gamma\varsigma kj}^{\zeta\xi} = Y_{\gamma+\varsigma}^{\zeta-\xi}\left(\theta_{kj}, \varphi_{kj}\right)(-1)^{(|\zeta|+|\xi|-|\zeta-\xi|)/2+\gamma}\left(\gamma+\varsigma-|\zeta-\xi|\right)!/\left(\gamma-|\zeta|\right)!\left(\varsigma+|\xi|\right)!$, θ_{kj}, φ_{kj} are the angular coordinates of the center of the j-th bubble in a spherical system with the origin at the center of the k-th bubble, $\alpha_{\gamma\varsigma\varrho}^{\zeta\xi\vartheta} = \frac{2\varrho+1}{4\pi}\frac{\left(\varrho-|\vartheta|\right)!}{\left(\varrho+|\vartheta|\right)!}\int_0^{2\pi}\int_0^{\pi}\sin\theta\, Y_\gamma^\zeta(\theta,\varphi)Y_\varsigma^\xi(\theta,\varphi)Y_\varrho^\vartheta(\theta,\varphi)\,d\theta\, d\varphi$, $\boldsymbol{\Theta}_{\gamma\varsigma\varrho}^p = p\boldsymbol{\alpha}_{\gamma\varsigma\varrho} + \boldsymbol{\beta}_{\gamma\varsigma\varrho}$, $\boldsymbol{\Upsilon}_{\gamma\varsigma\varrho}^p = p\boldsymbol{\alpha}_{\gamma\varsigma\varrho} - \boldsymbol{\beta}_{\gamma\varsigma\varrho}$, $\boldsymbol{\beta}_{\gamma\varsigma\varrho} = [\gamma(\gamma+1) + \varsigma(\varsigma+1) - \varrho(\varrho+1)]\boldsymbol{\alpha}_{\gamma\varsigma\varrho}/2$. Equations (1)-(3) are obtained of by the method of spherical functions [3, 4] under assumption that the bubbles are not very close to each other, and their nonsphericity and spatial displacement velocities are small. Moreover, equations (1)-(3) take into account the deformations of the central bubbles only in the form of the second harmonic.

4. Results of calculations

Figure 2 shows the change in the radius R_c of the central bubble in the linear configuration of three bubbles during one oscillation period of the liquid pressure p_∞. It can be seen that the bubble experiences strong radial expansion and contraction. Thus, at the stage of lowering the pressure of the liquid, it expands almost fourfold. The radii of other bubbles in this configuration, as well as the radii of all the bubbles in the planar configuration, vary approximately by the same law.

As a result of the interaction between the bubbles the surfaces of the central and other bubbles are deformed. Moreover, in all the configurations the deformations of the motionless central bubble appear lower than the deformations of the neighboring bubbles displacing to the central one. Figure 3 shows the changes of the mean-square dimensionless amplitude $\left|\varepsilon_{2c}\right| = \sqrt{0.2\sum_{m=-2}^{2}\left(\varepsilon_{2c}^{m}\right)^{2}}$ of the deformations of the surface of the central bubbles in the form of the second harmonic ($\varepsilon_{2c}^{m} = a_{2c}^{m} / R_{c}$) during one period of the liquid pressure oscillations for the both configurations under consideration. It can be seen that the maximum deformations of the central bubbles arise in the vicinity of the intense radial pulsations, while at the beginning and at the end of the period the bubbles are practically spherical. At that, the mean-square amplitude of deformations of the central bubbles in the linear and flat configurations change approximately identically.

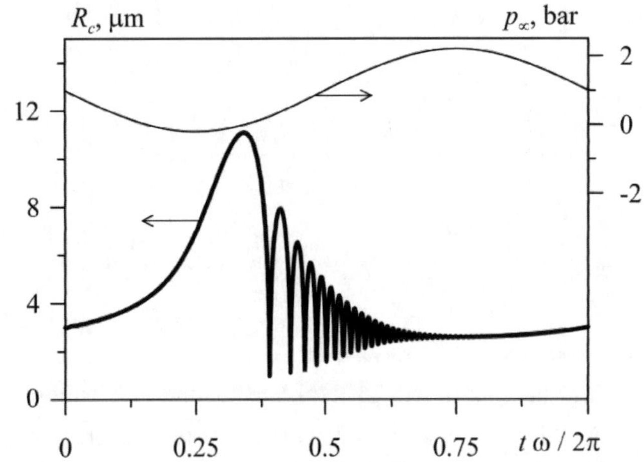

Figure 2. Change in the radius R_c of central bubbles (thick curve) during one period of the liquid pressure oscillations p_∞ (thin curve).

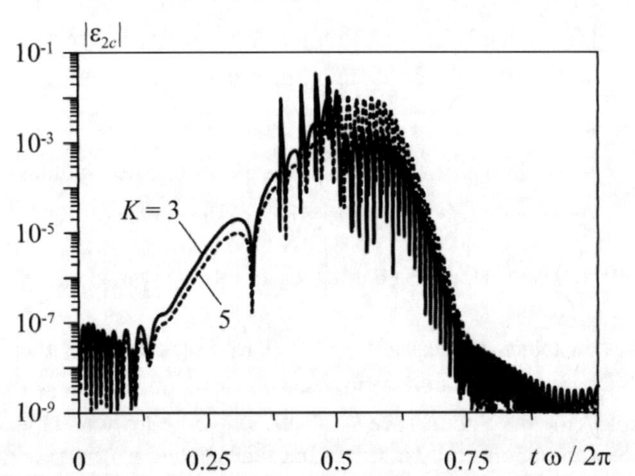

Figure 3. Change in the mean-square amplitude of the second harmonic, which determines the nonsphericity of the central bubbles in the configurations in question (figure 1).

References

[1] Doinikov A.A. 2004 *JASA* **116** no 2 pp 821–827
[2] Gubajdullin A.A. and Gubkin A.S. 2013 *Tyumen State University Herald. Physical and Mathematical Modeling. Oil, Gas, Energy* **7** pp 91–97

[3] Aganin A A and Davletshin A I 2009 *Matematicheskoe modelirovanie* **21** no 6 pp 89–102
[4] Aganin A A and Davletshin A I 2016 *Proceedings of the RAS Ufa Scientific Centre* **4** pp 9-13

IOP Conf. Series: Journal of Physics: Conf. Series **1058** (2018) 012068 doi:10.1088/1742-6596/1058/1/012068

Strong compression of bubbles in water, acetone and benzol

A A Aganin, D Yu Toporkov

Institute of Mechanics and Engineering - Subdivision of the Federal State Budgetary Institution of Science "Kazan Scientific Center of the Russian Academy of Sciences" (IME - Subdivision of FIC KazanSC of RAS) 2/31, Lobachevsky str., Kazan 420111 Russia

E-mail: top.dmtr@gmail.com

Abstract. Strong compression of the medium in cavitation bubbles in water, acetone and benzol realizing in the final stage of their collapse in the initially motionless liquid at a temperature of 293 K is considered. At the beginning of the collapse, the radius of the bubbles is 500 μm, the vapor in the bubbles is in its saturation state at the temperature of the surrounding liquid. The pressure of water and benzol is varied in the range $1 \leq p_0 \leq 100$ bar to determine its values at which the degree of compression of the medium inside the cavitation bubbles in these liquids is close to that achieved inside the cavitation bubble in acetone at the pressure $p_0 = 15$ bar. It is shown that in the case of the bubble in benzol such a degree of the vapor compression can be achieved at $p_0 = 4$ bar, whereas in the case of the bubble in water, even $p_0 = 100$ bar is not enough.

1. Introduction

One important feature of bubble dynamics in a liquid is the possibility of reaching high densities, pressures and temperatures in the final stage of the bubble collapse [1, 2]. The strong compression of the vapor-gas medium in a bubble can be gained by nearly homogeneous compression, by radially-converging isentropic (shockless) waves and by radially-converging shock waves [2, 3]. The greatest effect is achieved with compression by shock waves [2]. In particular, it was shown in [4] that the temperatures of the order of 10^7 K, the densities of the order of 1 g / cm^3 can be obtained at a liquid pressure of 15 bar and its temperature of 273 K in a small central region (200 nm in diameter) of the collapsing cavitation bubble in deuterated acetone.

This paper investigates the possibility of realizing the cavitation bubble collapse in water and benzol with attaining the same high degree of the vapor compression inside the bubble that is achieved in the case of the bubble in acetone collapsing at the room temperature (293 K) and the 15 bar pressure. At that, the room temperature of water and benzol is also taken, and their pressure is varied in the range of $1 \div 100$ bar so that the maximum values of the temperature and density inside their cavitation bubbles are close to those achieved in the case of the bubble in acetone. Water and benzol are considered due to the fact that they are widely used in various technologies. According to the criterion of [5], for producing shock waves inside a cavitation bubble, water (benzol) is less (more) suitable than acetone.

IOP Conf. Series: Journal of Physics: Conf. Series **1058** (2018) 012068 doi:10.1088/1742-6596/1058/1/012068

2. Problem statement

Collapse of a single cavitation bubble in an unlimited volume of liquid (acetone, water and benzol) is considered. At the beginning of the collapse (at time $t = 0$) the bubble is at rest, its radius is $R = R_0 = 500$ μm, the vapor and liquid temperature is $T = T_0 = 293$ K, the liquid pressure far from the bubble is $p = p_0$, the vapor pressure is equal to the saturation pressure p_S at temperature T_0. The pressure of water and benzol is varied in the range $1 \le p_0 \le 100$ bar so that the achieved compression degree of the medium inside the cavitation bubbles in these liquids is close to the degree achieved inside the cavitation bubble in acetone at $p_0 = 15$ bar. The initial bubble radius $R_0 = 500$ μm and the compressive liquid pressure $p_0 = 15$ bar are taken in accordance with the estimates [4] of the experimental data on bubble supercompression [1]. The main characteristics of the vapor compression degree inside a bubble are the levels of the maximum temperature and density values at the boundary of a small central region of the bubble with $r \le 0.25$ μm (the central super-hot core). This small area is excluded from consideration because the model of the present work is not adequate in this area since it does not take into account the effects of dissociation, ionization, nonsphericity, etc. However, these characteristics seem acceptable for the purposes of the present work.

The dynamics of the vapor and liquid is described by the equations [4, 6]:

$$\begin{aligned} &\frac{\partial}{\partial t}(\rho r^2)+\frac{\partial}{\partial r}(\rho w r^2)=0,\\ &\frac{\partial}{\partial t}(\rho w r^2)+\frac{\partial}{\partial r}(\rho w^2 r^2+pr^2)=2pr,\\ &\frac{\partial}{\partial t}(\rho e r^2)+\frac{\partial}{\partial r}[wr^2(\rho e+p)]=\frac{\partial}{\partial r}\left(r^2\kappa\frac{\partial T}{\partial r}\right). \end{aligned} \tag{1}$$

Here ρ is the density, w is the radial velocity, $e = U + w^2/2$ is the specific total energy, U is the specific internal energy, and κ is the thermal conductivity. The boundary conditions have the form [4, 6]:

$$\begin{aligned} &r\to\infty:\quad p=p_0,\quad T=T_0;\\ &r=R(t):\quad \dot R=w_l+\frac{j}{\rho_l}=w_g+\frac{j}{\rho_g},\quad p_l=p_g-\frac{4\mu_l w_l}{R}-\frac{2\sigma}{R},\\ &\kappa_l\left(\frac{\partial T}{\partial r}\right)_l-\kappa_g\left(\frac{\partial T}{\partial r}\right)_g=jl(p_g),\quad T_l=T_g. \end{aligned} \tag{2}$$

Here μ_l is the dynamic viscosity of the liquid, σ is the surface tension, l is the heat of vaporization, j is the evaporation and condensation velocity per unit surface area (determined by the Hertz-Knudsen-Langmuir formula [6]). The subscripts l and g indicate the correspondence to the liquid and the vapor, respectively.

The equations of the liquid and vapor states are taken as sums of the potential ($p^{(p)}$, $U^{(p)}$) and thermal ($p^{(T)}$, $U^{(T)}$) components of the pressure p and the internal energy U with the addition of a constant $U^{(ch)}$ in the equation for U

$$p(\rho,T)=p^{(p)}(\rho)+p^{(T)}(\rho,T),\quad U(\rho,T)=U^{(p)}(\rho)+U^{(T)}(T)+U^{(ch)}. \tag{3}$$

The density functions $p^{(p)}$, $U^{(p)}$ are presented in the form of the generalized Born-Mayer potential, the functions $p^{(T)}$ and $U^{(T)}$ are expressed as $p^{(T)}(\rho, T) = \rho\,\Gamma(\rho)\,U^{(T)}$, $U^{(T)}(T) = c_V T$ where Γ is the Grüneisen coefficient. The heat capacities c_V of the liquid and vapor are assumed to be constant. The model parameters for acetone are taken from [7], for water from [8] and for benzol from [9].

IOP Conf. Series: Journal of Physics: Conf. Series **1058** (2018) 012068 doi:10.1088/1742-6596/1058/1/012068

3. Results

A number of numerical experiments have been performed with various water and benzol pressure values in the range $1 \le p_0 \le 100$ bar, keeping the liquid temperature equal to 293 K. The results of those experiments have shown that to attain the vapor compression degree nearly as high as that achieved inside the cavitation bubble in acetone at $p_0 = 15$ bar, it is sufficient to take $p_0 = 4$ bar in the case of the bubble in benzol, and even $p_0 = 100$ bar is not enough in the case of the bubble in water.

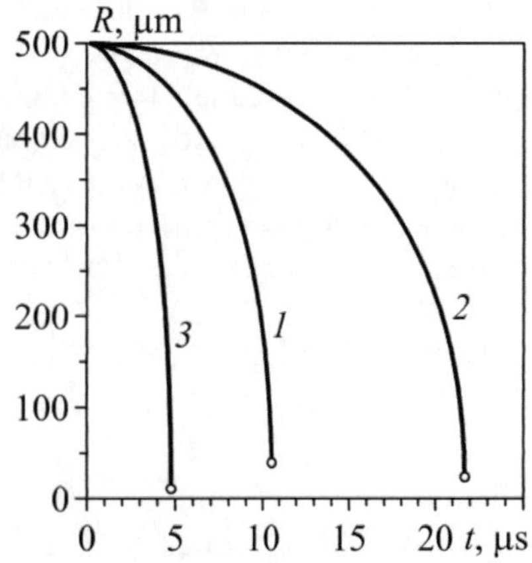

Figure 1. Variation of the radius R of the bubble during its collapse in acetone at $p_0 = 15$ bar (curve 1), in benzol at $p_0 = 4$ bar (curve 2) and in water at $p_0 = 100$ bar (curve 3). The circles indicate the minimum values of the bubble radii.

Figure 1 illustrates the change of the radii of the cavitation bubbles in acetone, water and benzol during their collapse at the pressure of acetone $p_0 = 15$ bar, water $p_0 = 100$ bar and benzol $p_0 = 4$ bar. It can be seen that duration of the bubble collapse in these three cases is very different. It is equal to 10.5 μs in the case of acetone, 4.8 μs in the case of water and 21.7 μs in the case of benzol. In the cases of water and benzol the bubble radii corresponding to the maximum values of the thermodynamic parameters attained inside the bubble are respectively 11 and 24 μm, which is significantly less than the similar radius (39 μm) in the case of acetone. The maximum velocities of collapse of the bubbles in acetone and benzol are approximately the same (about 600 m/s), whereas in the case of water it is much larger (3300 m/s).

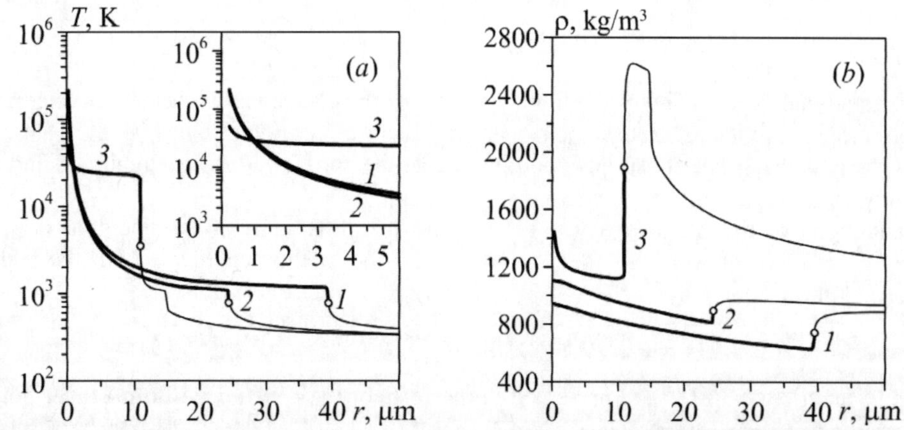

Figure 2. Radial distributions of the temperature (*a*) and density (*b*) in the cavitation bubble (at $r \ge 0.25$ μm) and the surrounding liquid in the cases of: acetone at $p_0 = 15$ bar (curves 1), benzol at $p_0 = 4$ bar (curves 2) and water at $p_0 = 100$ bar (curves 3) at the moment of extreme compression of the medium in the bubble. The circles indicate the values of the parameters on the bubble surface.

Figure 2 shows the spatial distributions of the temperature and density at the moment of reaching the maximum values of the thermodynamic parameters in the bubbles at the pressure of acetone $p_0 = 15$ bar, benzol $p_0 = 4$ bar and water $p_0 = 100$ bar. In all the three cases, the maximum values are reached when the shock waves are focused in the central region of the bubble. The case of the bubble collapse in water differs significantly from the others by that in the case of water the converging shock wave turns out to be initially very weak. After focusing at the bubble center and reflection from it, it transforms into a divergent one, which then reaches the bubble surface. As a result of their interaction, two waves appear: a shock wave converging toward the bubble center and a divergent compression wave propagating in the liquid. At the moment of focusing of this second convergent shock wave in the central region of the bubble, the thermodynamic parameters reach their maximum values. In contrast, during the collapse of the bubbles in acetone and benzol, the initially-arising convergent shock wave is sufficiently intense, so that its focusing in the central region of the bubble leads to the attainment of the maximum values of the thermodynamic parameters.

One can see in Figure 2 that in the case of water at $p_0 = 100$ bar, the temperature in the central region of the bubble is noticeably lower than in the cases of acetone at $p_0 = 15$ bar and benzol at $p_0 = 4$ bar. This is explained by the fact that, in comparison with water, acetone and, especially, benzol are much more suitable for production of convergent shock waves in collapsing cavitation bubbles. For example, a quite intense shock waves arise in the cases of acetone and benzol, starting from $p_0 \approx 3$-4 and 0.5 bar, respectively. In contrast, in the case of water, an increase of p_0 up to 40-50 bar leads to only an increase of the intensity of the convergent isentropic (shockless) compression wave arising in the bubble. No shock waves appear in this case. As a result, in the case of water, a substantially different density variation takes place. For example, with increasing p_0 to 100 bar the water vapor density maximum reaches values of the order of 1450 kg/m^3. In contrast, in the cases of acetone and benzol, due to the fact that the density increase on the shock wave is limited, the acetone and benzol vapor density maxima reach values of 940 kg/m^3 and 1100 kg/m^3, respectively. Moreover, these maxima are nearly constant in the range $3 < p_0 \leq 50$ bar in the case of acetone and in the range $0.5 < p_0 \leq 50$ bar in the case of benzol.

References

[1] Taleyarkhan R P, West C D, Cho J S, Lahey R T (Jr), Nigmatulin R I and Block R C 2002 *Science* **295** 1868
[2] Nigmatulin R I, Lahey R T (Jr), Taleyarkhan R P, West C D and Block R C 2014 *Physics-Uspekhi* **57** (9) 947
[3] Shaw S J and Spelt P D M 2010 *J. Fluid Mech.* **646** 363
[4] Nigmatulin R I, Akhatov I Sh, Topolnikov A S, Bolotnova R Kh, Vakhitova N K, Lahey R T (Jr) and Taleyarkhan R P 2005 *Phys. Fluids* **17** 107106
[5] Nigmatulin R I, Aganin A A, Toporkov D Yu and Ilgamov M A 2014 *Dokl. Phys.* **59** 431
[6] Nigmatulin R I 1990 *Dynamics of Multiphase Media* Vol. 1, 2. (Hemisphere, New York)
[7] Nigmatulin R I and Bolotnova R Kh 2007 *Dokl. Phys.* **52** (8) 442
[8] Nigmatulin R I and Bolotnova R Kh 2011 *High Temp.* **49** (2) 303
[9] Nigmatulin R I and Bolotnova R Kh 2017 *High Temp.* **55** (2) 199

IOP Conf. Series: Journal of Physics: Conf. Series **1058** (2018) 012069 doi:10.1088/1742-6596/1058/1/012069

The evolution of a gas bubble in a liquid near a flat wall

A A Aganin, L A Kosolapova, V G Malakhov

Institute of Mechanics and Engineering - Subdivision of the Federal State Budgetary Institution of Science "Kazan Scientific Center of the Russian Academy of Sciences" (IME - Subdivision of FIC KazanSC of RAS) 2/31, Lobachevsky str., Kazan 420111 Russia

E-mail: aganin54@mail.ru

Abstract. The axisymmetric dynamics of a cavitation bubble in a liquid near a flat solid wall is studied. The liquid is considered to be ideal, incompressible, its flow is potential, the gas in the bubble is homobaric. The Euler method is used for determining the position of the bubble surface and the velocity potential on it at every time moment and the boundary elements method is applied for determining the normal velocity component. The possibility of transforming an initially one-connected bubble into a toroidal one is taken into account. The process of expansion and subsequent contraction of a bubble in water under the influence of the excessive internal pressure in it under room conditions is considered. The shape of the bubble surface, the velocity and pressure fields in the surrounding liquid are determined. The results of calculations are given for the case when the ratio of the initial distance from the center of the bubble to the wall to the bubble maximum radius is 0.6, and also the results of comparing the calculations with known experimental data are given.

1. Introduction

Investigation of bubble dynamics near a wall is important for estimating the effect of cavitation on the surfaces of solids. A large number of publications are devoted to this topic, among which there are many papers in which the effect of the initial distance h_0 from the bubble center to the wall on the dynamics of a bubble during its expansion and subsequent compression is studied [1-3]. In most papers, the results are given for $\gamma = h_0 / R_{max} \geq 0.7$ [2, 3], where R_{max} is the maximum bubble radius reaching by the bubble in the course of its expansion in an unbounded volume of liquid At lower values of γ at the end of the expansion phase, a thin layer of liquid between the wall and the bubble surface is formed, which also persists on the toroidal phase of motion [4, 5]. The presence of such a layer complicates the implementation of numerical algorithms. Therefore, in a number of articles a thin layer of liquid is artificially eliminated by considering the bubble in contact with the wall [6, 7]. Under the assumption of the potentiality of the liquid flow, the most common technique used in calculations is to combine the time-step method with the boundary element method [1, 8-10]. The numerical instability of such an approach is eliminated by various smoothing procedures applied to the shape of the bubble surface and the potential values on it [11, 12]. In this paper, we present the results of calculations for the low-studied case $\gamma = 0.6$, obtained using the technique [8, 12].

2. Problem statement and computational technique

The process of expansion and subsequent contraction of a cavitation bubble in water near a flat solid wall under room conditions is considered (the liquid density is $\rho = 1000$ kg / m^3, the liquid pressure at a large distance from the bubble is $p_\infty = 1$ bar). At the moment of maximum expansion, the volume of the bubble V is equal to the volume of a sphere of radius $R_{max} = 1.45$ mm. It is assumed that at the initial time $t = 0$ the bubble has the form of a sphere of a small radius $R_0 = 0.25$ mm the center of which is distant from the wall by a distance $h_0 = 0.6R_{max}$, the pressure in the bubble is $p_{b0} = 88$ bar. The liquid is considered to be an ideal incompressible, its motion is potential, the surface tension is not taken into account. The pressure in the bubble varies according to the law

$$p_b = p_{b0}(V_0 / V)^\kappa,$$

where V_0 is the initial volume of the bubble, κ is the adiabatic exponent ($\kappa = 1.4$).

For the given data, two phases are distinguished in the evolution of the bubble at the wall. The first phase includes the expansion stage of the bubble and a part of the subsequent stage of its contraction up to the instant $t = t_c$, when the cumulative jet arising on the bubble surface impacts the layer of liquid between the bubble and the wall. The second phase includes the motion of the toroidal bubble formed as a result of the impact of the jet onto the liquid layer.

In accordance with this, the numerical solution algorithm is also divided into two stages. A detailed description of the calculation technique used in this paper is presented in [12]. It should be noted, that in [12], in contrast to other authors, smoothing by a cubic spline is used to eliminate the numerical instability of the functions that determine the bubble contour and the velocity potential value on it.

The following dimensionless quantities are utilized when presenting the results: $p^* = p / p_\infty$, $v^* = v / (p_\infty / \rho)^{1/2}$, $r^* = r / R_{max}$, $z^* = z / R_{max}$, $t^* = t / [R_{max}(\rho / p_\infty)^{1/2}]$, where p is the pressure, v is the velocity value, r, z are the radial and axial coordinates of the axisymmetric reference system with the origin on the wall and with the z axis passing orthogonally to the wall through the center of the bubble.

3. Results

To evaluate the reliability of the obtained results, a comparison is made with the experimental data [13] at four time moments both at the expansion stage and the stage of the subsequent contraction. The results of the comparison are shown in figure 1, where the upper row (*a*) presents the experimental data, the lower row (*b*) presents the numerical results. One can see their good qualitative agreement.

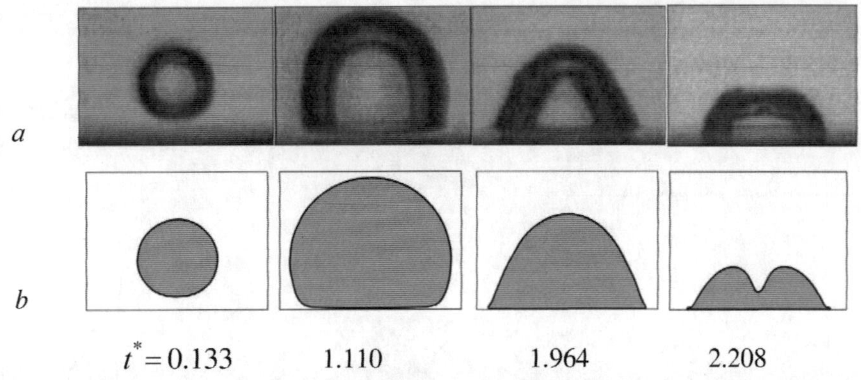

Figure 1. Comparison of the numerical results (*b*) of the present work with the experimental data [13] (*a*).

The results of calculations of the bubble motion in its first phase, *i.e.* prior to the moment t_c^* corresponding to the impact of the cumulative jet onto the thin liquid layer, are shown in figure 2. Figure 2*a* illustrates the change in the shape of the bubble surface during its expansion and contraction. In this figure, the first moment corresponds to the beginning of the process, the second moment corresponds to the maximum expansion of the bubble, the third and fourth moments correspond to the collapse with the formation of the cumulative jet and the fifth moment corresponds to the instant of the jet impact onto the thin liquid layer between the bubble and the wall. It can be seen that the part of the bubble surface near the wall at the moment of the maximum expansion of the bubble and then during its subsequent contraction is nearly flat. At that, the thickness of the liquid layer between the bubble and the wall at the moment t^*_c is approximately equal to $0.006R_{\max}$.

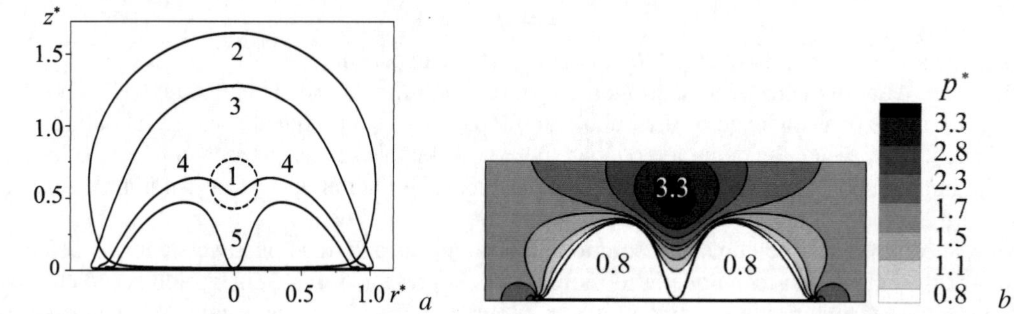

Figure 2. Bubble contours at five consecutive moments of time: 1 corresponds to $t^* = 0$, 2 to $t^* = 1.18$, 3 to $t^* = 1.89$, 4 to $t^* = 2.17$, 5 to $t^* = t^*_c = 2.24$ (*a*) and the pressure field in the liquid surrounding the bubble at the moment t^*_c (*b*)

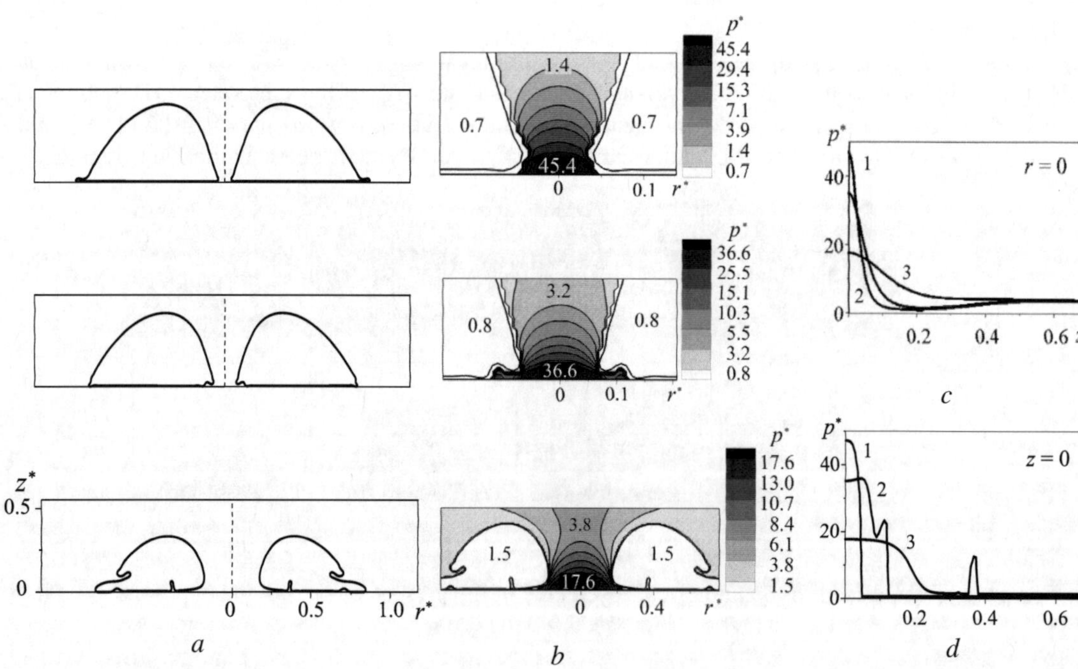

Figure 3. The contour of the bubble at the toroidal phase of motion (*a*), the pressure field in the liquid in the area of contact between the jet and the layer (*b*), the pressure profiles along the symmetry axis (*c*) and the radial pressure profiles on the wall (*d*) at three moments of time: 1 corresponds to $t^* - t^*_c = 0.0034$, 2 to $t^* - t^*_c = 0.0136$, 3 to $t^* - t^*_c = 0.069$

Figure 2*b* shows the pressure field in the liquid surrounding the bubble at the end of the first phase. It is seen that the pressure in the vicinity of the jet end is approximately equal to the pressure in the bubble $p^* = 0.8$, and the largest pressure $p^* = 3.3$ is observed at the base of the jet. The jet end velocity in the final of the first phase is $v^* = 7.04$.

The results of calculations on the toroidal phase of motion are shown in figure 3. It can be seen (figure 3*a*), that at the beginning of this phase the part of the bubble surface near the wall, as in the first phase, remains practically flat. As a result of the jet impact onto the liquid layer, a thin ring-like liquid splash directed into the bubble interior appears on the bubble surface at the impact region periphery [2]. With time it moves away from the axis of symmetry and its amplitude increases. Figure 3*b* illustrates the changes of the liquid pressure field in the vicinity of the torus hole. It is seen that with time the torus hole increases, the maximum pressure on the wall near the axis of symmetry decreases from 45.4 to 17.6. The liquid pressure profiles along the symmetry axis are shown in figure 3*c*. It can be seen that the maximum pressure due to the jet impact is realized on the wall and rapidly decreases with time and with increasing z^*. The radial pressure profiles depicted in figure 3*d* show that the greatest pressure observed at the axis of symmetry, with time, as the liquid splash on the bubble surface develops, becomes comparable with the pressure in the region of the splash ($t^* - t^*_c =$ 0.069).

4. Conclusion

Using the time-stepping method and the boundary element method, the dynamics of a cavitation bubble in water near a solid wall has been studied in the case the pressure in the bubble at the initial instant of time considerably exceeds the pressure in liquid and the thickness of the liquid layer between the bubble and the wall is 0.6 of the radius of the bubble at the moment of its maximum expansion. The numerical results have been compared with the experimental data by other authors. Good agreement is obtained. It is shown that at the beginning of the impact of the cumulative jet, a pronounced pressure maximum at the wall arises in the center of the region of action. Over time, this maximum decreases, becoming comparable with the peripheral local maximum in the region of the liquid splash into the bubble.

References

[1] Blake J R, Taib B B, Doherty G 1986 Journal of Fluid Mechanics **170** 479
[2] Tong R P, Schiffers W P, Shaw S J, Blake J R, Emmony D C 1999 Journal of Fluid Mechanics **380** 339
[3] Pearson A, Blake J R, Otto S R 2004 Journal of Engineering Mathematics **48** 391
[4] Wang Q X, Yeo K S, Khoo B C, Lam K Y 2005 Theoretical and Computational Fluid Dynamics **19** No 5 303.
[5] Yang Y 2014 Experimental and numerical investigations on cavitation bubble dynamics near a solid boundary Doctoral thesis (Syngapore) p 197
[6] Ni B Y, Zhang A M, Wu G X 2015 Journal of Fluids Engineering **137** 031206
[7] Wang Q X, Liu W, Zhang A M, Sui Y Interface Focus 5: 20150048
[8] Aganin A A, Ilgamov M A, Kosolapova L A, Malakhov V G 2016 Thermophysics and Aeromechanics **23** No 2 211
[9] Voinov O V, Voinov V V 1975 Soviet Physics Doklady **20** 179
[10] Best J P 1994 Bubble Dynamics and Interface Phenomena (Dordrecht: Kluwer Academic Publishers) 405
[11] Best J P, Kucera A 1992 Journal of Fluid Mechanics **245** 137

[12] Aganin A A, Kosolapova L A, Malakhov V G 2018 Mathematical Models and Computer Simulations **10** No 1 89
[13] Philipp A, Lauterborn W 1998 Journal of Fluid Mechanics **361** 75

Effect of wetting a wall on the impact of a liquid jet

A A Aganin and T S Guseva

Institute of Mechanics and Engineering - Subdivision of the Federal State Budgetary Institution of Science "Kazan Scientific Center of the Russian Academy of Sciences", 2/31, Lobachevsky str., Kazan 420111 Russia

E-mail: ts.guseva@mail.ru

Abstract. The influence of a thin layer of water wetting a wall on the impact of a water jet is studied. The jet velocity is 250 m/s. The end of the jet is somewhat blunter than the hemispherical one. The thickness of the wetting layer is 4% of the jet radius so that the level of the pressure on the wall in the center of the impact area is approximately equal to that arising when hitting a dry wall. A numerical technique based on the approach without explicit separation of the interphase boundary is applied. The CUP-CUP method is used for integrating the Euler equations. An adaptive soroban-grid is utilized. It is shown that the presence of the wetting layer mainly manifests itself in a significant decrease of the pressure peaks realized in the narrow peripheral area of the impact zone on the wall.

1. Introduction

The problem of the high-speed impact of liquid mass (jet or drop) on a wall covered with a thin layer of liquid is important for many applications. Such impacts can occur during the operation of steam turbines, the flight of aircraft in rain conditions, the collapse of cavitation bubbles near the surfaces of bodies [1]. Jet impacts of cavitation nature are considered to be one of the main causes of the well-known phenomenon of cavitation destruction of body surfaces [1-3].

To date, the focus has been on studying the liquid impact on a dry wall (for example, [4, 5]). In the present work, the numerical investigation of the effect of wall wetting on the impact of a high-speed liquid jet is investigated. The case of a small thickness of the wetting layer is under consideration, in which the damping effect of the layer is relatively small. In particular, with such a thickness of the wetting layer, the level of the load on the wall in the center of the impact is comparable to what is realized when hitting a dry wall.

The important features of the considered jet impact on a wall are the shock waves arising in the liquid, their interaction with the free liquid surface leading to its large and rapid deformations, the high-speed lateral spreading of liquid with the possible formation of the smallest drops and bubbles [4]. All the features should be taken into account when choosing a method of numerical modeling. In this paper, we use the CIP-CUP (Constrained Interpolation Profile - Combined Unified Procedure) [6] method in combination with the approach without explicit treatment of the interphase boundary. The CIP-CUP method is quite effective in simulating the shock waves of moderate intensity and the contact boundaries with a large difference of the acoustic impedance (including boundaries of the gas-liquid type) [7]. Unstructured dynamically adaptive soroban-grids [8] are used to increase the efficiency of calculations.

2. Mathematical model

The impact of a high-speed jet of liquid (water) on a fixed flat rigid wall uniformly wetted with a thin layer of the same liquid is considered (Figure 1). The jet is surrounded by air under room conditions. The jet velocity is $V = 250$ m / s, the wetting layer thickness is $d = 0.04\ R$ where R is the jet radius, the undisturbed liquid pressure in the jet, the ambient air and the wetting layer is $p_0 = 1$ bar.

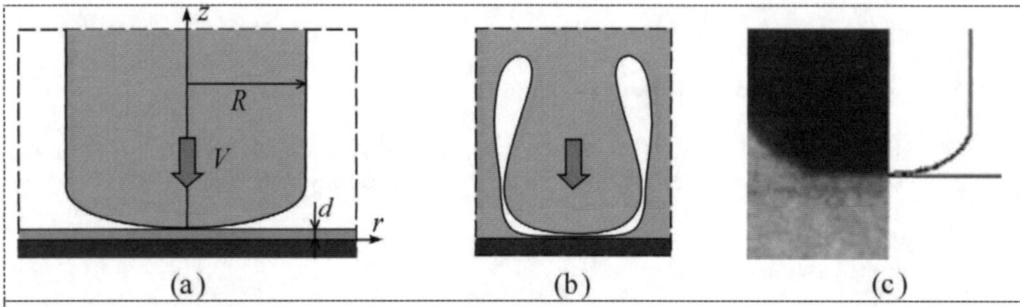

Figure 1. Schematic of the impact of the jet of the present work (a), the shape of the jet obtained in [9] by calculation of the collapse of a bubble near a wall by boundary element method (b) and the shape of the jet taken in [1] in the numerical simulation of the experiments on the impact of the jet on the surface of the liquid (c).

The jet is supposed to be a cylindrical liquid column with a rounded end, the shape of which is described by the equation

$$z = \alpha\left(R - \sqrt{R^2 - r^2}\right) \tag{1}$$

where r, z are the radial and axial cylindrical coordinates, α is a parameter determining the roundness of the shape. It is taken that α=0.25 (Figure 1 a), which approximately corresponds to the ends of the jets in the works [9, 1] (Figure 1 b, c).

The influence of the thin liquid layer covering a wall on the impact of the jet is studied. For this purpose, a comparison is made with the case of impact on the dry wall.

3. Results

Figure 2 illustrates the features of the jet impact on the dry wall. At the beginning of the impact, a shock wave occurs travelling up the jet. Initially, the edge of this wave coincides with the boundary of the rapidly expanding area of contact between the jet and the wall, on the periphery of which the liquid pressure maximum is localized. Then the edge of the shock wave detaches from the wall and goes up the free surface of the jet, a high-speed lateral liquid spreading along the wall arises in the form of a thin film. The interaction of the shock wave with the jet surface leads to the formation of rarefaction waves. Focusing of some of those waves generated on the concave part of the jet boundary results in the appearance of a metastable zone in the vicinity of the jet surface. The metastable zone is then greatly increased, covering almost all the cross-section of the jet.

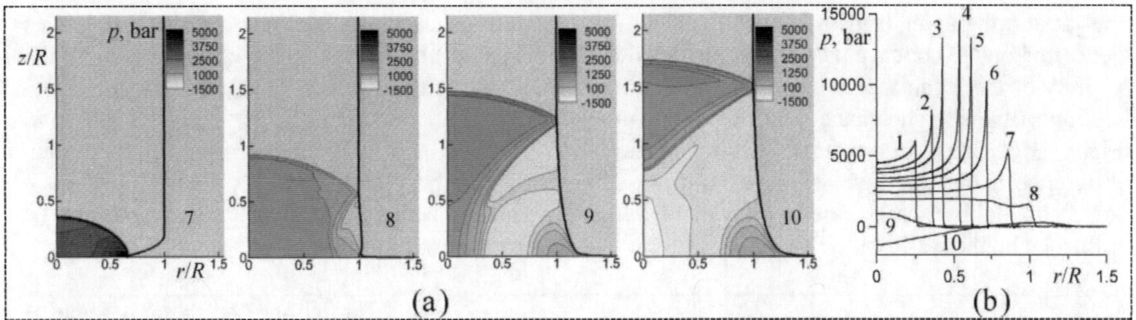

Figure 2. Impact of the jet on the dry wall: variation of the pressure field in the jet (a) and the pressure on the wall (b) during the impact. Numbers 1-10 correspond to the time moments t_1*- t_{10}* = 0.062 (number 1), 0.134 (2), 0.19 (3), 0.3 (4), 0.4 (5), 0.55(6), 0.8 (7), 1.2 (8), 2 (9), 2.4 (10), $t^*=t/(RD^{-1})$, $D° = C + \lambda V$ is the shock wave velocity, C is the undisturbed speed of sound, $\lambda \approx 2$ for water.

It can be seen that the pressure in the metastable zone of the jet, especially in the vicinity of the axis of symmetry, takes large negative values. In reality, this would lead to cavitation, which is not taken into account in the model of this paper. One may expect that the presence of the cavitation region will significantly influence on the subsequent dynamics of the liquid in the vicinity of the wall, therefore the subsequent process is not considered in this paper. The change of the load on the wall in the initial most intense stage of the jet impact is shown in Figure 2 b. It can be seen that in the center of the impact region the pressure decreases monotonically, whereas on the periphery of that region it first grows quite strongly, and then falls. The radial pressure distribution on the wall up to the moment t_7 is inhomogeneous. The inhomogeneity increases to the moment t_4, after which it decreases quite quickly.

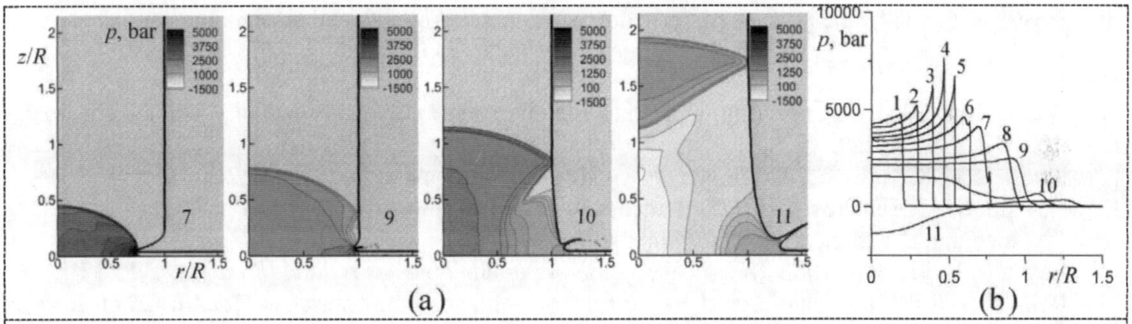

Figure 3. The same as in Figure 2, but for the impact of the jet on the wetted wall. Numbers 1-11 correspond to the time moments t_1* - t_{11}*: 0.08 (number 1), 0.126 (2), 0.2 (3), 0.26 (4), 0.32 (5), 0.4 (6), 0.56 (7), 0.84 (8), 1 (9), 1.48 (10), 2.76 (11).

Figure 3 illustrates the process of the impact of the jet on the wetted wall. In this case, two shock waves arise, the first of which propagates into the area of the jet, while the second one travels into the area of the layer. The edges of these waves are attached to the boundary of the rapidly expanding area of contact between the jet and the layer, where the liquid pressure maximum is localized. Soon the second shock wave reaches the wall, generating a third shock wave travelling upward. By the time t_7, the front of the reflected shock wave in the jet merges with the front of the first wave, and it merges with the front of the second wave in the wetting layer. The edges of the shock waves formed by the merging detach from the boundary of the jet-layer contact area and propagate along the surfaces of the jet and the layer, respectively. At the periphery of the jet-layer contact area, a thin splash of liquid is formed, which eventually becomes more pronounced (moments t_9-t_{11}). Then the process develops in many ways similar to the case of the dry wall. The interaction of the shock waves with the surfaces of the jet and the layer leads to the formation of rarefaction waves. Focusing of some of those waves generated on the concave part of the jet boundary results to the appearance of a metastable zone in the vicinity of the jet surface. From then on, the metastable zone is greatly increased, covering the cross-section of the jet. The change of the load on the wall in the initial stage of the jet impact is shown in Figure 3 b. One can see that it is largely similar to the case of impact on the dry wall. The main difference is that the level of maximum pressure at the periphery of the impact area is now much lower. In addition, in the case of the wetted wall, the peripheral peaks decrease faster $(t > t_6)$ due to the influence on the wall of the rarefaction wave coming from the layer surface.

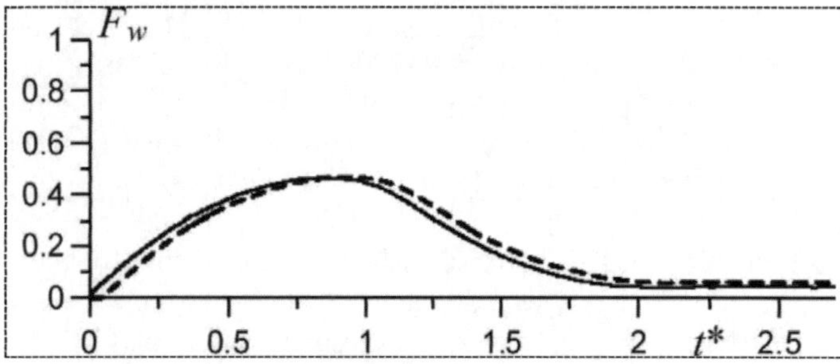

Figure 4. The integral force of the jet action on the wall at the initial most intense stage in the cases of impact onto the dry (solid line) and wetted (dashed line) walls.

Figure 4 demonstrates the time dependences of the integral force of the jet action on the wall determined by the expression

$$F_w = \frac{2}{(p_{wh} - p_0)R^2} \int_0^\infty [\max(p,0) - p_0] r dr \qquad (2)$$

The initial stage is presented, where the action is most significant. It is seen that the influence of the wall wetting on the integral action is insignificant: the difference in the curves is mainly determined by the time delay caused by the passage of the shock wave incident on the wall through the wetting layer.

References

[1] Bourne N K 2005 On impacting liquid jets and drops onto polymethylmethacrylate targets *Proc. R. Soc. A.* **461** 1129
[2] Kornfeld M, Suvorov L 1944 On the destructive action of cavitation *J. Appl. Phys.* **15** 495
[3] Hawker N A, Ventikos Y 2012 Interaction of a strong shockwave with a gas bubble in a liquid medium: a numerical study *J. Fluid Mech.* **701** 59
[4] Lesser M B, Field J E 1983 The impact of compressible liquids *Ann. Rev. Fluid Mech.* **15** 97
[5] Rein M 1993 Phenomena of liquid drop impact on solid and liquid surfaces *Fluid dynamics Research* **12** 61
[6] Yabe T, Wang P Y 1991 Unified numerical procedure for compressible and incompressible fluid *J. Phys. Soc. Japan* **60**(7) 2105
[7] Yabe T, Xiao F, Utsumi T 2001 The constrained interpolation profile method for multiphase analysis *J. Comput. Phys.* **169**(2) 556
[8] Takizawa K, Yabe T, Tsugawa Y, Tezduyar T E, Mizoe H 2007 Computation of free-surface flows and fluid-object interactions with the CIP method based on adaptive meshless Soroban grids *Comput. Mech.* **40** 167
[9] Aganin A A, Ilgamov M A, Kosolapova L A, Malakhov V G 2016 Dynamics of a cavitation bubble near a solid wall *Thermophysics and Aeromechanics* **23**(2) 211.

Technology of growing silicon nanotubes

B A Timerkaev[1], A A Andreeva[2]

[1]Kazan National Research Technical University named by A.N.Tupolev, 420111, K.Marx str., 10, Kazan, Russian Federation
[2]Kazan Federal University, Institute of Physics, 420008, Kremlevskaya St, 18, Kazan, Russian Federation

E-mail: btimerkaev@gmail.com

Abstract. The paper presents the results of studies on silicon nanotubes production in an arc discharge. The features of the experiments, the description of the experimental setup are given. A description of the obtained nanostructures studied on an electron microscope is presented. It is unequivocal to assert that the obtained silicon nanotubes are single-walled, it is not yet possible. But the fact of obtaining silicon nanotubes is indisputable.

Keywords: vacuum, silicon, nanotubes, arc discharge

1. Introduction

Nanostructured semiconductors are key materials of modern electronics. Silicon nanostructures are considered promising candidates in nanoelectronics, optoelectronics, sensorics, biology, medicine and other fields of science and technology in connection with their unique physical properties, simple and economical methods of their obtaining [1-4].

The available technological methods for depositing silicon-based coatings are subdivided into chemical and physical methods [5-6]. In the described work, a physical method of coating deposition was used, namely, the growth of silicon nanostructures in an electric arc discharge. Vacuum arc evaporation of silicon is considered to be one of the promising methods of deposition of coatings due to low energy capacities of processes (50 eV / atom), high coating characteristics and the ability to flexibly manipulate coating parameters [7]. High adhesion (up to 250 MPa), high film growth rate (up to 1.5 μm / min) and high productivity are also provided.

During the recent years, reports where an arc is described on polycrystalline silicon were published [8]. However, a large concentration of power at cathode leads to its cracking and destruction. In the research paper [9], the arc burned on a polycrystalline cathode no more than 30 s. In Ref. [10] after work more than 2 minutes, melting of the working surface and destruction of a part of the cathode were observed.

2. Experiment

In this paper, an attempt is made to create silicon nanomaterials in a low-current argon arc. To obtain silicon nanostructures in an arc discharge, an experimental setup was assembled. Silicon and tungsten were used as electrodes. The arc discharge between the electrodes was ignited in argon at a pressure of 500 Torr. The experimental setup consists of a water-cooled vacuum chamber of a cylindrical shape (figure 1), a vacuum and gas supply system, a source of electrical power, and also measuring instruments.

During the preparation of the experiment, air is evacuated from the vacuum chamber and then the chamber is filled with argon up to 500 Torr.

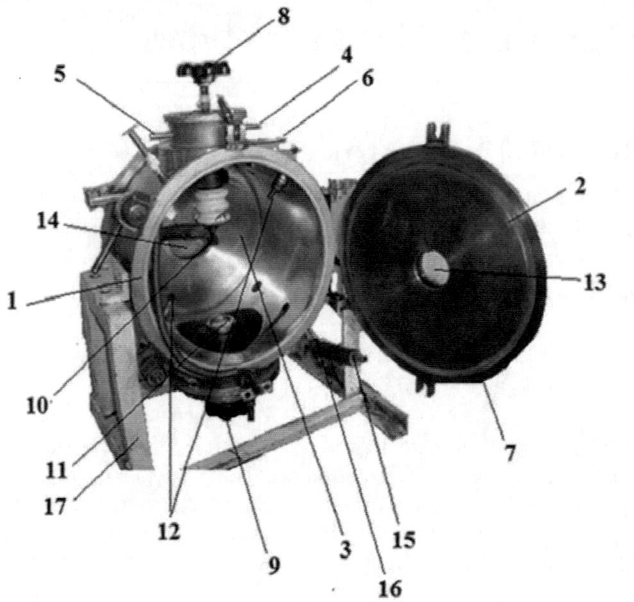

Figure 1. Vacuum chamber. 1 - cylindrical chamber; 2, 3 - the door; 4, 5, 6 - pipes for supplying the cooled liquid; 7 - a branch pipe for drainage of a liquid; 8, 9 -regulators of the distance between the electrodes; 10 - electrode retainer; 11 - stand for the substrate; 12 - current leads; 13, 14 - inspection windows; 15 - a branch pipe for gas supply; 16 - a branch pipe for gas extraction; 17 - support for the camera.

The pressure in the vacuum chamber was measured with a U-shaped manometer, the current was fixed with an ammeter. In order to ignite the arc discharge into the chamber between the electrodes, the RNO-250-5 transformer used for smooth regulation of the AC voltage with a frequency of 50 Hz was used, the maximum current of the regulator at the output was 12 A / 20 A; power - up to 5 kW. The ballast was connected in series to the transformer to ensure a stable discharge. The surface of the electrodes before the experiment is thoroughly polished to a mirror finish, in order to reduce the amount of oxides. The materials of the electrodes were tungsten and silicon, the interelectrode distance was 1 cm (figure 2), the voltage is applied to the electrodes on the order of 80 V. Since the resistance of silicon at room temperatures is large enough, preheating of silicon is required to ignite the microarc. For this purpose, the electrodes are closed for a short time. When the current flows, silicon is heated, its conductivity increases, and when the contacts are opened, a micro-arc with a current of the order of 8 A is formed. In the process of burning the arc, a built-up edge is formed on the surface of the tungsten. This build-up was examined using a scanning electron microscope AURIGA CARL ZEISS with an allowable magnification of 300 to 500 000 times and a resolution of about 2.5 nm.

IOP Conf. Series: Journal of Physics: Conf. Series **1058** (2018) 012071 doi:10.1088/1742-6596/1058/1/012071

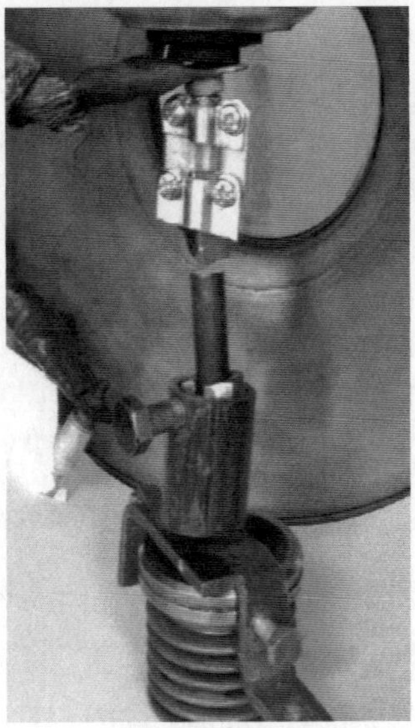

Figure 2. Fixation of the silicon anode (from above) and tungsten cathode in the working unit (bottom).

Figure 3 shows photographs of the build-up from the cathode, obtained with an electron microscope. A large number of formless tubular formations of various sizes, intertwining with each other and curling along the entire length.

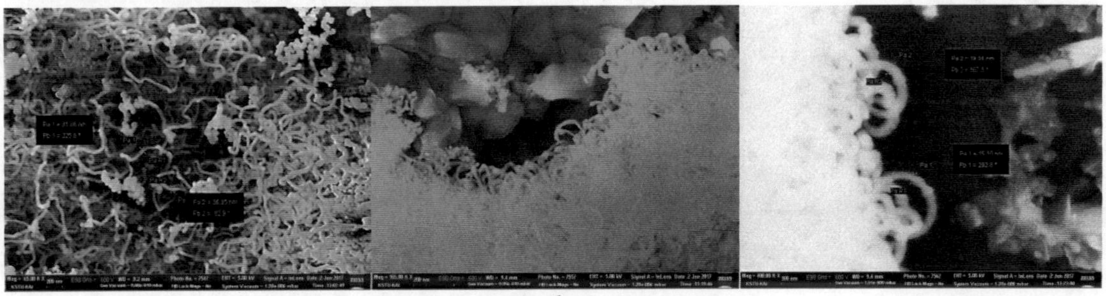

a b c

Fig.3. Electron microscopic image in magnification:
a-65,000 times, b-165,000 times, c-400,000 times

3. Results

Figure 3a shows that the winding tubes have approximately the same diameters, of the order of 30-37 nm. In addition to these tubes of other, clearly distinguishable formations, this increase is not observed. In the photographs, 3 b, we can see the same structures of climbing tubes, intertwined with each other, but already having the form of cylinders, with a diameter of 15-19 nm. It is unequivocal to assert that the obtained silicon nanotubes are single-walled, it is not yet possible. But the fact of obtaining silicon nanotubes is indisputable.

References

[1] Timerkaev B A, Sofronitskiy A O and Andreeva A A 2016 *J. Phys.: Conf. Series* **669** 012062

[2] Galeev I G and Asadullin T Ya 2016 *J. Phys.: Conf. Series* **669** 012015

[3] Galeev I G and Asadullin T Ya 2017 *J. Phys.: Conf. Series* **789** 012003

[4] Galeev I G and Asadullin T Ya 2017 *J. Phys.: Conf. Series* **789** 012012

[5] Timerkaev, B.A., Melnik, K.S. Plasma deposition of silicon nanostructures. *2016, Science in motion: from reflection to creation of reality Materials of the All-Russian Scientific and Practical Conference,* p. 190-193

[6] Timerkaev, B.A., Melnik, K.S. Synthesis of silicon nanostructures in a glow discharge. 2016, *Professional communications in the scientific community - the factor of ensuring the quality of research materials of the II All-Russian Scientific and Practical Conference with foreign participants, Kazan National Research Technical University, A.N. Tupolev, Almetyevsky branch*, p. 15-17

[7] Andreeva A.A., Sofronitsky A.O., Timerkaev B.A. Creation of carbon nanotubes in a high-voltage discharge. 2016, *Professional communications in the scientific environment - the factor of ensuring the quality of research materials of the II All-Russian Scientific and Practical Conference with foreign participants, Kazan National Research Technical University, A.N. Tupolev, Almetyevsky branch,* p. 22-24

[8] Timerkaev, B.A., Melnik, K.S., Zalyaliev, B.R. Application of silicon nanocoatings in a magnetron discharge. *2017, Bulletin of the Kazan State Technical University, A.N. Tupolev, Vol. 73, № 2,* p. 32-36

[9] Kesaev, I.G., Pashkova, V.V. Electromagnetic fixation of a cathode spot. *1959, Journal of Technical Physics, T. 29, № 3,* p. 287-298

[10] Dukhopelnikov, D.V., Zhukov, A.V., Kostin, A.A., Yurchenko A.A. Control of the motion of a cathode spot in linear vacuum-arc evaporators. *2005, Strengthening technologies and coatings, № 11, p. 45-49*

IOP Conf. Series: Journal of Physics: Conf. Series **1058** (2018) 012072 doi:10.1088/1742-6596/1058/1/012072

Effect of vapor concentration in bubbles on the acoustic waves propagation in bubbly liquids

D A Gubaidullin[1], A A Nikiforov[1]

[1]Institute Institute of Mechanics and Engineering, Federal Research Center KazSC, Russian Academy of Sciences, 2/31 Lobachevsky str., Kazan 420111, Russia

E-mail: anikiforov1@yandex.ru

Abstract. This article theoretically investigates the acoustic waves propagation in bubbly liquids in the presence of diffusion mass transfer. Numerical calculations of the dynamics of pressure pulse in bubbly liquids are performed for different values of the vapor concentration. An increase the vapor concentration in the bubbles leads to a decrease in the propagation velocity of the low-frequency harmonics that make up the pulse and to an increase their attenuation due to the nature of the dependence of the phase velocity and attenuation coefficient on the vapor content in the gas bubbles.

Introduction

A propagation of acoustic waves in a liquid with gas bubbles is studied intensively both theoretically and experimentally. Nigmatulin [1], Temkin [2], Leighton [3] considered various basic aspects of acoustics of multiphase flows. Gumerov [4], Hao and Prosperetti [5] obtained the results of study of the vapor bubble dynamics in liquid. Nigmatulin et al. [6] studied the small radial oscillations of vapor-gas bubbles in liquid under the influence of an acoustic field. They showed that capillary effects and phase transitions lead to the new resonant frequency of small bubbles. This frequency is different from the Minnaert frequency. Vakhitova and Shagapov [7] investigated the propagation of small disturbances in two-component two-phase bubbling media with a three-temperature model. They found that heat and mass transfer rather than an interphase friction mainly determine dispersion.

In this work the acoustic waves propagation in bubbly liquids is theoretically investigated in the presence of diffusion mass transfer.

Dispersion relation

Let us consider a plane one-dimensional flow of a two-fractional mixture of liquid with vapor-gas bubbles with phase transitions in the acoustic field. The volume content of the bubbles, α_2, and the liquid, α_1, are determined by

$$\alpha_2 = \frac{4}{3}\pi a^3 n, \quad \alpha_1 + \alpha_2 = 1,$$

where n is the bubbles number in the unit of volume, subscripts 1 and 2 refer to the parameters of the liquid and bubbles, respectively. The reduced densities ρ_j are given by

$$\rho_1 = \rho_1^\circ \alpha_1, \quad \rho_2 = \rho_G + \rho_V = \rho_G^\circ \alpha_2 + \rho_V^\circ \alpha_2$$

where ρ_j° is the true densities of the liquid ($j=1$), the vapor ($i=V$) and the gas in bubbles ($j=G$), respectively.

The dispersion relation, connecting the complex wave number K_* and oscillation frequency ω, has the following form:

$$\left(\frac{K_*}{\omega}\right)^2 = V(\omega)D(\omega). \tag{1}$$

$$V(\omega) = 1 + \alpha_{20}(m^\circ - 1)\frac{1 - i\omega\tau_v\left(\frac{1}{2} + \frac{\alpha_{10}}{m^\circ}\right)}{1 - i\omega\tau_v\left(\frac{1}{2} + \frac{\alpha_{10}}{m^\circ}(\alpha_{20} + m^\circ\alpha_{10})\right)},$$

$$D(\omega) = \frac{\alpha_{10}}{C_1^2} + \alpha_{20}\frac{1}{H_R + H_K}, \quad H_R = \frac{a_0^2(i\omega)}{6}\left[\frac{i\omega\tau_R - 2}{2\tau_R - t_A(i\omega\tau_R - 2)}\right],$$

$$H_K = A_1\frac{U_2H_1H_2 - U_4H_1 - \frac{c_1}{c_2}\frac{\tau_{T2}}{\tau_{T1}}U_1H_2}{U_2H_1H_3 - U_3U_4H_1 + \frac{c_1}{c_2}\frac{\tau_{T2}}{\tau_{T1}}(U_4G - U_1H_3)},$$

$$H_1 = \frac{i\omega\tau_{T1}}{m} - 1, \quad H_2 = i\omega\tau_{T2} - 1, \quad H_3 = i\omega\tau_{T2}(1 - R_0) - 1 + G,$$

$$U_1 = E(i\omega\tau_m - 1), \quad U_2 = -\frac{l_0 i\omega\tau_{T2}}{T_0c_2(1 - k_{V0})} + b_1 - U_1\left(1 + \frac{c_1}{c_2}\frac{\tau_{T2}}{\tau_{T1}}\right),$$

$$U_3 = 1 - G\left(1 + \frac{c_1}{c_2}\frac{\tau_{T2}}{\tau_{T1}}\right), \quad U_4 = b_1H_2 + U_1, \quad m^\circ = \frac{\rho_{20}^\circ}{\rho_{10}^\circ}, \quad m = \frac{\rho_{20}}{\rho_{10}},$$

$$\tau_R = \frac{a_0^2}{8\nu_1}, \quad t_A = \frac{a_0}{C_1\alpha_{10}^{1/3}}, \quad \tau_v = \frac{2a_0^2}{9\nu_1}, \quad \tau_{Ti} = \frac{4c_i\rho_{20}^\circ a_0^2}{3Nu_i\lambda_i}, \quad \tau_m = \frac{2a_0^2}{3Sh_1D_1}.$$

Here, a is the bubble radius, l is the specific heat of vaporization, k_i is the mass concentration of the i-th component of the disperse phase, c is the specific heat, ν_1 is the kinematic viscosity of the liquid, Nu_i is the Nusselt number for heat transfer of the phase surface with the i-th phase, λ_i is the coefficient of thermal conductivity of the i-th phase (i = 1,2), Sh_i is the dimensionless mass transfer coefficient (Sherwood number), D_1 is the diffusion coefficient, τ is the relaxation times. The index Σ refer to the phase interface parameters, index 0 - initial unperturbed state.

Results

The calculations are performed in water with vapor-air bubbles with the following values: $\alpha_{20} = 0.001$ $a_0 = 1\,\text{mm}$, $p_0 = 0.1\,\text{MPa}$. Figures 1 and 2 show the dependence of the attenuation coefficient and the phase velocity on the frequency for water with vapor-air bubbles at different k_{V0}. The calculations were carried out using the dispersion relation (1). The effect of a change in the vapor concentration in bubbles on the phase velocity is substantially manifested only in the low-frequency region. At low frequencies, the phase velocity decreases with respect to the value calculated for gas bubbles when the effects of mass transfer are neglected. The attenuation coefficient in this case demonstrates the growth with increasing vapor concentration k_{V0}.

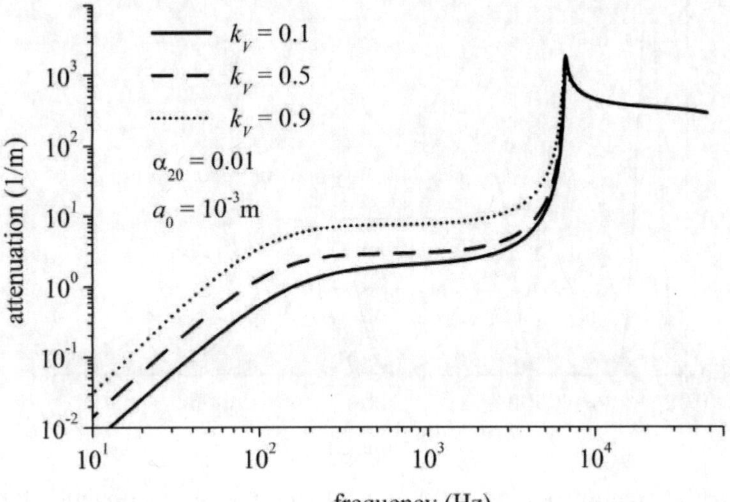

Figure 1. Frequency dependences of the attenuation coefficient in water with vapor-air bubbles

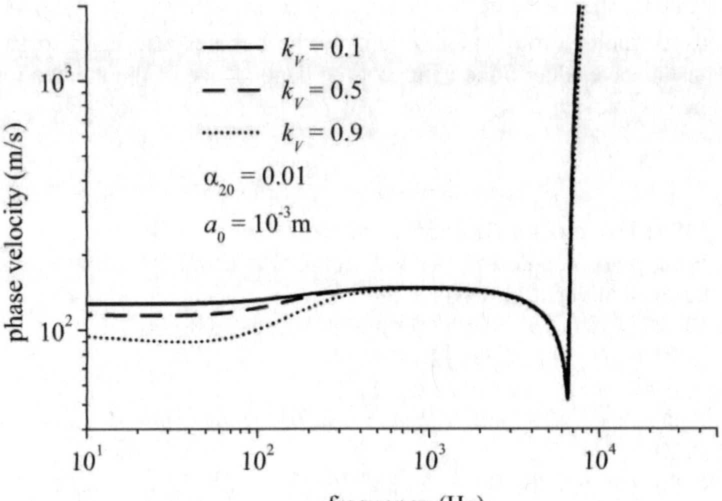

Figure 2. Frequency dependences of the phase velocity in water with vapor-air bubbles

Using the dispersion relation (1) and the fast Fourier transform algorithm, the evolution of pressure pulse is calculated. The initial form of pulse is described by the function

$$p(0,t) = \exp[-((t-t_*)/N)^2]$$

where t_* is a half of the pulse duration and N is the parameter determining the width of the pulse peak. The profiles, shown in figure 3 are calculated for the distances of 2 m and 4 m from pulse initiation point.

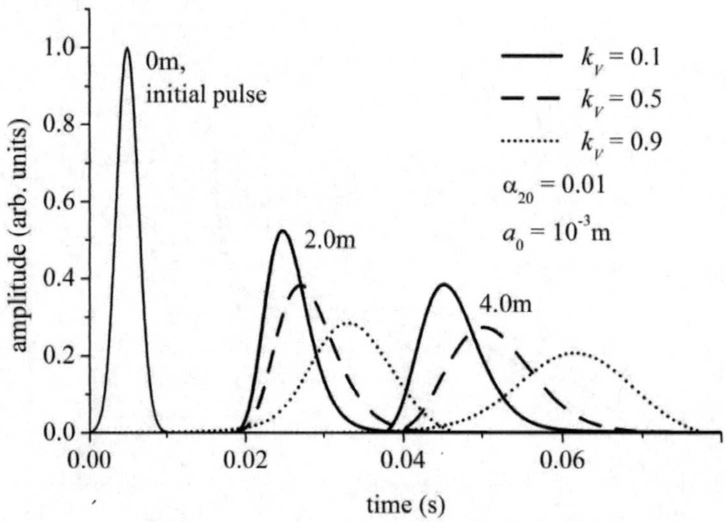

Figure 3. Evolution of pressure pulse in water with vapor-air bubbles for different k_{V0}

An increase the vapor concentration in the bubbles leads to a decrease in the propagation velocity of the low-frequency harmonics that make up the pulse. As a result, with an increase in k_{V0}, the original waveform is distorted. An increase the vapor content in the bubbles also leads to an increase in the attenuation of the acoustic waves due to the nature of the dependence of the attenuation coefficient on the vapor content in the gas bubbles.

References

[1] Nigmatulin R I 1991 *Dynamics of Multiphase Media* (New York: Hemisphere)
[2] Temkin S 2005 *Suspension Acoustics: An Introduction to the Physics of Suspensions* (New York : Cambridge University Press)
[3] Leighton T G 1994 *The Acoustic Bubble* (London: Academic Press)
[4] Gumerov N A 2000 *Physics of Fluids* **12** 71
[5] Hao Y, Prosperetti A 1999 *Physics of Fluids* **11** 2008
[6] Nigmatulin R I, Khabeev N S and Nagiev F B 1981 *International Journal Heat and Mass Transfer* **24** 1033
[7] Vakhitova N K, Shagapov V Sh 1984 *Journal of Applied Mechanics and Technical Physics* **25** 682

IOP Conf. Series: Journal of Physics: Conf. Series **1058** (2018) 012073 doi:10.1088/1742-6596/1058/1/012073

Resistance of the powder coatings obtained in the electrostatic field of the corona discharge to the static action of liquids

Marsel Fazlyyyakhmatov, Alexandra Teveleva and Nail Kashapov
Kazan (Volga Region) Federal University, 18 Kremlyovskaya str., Kazan 420008, Russian Federation

E-mail: mfazlyjy@kpfu.ru

Abstract. The results of an experimental study of the static action of liquids on polymer-powder coatings obtained in the electrostatic field of the corona discharge are presented. Epoxy, epoxy-polyester, polyester, polyurethane and silicone powder materials have been studied. It has been shown experimentally that epoxy-polyester coating EP111095G is resistant, while epoxy EX611434SG, polyester PD810119G, polyester PD510226G and polyurethane PD010186G coatings are relatively resistant.

1. Introduction

Electrostatic powder coating was first introduced to the finishing industry in early 1960s [1]. In recent years, the polymer powder coatings are widely used in various industries. Major application area is obtaining protective, functional and decorative coatings in the finishing industry. The electrostatic coating of powders for the finishing industry is an important and increasing commercial business [2].

Several ways of producing powder coatings are existing. The most often used at present is a method of applying powder coatings in the electrostatic field [3]. A large number of spraying devices both by tribostatic [4-5] and electrostatic methods have been developed [6]. The great advantages and most commonly used at present is the spraying of powder coatings with spray guns with external powder charging [2]. Often this method is called “the production of coatings in corona discharge plasma” [7].

Powder paints are multicomponent disperse systems consisting of solid particles – film-forming basis and air-separating medium [8]. The main advantage of powder coatings is the absence of solvents. Therefore, the physical, mechanical and dielectric properties of powder coatings are significantly higher than for paint and varnish materials, in particular, their protective properties [9].

There are two types of polymer powder coatings: thermosets and thermoplastics. For obtaining protective coatings in recent years in most cases, are used thermosetting polymer powders. These coatings are formed by fusion of the powder particles and further chemical transformations.

By type of film former, the most common thermosetting compositions are epoxy, polyester, epoxy-polyester, polyurethane and polyacrylate [10, 11].

In [12, 13] was proposed to use powder coatings for the protection and recovery of medical ultrasound sensors.

Ultrasound is the most popular method of non-destructive testing and diagnostics in various fields including medicine. In [14, 15] are shown results of research and development of new methods and equipment for non-destructive testing multiphase and biological objects.

Ultrasonic sensors are used for measurements and diagnostics in various liquids, aggressive gases and multiphase media (e.g., drilling mud, etc.). Thereby, it is important to use protective coatings that are resistant to contaminants and prevent the destruction of the sensor.

In the process of diagnosis liquid and solid objects, the close contact of the ultrasonic transducer with the test object is necessary. This requires to use different protective coatings and prisms. They prevent the damage of sensor and protect it from aggressive environments. Particularly in the medicine, copolymers of vinyl chloride and ethyl acrylate are applied. However, the protective coatings are abraded and are damaged, resulting in a failure of the transducer. To solve this problem, the applying of polymer powder coating in the electrostatic field are proposed [12, 13, 16, 17].

The main task of this work is to study powder coatings for resistance to aggressive liquids.

2. Materials and methods

For most tasks of ultrasound diagnostics and probing the frequency range 0.5-15 MHz is used. Ultrasonic vibrations in this frequency range provide sufficient resolution in the low-level vibration damping. In previous paper [17] was shown that the thickness of the matching layer the piezoceramic at these frequencies is 10-300 um.

The studies [12] of polyether PD510226G (Inwer, Italy) and epoxy-polyether EP110022G (Inwer, Italy) powder coatings properties was showed the suitability of using this types of coatings to the surface of the ultrasonic transducers. Such parameters as impact strength, flexibility, adhesion, gloss, hardness and wear resistance was studied.

In this paper we studied powder coatings (Inwer, Italy): epoxy EX611434SG (blue, 5019), epoxy-polyester EP111095G (white, 9016), polyester PD810119G (grey light, 7035), polyester PD510226G (green, 6029), polyurethane PD010186G (colourless varnish), silicone PD208120MRT (black matte). Colours are indicated in brackets according to the RAL colour catalogue. All powder coatings meet safety requirements due to the absence of organic solvents and heavy metals.

The substrate samples for studies are chosen according to ISO 1514:2016. The steel plates with dimensions of 10×5×0.2 sm are used in this study.

The surface preparing process was discussed in [13] and it is carried out according to ISO 1513:2010. It should be noted that surface must be clean and dry.

Spraying the coating is carried out using a spray gun START-50 (Radar, Russia) with voltage minus 30 kV on the corona discharge electrode. The measurement of the potential electrostatic field during application polymer powder coatings on metallic substrates using different speeds of micro-particles and the distance to the substrate was carried out [18]. Spraying is carried out in the chamber of coating powder materials at the samples.

Experimental studies in [17] were shown optimum operating parameters of the system of homogeneous coating polymer powder coatings. Based on these data the following parameters are chosen: time of spraying – 6 sec., distance from gun to substrate – 15 sm., air pressure – 200 kPa. The homogeneity of the obtained coatings was shown in [19, 20].

Baking was carried out in a drying chamber. In this work, we used a drying chamber ES 4610 (Ekohim, St. Petersburg, Russia).

For polymer coatings based on epoxy and silicone resins, the rejection time is 15 minutes at 200 °C, based on polyester and polyurethane – 15 minutes at 180 °C.

There are several internationally recognised standards for assessing chemical resistance. All standards are grouped in section 87.040 Paints and varnishes of the International Organization for Standardization (ISO). Immersion method is described in the standard ISO 2812-1:2017, method using an absorbent medium in the ISO 2812-3:2017, spotting method in the ISO 2812-4:2017. In the Russian Federation, the standard GOST 9.403-80 "The uniform system of protection against corrosion and aging. Paint and varnish coatings. Test methods for resistance to static effects of liquids" is applied.

In this work, tests of coatings were carried out in accordance with GOST 9.403-80. The method B was used (with a high degree corresponds to ISO 2812-3:2017). In this method, 10 drops of liquid are applied to a horizontally placed testing plate at a distance of 20 mm from the edge of the plate. The distance between the droplet centres should not be less than 20 mm. The tests were carried out for 1 hour under atmospheric conditions at air temperature (20 ± 2) °C and relative humidity (65 ± 5) %.

The following liquids were used for the tests: acetic acid according to GOST 61-75 (CH_3COOH), 8% solution; sulphuric acid according to GOST 4204-77 (H_2SO_4), 25% solution; acid hydrochloric in accordance with GOST 3118-77 (HCl), 25% solution; nitric acid according to GOST 4461-77 (HNO_3), 25% solution; sodium hydroxide according to GOST 4328-77 (NaOH), 3%, 5% and 10% solutions; sodium chloride according to GOST 4233-77 (NaCl), 5% solution; mineral oil according to GOST 20799-88 and petrol.

3. Experimental results and discussion

After the tests, the decorative and protective properties are determined by comparison with the control sample using a magnifying glass (change in gloss, shade, appearance of bubbles, peeling, wrinkling, corrosion, etc.). The physical and mechanical properties of the coating are also determined (adhesion, bending elasticity, impact strength). The coating is considered stable if after the tests all characteristics correspond to the initial values.

Table 1 shows the test results. The symbol "+" marks the resistance of the coating, the symbol "–", if there are changes in any characteristics of the coatings.

Table 1. Test results for coatings for resistance to liquids.

Powder coating	Liquid							
	CH_3COOH	H_2SO_4	HCl	HNO_3	NaOH	NaCl	Mineral oil	Petrol
EX611434SG	–	+	+	–	+	+	+	+
EP111095G	+	+	+	+	+	+	+	+
PD810119G	+	–	+	–	+	+	+	+
PD510226G	+	+	+	–	+	+	+	+
PD010186G	+	+	+	–	+	+	+	+
PD208120MRT	+	+	–	–	–	+	+	+

3.1. Acetic acid.

Influence of the 8% acetic acid solution did not cause any changes in all samples, except the epoxy coating EX611434SG. On this sample, there was an insignificant colour change, i.e. sombreness. Also, the gloss of the coating was changed. No changes in the physical and mechanical properties were observed in any sample. Also, no changes were observed on the metal substrate surface. Tests with less concentrated solutions were not performed.

3.2. Sulphuric acid.

Influence of the 25% sulphuric acid solution did not cause any changes in all samples, except the polyester coating PD810119G. On this sample, there was a small spot and the shine of the coating was changed. No changes in the physical and mechanical properties were observed in any sample. Also, no changes were observed on the metal substrate surface.

3.3. Hydrochloric acid.

Influence of a 25% hydrochloric acid solution did not cause any changes in all samples except the silicone coating PD208120MRT. There was a complete destruction of the coating. During the test, a chemical reaction occurred with the formation of gas bubbles. On the plate surface were formed traces of corrosion and rust. The remaining samples showed no changes in the in the physical and mechanical properties. Tests with less concentrated solutions were not performed.

3.4. Nitric acid.

Influence of the 25% nitric acid solution resulted in colour change of the EX611434SG epoxy coating. On the epoxy polyester coating EP111095G, no changes were detected. The sample of polyester coating PD810119G changed colour (yellowed). Gloss change occurred on the samples of polyester PD510226G and polyurethane PD010186G coatings. Changes in the physical and mechanical properties of the above samples were not observed, except for the sample PD810119G. This sample deteriorated all the physical and mechanical properties (adhesion, impact strength, flexural elasticity). No changes were observed on the metal substrate surface in any of the above samples.

The sample of silicone coating PD208120MRT was destroyed, rust formed on the metal surface.

3.5. Sodium hydroxide.

Influence of 3%, 5% and 10% sodium hydroxide solutions did not cause any changes in all samples except the silicone coating PD208120MRT. Also, in all samples, except PD208120MRT, no changes in the physical and mechanical properties and changes on the metal substrate surface were observed.

Influence of the 3% sodium hydroxide solution resulted in a slight peeling, 5% solution to a significant peeling, 10% solution to complete peeling of the PD208120MRT coating. Tests with more concentrated solutions were not performed.

3.6. Sodium chloride, mineral oil and petrol

Influence of the 5% sodium chloride solution, mineral oil and petrol did not lead to any changes in any sample.

4. Conclusions

The results of an experimental study of the static action of liquids on polymer-powder coatings obtained in the electrostatic field of the corona discharge are presented in this article. Epoxy, epoxy-polyester, polyester, polyurethane and silicone powder materials have been studied. It has been shown experimentally that epoxy-polyester coating EP111095G is resistant, while epoxy EX611434SG, polyester PD810119G, polyester PD510226G and polyurethane PD010186G coatings are relatively resistant.

References

[1] Janardhanan S, Zvonkina I and Soucek M D 2016 Powder coating technology *Kirk-Othmer Encyclopedia of Chemical Technology* (New York: John Wiley & Sons) pp. 1-16
[2] Dastoori K, Makin B and Telford J 2001 *J. Electrostat.* **51-52** 545-551
[3] Bailey A G 1998 *J. Electrostat.* **45** 85-120
[4] Kleber W and Makin B 1998 *Part. Sci. Technol.* **16(1)** 43-53
[5] Boncza-Tomaszewski Z and Penczek P 2002 *Macromol. Symp.* **187** 417-425
[6] Kleber W 1993 *J. Electrostat.* **30** 393-402
[7] Wu S 1976 *Polym.-Plast. Technol. Eng.* **7(2)** 119-220
[8] Yakovlev A D 1987 *Poroshkovye kraski* [Powder Coatings] (Leningrad: Khimiya) p 216
[9] Polyakova K K and Payma V I 1972 Tehnologiya i oborudovanie dlya naneseniya poroshkovyh polimernyh pokrytiy [Technology and equipment for applying polymer powder coatings] (Moscow: Mashinostroenie) p 136
[10] Bosshard G 1982 *Prog. Org. Coat.* **10** 205-212
[11] Weiss K D 1997 *Prog. Polym. Sci.* **22** 203-245
[12] Gavrilova V A, Kashapov N F and Kashapov R N 2011 *Biomedical Engineering* **45** 198-200
[13] Gavrilova V A, Fazlyyyakhmatov M G and Kashapov N F 2013 *Repair, Rebuild, Modernization* **12** 13-16
[14] Denisov E S, Temyanov B K, Sagdiev R K and Fazlyyyakhmatov M G 2014 *IOP Conf. Series: Mater. Sci. Eng.* **69** 012014

[15] Sagdiev R K, Denisov E S, Evdokimov Yu K, Fazlyyyakhmatov M G and Kashapov N F 2014 *IOP: Conf. Ser. Mater. Sci. Eng.* **69** 012012
[16] Fazlyyyakhmatov M and Kashapov N 2014 *High Temperature Material Processes* **18(4)** 273-279
[17] Gavrilova V A, Fazlyyyakhmatov M G and Kashapov N F 2013 *J. Phys.: Conf. Series* **479** 012010
[18] Gavrilova V A, Fazlyyyakhmatov M G and Kashapov N F 2014 *Russian Physics Journal* **57 (3/3)** 114-118
[19] Gavrilova V A, Fazlyyyakhmatov M G and Kashapov N F 2014 *J. Phys.: Conf. Series* **567** 012023
[20] Fazlyyyakhmatov M 2017 *J. Phys.: Conf. Series* **789** 012005

IOP Conf. Series: Journal of Physics: Conf. Series **1058** (2018) 012074 doi:10.1088/1742-6596/1058/1/012074

Increase of the resource of machine parts working by combined methods using plasma electrolytic oxidation

A V Kolomeichenko[1], S N Sharifullin[2, 3], Yu A Kuznetsov[1], V N Logachev[1], N V Titov[1]

[1]Orel State Agrarian University. N.V. Parakhina
[2]Kazan Federal University
[3]Kazan State Agrarian University

E-mail: saidchist@mail.ru

Annotation. The paper presents the results of a study of the physicomechanical properties of oxide-ceramic coatings formed by plasma electrolytic oxidation on various aluminum alloys. It has been found that filling the pores of an oxide-ceramic coating with oil or applying a copper layer to its surface with a friction-mechanical method increases the wear resistance of movable joints of machine parts by 1.7 times and 4.5 times, respectively. Based on the complex of studies carried out, combined methods are proposed that significantly increase the life of reinforced parts of machines in operation.

1. Introduction

During operation 85...90% of cars goes down not because of broken parts, and due to the wear of their working surfaces. To increase the resource it is necessary to create hardened layers with high physical and mechanical properties on wear surfaces of parts. One of the methods that allow to realize this is electro-spark and plasma – electrolytic [1-19]. Among these methods, plasma-electrolytic oxidation (PEO) is the most common. The method of PEO does not have many disadvantages inherent in other technologies of surface hardening of parts. The main advantages of PEO include: obtaining multifunctional coatings with high physical and mechanical properties of the specified composition, structure and thickness, the availability of chemical reagents, environmental friendliness of the process and the lack of special treatment facilities using silicate-alkaline electrolytes. Formed coatings have high adhesion strength to the substrate material, hardness, wear resistance, corrosion resistance, heat resistance and controlled porosity.

At the same time, the coatings formed by PEO are not devoid of drawbacks. Firstly, due to the limited thickness only minor wear of parts (up to 0.10...0.13 mm) can be compensated with their help. Secondly, the peo technology is comparatively well developed for a group of so – called valve metals and their alloys-aluminum, titanium, magnesium, tantalum, etc., while a significant proportion of machine parts are made of steels and cast irons.

For elimination of the specified shortcomings it is perspective to use the combined technologies of increase of durability including increment of working surfaces of details for the purpose of compensation of their wear by one of the known methods, hardening of the restored surfaces of PEO and the subsequent modification of the hardening covering.

2. Methods of research

The thickness of the coating formed by PEO was measured by eddy current thickness gauge VT-201 with an error δ=±(0,03 S+1,0), microns, where S is the measured value, microns. The hardness of coatings formed by PEO was measured on the computerized hardness testing PMT-3M-01 according to GOST 9450-76. The evaluation of the thermal conductivity of the coating formed by PEO was carried out with a monotone one-sided heating of the sample with an oxide coating on a dynamic λ-calorimeter, allowing an error of not more than 10% in the temperature range -196...377°C to measure the thermal conductivity of materials with λ = 0,1...5 W/(m·°C). When estimating the specific heat of the coating formed by PEO, a comparative method was used, realized by means of a dynamic C-calorimeter with a heat meter and an adiabatic shell. The through porosity of the coatings was determined by the color defectoscopy method. After PEO and removing the loose coating layer, the samples were washed in cold running water, then immersed for 30 seconds in a warm (35 ... 40 ° C) solution of caustic soda (20 g / l) and for 1 minute in a clarification solution (nitric and hydrofluoric acid in the ratio 1: 1). For pore staining the samples were placed for 5 min in a solution containing 20 g/l of copper sulfate and 20 ml/l of hydrochloric acid. Washed and dried samples with areas of pink spots of copper on the coating were investigated under a microscope, 20x. The through porosity of the coatings was determined as the percentage ratio of the areas of the painted areas to the area of the site under consideration.

Wear resistance was determined in the conditions of boundary lubrication in accordance with GOST 23.224-86 in group A on the friction machine SMT-1M, reproducing the friction scheme when rotating the counterbody in the form of a sleeve relative to the fixed sample in the form of a ring. Samples for research were made of aluminum alloys Ak7ch, AO3-7, D16, PA12, AK9M2, AMg6, AK5. After that, on their surfaces by PEO coatings were formed, which were then polished to remove the loose layer and roughness Ra=1,20...1,25 microns. Further, a copper layer was applied to the surface of oxide-ceramic coating by porosity of 3...4% by friction-mechanical method. The hardened layer with porosity of 14 ... 15% was filled with oil. The porosity of the coatings was changed by varying the temperature of the electrolyte. The material for the manufacture of the counterbody was steel 45, hardened to HRC 40...45 and having a working surface roughness Ra=0,63 mm. When forming coatings by PEO method, an installation operating at an alternating electric current of industrial frequency of 50 Hz, from a network of 380 V voltage, providing an anode-cathode formation of coatings with a ratio of the amplitude of the cathode and anode currents I_k / I_A=1.0, was used. In the formation of coatings used silicate-alkaline electrolyte type "CON-Na_2SiO_3". The content of potassium hydroxide CON in the electrolyte was 3 g/l, sodium liquid glass Na_2SiO_3 – 10 g / l, distilled water – the rest. PEO modes: current-20 A / dm2, the duration of oxidation – 2 h, the electrolyte temperature ranged from 20 ^{0}C to 60 ^{0}C.

Tested the following connection: "steel – coating formed by PEO, porosity 3...4%, steel – coating formed MAO, porosity 14...15%", "steel – coating formed MAO, porosity 14...15% filled with oil", "steel – coating formed MAO, porosity 3...4% with copper layer". When applying the copper layer, the contact pressure was-15 MPa, sliding speed-0.5 m / s, the thickness of the rubbing plate of copper – 0.05...0.10 mm, the duration of application – 230...240 s, lubrication of the contact surfaces – technical glycerin. The exact oil flow was regulated by a metering valve. To accelerate wear, an abrasive was added to the oil, which was used as quartz sand, corresponding to GOST 2138-91, dispersion of 5 microns. The abrasive concentration was 3±0,5% by weight of oil. Tests were carried out at room temperature. Before carrying out researches carried out an operating time of the tested mobile connection at smooth change of loading from 20 to 200 N with an interval of 20 n which completion was fixed on stabilization of coefficient of friction in a contact zone. The loading was carried out through the lever system of the friction machine. Wear was determined by changing the mass of samples and contrabass with the use of analytical weights ADV-200M. The error in weighing was not more than 0.002 g.

Before measurements, the sample and the counterbody was thoroughly washed in benzene for 10 minutes, then dried in a drying Cabinet SNOL-3,5 at a temperature of 60...70 °C for 40 minutes, followed by air cooling. After the measurements, the test sample and the counter-body were placed on the friction machine according to the coincidence of the pre-applied marks. The wear was determined after every 10 hours of testing. The total test duration of each mobile unit was 60 hours. All experiments were carried out with a fivefold repetition.

3. Research results and their discussion

Physical and mechanical properties of coatings formed by PEO obtained in the course of research on various materials are presented in the Table 1. Studies have shown that the thickness and hardness of coatings on foundry, antifriction alloy and surfaces obtained using welding wires, rods and solders is lower than on the deformable alloy D16. This is consistent with the research of other scientists and is explained by the different chemical composition of oxidized aluminum alloys, which affects both the thickness of the hardened layer and its hardness, which is provided by the phase composition of the coating [3, 10, 13, 14].

The results of the wear tests showed the following. The wear and tear of movable joints "steel - coating formed by PEO, porosity 3 ... 4%" and "steel - coating formed by PEO porosity 14 ... 15%" differs insignificantly (no more than 2 ... 3%) and accordingly makes from 15,5 to 14.9 mg (Figure 1).

Obviously, the excess porosity (14 ... 15%) does not have a significant effect on wear in the initial period of work. The movable compound "steel - coating, formed by PEO, porosity 14 ... 15% filled

Table 1. – Physical and mechanical properties of coatings formed by PEO

Indicator	De-for-mable alloy D16	Foundry alloy Ak7ch	Foun-dry alloy Ak7ch after com-pres-sion	Antifriction alloy AO3-7 after compression	The aluminum-containing solder ПA12	Aluminium-containing welding wire Svamg6, Svak5 and rod AK9M2
Thickness, μm	150	120	120	120	120	120
Hardness, GPa	22	9	10	12	8	11 10 9
Coefficient of thermal conductivity, W/m·°C	when the temperature varies from 0 to 380 ° C, it varies within the interval 0.42 .... 1.12					
Specific heat capacity, j/kg·°C	when the temperature varies from 0 to 380 ° C, changes in the range 800 1300					

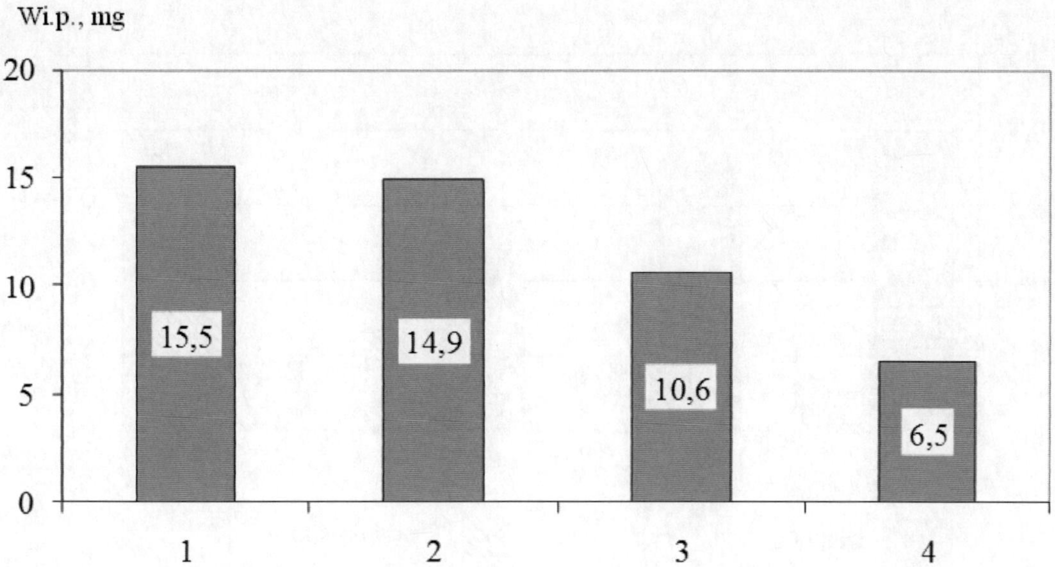

Figure 1 - Ratio of wear of movable joints in the initial period of work $W_{i.p.}$: "steel-coating formed by PEO, porosity 3 ... 4%" (1), "steel - coating formed by PEO, porosity 14 ... 15%" (2), "steel - coating formed by PEO, porosity 14 ... 15% filled with oil "(3)," steel - coating formed by PEO porosity 3 ... 4% with a copper layer "(4)

with oil", this figure is reduced and is 10.6 mg. This is facilitated by the formation of an oil film between the test samples. The oil in this case protrudes from the pores of the coating due to a sharp increase in temperature in the friction zone at the beginning of the test and helps to reduce the wear rate. The minimum wear in the initial period of operation (6.5 mg) is provided by the movable compound "steel - coating formed by PEO, porosity 3 ... 4% with a copper layer", which is facilitated by a copper layer that reduces the coefficient of friction between the sample and the counterbody.

The lowest wear rate after running-in was observed in the movable compound "steel - coating formed by PEO, porosity 3 ... 4% with a copper layer" (Figuer 2). It was 6.7 times lower than the mobile compound "steel - coating formed by PEO, porosity 3 ... 4%." The results obtained are consistent with the studies of other authors who found that the formation of a PEO-coating compound on one of the parts leads to intensive wear of the response part of the compound [15-18]. Filling the coatings with a porosity of 14 ... 15% with oil reduced the wear rate by 1.6 times. The presence of a porosity coating of 14 ... 15%, compared to a porosity of 3 ... 4%, had no negative effect on the wear rate of the test compounds.

IOP Conf. Series: Journal of Physics: Conf. Series **1058** (2018) 012074 doi:10.1088/1742-6596/1058/1/012074

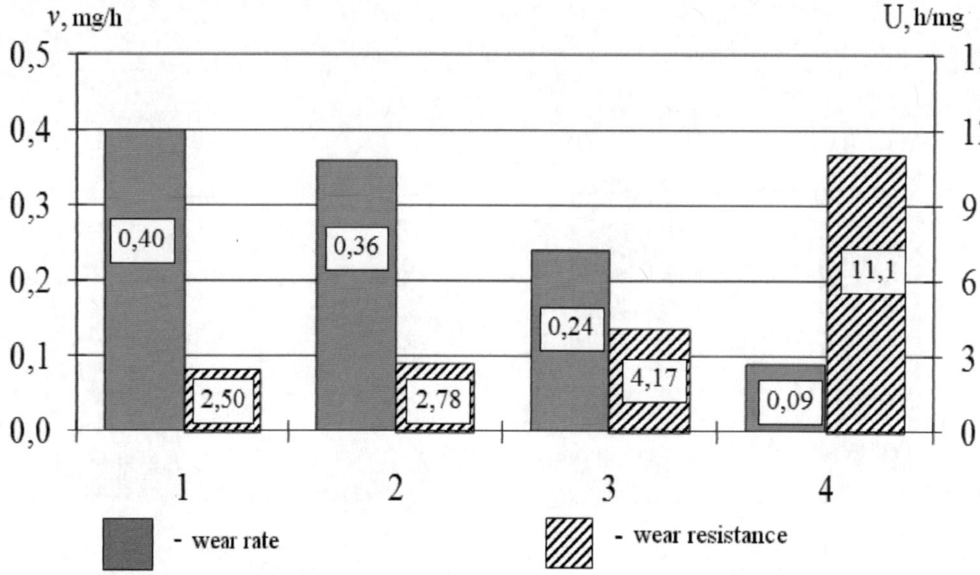

Figure 2 - Ratio of wear rates v and wear resistance of U mobile joints: "steel - coating formed by PEO, porosity 3 ... 4%" (1), "steel - coating formed by PEO, porosity 14 ... 15%" (2), "steel - coating formed by PEO, porosity 14 ... 15% filled with oil "(3)," steel - coating formed by PEO, porosity 3 ... 4% with a copper layer "(4).

The wear resistance of the investigated mobile compounds was:
1) "steel - a coating formed by PEO, porosity of 3 ... 4%":

$$U = \frac{1}{0{,}40} = 2{,}5\ h/mg$$

2) "steel - coating formed by PEO, porosity 14 ... 15%":

$$U = \frac{1}{0{,}36} = 2{,}78\ h/mg$$

3) "steel - coating formed by PEO, porosity 14 ... 15% filled with oil":

$$U = \frac{1}{0{,}24} = 4{,}17\ h/mg$$

4) "steel - coating, formed PEO, porosity 3 ... 4% with a copper layer":

$$U = \frac{1}{0{,}09} = 11{,}1\ h/mg$$

Thus, the pore filling of the oxide-ceramic coating with oil or the application of a copper layer on its surface increases the wear resistance of movable joints, respectively, by 1.7 times and 4.5 times in comparison with the conventional coating formed by PEO, porosity of 3 ... 4%. Therefore, it is advisable to use such technological methods as finishing operations when restoring worn working surfaces of parts with reinforcement of PEO, since not only wear in the initial period of work depends on them, but also the wear rate of the movable joint during the operation of the machine.

On the basis of the research, combined methods for increasing the life of machine parts using PEO have been developed. The parts are first cleaned and defective. Then, depending on the wear, the area of damage, the size and type of the renewable working surfaces, the material of the parts, and the availability of the necessary technological equipment at a particular plant, they perform their restoration. To compensate for wear, you can use: surfacing, spraying, soldering, plastic deformation, an additional repair part. Further, depending on the required resource of the mobile connection after repair, the restored working surfaces strengthen the PEO and modify the oxide-ceramic coating by one of the methods considered in the article.

After performing the recovery operations, the parts are degreased in an aqueous solution containing 5 ... 10 g / l NaOH, 40..50 g / l Na^3PO^4 and 3 ... 5 g / l Na^2SiO^3, at a temperature of 60 ... 70 ° C for 1.5 min. Then rinse with water, heated to a temperature of 40 ... 50 ° C, for 3 ... 5 minutes. Prefabricated parts are mounted in special devices, install electrodes, mount on a suspension bracket and place in an electrolytic bath [11]. When using hollow electrodes with ceramic sprayers, compressed air is supplied to them. The use of such electrodes promotes the initiation of burning of microarc discharges and the mixing of electrolyte in PEO, which provides an improvement in the quality and increase in the thickness and hardness of the strengthening oxide-ceramic coating [3, 11].

Oxidation of parts is advisable to implement in devices that are simple to manufacture. To prevent the electrolyte from falling into the PEO under protective devices and cases, which contributes to damage to the non-hardened surfaces of the parts, a silicone sealant is used.

PEO is carried out in a silicate-alkaline electrolyte of the "KOH-Na_2SiO_3" type, the rest - distilled water. Oxidation must be started at a current density of 30 ... 35 A / dm_2, and after the process is exited to a mode, reduce it to a working one - 20 A / dm_2. The duration of oxidation is usually 1.5 ... 2 hours. Depending on the purpose of the coating, the electrolyte temperature in the cooler is regulated. When coatings are used as anticorrosive coatings and to increase the wear resistance of movable joints by means of a copper layer, their porosity should be 3 ... 4%. To increase the wear resistance of mobile compounds due to the filling of coatings with oil their porosity should be 14 ... 15%. The porosity of the coatings is regulated by the temperature of the electrolyte and the modes of PEO. After the PEO, the parts are removed from the demountable blocks, washed with running water at room temperature, dried and the coating obtained is monitored.

To remove the technology layer of the coating and give the desired time-mers and geometric shapes hardened working surfaces of parts subjected to mechanical processing. At the same time, an elastic abrasive tool is used, consisting of petals of a sandpaper sandwiched between two discs. The processing is carried out by the peripheral part of the petals of the skin when the tool rotates.

Further on the part with a hardening coating formed by PEO, a porosity of 3 ... 4%, operating under conditions of boundary lubrication or friction without lubricant, a copper layer is applied by a friction-mechanical method with a special device [9, 11].

The reinforcing coating formed by PEO, the porosity of 14 ... 15% of the working surfaces of parts operating under hydrodynamic or semi-liquid (mixed) lubricants is filled with oil. For this purpose, the parts are laid in a wire basket and lowered into a bath with a spindle AU GOST 1642-75 or transformer GOST 10121-76 heated with oil (or a mixture of oil with 5 ... 8% graphite or molybdenum disulfide) heated to 100 ... 110°C and kept 1, 5 ... 2.0 h. After completing the oil coating, the parts are taken out of the bath, the residual oil is drained and dried dry.

When choosing any of the combined methods, it should be borne in mind that during operation of the parts, the contact pressure on the coating should not exceed 15 MPa. Otherwise, because of the possible deformation of the metal base, the hardened layer may break down [9, 11].

4.The conclusion

On the basis of the results of the conducted studies, it was established that the greatest thickness and hardness of oxide-ceramic coatings formed by PEO is ensured by using a deformable aluminum alloy D16. Filling the pores of the oxide-ceramic coating with oil or applying a copper layer to its surface with a frictional-mechanical method increases the wear resistance of movable joints of machine parts by 1.7 times and 4.5 times, respectively. Based on the results of the studies, combined methods have been suggested that allow significantly increasing the life of reinforced parts of machines in operation.

IOP Conf. Series: Journal of Physics: Conf. Series **1058** (2018) 012074 doi:10.1088/1742-6596/1058/1/012074

References

[1] Kolomeichenko A V, Kuznetsov I S, Izmaylov A Yu, Solovyev R Yu, Sharifullin S N 2017 *IOP Conf. Series: Mater. Sci. Eng.* **240** 012047.

[2] Kosenko A. Investigation of the thickness of coatings obtained by plasma electrolytic oxidation on aluminum alloys in an electrolyte of the "KON-H3VO3" type /A. Kosenko, Yu.A. Kuznetsov, A.V. Kolomeichenko, N.V. Titov, R.Yu. Soloviev, S.N. Sharifullin //Collected theses of the All-Russian (with international participation) conference "Low-temperature plasma physics" FNTP-2017, June 5 - 9, 2017 - Kazan, 2017. - P. 175.

[3] Suminov I.V. Plasma-electrolytic modification of the surface of metals and alloys. Suminov, P.N. Belkin, A.V. Eppelfeld et al. - M .: Technosphere. - 2011. - 512 p.

[4] Minaev A.N. Composite coatings formed by plasma electrolytic oxidation /Minaev, S.V. Gnedenkov, S.L. Sinebryukhov et al. //Corrosion: materials, protection. - 2011. - No. 3. - Pp. 1-10.

[5] Shatrov A.S. Protective oxide-ceramic coatings formed on aluminum alloys by the method of plasma electrolytic oxidation /AC. Shatrov //Corrosion: materials, protection. - 2012. - № 4. - P. 28 - 37.

[6] Dudareva N.Yu. Adhesion strength of a plasma electrolytic coating formed on a high-silicon alloy by the microarc oxidation method /N.Yu. Dudareva, R.V. Kalshchikov, F.F. Musin et al. //Bulletin of the Irkutsk State Technical University. - 2014. - No. 10 (93). - Pp. 38 - 43.

[7] Egorkin VS Formation of protective coatings on aluminum and steel by the method of plasma electrolytic oxidation /V.S. Egorkin, A.V. Puz, OA Khrisanfova et al. //Bulletin of the Engineering School of the Far Eastern Federal University. - 2011. - No. 1 (6). - Pp. 58 - 75.

[8] Dehnavi V., Effect of duty cycle and applied current frequency on plasma electrolytic oxidation (PEO) coating growth behavior /Dehnavi V., Li Luan V., Shoesmith D. Liu X., Rohani S. //Surface & Coatings Technology. - 2013. - Vol. 226. - Pp. 100 - 107.

[9] Basinyuk V.L. Method of friction-mechanical formation of antifriction coatings on Al2O3 /VL. Basinyuk, A.V. Kolomeichenko, V.A. Kukareko et al. //Friction and wear. - 2005. - T. 26. - No. 5. - Pp. 530-538.

[10] Dunkin ON Influence of microarc oxidation parameters on the properties of coatings formed on aluminum alloys /O.N. Dunkin, A.P. Efremov, B.L. Crete //Physics and chemistry of material processing. - 2000. - №2. - Pp. 49 - 53.

[11] Kolomeichenko A.V. Technologies of restoration and strengthening of details of agricultural machinery by microarc oxidation. Textbook /AV. Kolomeichenko, N.V. Titov, V.N. Logachev et al. - Eagle: Ed. Eagle GAU, 2013. - 131 p.

[12] Saifutdinov A I, Fairushin I I and Kashapov N F 2016 *JETP Lett.* **104** 180–185

[13] Butyagin P. Microarc oxidation. Properties of coatings /P. Butyagin , T. Unilateral, S. Safronova //Nanoindustry. - 2013. - No. 5. - Pp. 48 - 51.

[14] Kuznetsov Yu.A. Restoration and strengthening of parts of APC machines and equipment by microarc oxidation. - Author's abstract. diss. Doct. tech. sciences. - M :, 2006. 35 p.

[15] Kuchmin I B, Nechaev G G and Solovyov N D 2015 *Physics and Chemistry of Materials Processing* **4** 45–49

[16] Prozhega M.V. Investigation of the influence of the substrate material on the wear resistance of MDO coatings. /M.V. Prozhega, N.N. Smirnov, O. Somov. et al. //Friction and lubrication in machines and mechanisms. - 2015. - No. 2. - P. 44 - 48.

[17] Kuznetsov Yu.A. Anti-wear properties of ceramic coatings obtained by microarc oxidation. Kuznetsov // Mechanization and electrification of agriculture. - 2004. - No. 6. - P. 28.

[18] Kozlov A.V. Increase of wear resistance of machine parts by microarc oxidation with subsequent modification of coating. - Author's abstract. diss. Cand. tech. sciences. - Ufa, 2014. - 16 p.

[19] Volkhin A.M. Tribotechnical characteristics of composite porous MDO coatings impregnated with ultrahigh molecular weight polyethylene. - Author's abstract. diss. Cand. tech. sciences. - M., 2013. - 26 p.

IOP Conf. Series: Journal of Physics: Conf. Series **1058** (2018) 012075 doi:10.1088/1742-6596/1058/1/012075

Finished plasma strengthening and restoration of fuel equipment details

P A Topolyansky[1], S N Sharifullin[2,3], N R Adigamov[3], S A Ermakov[1], A P Topolyansky[1]
[1]Research-and-production firm "Plazmacenter"
[2]Kazan (Privolzhsky) Federal University
[3]Kazan State Agrarian University

E-mail: saidchist@mail.ru

Abstract. The results of the investigation of the physical and mechanical properties of diamond-like coatings of the DLCPateks type (a-C: H / a-SiOCN) obtained on friction surfaces by transporting the atomic and molecular flux of vapor particles of liquid chemical compounds by a plasma jet of an arc plasma torch of atmospheric pressure are presented. The layer formed on the working surfaces is a non-metallic amorphous multilayer coating with a low coefficient of friction, increased microhardness, chemical inertness, hydrophilicity, high heat resistance and dielectric characteristics. To minimize the possible defectiveness of the main material, it is proposed to apply thin-film coatings to them at the final stage of manufacturing fuel equipment parts.

1. Introduction.

Recently, there has been a demand for technologies to strengthen and restore wear parts of fuel equipment for diesel engines that can extend their life, save financial, material and time costs of the consumer. In connection with this, the development of effective technologies for hardening and restoring fuel equipment parts is a promising direction of research [1 - 20]. This is also facilitated by the widespread use of machinery and equipment with foreign fuel equipment, for which the use of repair technologies is particularly economically justified.

One of the leaders in the diesel engine industry is the common rail injection fuel system manufactured by Bosch, Delphi, Denso, Siemens, etc. Almost every second diesel engine is equipped with such systems, therefore the development and research of technologies for the recovery of wear parts of this equipment is also actual task.

The main load on the parts that work under friction and wear conditions is perceived by a thin surface layer, and therefore wear resistant thin-film coatings are most often used. For a rational choice of coating materials and technologies for their application in order to increase the resource, reliability, and restoration of precision parts of fuel equipment, it is advisable to consider the conditions of their operation, the causes of wear and tear, to develop a methodology for choosing the optimal coating and the technology for applying it, and to investigate specific properties of coatings.

Coating technologies for hardening and restoring fuel equipment parts can be conditionally divided into three groups depending on the hardness of the created surface layer:

1. Hardness $\leq$ 5 GPa (less hardness of the base material):

IOP Conf. Series: Journal of Physics: Conf. Series **1058** (2018) 012075 doi:10.1088/1742-6596/1058/1/012075

- friction-mechanical brass; - alumochromophosphation; - ion-plasma spraying of the TiN-Cu-MoS_2 coating; - electrochemical-mechanical treatment using a renovation fluid in the form of a solution of polyethylene glycol with zinc; - electrospark deposition of copper-graphite coatings; - application of fluoroorganic surfactants (epilam);

- application of metal-ceramic coating from natural layered silicates on the basis of geoactivators (vermiculite, serpentinite, etc.).

2. Hardness 5-8 GPa (of the order of hardness of the base material):

- electrolytic chromium plating; - diffusion chrome plating; - carbonitriding;

- electrolytic and gaseous chromium by thermal decomposition of chromium hexacarbonyl; - sulfochroming; - carbonitriding with the introduction of an activating coating.

3. Hardness $\geq$ 11-12 GPa (exceeding the hardness of abrasive particles): - ion-plasma spraying of TiN coatings; - Electrolytic chromium plating with additional modification by ultradisperse diamonds; - Electrolytic chromium plating with dielectric filler (aluminum oxide); - ion-plasma spraying of nanostructured TiN coatings; - coatings applied using physical (PVD) and chemical (CVD) vapor deposition coatings; - finishing plasma hardening with application of multi-layer wear-resistant coatings.

The technologies discussed above, as well as structural, technological and operational factors associated with wear resistance, have allowed us to formulate a methodology for selecting the optimal coating process for the purpose of hardening and restoring the details of fuel equipment. In accordance with the considered concept of choosing the technology for improving the durability of fuel equipment parts, we consider the final plasma hardening (FPH) with the application of a diamond-like coating DLCPateks (a-C: H / a-SiOCN). The FPH process is based on the decomposition of vapors of liquid chemical compounds introduced into the arc discharge plasma and the formation of an atomic and molecular particle flux in a plasma jet reactor. Heating of articles with FPH does not exceed 150 ° C. As a result, a nonmetallic amorphous multilayer coating with low friction coefficient, increased microhardness, chemical inertness, hydrophilicity, high heat resistance and dielectric characteristics is formed on the working surfaces on the working surfaces.

The FPH also cleanses parts from process contaminants (microparticles, particles of processed and cutting material, metal oxides, components of lapping pastes, oil residues, solvents, fats, technological solutions, dust, sand, moisture, decomposition products of working and process fluids, microorganisms and products of their activities).

The analysis shows that about 80% of the plunger pairs initially have an increased clearance between the plunger and the bushing. Therefore, during the replacement of parts during the repair of fuel equipment, it is also advisable to produce their FPH with coating on the working surfaces of the DLCPateks coating, which will reduce the initial gap and will prevent the working surfaces from grasping.

2.Methods of conducting research.

When studying the properties of DLCPateks coated with FPH technology, the heat-treated steel ShKh15 was used as the substrate material. The thickness of the applied coating was of the order of 1 μm. To determine the coefficient of friction of the DLCPateks coating, the tests were carried out according to the "ball-disk" scheme using balls with a diameter of 3 mm made of silicon nitride Si_3N_4. The load on the counterbody was 5 N. The linear slip velocity was 10 cm /s. The friction path is 80-100 m. Nissan Oil SAE 5W-40 engine oil was used in the tests. Amorphous coating studies were carried out using a transmission electron microscope JEM 2100 (JEOL, Japan). To measure roughness parameters according to EN ISO 13565-2: 1996, the "Profile" measuring and computing complex was used. The adhesion of the DLCPateks coating to SHKH15 steel was determined by the scratch test method with the determination of the fracture initiation load at longitudinal displacement and its variable force on the diamond indentor.

3. The results of the research and their discussion.

Analysis of the results of nanoindentation revealed that with increasing load and, accordingly, with increasing contact depth, the hardness decreases, which characterizes the DLCPateks coating as gradient. Within one load, the hardness varies from 14 GPa to 23 GPa, which is typical of multicomponent coatings. The average properties of the DLCPateks coating: nanohardness is 18 GPa, Young's modulus is 127 GPa, elastic recovery is 87%. In Fig. 1 is a diagram of the indentation of DLCPateks coverage. The resistance of the surface layer to elastic deformation (plasticity index) of H_{IT} / E_r is 0.14. The Young's modulus of SHX15 is 211 GPa. High value of plasticity index provides an increased life in the conditions of cyclic loads, and the proximity of the modulus of elasticity of the coating and the substrate contributes to the reduction of technological stresses on the interface and the increase of adhesive strength. In Fig. 2 is a diagram characterizing the decreasing character of the change in the coefficient of friction (lower curve) with the time of DLCPateks coating. The average value of the friction coefficient is 0.024.

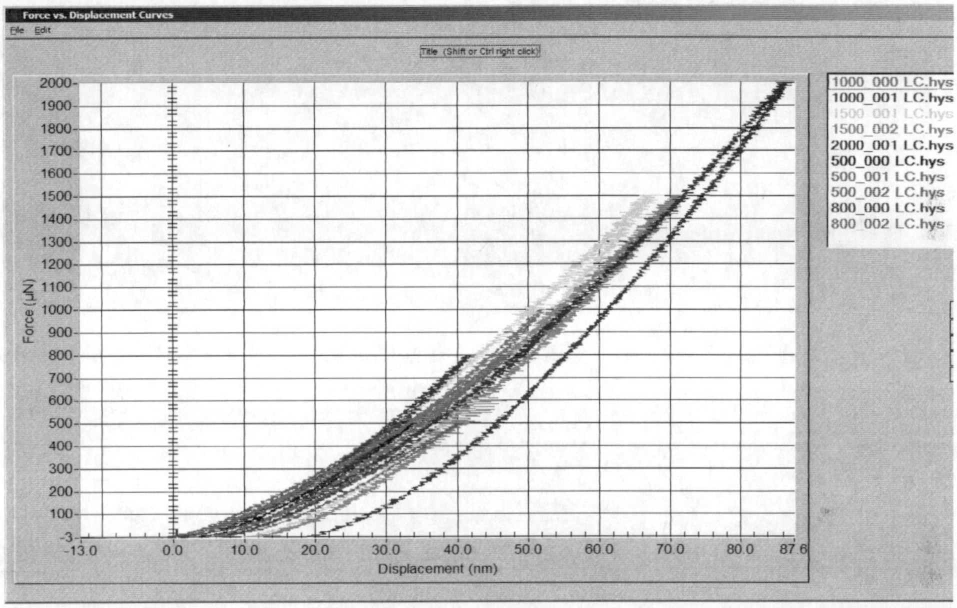

Fig. 1. F-h indentation diagrams for different loads of the DLCPateks coating (TI 750Ubi, Hysitron, USA)

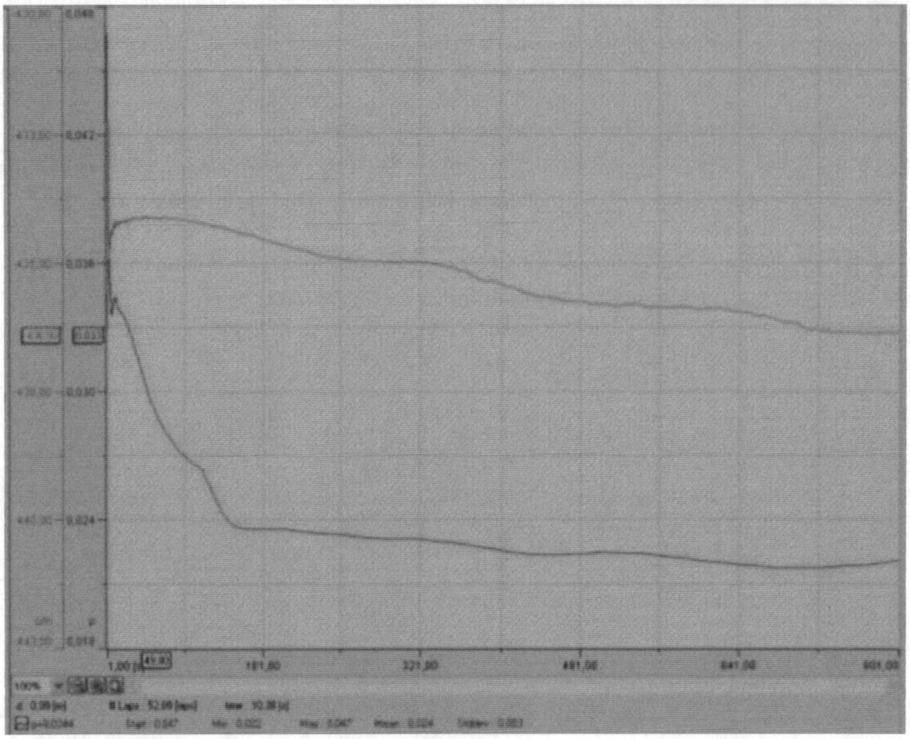

Fig. 2. Change in friction coefficient with coating time DLCPateks (TRB-S-DE, CSM-Instruments, Switzerland)

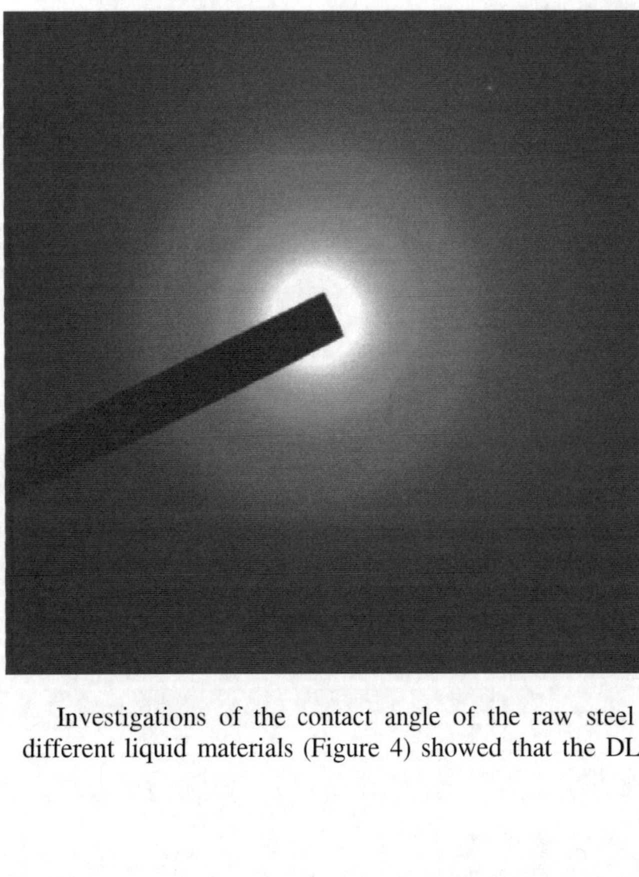

Studies with the JEM 2100 transmission electron microscope (JEOL, Japan) showed that the DLCPateks coating is amorphous (Figure 3) and represents a 60-100 Å structure. The amorphous (glassy) state of the coating material (in contrast to the crystalline state) is characterized by the absence of grain boundaries and dislocation defects, ensures the maximum effective filling of the cavities of the substrate profile.

Fig. 3. Electronogram of DLCPateks covering its amorphous character

Investigations of the contact angle of the raw steel SHKH15 without coating and coating for different liquid materials (Figure 4) showed that the DLCPateks coating provides more hydrophilic

IOP Conf. Series: Journal of Physics: Conf. Series **1058** (2018) 012075 doi:10.1088/1742-6596/1058/1/012075

properties of the surface (has a smaller angle of wetting). This leads to an increase in surface energy, which increases the wettability of the fuel and, accordingly, its lubricity.

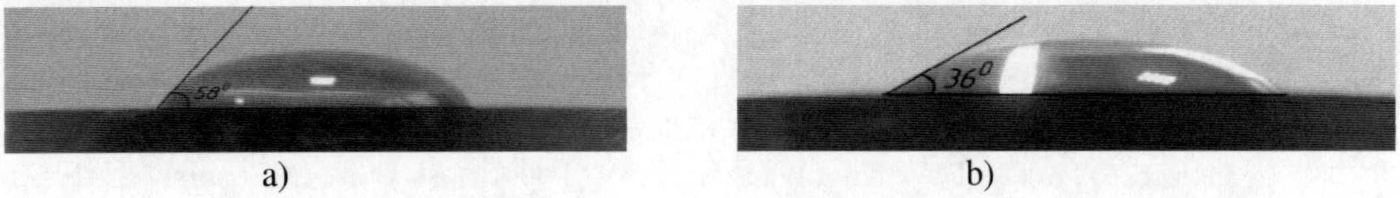

Fig. 4. The contact angle of wetting, where a) - is uncoated; b) - coated with DLCPateks (OCA 15EC, DataPhysics Instruments GmbH, Germany)

The adhesion of the coating DLCPateks to steel SHKH15 was determined by the method of scrap testing in accordance with ISO 20502: 2005. The measured load at which the depth of occurrence of the indenter in the composition "coating-substrate" ceases to grow smoothly, indicates the moment of destruction of the coating and characterizes its adhesion to the substrate. The loading of the DLCPateks coating with a thickness of 1.2 μm was 20 mN (Figure 5).

After coating with DLCPateks, the surface roughness parameters are improved. To measure roughness parameters according to EN ISO 13565-2: 1996, the "Profile" measuring and computing complex was used. When comparing the reference curves after the FPU, the parameter R_{pk}, which characterizes the height of the projections that wear out during run-in, decreased by 1.3 times; The parameter R_k, which characterizes the basis of the profile, decreased by 3.9 times.

Investigations of the three-dimensional surface topography using MarSurf WS1 from Mahr GmbH (Germany) transition zones DLCPateks coating SHKH15 steel backing also showed that after the coating is applied, the deep depressions of the surface relief are healed.

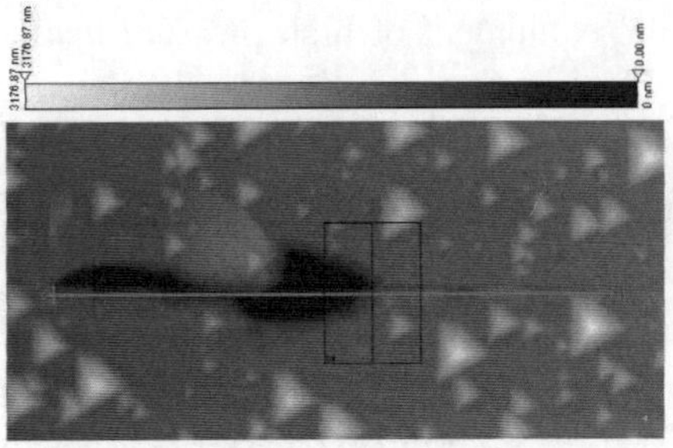

Fig. 5. Three-dimensional surface relief and scratches, obtained by the scanning probe microscopy of the DLCPateks coating in determining the fracture load

When studying the electrical properties of the DLCPateks coating, its specific electrical resistance is determined, which is 106 Ohm · m. These dielectric characteristics of the coating should ensure the absence of electrochemical and electromechanical phenomena in friction.

Experimental bench tests to meet the tolerances of test plans have proved the possibility of using FPH technology with DLCPateks coating when restoring rod and multiplier valves, spray nozzles and Common Rail fuel injector valves. Currently, the plunger-type parts and the DLCPateks-coated prism are being tested at the UK MC "Altai Precision Products Plant" (Barnaul) on a non-motorized stand Hartridge Cri-PC.

4. Conclusions.

1. The most promising ways to harden and restore parts of fuel equipment are processes that exclude subsequent abrasive processing, for example, the FPH process with the application of a diamond-like coating of DLCPateks.

2. The effectiveness of the diamond-like coating of DLCPateks is determined by its amorphous state, hydrophilic properties, optimal physicomechanical, tribological and dielectric characteristics, the creation of compressive stresses on the surface, the reduction of roughness parameters, and increased adhesion.

References.

[1] Topolyansky P.A. Application of international standards to assess the properties of surfaces subject to wear /P.A. Topolyansky, A.P. Topolyansky, N.A. Sosnin, S.A. Ermakov. //Technology of mechanical engineering. - 2010. - No. 6. - Pp. 56 - 60.

[2] Kanaev A.T. Certification of materials and coatings for micro abrasive wear parameters /A.T. Kanaev, P.A. Topolyansky, S.A. Ermakov, A.P. Topolyansky. //Bulletin of Science of the Kazakh Agrotechnical University. S. Seifullin. - 2017. - No. 2 (93). - Pp. 111 - 119.

[3] Topolyansky P.A. Testing of tribological coatings for micro abrasive wear /P.A. Topolyansky, S.A. Ermakov, A.P. Topolyansky. // Mechanics and tribology of transport systems. - Rostov-on-Don, on 8-10.11.2016: in 2 tons - Rostov-on-Don: FGBOU IN RGUPS. - 2016. - T. 2. - P. 217 - 223.

[4] Adigamov N.R. Plasma technologies in increasing the efficiency of high-pressure fuel pumps of diesel engines /N.R. Adigamov, V.P. Lyalakin, R.Yu. Soloviev, S.N. Sharifullin. //Welding production. - 2016. - No. 2. - Pp. 49 - 51.

[5] Sharifullin S N and Pirogova A S 2017 *J. Phys.: Conf. Series*. **789** 012051

[6] Solovev R Y, Sharifullin S N and Adigamov N R 2016 *J. Phys.: Conf. Series* **669** 012050

[7] Sharifullin S N and Dunayev AV 2016 *IOP Conf. Series: Mater. Sci. Eng.* **134** 012025.

[8] Dounaev, A., Sharifullin, S. Friction surfaces modification using tribo-compounds (2014) World Applied Sciences Journal, 31 (2), pp. 272-276.

[9] Sharifullin S N, Adigamov N R, Adigamov N N, Solovev R Y and Arakcheeva K S 2016 *J. Phys.: Conf. Series* **669** 012049

[10] Kadyrmetov A M, Sharifullin S N and Maltsev A F 2016 *J. Phys.: Conf. Series* **134** 012009

[11] Sagin S.V. Tribotechnical processes occurring in the fuel equipment of high pressure marine diesel engines /S.V. Sagin. //Glimpses of technology. - 2013. – No. 2. – Pp. 33 – 42.

[12] Bodrov A.S. Investigation of the causes of failure of the needle of the atomizer of diesel fuel injectors /A.S. Bodrov, A.A. Katunin, A.S. Markin and etc. // Problems of research of systems and means of motor transport. Materials MNTC. - 2017. – Vol. 1. – Tula. – Pp. 34 – 40.

[13] Timofeev S.S. increase of wear resistance of plunger pairs. Surface engineering and renovation of products /S.S. Timofeev. //Proceedings of 10-th international science-technical conference. May 24-28, 2010. -Kyiv: ATM of Ukraine. - 2010. – Pp. 194 – 196.

[14] Timofeev S.S. Improvement of wear resistance of parts of fuel equipment of diesel engines /S.S. Timofeev, A.Y. Movshovich, V.N. Ostapchuk/ //surface Engineering and renovation of products. Materials of the 6th international scientific and technical conference. 30.05-01.06. - 2006. Yalta-Kiev. – Pp. 143-145.

[15] Lebedev A.T. Recovery of the plunger /A.T. Lebedev, P.A. Lebedev. // Mechanization and electrification of agriculture. - 2010. – No. 1. Pp. 23 – 24

[16] Gaidar S.M. Restoration of the operating parameters of plunger pairs and improving the reliability of the fuel system of diesel engines /S.M. Gaidar. //The international technical-economic journal. - 2010. – No. 1. – Pp. 54 – 58.

[17] Skoblo T.S. A study of the influence of ultradispersed diamonds with the restoration and hardening of parts of the nodes of fuel equipment /T.S., Skoblo, O. Y. Klochko, A.I., Sidorenko, A.V. Plugatarev, A.K. Olejnik. //Problems of Tribology. - 2015. – No. 1 – Pp. 106 – 111.

[18] Jackin C. S. Increase of adhesion of electroplated composite coatings used in the restoration plunger pairs fuel pump /S.U. Jacquin, M.N. Krasnova, N.A. Pen'kov, A.I. Krasnov. – Proceedings of GOSNITI. - Moscow: - 2015. - T. 119. - H 1. – S. 54 – 60.
[19] Gabitov I.I. Hardening of valve assemblies electro-hydraulic injectors of diesel engines at repair /I.I. Gabitov, R.N. Saifullin, A.R. Valiev. //Hardening technology and coatings. - 2017. – No. 7. –Pp. 328 – 335.
[20] Hershberger J. Evaluation of DLC coatings for spark-ignited, direct-injected fuel systems /Hershberger J., Ozturk O., Ajayi O.O., Woodford J.B., Erdemir A., Erck R.A., Fenske G.R. //Surface and Coatings Technology. – Vol. 179 (2004). – Pp. 237 – 244.

IOP Conf. Series: Journal of Physics: Conf. Series **1058** (2018) 012076 doi:10.1088/1742-6596/1058/1/012076

Modeling of radio frequency argon dielectric barrier discharge at atmospheric pressure in argon

A A Saifutdinova
A.N.Tupolev Kazan National Investigation Technical University

E-mail:aliya_2007@list.ru

Abstract. In this work modeling of atmospheric pressure barrier discharge in argon for rf-voltage signal applied to the electrodes was conducted. The model included balance equations for the densities of charged (electrons, ions) and the excited particles, the electron energy density, and the Poisson equation for the electric potential. The fluxes of charged and excited particles (electrons, ions) and electron energy were given in the drift-diffusion form. Distributions of spatiotemporal parameters of the discharge were obtained.

1. Introduction

Atmospheric pressure dielectric barrier discharges (DBD) have recently commanded much attention, fuelled by their promise to rival the widely used low-pressure glow discharges and their facilitation of numerous applications through the removal of the usually indispensable vacuum chamber [1,2]. They are capacitive nonthermal plasmas generated between two parallel electrodes that are either metallic [3,4] or coated with a dielectric layer[1,5], and their generation has been achieved over a very wide spectrum from kilohertz[1,5] and megahertz [3,7] to microwave [8]. They can also be generated as surface-wave discharges [9]. Today atmospheric pressure DBD are finding wide ranging applications including etching, deposition, surface modification, and sterilization [1,2]. The progress of their fundamental understanding, on the other hand, lags markedly behind their technological advancement largely due to comparatively few theoretical studies so far [10–13].

In this paper, atmospheric pressure (DBD) in argon initiated by the applied rf-voltage is numerically investigated.

2. Description of the Mathematical Model

The spatiotemporal characteristics of the DBD were determined using an extended hydrodynamic model describing the parameters of the gas-discharge plasma [17]. The model is based on the drift-diffusion equations and includes k equations for the densities of charged (electrons, ions) and excited particles n_k and the balance equation for the electron energy density n_ε. The self-consistent electric field is determined from Poisson's equation for the potential φ. The system of equations has the form

$$\frac{\partial n_k}{\partial t} + \nabla \cdot \mathbf{\Gamma}_k + (\mathbf{u} \cdot \nabla) n_k = \sum_j c_{k,j} R_j, \quad (1)$$

$$\frac{3}{2}\frac{\partial n_\varepsilon}{\partial t} + \nabla \cdot \mathbf{Q} + (\mathbf{u} \cdot \nabla)\, n_\varepsilon = \\ = -\, q_e \mathbf{\Gamma}_e \cdot \mathbf{E} - S_{el} - \sum_j \Delta\varepsilon_j R_j, \tag{2}$$

$$\Delta\varphi = -\frac{q_e}{\varepsilon_0}\left(\sum_{k=1}^{N} Z_k n_k\right), \quad \mathbf{E} = -\nabla\varphi. \tag{3}$$

Here, q_e is the charge of an electron, ε_0 is the permittivity of vacuum, Z_k is the charge of a particle of species k, $\mathbf{E}$ is the electric field, $n_\varepsilon = n_e\bar{\varepsilon}$ is the average electron energy density, and $\bar{\varepsilon}$ is the average electron energy. The flux densities of charged particles $\mathbf{\Gamma}_k$, as well as the electron energy flux density $\mathbf{Q}$, are described in the drift-diffusion approximation,

$$\mathbf{\Gamma}_k = -D_k \nabla n_k + z_k \mu_k \mathbf{E} n_k, \tag{4}$$

$$\mathbf{Q} = -D_\varepsilon \nabla n_\varepsilon - \mu_\varepsilon \mathbf{E} n_\varepsilon. \tag{5}$$

Here, μ_k and D_k are the mobilities and diffusion coefficients of charged particles, respectively; $\mu_\varepsilon = 5/3\mu_e$; and $D_\varepsilon = 2/3\mu_\varepsilon T_e$, where μ_e is the electron mobility and $T_e = 2/3\bar{\varepsilon}$ is the electron temperature defined as 2/3 of the average electron energy.

The right-hand side of Eq. (2) is determined by the reactions occurring in the discharge. Each reaction makes a positive contribution to the source function if it results in the production of the corresponding particle species and a negative contribution if the particles of this species disappear. In this equation, j is the reaction index (28 plasma chemical reactions in total were taken into account) and $c_{k,j}$ is the number of particles of the kth species produced in the jth reaction ($c_{k,j}$ can be either positive or negative). The reaction rates R_j are determined by the rate constants $k_j(T_e)$ of the corresponding processes and are proportional to the product of the densities of the reacting species, $R_j = k_j(T_e) \cdot \prod_{k=1}^{N} n_k$, where $N = 2$ for reactions involving two species and $N = 3$ for reactions involving three species. The first summand on the right-hand side of Eq. (2) describes Joule heating of electrons in the electric field. The second summand, $S_{el} = (3/2)\delta\nu_{el} n_e (T_e - T)$ describes energy exchange between electrons and neutral gas particles due to elastic collisions. Here, T is the gas temperature, which is assumed to be equal to room temperature ($T = 293.15$ K), and $\delta = 2m_e/M_a$ is the doubled ratio of the electron mass m_e to the gas atom mass M_a. The third summand takes into account inelastic energy losses of the electron gas. Here, $\Delta\varepsilon_j$ is the energy lost (or acquired if $\Delta\varepsilon_j < 0$) by an electron in a given reaction and R_j is the reaction rate, which is determined by the rate constant of the corresponding inelastic process, $R_j = k_j(T_e) n_e n_n$, where n_e is the electron density and n_n is the density of neutral particles.

The set of plasmachemical reactions used to describe a discharge in argon is presented in Table 1. It takes into account three effective excited atomic levels (metastable Ar^m, resonance Ar^r, and excited Ar* ones), two effective excimer levels (Ar_2^* and Ar_2^{**}), and three positive ion species (Ar^+, Ar_2^+, and Ar_3^+). In addition, the effective process $e + \mathrm{Ar} \leftrightarrow e + \mathrm{Ar}^{\mathrm{eff}}$ with a threshold energy of $\Delta\varepsilon = 13.9$ eV was introduced to correctly take into account inelastic energy losses of the electron gas in Eq. (2). The cross section for this process was found similarly to [18] by subtracting the cross sections for individual processes of excitation of high-lying levels (above 13.9 eV) [17] from the total excitation cross section taken from.

The coefficients of electron mobility and diffusion, as well as the rate constants of elastic and inelastic electron-impact processes, were calculated using the electron energy distribution function (EEDF). The EEDF was found by solving the Boltzmann kinetic equation in the two-term (f_0, f_1) approximation under the assumption of a local dependence on the electric field strength. The kinetic equation took into account electron heating in the longitudinal electric field and the change in the electron energy due to elastic, electron–electron, and inelastic collisions.

All constants, transport coefficients (6), and the effective temperature were first calculated as functions of E/N. Then, by virtue of the monotonic dependence $T_e(E/N)$, the constants and transport coefficients were redefined as functions of the effective temperature. The distribution of the latter was found from the electron thermal balance equation (2), in which not only volumetric processes but also heat conduction transfer were taken into account. Therefore, there was a spatial shift between the distributions of the field, electron density, and electron temperature. In other words, the profile $T_e(x)$ and, accordingly, the profile of the electron-impact ionization rate were spread over the electron thermal relaxation length; i.e., there appeared a nonlocal dependence of the discharge parameters on the electric field.

The ion mobilities $\mu_{Ar^+}p = 6.86\times10^{-4}$ m^2/(V s), $\mu_{Ar_2^+}p = 5.14\times10^{-4}$ m^2/(V s), and $\mu_{Ar_3^+}p = 4.84\times10^{-4}$ m^2/(V s), where p is the gas pressure, were taken from, while the diffusion coefficients were found from the Einstein relation.

The boundary conditions for the electron number density, electron energy density, and densities of ions and excited particles were specified at the boundaries of the discharge volume (i.e., at the inner surfaces of the dielectrics) as follows:

$$-\mathbf{n}\cdot\mathbf{\Gamma_e} = -\alpha n_e\mu_e\mathbf{E}\cdot\mathbf{n} + \qquad (7)$$

$$+\frac{1}{2}\nu_{th}n_e - (1-\alpha)\sum_p \gamma_p(\mathbf{\Gamma_p}\cdot\mathbf{n}),$$

$$-\mathbf{n}\cdot\mathbf{Q} = -\alpha n_\varepsilon\mu_\varepsilon\mathbf{E}\cdot\mathbf{n} + \qquad (8)$$

$$+\frac{5}{6}\nu_{e\,th}n_\varepsilon - (1-\alpha)\sum_p \gamma_p\bar{\varepsilon}_p(\mathbf{\Gamma_p}\cdot\mathbf{n}),$$

$$-\mathbf{n}\cdot\mathbf{\Gamma}_k = \left(\frac{1}{4}\nu_{th}n_e\right) - \alpha\mu_k n_k\mathbf{E}\cdot\mathbf{n}, \qquad (9)$$

where $\mathbf{n}$ is the unit normal vector; $\gamma_p = 0.02$ is the secondary electron emission yield from the dielectric surface under bombardment by particles of the pth species; $\mathbf{\Gamma_p}$ is the flux of particles of the pth species onto the electrode; $\bar{\varepsilon}_p$ is the average energy of electrons emitted due to bombardment by particles of the pth species; and $\nu_{e\,th} = \sqrt{8k_BT_e/\pi m_e}$ and $v_{th} = \sqrt{8k_BT_k/\pi M}$ are the thermal velocities of electrons and heavy plasma particles, respectively (here, T_e is expressed in eV and it is set $T_k = T_g$). It is assumed that $\alpha = 1$ if the drift component of the flux of charged particles of a given species, $z_kn_k\mu_k\mathbf{E}$, is directed to the wall (i.e., if $z_k\mu_kn_k\mathbf{n}\cdot\mathbf{E} > 0$) and $\alpha = 0$ in the opposite case. Here, the heavy particle (an ion or an excited particle) of a given species is assumed to lose its charge and decay (in our case, into Ar atoms) on the dielectric surface, i.e., a set of surface reactions is specified.

The potentials of the metal electrodes (adjacent to the outer boundaries of the dielectric) were specified as the boundary conditions for Poisson's equation:

$$\varphi = 0 \qquad (10)$$

at the grounded electrode and

$$\varphi = V(t) \qquad (11)$$

at the powered electrode.

At the interface between the plasma and dielectric, the discharge is accumulated at the latter. For the jump of the normal component of the electric induction at the dielectric surface, $\mathbf{D_1} - \mathbf{D_2}$, we have

$$\mathbf{n}\cdot(\mathbf{D_1} - \mathbf{D_2}) = \varepsilon_1\varepsilon_0\mathbf{n}\cdot\mathbf{E}_1 - \varepsilon_2\varepsilon_0\mathbf{n}\cdot\mathbf{E}_2 = \sigma, \qquad (12)$$

where ε_1 and ε_2 are the relative permittivities of the plasma and dielectric, respectively; ε_0 is the permittivity of vacuum; E_2 and E_1 are the normal components of the electric field in the dielectric and plasma, respectively; and σ is the surface charge density accumulated on the barrier surface (it is defined only at the plasma–dielectric interface; in the one-dimensional case, it consists of two points).

For the surface charge density σ, the following equation is solved separately:

$$\frac{d\sigma}{dt} = -\mathbf{n} \cdot e\left(\sum_{k} z_k \mathbf{\Gamma}_k - \mathbf{\Gamma}_\mathbf{e}\right), \tag{13}$$

where the sum over k is extended to all ion species.

3. Numerical simulation and results

To model the spatiotemporal behavior of a dielectric barrier rf-discharge, a one-dimensional computation domain was defined. The discharge gap length was specified to be equal to L = 1 mm, while the thicknesses of the dielectric barriers were assumed to be d = 1 mm. The relative permittivity was set at 4, which corresponds to quartz. The feeding voltage amplitude was 1.5 kV.

Figure 1 *a* and *b* present the spatial distribution of the electron number density and electron temperature (along the horizontal axis) extruded in time for one period of the external excitation along the vertical axis. These results show that this discharge is similar to lower pressure RF capacitively coupled discharges. The electron density is static in time in the discharge center. However, near the wall the electron density evolves in time driven by the applied time–varying electric field. The electron temperature near the wall is also strongly time modulated attaining peak electron temperatures of 6.9 eV. In the plasma bulk the electron temperature is much cooler than near the walls and is much less modulated since the electric field is much less intense than in the sheaths.

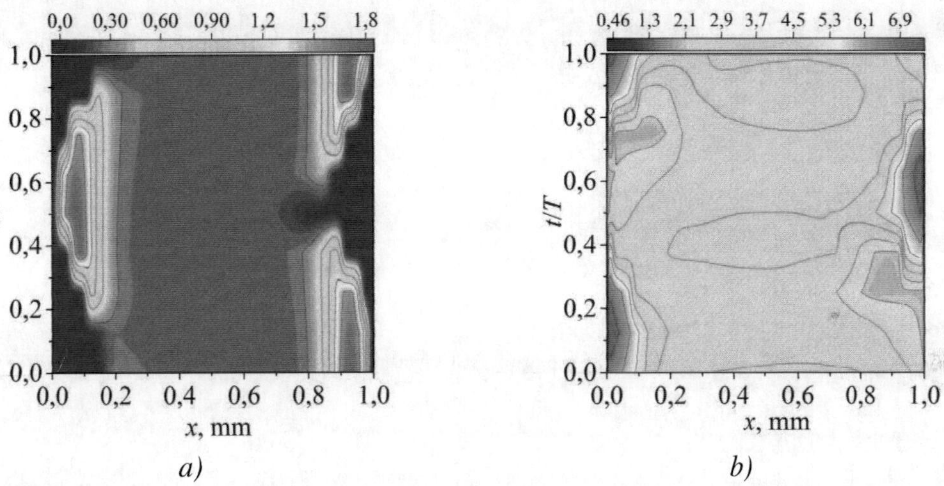

Figure 1. *a)* Electron number density and *b)* electron temperature spatial and time evolution.

4. Conclusions

Based on the one-dimensional extended fluid model, the characteristics of dielectric barrier discharge excited rf voltage are numerically investigated in atmospheric pressure argon. All the main discharge characteristics were obtained.

Acknowledgments

This work was supported by the RFBR, project no. 18-31-00098 mol_a.

References

[1] J R Roth, Industrial Plasma Engineering, Principles Vol. 1 (Institute of Physics, Bristol, 1995), pp. 453–461.
[2] E E Kunhardt, *IEEE Trans. Plasma Sci.* **28**, 189 (2000).
[3] J. Park, I. Henins, H W Herrmann, G S Selwyn, J Y Jeong, R F Hicks, D Shim, and C S Chang 2000 *Appl. Phys. Lett.* **76**, 288.

[4] R. H. Stark and K. H. Schoenbach, *J. Appl. Phys* 1999 **85**, 2075.
[5] S Kanazawa, M. Kogoma, T. Moriwaki, and S. Okazaki, *J. Phys.* 1988 *D* **21**, 838.
[6] A I Saifutdinov, A A Saifutdinova and N F Kashapov 2016 *Journal of Physics: Conference Series* **669** 012044.
[7] H Koinuma, H. Ohkubo, T Hashimoto, K Inomata, T Shiraishi, A Miyanaga, and S Hayashi, *Appl. Phys. Lett*. 1992 **60**, 816.
[8] M J Shenton and G. C. Stevens, *IEEE Trans. Plasma Sci*. 2002 **30**, 184.
[9] M Moisan, Z Zakrzewski, R Etmadi, and J C Rostaing, *J. Appl. Phys*. 1998 **83**, 5691.
[10] G M Petrov et al., *Plasma Chem. Plasma Process*. 2000 **20**, 183.
[11] X Yuan and L L Raja, *IEEE Trans. Plasma Sci*. 2003 **31**, 495.
[12] M. G. Kong and X. T. Deng, *IEEE Trans. Plasma Sci*. 2003 **31**, 7.
[13] Yu B Golubovskii, V A Maiorov, J Behnke, and J F Behnke, *J. Phys. D* 2003 **36**, 975.
[14] A I Saifutdinov, A A Saifutdinova, and B A Timerkaev 2018 *Plasma Physics Repotrs* **44**, 3.

On the question of structure of ZnO thin films formed by IBAD and subsequently implanted with silver ions

N M Lyadov[1], R N Kashapov[1,2], I R Vakhitov[3], A I Gumarov[3*], V A Shustov[1] and I A Faizrakhmanov[1]

[1] Zavoisky Physical-Technical Institute, FRC Kazan Scientific Center of RAS
[2] Engineering Institute, Kazan Federal University, Kazan 420008, Russia
[3] Institute of Physics, Kazan Federal University, Kazan 420008, Russia

*E-mail: amir@gumarov.ru

Abstract. Nanocrystalline ZnO thin films with a thickness of ~ 235 nm were synthesized by ion beam-assisted deposition (IBAD) technique using a metal target of zinc and oxygen (O_2) as a reactive gas. The near-surface region of the synthesized films was subsequently implanted with 30 keV Ag^+ ions in the fluence range of $(0.25\text{-}1)\times10^{17}$ ion/cm^2 at high ion current density of 12 μA/cm^2. The structure parameters and morphology of as-deposited and subsequently implanted with silver ions ZnO films were investigated by X-ray diffraction and scanning electron microscopy techniques. It was found that the as-deposited ZnO films have inhomogeneous structure, which consists of nanocrystallites and disordered amorphous phase. The nanocrystallites of the obtained ZnO thin films have values of lattice parameters higher than for a bulk ZnO. Subsequent implantation with silver ions leads to a significant radiation heating and microstress relaxation of the film as well as to an increase in the size of nanocrystallites due to the amorphous phase.

1. Introduction

Zinc oxide is a n-type semiconductor with direct bandgap of ~3.4 eV and high exciton binding energy of 60 meV at room temperature [1]. These properties make ZnO an advanced material for optoelectronic devices creation based on it. In addition, zinc oxide has a high optical transparency (T ~ 90%), low resistivity and high radiation resistance [2]. Currently, ZnO is widely used in solar cells as planar transparent electrodes [3,4]. From a practical point of view, such a successful combination of physical-chemical properties of ZnO makes it promising to create UV, blue, white and green light-emitting diodes, optical and gas sensors, HUD-displays etc [5,6]. Additional doping with various impurities is carried out in order to expand the range of ZnO thin films application. In particular, doping of N or Ag is used to change the type of charge carriers from n-type to p-type, for instance [7-9]. Ion implantation is one of the most precise and widely used methods for introduction of impurities into semiconductor materials among other technics of doping. It is well known, that ion implantation leads to structural irradiation-induced defects and even to amorphization of irradiated layer [10]. Since ZnO is a radiation-resistant material [1,2] and does not amorphize even

IOP Conf. Series: Journal of Physics: Conf. Series **1058** (2018) 012077 doi:10.1088/1742-6596/1058/1/012077

at high fluence and irradiation energies, so study of the effect of ion implantation on the structural properties of ZnO is an actual topic. In addition, methods and conditions of film synthesis significantly affect the structure [11]. Energy of condensing atoms on a substrate is three orders of magnitude higher for ion beam synthesis of thin films in comparison with other methods, which can lead to the implantation of target atoms into the film during the growth process. Thus, IBAD is a promising technique for thin films synthesis with attractive physical-chemical properties associated with features of the structure during film formation [12,13]. The main aim of this work was to study the structural properties of ZnO thin films obtained by the IBAD technique, as well as to investigate the effect of subsequent silver ion implantation.

2. Experimental details

ZnO thin films were deposited by the IBAD technique onto plane-parallel plates of quartz glass with a diameter of 20 mm and a thickness of 2 mm. Chemically pure Xe (99.99% purity) was used as a working gas. The sputtered target was a metal disk of Zn (99.99% purity) with a diameter of 100 mm and a thickness of 4 mm. The oxygen (99.99% purity) was used as a reactive gas. The pressure during deposition was 2×10^{-2} Pa. The deposition time was 45 min. Subsequent implantation of 30 keV silver ions into the synthesized films was carried out by using ILU-3 ion accelerator in the fluence range of $(0.25\text{-}1)\times10^{17}$ ion/cm^2 at ion current density of 12 μA/cm^2. The morphology and thickness of the films were determined using a scanning electron microscopy (SEM, Zeiss EVO 50 XVP) technique. The elemental composition of films was investigated by energy-dispersive X-ray spectrometer (EDX, Inca Energy-350) built-in into the vacuum chamber of SEM and X-ray photoelectron spectroscopy setup (XPS, by SPECS). The crystal structure and the phase composition of the films were studied using X-ray diffractometer (XRD, Dron 7) with an X-ray source of Cu-Kα (0.15406 nm).

3. As-deposited ZnO films

Figure 1a shows SEM cross-section of the as-deposited ZnO film. The thickness of the film was obtained from SEM image and was about ~235 nm. The calculated deposition rate was 6 nm/min. The surface morphology of the as-deposited ZnO films is smooth without noticeable features even at high magnification (Fig. 1b).

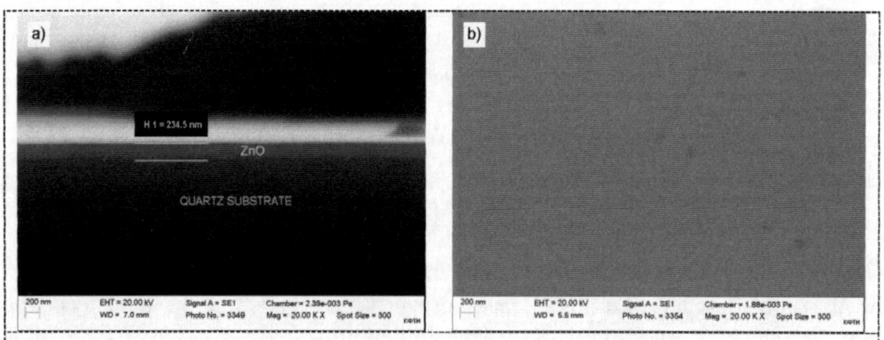

Figure 1. SEM images of as-deposited ZnO film by IBAD technique on quartz glass substrate: a) – cross-section and b) – morphology of the surface.

Elemental analysis of the as-deposited ZnO films was performed using 5 keV electron energy probing beam. Low energy of the beam gave an option to exclude the signal from the substrate (Si). It was found that composition of the obtained films was 49 at.% and 51 at.% for Zn and O, respectively. Elemental composition of the as-deposited ZnO thin films was obtained by EDX and is in a good agreement with XPS data.

X-ray diffraction analysis has shown that ZnO thin films have a polycrystalline wurtzite type structure (Fig. 2). Positions and intensity of reflexes for the powder diffraction pattern of ZnO are shown with blue lines on Fig. 2 [14]. Intensity of the reflection from the (002) plane is much larger than intensities of other reflections. This indicates an orientated growth of crystallites. We calculated the lattice parameters using the analytic expression in quadratic form for the hexagonal crystal system:

$$Q = \frac{1}{d^2} = \frac{1}{a^2}\left[\frac{4}{3}(h^2 + k^2 + hk) + \frac{l^2}{\left(\frac{c}{a}\right)^2}\right] = \frac{1}{a^2}\left[\frac{4}{3}s + \frac{l^2}{\left(\frac{c}{a}\right)^2}\right], \tag{1}$$

where $s = h^2 + k^2 + hk = k^2 + i^2 + ki = i^2 + h^2 + ih, i = -(h + k), h, k$ and l are Miller indices.

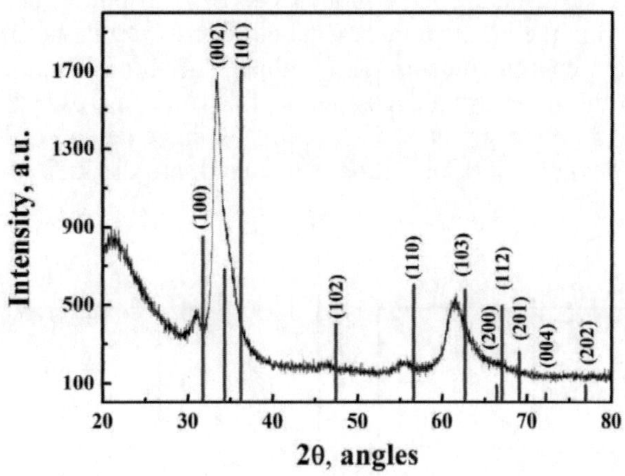

Figure 2. X-ray pattern of as-deposited ZnO thin film in the range of angles 2θ.

Table 1. Structural parameters of as-deposited ZnO thin films by IBAD technique on quartz glass

(hkl)	2θ	d_{exp} (Å)	d_{lit} (Å)	Δd (%)	a=b (Å)	c (Å)
(100)	31,00°	2,882	2,814	2,4	3,327	-
(002)	33,40°	2,681	2,613	2,6	-	5,361
(110)	55,56°	1,652	1,625	1,7	3,305	-
(103)	61,45°	1,507	1,481	1,7	3,285	5,428

Crystal lattice parameters of the as-deposited ZnO films obtained from the diffractogram (Fig. 2) are represented in Table 1. The lattice parameters a and c have values of 3.327 Å and 5.361 Å, respectively, which are much larger than for bulk ZnO (3.250 Å and 5.250 Å). Such increase in values of unit cell parameters may be due to the tensile microstresses that arise from implantation of zinc atoms into the interstitial positions during the growth of the ZnO film.

Let us consider the possibility of this process. In radiation physics, the maximum energy transferred by the incident ion to the target atom in the case of an elastic collision is determined by the following expression:

$$T_{max}(E) = \frac{4M_1M_2}{(M_1+M_2)^2} \cdot E, \tag{2}$$

where M_1 and M_2 are the masses of the incident ion and the target atom, respectively, and E is the energy of the incident atom. T_{max} determines the maximum possible energy of sputtered zinc atoms, which come to the surface of the deposited film and lead to the structural deformation of ZnO nanocrystallites. Standard atomic weights of Xe and Zn are 131.29 and 65.38, respectively. The energy of the primary beam (primary ions) is 1 keV. Thus, according to expression (2), T_{max} in our case is equal to 0.887 keV.

In the case of ion sputtering, the energy distribution of sputtered zinc atoms (E_{Zn}) is given by:

$$\frac{dN}{dE} \sim \frac{E_{Zn}}{(E_{Zn}+E_b)^3} \tag{3}$$

This distribution has a maximum at the energy of sputtered zinc atoms $E_{Zn} \cong E_b/2$, where E_b is the binding energy of surface atoms estimated as ~$4E_s$ (E_s is the energy of sublimation). Sublimation energy for zinc has a value of 1.356 eV/atom [15]. Therefore, E_b will have a value of ~ $4E_s$ = 5.424 eV. According to the obtained data, we have plotted energy distribution of condensing zinc atoms (Fig. 3). As it can be seen from the Figure 3 there is a high-energy tail in the energy distribution of the condensing zinc atoms. High-energy zinc atoms from this distribution region can be implanted into a near-surface thin region of the film during deposition and produce microstresses in it. Magnitude and sign of microstresses depend on the structure of the film. We have determined the average projection range R_p and the standard deviation ΔR_p of Zn ions implanted into the ZnO matrix using the "SRIM" algorithm [16].

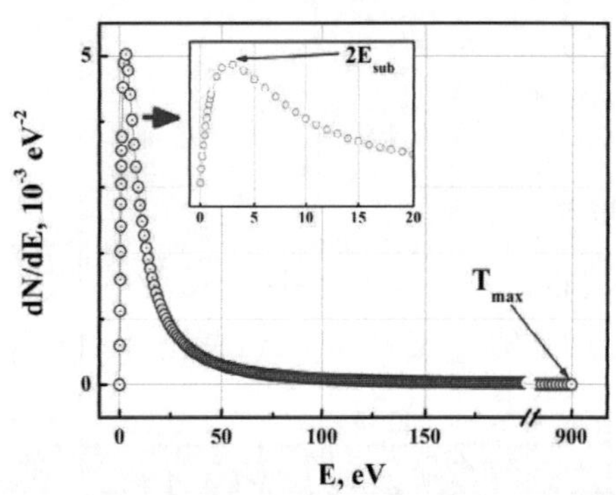

Figure 3. Energy distribution of the condensing zinc atoms at ES = 5,424 eV and Tmax = 887 eV.

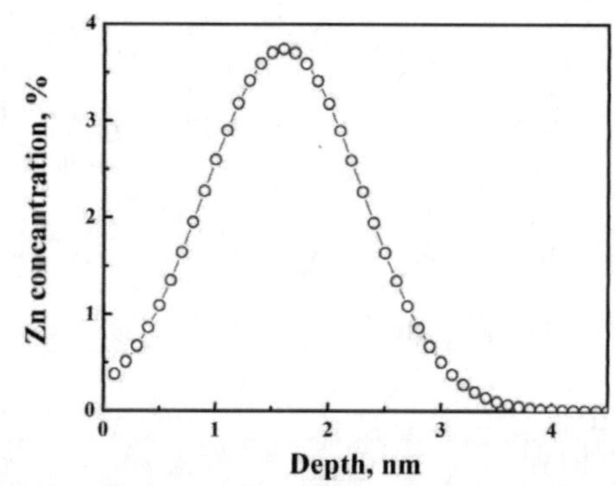

Figure 4. Depth distribution profile of Zn ions implanted into the ZnO matrix with energy Tmax = 0.8 keV.

Further, we have plotted zinc distribution profile according to the formula taken from Ref. [17] (see Fig. 4). The energy of the Zn ions was taken T_{max} = 0.8 keV and the radiation fluence was equals 1×10^{15} ion/cm^2. The obtained profile has the Gaussian-shape distribution form with Rp ≈ 1.6 nm and ΔRp ≈ 0.73 nm. Consequently, the maximum depth distribution of zinc ions is equal to Rp + 2ΔRp = 3.06 nm for T_{max} = 0.8 keV. Thus, we can conclude that a small part of the high-energy sputtered zinc atoms were implanted into the surface layer during the deposition of ZnO films. Possibly, zinc atoms implantation leads to the structure deformation of ZnO nanocrystallites. We assume that this process can explain the observed shift of the (002) reflection from ZnO nanocrystallites to the small-angle region of the diffractogram (Fig. 5a). At the same time, the wide arm relating to the amorphous component of ZnO remains at the standard position (002). We note that a shift in the position of reflexes is usually observed from both the nanocrystalline and amorphous components in the case of metal films [18]. The observed changes in the structural parameters of the lattice of ZnO nanocrystallites are about 2-2.5% (see Table 1), which are significantly exceed the limit of elastic deformation of solids.

4. As-implanted with silver ions

Figure 5a shows a decomposition of reflex (002) for the as-deposited ZnO film by two components. The first narrower component (blue line) can be attributed to the nanocrystallites of ZnO. The second component (pink line) is assigned to the intercrystalline amorphous region with crystal lattice parameters corresponding to the literature data for ZnO.

Silver ion implantation leads to a sequential change in the structure of films, which is reflected in the change of the reflex (002) shape (Fig. 5b and Fig. 5c). In particular, quantitative content of 1 and 2 components of as-implanted ZnO film decreases from 40 and 60% to 17 and 38%, respectively (see Table 2) for the minimal radiation fluence (0.25×10^{17} ion/cm^2) in our experiments.

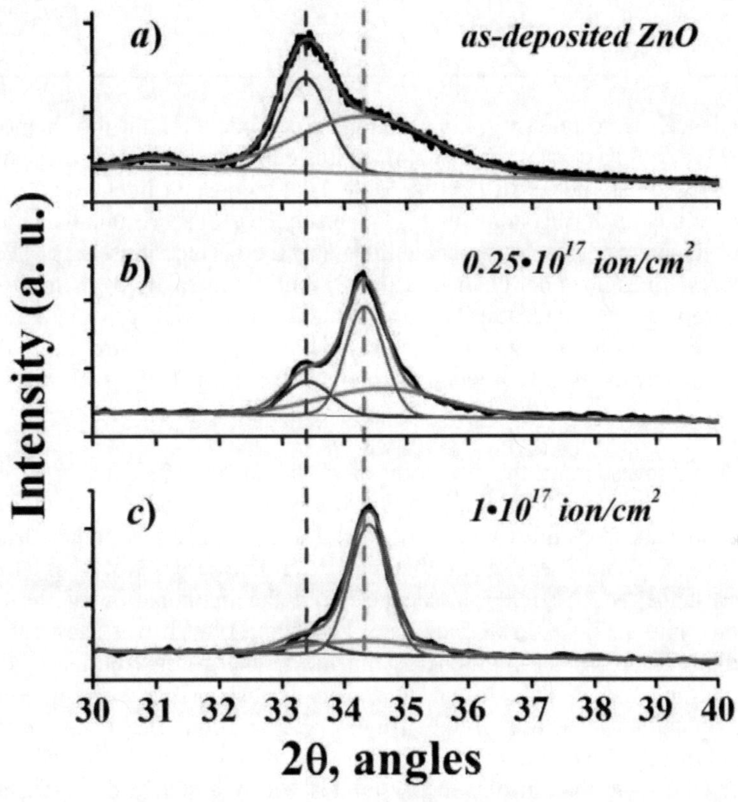

Figure 5. Decomposition of the reflex (002) by components for as-deposited and subsequently implanted with silver ions ZnO films. The blue, pink and green lines correspond to the 1, 2 and 3 components on the diffractogram.

The formation of the third component (green line), corresponds to ZnO crystallites with a standard lattice parameters. The content of the third component increases to 68% at the maximum implantation fluence of 1×10^{17} ion/cm^2. This indicates a partial recovery of the crystal structure. The content of "deformed" nanocrystallites and amorphous component decrease to 8% and 24%, respectively. Obviously, the observed changes in the structure of the films are due to their thermal annealing during implantation process. Thus, at the high ion current density, the irradiated film is heated because of the low thermal conductivity of the quartz glass substrate.

Table 2. Decomposition parameters of the reflex (002) for as-deposited and subsequently implanted with silver ions ZnO film.

Components (%)	FWHM (2θ)	d (Å)	Δd (%)	c (Å)	Center (2θ)	ΔL (2θ)
as-deposited ZnO						
1 (40 %)	0.894	2.683	2.7 %	5.367	33.36	0.97
2 (60 %)	2.44	2.610	0.1 %	5.220	34.33	
ZnO:Ag (Fluence – $0.25{\cdot}10^{17}$ ion/cm^2)						
1 (17 %)	0.894	2.679	2.5 %	5.359	33.41	1.02
2 (38 %)	2.44	2.602	0.4 %	5.205	34.43	
3 (45 %)	0.724	2.611	0.07 %	5.222	34.32	
ZnO:Ag (Fluence – $1{\cdot}10^{17}$ ion/cm^2)						
1 (8 %)	0.894	2.681	2.6 %	5.361	33.40	0.99
2 (24 %)	2.44	2.605	0.3 %	5.211	34.39	
3 (68 %)	0.731	2.605	0.3 %	5.211	34.39	

Finally, dependence of the intensity and position of component 1 on the fluence of implantation was investigated. One could expect a gradual shift of this component towards the standard value of the diffraction angle 2θ (see parameter ΔL (2θ) in Table 2) at a constant intensity, which corresponds to the gradual stress decrease during annealing. Conversely, we observe another situation, which we associate with the nanometer size of the crystallites and the surface closeness. The value of 2θ for "deformed" nanocrystallites does not change (Table 2), but the intensity of the reflex decreases while the intensity of component 3 increases. The interstitial atoms, which produce stresses, can quickly diffuse to the surface due to the small size of the crystallites. Therefore, stresses in the nanocrystallite decrease to zero and intensity of the component 3 increases as a result of a such rapid process.

5. Conclusions

In summary, we have investigated the structural properties of the ZnO thin films formed by IBAD technique and the effect of subsequent implantation with silver ions. The structure of the as-deposited ZnO film consists of nanocrystallites with lattice parameters of 2-2.5% higher than standard values and amorphous regions. The mechanism that can lead to increase of the lattice parameter of nanocrystallites is suggested. Also it is associated with the implantation of the most high-energy sputtered zinc atoms into the near-surface region of a growing film. It was shown that such process is possible due to the presence of the high-energy tail in the energy distribution of the deposited zinc atoms. Silver ions implantation at high current density is responsible for radiation heating and the recovery of the crystalline structure of the irradiated ZnO film. The quantitative content of the "deformed" nanocrystalline and amorphous components of the thin film decreases as a result of heating. The later lead to formation of nanocrystallites with the standard crystal lattice parameters. Thus the content of nanocrystallites prevails at the maximum fluence of silver ion implantation.

Acknowledgments

The authors are grateful to V F Valeev and V I Nuzhdin (Zavoisky Physical-Technical Institute, FRC "*Kazan Scientific Center of RAS*") for ion implantation experiment. A.I. Gumarov is grateful to the research grant of Kazan Federal University for supporting studies of element composition using XPS setup.

References

[1] Özgür Ü, Alivov Ya I, Liu C, Teke A, Reshchikov M, Doğan S, Avrutin V, Cho C, Morkoc H 2005 J. Appl. Phys. 98 041301

[2] Jagadish C, Pearton S 2006 *Zinc Oxide Bulk, Thin Films and Nanostructures: Processing, Properties and Applications* (Elsevier Ltd) p 600
[3] Chen J, Chen D, Zhou Y, Li W, Ren Y, Hu L 2014 *Materials Letters* 117 162
[4] Hirahara N, Onwona-Agyeman B, Nakao M 2012 *Thin Solid Films* 520 (6) 2123
[5] Lyadov N M, Gumarov A I, Kashapov R N, Noskov A I, Valeev V F, Nuzhdin V I, Bazarov V V, Khaibullin R I, Faizrakhmanov I A 2016 *Semiconductors* 50 (1) 44
[6] George R S Andrade, Cristiane C Nascimento, Zenon M Lima, Erico Teixeira-Neto, Luiz P Costa, Iara F Gimenez 2016 *Applied Surface Science* 399 573
[7] Mezdrogina M M, Vinogradov A Ya, Kuzmin R V, Levitski V S, Kozanova Yu V, Lyanguzov N V, and Chukichev M V 2016 *Semiconductors* 50 (10) 1304
[8] Georgobiani A N, Gruzintsev A N, Volkov V T, Vorobiev M O, Demin V I, Dravin V A 2003 *Nucl. Inst. and Meth. in Phys. Res. A.* 514 117
[9] Michelle A Myers, Joon Hwan Lee, Zhenxing Bi and Haiyan Wang 2012 *J. Phys.: Condens. Matter.* 24 145802
[10] Bazarov V V, Nuzhdin V I, Valeev V F, Stepanov A L, 2018 *Vacuum* 148 254
[11] Gleiter H 2000 *Acta mater.* 48 1
[12] Faizrakhmanov I A, Bazarov V V, Zhikharev V A, Stepanov A L, Khaibullin I B 2001 *Vacuum* 62 15
[13] Faizrakhmanov I A, Bazarov V V, Zhikharev V A, Khaibullin I B 1999 *Nucl. Instr. and Meth. in Phys. Res. B* 148 669
[14] Mirkin L I 1961 *Handbook of X-ray Analysis of Polycrystalline Materials* (Consultants Bureau, New York) p 731
[15] Chicherskaya A L, Pupyshev A A 2015 *Analytics and Control* 19 (3) 230
[16] Ziegler J F, Biersack J P, Littmark U 1985 *The Stopping and Range of Ions in Solids* (Pergamon Press, New York) (SRIM-2008 software at http://www.srim.org/)
[17] Batalov R I, Vorobev V V, Nuzhdin V I, Valeev V F, Bayazitov R M, Lyadov N M, Osin Yu N, and Stepanov A L 2016 *Tech. Phys.* 61(12) 1871
[18] Lyadov N M, Bazarov V V, Vagizov FG, Vakhitov I R, Dulov E N, Kashapov R N, Noskov A I, Khaibullin R I, Shustov V A, Faizrakhmanov I A 2016 *Applied Surface Science* 378 114